Biology: The Study of Life

Ruth Bernstein • **Stephen Bernstein**

University of Colorado, Boulder

 Harcourt Brace Jovanovich, Inc.

New York   San Diego   Chicago   San Francisco   Atlanta
London   Sydney   Toronto

Requests for permission to make copies of any part of the work should be mailed to: Permissions, Harcourt Brace Jovanovich, Publishers, 757 Third Avenue, New York, NY 10017.

Printed in the United States of America.

Library of Congress Catalog Card Number: 81-84968

ISBN: 0-15-505440-6

Illustration and Photo Credits

Cover photograph © Animals Animals / Dan Suzio

Part Openers
Part 1: © Arch. Phot., Paris/SPADEM/VAGA, 1981 **Part 2:** © Peter Butler/ Photo Researchers, Inc. **Part 3:** Rich Clarkson/Sports Illustrated © Time, Inc. **Part 4:** Exerpted from the book *A Child is Born* by Lennart Nilsson. English translation copyright © 1966, 1977 by Dell Publishing Co., Inc. Originally published in Swedish under the title *Ett Barn Blir Till* by Albert Bonniers Forlag. Copyright © 1965 by Albert Bonniers Forlag, Stockholm. Revised edition copyright © 1976 by Lennart Nilsson, Mirjam Furuhjelm, Axel Ingerlman-Sundberg, Cales Wirson. Used by permission of Delacorte Press/Seymour Lawrence. **Part 5:** Keith Axelson. Photograph won first place in *Natural History Magazine* 1981 competition.

Chapter 1
1–1(a) By Robert Campbell © National Geographic Society. **(b)** © George Holton/ Photo Researchers, Inc. **(c)** M. H. Luque. **1–2** Al Giddings/Ocean Films, Ltd. **1–3(a)** Lynn McLaren/Photo Researchers, Inc. **(b)** Copyright by the American Dental Association. Reprinted by permission. **(c)** Photo by Dr. Robley C. Williams/The National Foundation. **1–4** Photo by Stephen Bernstein. **1–5** Palomar Observatory; California Institute of Technology. **1–6** NASA photo. **1–7** NASA photo. **1–8** Adapted from *Biology Today*, 1972, Random House, Fig. 33–11, p. 667. **1–15** The Bettmann Archive, Inc.

Chapter 2
2–2 Art by Helmut K. Wimmer, American Museum—Hayden Planetarium. **2–3** From *Red Giants and White Dwarfs* by Robert Jastrow. New American Library, p. 95.

Chapter 3
3–1(a) © Photo Researchers, Inc. **(b)** © Eric V. Gravé/Photo Researchers, Inc. **(c)** Photo by Hugh Spencer from National Audubon Society/Photo Researchers, Inc. **(d)** © Eric V. Gravé/Photo Researchers, Inc. **(e)** R. Krauft–Biology Media 1979/Photo Researchers, Inc. **3–11(b)** Watson and Crick in front of the Model, photo by A. C. Barrington Brown. From J. D. Watson, *The Double Helix*, Atheneum, New York, p. 215. © 1968 by J. D. Watson.

Chapter 4
4–1 Wide World Photos.

Chapter 5

5–2 Cell Research Institute, University of Texas at Austin. 5–3 Runk/Schoenberger from Grant Heilman. 5–4 From Bloom, W., and Fawcett, D. W.: *A Textbook of Histology* 10 Edition. Philadelphia, W. B. Saunders Company 1975. 5–5 Ursula Goodenough, Washington University. 5–6(a) K. R. Porter. (b) Drs. David Green and Humberto Fernandez–Moran. 5–8 L. Kay Shumway. 5–10 From "The Ground Substance of the Living Cell" by Keith R. Porter and Jonathan B. Tucker. Copyright © March, 1981 by Scientific American, Inc. All rights reserved. 5–11 H. J. Arnott and W. R. Fagerberg, University of Texas at Arlington. 5–12 Dr. Eva Frei and Professor R. D. Preston, F.R.S. 5–13 Industrial Genetics, Inc.

Chapter 6

6–3 Courtesy of A. H. Sparrow and R. F. Smith, Brookhaven National Laboratory. 6–4 A. S. Bajer and J. Mole–Bajer.

Chapter 7

7–1 Lowell Georgia. 7–2 Runk/Schoenberger from Grant Heilman. 7–3 Courtesy of the Cytogenics Laboratory, University of California, San Francisco. 7–8 The Bettmann Archive, Inc. 7–19(a) Photography by Runk/Schoenberger from Grant Heilman. (b) and (c) New Jersey Medical School. 7–20 Evelyn M. Shafer.

Chapter 8

8–1 The Bettmann Archive, Inc. 8–3(a) Suzanne Szasz. (b) © Hans Pfletschinger/Peter Arnold, Inc. (b) © 1978 Laimute E. Druskis/Taurus Photos. 8–6 The Alan Fletcher Research Station, Dept. of Lands, Queensland, Australia. 8–8 © 1979 Dunstan–BSL from Black Star. 8–9 M. W. F. Tweedie/Photo Researchers, Inc. 8–10 © Tom McHugh/Photo Researchers, Inc. 8–12(a) © Ingeborg Lippman/Magnum. (b) © Len Rue, Jr./Photo Researchers, Inc.

Chapter 9

9–1 Adapted from *Living in the Environment*, Second Edition, by G. Tyler Miller, Jr.© 1979 by Wadsworth, Inc. Reprinted by permission of Wadsworth Publishing Company, Belmont, California 94002. 9–2 Adapted from W. H. Dowdeswell, *Animal Ecology*, 1961, Harper & Row. 9–4 Joseph R. Jehl, Jr. 9–7 Corvallis Environmental Research Laboratory. 9–10 Adapted from R. H. MacArthur, *Ecology*, 1958, Vol. 39, pp. 599–619, Duke University Press. 9–11 Figure 24–4 (p. 563) from *Biology* by Richard A. Goldsby, copyright © 1976 by Harper & Row, Publishers, Inc. Reprinted by permission of the publisher. 9–12 Right: © Animals Animals/Stan Schroeder. Left: © Grant Heilman, photo by S. Rannels. 9–13 Adapted from H. B. Cott, *Adaptive Coloration in Animals*, 1940, Methuen & Company. 9–14 Adapted from R. E. Hall, K. R. Kelson, *The Mammals of North America*, Ronald Press, 1959. 9–16 Adapted from Martin L. Cody, *Competition and the Structure of Bird Communities*. Copyright © 1974 by Princeton University Press, Figure 1e, p. 13. 9–17 Adapted from D. W. Johnston, E. P. Odum, *Ecology*, 1956, Vol. 37, Duke University Press. 9–18 Adapted from B. J. Meggers and E. S. Ayensu, *Tropical Forest Ecosystems in Africa and South America*, Smithsonian Institution Press. 9–20 © Lee Boltin.

Chapter 10

10–1 Dr. Alec N. Broers and Dr. Barbara J. Panessa, IBM, Thomas J. Watson Research Center. 10–3(a) Left: © Eric V. Gravé/Photo Researchers, Inc. Right: © Jeanne White from National Audubon Society/Photo Researchers, Inc. (b) © Eric V. Gravé/Photo Researchers, Inc. (c) © 1981 Eric. V. Gravé/Photo Researchers, Inc. (d) Left: © 1981 Eric V. Gravé/Photo Researchers, Inc. Right: © 1981 Eric V. Gravé/Photo Researchers, Inc. 10–11 Courtesy American Museum of Natural History. 10–18 Courtesy Field Museum of Natural History, Chicago. 10–26 Adapted from V. Grant & K. Grant, *Flower Pollination in the Phlox Family*, Columbia University Press, 1965. 10–27 © New York Zoological Society photo. 10–29 Adapted from *Invitation to Biology*, Third Edition, by Helena Curtis and N. Sue Barnes, 1981, p. 620, Worth Publishers, Inc. 10–31 © Russ Kinne, 1973/Photo Researchers, Inc. 10–32(a) © A. W. Ambler from National Audubon Society/Photo Researchers, Inc. (b) San Diego Zoo–Ron Garrison/Photo Researchers, Inc. 10–33 From "The Casts of Fossil Hominid Brains" by Ralph L. Halloway. Copyright © July, 1974 by Scientific American, Inc. All rights reserved. 10–35 Gibbon, courtesy Animal Talent Scouts, Inc. 10–36(a) Ylla/Photo Researchers, Inc. (b) © Akira Uchiyamaw/Photo Researchers, Inc. 10–38(b) Hans Wendt. 10–40 Trustees of the British Museum (Natural History). 10–41(a) Irven DeVore/Anthro Photo. (b) Line drawings adapted from *Life Before Man*, Time–Life Books, New York, 1972. (b) Photographs, Nina Leen, *Life Magazine*, © Time, Inc. 10–42 Photographs courtesy of the American Museum of Natural History.

Chapter 11

11–1 Stephen Bernstein. 11–2(a) Courtesy Great Lakes Fishery Commission. (b) Michigan Department of Natural Resources. 11–3(a) Left: Eli Lilly and Company. Right: Photo courtesy of Fisher Scientific Co., Educational Materials Division. (b) Norma J. Lang, Department of Botany, U.C. Davis/Biological Photo Service. (c) Time Magazine. 11–4(a) Left: © Eric V. Gravé/Photo Researchers, Inc. Right: © Eric V. Gravé/Photo Researchers, Inc. (b) Left: © Eric V. Gravé/Photo Researchers, Inc. Right: © J. Robert Waaland/Biological Photo Service. 11–5 Adapted from G. C. Stephens, B. B. North, *Biology*, 1974, John Wiley and Sons. 11–8 Adapted from Kimball, *Biology*, © 1974, Addison–Wesley, Reading, MA, Figure 13–13. Reprinted with permission. 11–9 Photography by Hal Harrison/Grant Heilman. 11–10 Adapted from "The Thermostat of Vertebrate Animals," by H. Craig Heller, Larry I. Cranshaw and Harold T. Hammel. Copyright © August, 1978 by *Scientific American*. All rights reserved. 11–11 Adapted from G. C. Stephens, B. B. North, *Biology*, 1974, Figure 3–26, John Wiley and Sons. 11–13(a) Left: Photo by Harley Barnhart from *Smithsonian*, 1980, May, Vol. II, No. 2, pp. 96–105. Right: Kit Scates. (b) © Biology Media/Photo Researchers, Inc. (c) Runk/Schoenberger from Grant Heilman Photography. 11–14 David Pramer, Rutgers University. 11–15(a) © L. West/Photo Researchers, Inc. (b) Art adapted from *Biology* by G. C. Stephens and B. North, John Wiley and Sons, Inc., 1974, p. 255, Figure 9–6(b). (c) © Ken Brate/Photo Researchers, Inc. (d) © John Bova/Photo Researchers, Inc. 11–16 Photo courtesy Dr. Robert D. Lumsden.

Chapter 12

12–1 Walter Iooss, Jr./*Sports Illustrated*. 12–4(a) Adapted from Stephens & North, *Biology*, 1974, Figure 11–22(b), John Wiley and Sons. (b) Adapted from Stephens & North, *Biology*, 1974, Figure 12–44fi John Wiley and Sons. 12–5(a) Art adapted from Vander, A. J., Sherman, J. H., and Luciano, D. S., *Human Physiology*, McGraw-Hill, 1970, Figure 6–7, p. 165. (b) Art adapted from Kimball, *Biology*, © 1974 Addison-Wesley, Reading, MA, Figure 7–7. Reprinted with permission. (12–7) Adapted from J. D. Ebert, *Biology*, 1973, Holt, Rinehart and Winston, Figure 15–7, p. 459. 12–8 Reprinted by permission of *American Scientist*, Journal of Sigma Chi. 12–10 Adapted from Stephens & North, *Biology*, 1974, Figure 11–26, p. 338, John Wiley and Sons. 12–11 Art adapted from Vaner, A. J., Sherman, J. H., and Luciano, D. S., *Human Physiology*, McGraw-Hill, 1970, Figure 16–24, p. 505 12–13(a) U.P.I. (b) From *The Pituitary Body and Its Disorders* by H. Cushing, J. B. Lippincott, 1912. 12–19 Adapted from P. M. Ray, *The Living Plant*, 1963, Holt, Rinehart and Winston, Figure 1–1, p. 4 12–20(b) Photograph by Louis Mantonyi from P. M. Ray, *The Living Plant*, 1963, Holt, Rinehart and Winston, Figure 7–2, p. 81 12–21 Adapted from P. M. Ray, *The Living Plant*, 1963, Holt, Rinehart and Winston, Figure 7–2, p. 81. 12–24 Graph adapted from Frank B. Salisbury, *The Flowering Process*, 1963, Pergamon Press, Inc. 12–26 With permission from *Natural History*, October, 1973. Copyright by The American Museum of Natural History.

Chapter 13

13–1 Adapted from *Biology*, Second Edition, by Richard A. Goldsby, Figure 20–1 (p. 506). Copyright © 1979 by Harper & Row, Publishers, Inc. Reprinted by permission of the publisher. 13–4 Adapted from a A. J. Vander, et al., *Human Physiology: The Mechanisms of Body Functions*, 1970, McGraw-Hill, Figure 5–22a, p. 149. 13–5 From "Myelin" by Pierre Morell and William F. Norton. Copyright © May, 1980 by Scientific American, Inc. All rights reserved. 13–6 Adapted from *Nerve, Muscle and Synapse* by Bernard Katz, McGraw-Hill, 1966, p. 8, Figure 2. 13–7 Adapted from *Biology* by Richard A. Goldsby, Harper & Row, New York, 1979, p. 514, Figure 20.6. 13–8(a) Adapted from *Being Human* by David W. Deamer, 1981, Holt, Rinehart and Winston. (b) Professor John E. Heuser, Washington University School of Medicine, St. Louis, MO. 13–9(a) Figure 20.6 (p. 514) from *Biology*, Second Edition, by Richard A. Goldsby, copyright © 1979 by Harper & Row, Publishers, Inc. Reprinted by permission of the publisher. (b) By E. R. Lewis and Y. Y. Zeevi in *The Scanning Electron Microscope*, by Everhart and Hayes; also in Ebert, et al., 1973, *Biology*, Holt, Rinehart and Winston, New York, Figure 13–3, p. 404. 13–11 Adapted from A. J. Vander, et al., *Human Physiology: The Mechanisms of Body Function*, 1970, McGraw-Hill, Figure 16–13, p. 511. 13–12 Adapted from N. R. Carlson, *Physiology of Behavior*, 1977, Allyn and Bacon, Figure 8–2, p. 161. 13–13 Adapted from *Biology*, Second Edition, by Richard A. Goldsby, Figure 20–11, p. 522. Copyright © 1979 by Harper & Row, Publishers, Inc. Reprinted by permission of the publisher. 13–14(a) Adapted from N. R. Carlson, *Physiology of Behavior*, 1977, Allyn and Bacon, Figure 8–5, p. 164. (b) Micrograph by E. R. Lewis, Y. Y. Zeevi and F. S. Werblin. 13–15(a) William H. Amos. (b) and (c) Kimball, *Biology*, © 1974, Addison–Wesley, Reading, MA 01867, Figure 26–2. Reprinted with permission. 13–16 Thomas Eisner. 13–17 Ritchie, *Biology* © 1979, Addison–Wesley, Reading, MA 01867, Figure 20–19. Reprinted with permission. 13–18(a) Adapted and reproduced by permission of Harcourt Brace Jovanovich, Inc. from *Health*, Second Edition, by B. A. Kogan © 1970, Figure 4–5. (b) Adapted from W. T. Keeton, *Elements of Biological Science*, Second Edition, 1973, W. W. Norton, Figure 10–26, p. 225. (c) and (d) A. J. Vander et al., *Human Physiology: The Mechanisms of Body Functions*, 1970,

Page 629 constitutes a continuation of the copyright page.

Preface

The exploration of the universe is the most profound and challenging adventure of human existence. Part of this adventure—the study of biology—focuses on the world of life that exists precariously in the thin, delicate biosphere resting on the crust of the earth. In this book, we will examine the biological principles that apply to all living organisms and then unravel the specific interconnections between plant and animal life, placing a particular emphasis on human life.

The first goal of this text is to review the more significant discoveries of the science of biology. We are living in what has been called "the golden age of biology." Scientists are just beginning to enter some of the more mysterious frontiers of the world of life, and every mystery that they solve reveals a deeper layer of mystery. The science of biology is one of the greatest achievements of the human mind, as well as an important part of our cultural heritage as citizens of twentieth-century civilization. The study of life is also a beautiful and wondrous experience. As the great nineteenth-century biologist Thomas Henry Huxley put it: "In travelling from one end to the other of the scale of life, we are taught one lesson, that living nature is not a mechanism but a poem."

The second goal of this text is to show how the discoveries of biology relate to our lives. As individuals and as members of the human community, we are continually faced with a bewildering array of problems, which range from choosing the most effective birth-control method to understanding global crises. To understand such problems as energy shortages, environmental pollution, starvation, and war, we need solid and reliable information about their origins and their possible solutions. The science of biology provides us with some of this information, and throughout the text this information is applied to the problems and crises encountered in modern life.

Biology: The Study of Life is divided into five parts. Part 1 (Chapter 1) is a discussion of the goals and methods of science. The intent of this introduction is to demonstrate that science is not a foreign or incomprehensible subject. Part 2 (Chapters 2–10) explains the mechanisms of evolution and traces the changes in life forms from their probable origins some 3.5 billion years ago to the emergence of human beings and the development of modern civilizations. Part 3 (Chapters 11–20) concentrates on the individual organism and deals specifically with the inherited capacities that give each organism a chance to survive to maturity and to reproduce in a hostile environment. The structures and functions of many types of organisms—from minute unicellular protozoans to elephants and redwood trees—are examined in these chapters, but the central focus is on the individual human being. Part 4 (Chapters 21–23) considers the orderly sequence of developmental changes that occur in each life and follows the human individual through each stage of existence, from conception and birth through maturity and old age to death. Finally, Part 5 (Chapter 24) investigates the biological origins and possible solutions to the crises and problems present in modern society.

Throughout the book, we gradually develop an integrated body of knowledge. Some of the earlier chapters, which emphasize the basic principles of biology, may seem far removed from the later chapters, where applications of these principles are made to modern life. However, the material in these earlier chapters is essential to an understanding of the later material. An understanding of infectious diseases (Chapter 18) or the proper diet for physical well-being (Chapter 16) is based on an understanding of how cells function (Chapters 4 and 5). An understanding of genetics (Chapter 7) is essential to an understanding of the mechanisms of sex and reproduction (Chapter 23). Real understanding is never easy to acquire, but we have tried throughout the text to make understanding as easy and enjoyable as possible.

We would like to thank the following reviewers for their contributions to the text: Marvin Druger, Syracuse University; Curtis Eklund, The University of Texas at El Paso; John E. Frey, Mankato State University; Judith Goodenough, University of Massachusetts; Albert J. Grennan, San Diego Mesa College; Gordon Locklear, Chabot College; William H. Mason, Auburn University; Richard E. Pieper, University of Southern California; and Kareen Sturgeon, University of Colorado.

We would also like to express our gratitude to the members of the editorial and production staffs of Harcourt Brace Jovanovich who have made major contributions to this book: Bill Bryden, Geri Davis, Sue Lasbury (especially for her photographic research), Ann Nemanich, Michelle Pinney, André Spencer, Emily Thompson, Fran Wager, and, in particular, our editor, Mary George.

Ruth Bernstein
Stephen Bernstein

Contents

Part 3

The Individual Life: Adaptations for Survival

11 Organizations for Survival   243

Part 4
The Individual Life: Maturation, Reproduction, and Aging

21 Patterns of Life History 541

22 Reproduction of the Individual: From Cell to Organism 553

An Introduction to Science

1

Science: Understanding the Past and Predicting the Future

You are about to enter the world of biology. The goal of this first chapter is to prepare you for the experience—to take you behind the scenes to show you how scientific knowledge is acquired and how it is similar to but uniquely different from other forms of knowledge.

Science: The Open Frontier

Science has developed in response to the basic human characteristic of curiosity— the drive to explore and to understand the unknown. This drive has always taken some men and women in each generation one step farther into the unknown than anyone has gone before. In our time, it has produced astronauts and cosmonauts— the first humans to enter outer space. Before them, in an unbroken chain stretching back through history, were the first to fly, the first to enter and explore the ocean depths, the first to set sail in frail wooden ships to discover unexplored continents, and many other familiar, yet incredible, adventurers in human history.

To most of us, the great adventure of exploration may seem beyond reach; few of us will travel into space, and the last frontiers on earth have already been conquered. But a frontier can be a state of mind—a sense of being somewhere that no human has ever been before. Scientists are continually discovering new frontiers and are currently exploring aspects of the earth and the universe that were not even known to exist at the time you were born.

The world you will encounter in this book includes structures many thousands of times smaller than the limits of human vision and extends out into the universe for billions of miles and back in time for billions of years. The goal of science is the complete understanding of this entire vastness of space and time.

Humans have always been interested in the natural history of plants and animals and have described and documented the lives of many organisms in great detail. Biologists are now studying living organisms in their natural settings everywhere on earth that is capable of supporting life. Biologists interested in the social behavior of animals are living with the mountain gorillas of Africa, the

Figure 1-1
Investigating Animal Social Behavior
(a) The mountain gorilla in Africa; (b) king penguins in Georgia (U.S.S.R.); (c) the Alaskan brown bear.

(a)

(b)

(c)

brown bears of Alaska, and are studying king penguins in the U.S.S.R. (Figure 1-1). They are also flying small planes, tracking migratory birds to discover what mechanisms permit birds to navigate accurately over land and water for thousands of miles. Biologists can also be found diving under the ice in the Arctic Ocean and in the Gulf Stream of the Bahama Islands to investigate the complex relationships between multitudes of diverse sea creatures (Figure 1-2).

Until comparatively recent times, direct observation of the world of life was almost exclusively limited to large-scale, surface phenomena. However, the technological explosion in the twentieth century has extended the limits of direct observation to the basic microstructures of life (Figure 1-3) and into the flow of chemical and electrical activity that is life itself. Scientists are now examining the molecular building blocks of heredity—the blueprints that determine how a complete organism develops from the union of a sperm and an egg. These observations may someday lead to the correction of hereditary defects that cause such afflictions as diabetes and some forms of mental retardation before these tragedies occur. Other scientists are studying the development of learning in the brain—not only in the large brains of humans and monkeys, but also in the tiny brains of ants (Figure 1-4), locusts, and sea snails. Probes sensitive to minute currents of electrical activity are inserted into the brain, and changes in this activity are monitored as the animals learn to solve problems. These are just a few examples of recent explorations into the microscopic world of life.

To go back in time, twentieth-century technology offers techniques for dating the ages of the fossilized remains of plants and animals that have been embedded

Figure 1-2
Undersea Research
Scuba-diving researcher investigating small oceanic organisms in the Gulf Stream near Bimini in the Bahama Islands.

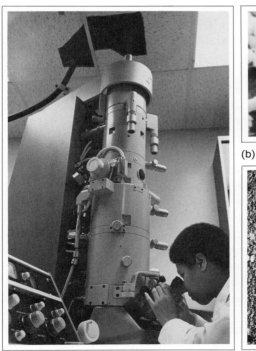

(a)

(b)

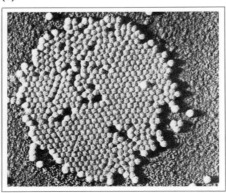

(c)

Figure 1-3
Observing the Microstructures of Life

(a) The scanning electron microscope can show tiny structures enlarged as much as 300,000 times. (b) Plaque on human teeth, formed of corncob-like clusters of bacteria ($\times 8,500$). (c) The virus that causes polio ($\times 53,880$).

Figure 1-4
The Brain of an Ant
A slice through the interior of an ant
brain 1 millimeter in width.

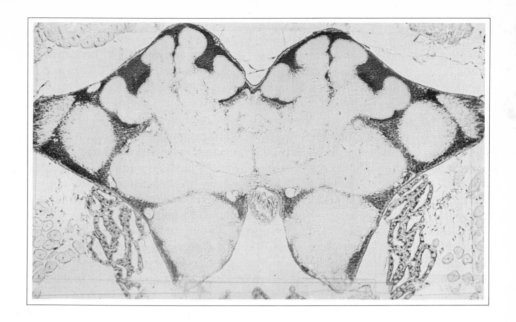

for millions of years in the rock layers of the earth's crust. These techniques are now so accurate that it may be possible to trace the emergence and extinction of ancient organisms back to the origins of life on earth. Not only have scientists discovered when human beings first appeared on earth and how they developed into our modern form, but they have also inferred such details about early human life as the time when early men and women first began to use fire for heating and cooking, the state of their health, and the types of weather conditions they endured.

Scientists are also extending their observations out into space beyond the planet earth. As yet, there is no evidence that life exists elsewhere in the universe. However, many scientists are convinced that life has originated at different times in separate parts of the universe, that other forms of intelligent life comparable to human intelligent life may have evolved elsewhere, and that extraterrestrial civilizations much older than human civilizations may very well exist.

The earth is a planet of a star (the sun) that is only one of the billions of stars in our galaxy. A very large optical telescope, such as the one at Mount Palomar in California, can sight as many as 10 billion galaxies within its range (Figure 1-5). The universe is so enormous that scientists believe planets similar to the earth may have undergone comparable developmental stages. In 1971, an international conference in the Soviet Union devoted to the topic of communication with extraterrestrial intelligence concluded that intelligent civilizations may exist on as many as 1 million planets in our galaxy alone. The world of life may indeed extend throughout the entire universe!

We have now entered the era of space exploration. Humans have already walked on the surface of the moon, 237,000 miles from the earth, searching for

Figure 1-5

The Nearest Galaxy to Our Own—The Great Spiral Galaxy in the Constellation *Andromeda*

Astronomers measure distance in light-years (a *light-year* is the distance that a ray of light moving at 186,000 miles per second travels in one year). The Andromeda galaxy is 2 million light-years (millions of billions of miles) distant from our galaxy. The small points of light in this photograph are individual stars within our own galaxy.

Figure 1-6
Astronaut Harrison Schmitt,
a Member of the Apollo 17
Mission, Taking Rock
Samples on the Moon

clues to the age and development of our solar system and the universe (Figure 1-6). Unmanned space probes have traveled millions of miles to Mars, Mercury, and Venus to send back televised pictures and other data concerning the atmosphere and geology of these planets. This information may provide evidence that some form of life exists on these planets. The space probe Pioneer 10 has journeyed into deep space carrying a message designed to inform possible extraterrestrial civilizations that the planet earth is home to intelligent life (Figure 1-7).

The exploration of life on earth and the search for intelligent life elsewhere in the universe is such a great task that it will probably never be fully accomplished. The human mind will always face the challenge of an unconquered frontier.

The Universe Is Comprehensible

Curiosity motivates scientists to explore the unknown. They strive to accomplish this difficult task because they believe that the universe is comprehensible, that the vast complexity of space and time can be understood, that the objects and events in the universe can be related to one another in a systematic and orderly way. Philosophers of science have called these relationships the "laws of nature." The goal of scientists is to isolate and describe these laws.

The laws are workable. Laws of nature can be demonstrated in every area of daily life. Homes can be heated with gas because laws exist that describe how pressures and temperatures affect all gases. The food that is available in supermarkets has been produced by utilizing the laws that govern the growth and development of plants and animals. And the next time you are flying in a plane, consider that the plane is up there and will come down again safely because there are laws that govern moving bodies.

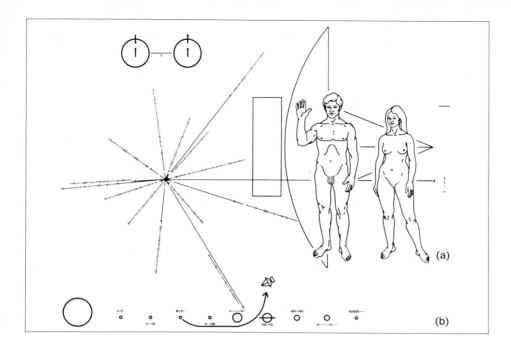

The fact that these laws work provides evidence that there is order in nature. However, as you will see in this chapter, the human interpretation of this order is imperfect and subject to change. Scientists examine both direct and indirect information to determine if patterns of organization indicate that there are underlying relationships between objects and events. When a relationship appears to exist, scientists test and retest it, making more observations to see if the human interpretation conforms to reality. Science is a long, gradual process of isolating and improving our understanding of the relationships that exist in nature.

During the process of discovery, scientists concentrate on three important types of relationship: common properties, levels of organization, and cause-and-effect relationships.

Common Properties

The billions of diverse objects and events in the universe can be divided into groups that have *common properties.* For example, although there are many kinds of birds of different sizes, colors, and shapes, all birds share certain unique features, such as having feathers and hatching their young from eggs.

Laws also describe properties that are shared by large groups of seemingly different objects and events. The same force—gravity—holds objects, such as humans, to the surface of the earth, the earth to the sun in the solar system, and the sun to its galaxy in the universe. The laws of gravity describe the characteristics of this force. Similarly, the ways in which all plants and animals reproduce and transmit hereditary information have certain common properties that are described by the laws of heredity.

In our sophisticated twentieth century, it is easy to forget that the ability to organize very different objects and events on the basis of common properties is a great human achievement. The English mathematician and philosopher Jacob Bronowski reminds us that

> Nature does not provide identical objects; on the contrary, these are always human creations. What nature provides is a tree full of apples which are all recognizably alike and yet are not identical: small apples and large ones, red ones and pale ones, apples with maggots and apples without. To make a statement about all these apples together, and about crabapples, Orange Pippins, and Beauties of Bath, is the whole basis of reasoning.*

Levels of Organization

Objects do not occur in the universe in isolation; each object is a part of a large grouping of objects. In Figure 1-8, a grasshopper is shown as one part of a community of animals living in a saltwater marsh. A human observer of this community can see that it is composed of various *levels of organization* (orderly, systematic relationships). The organization shown in the figure is a *food web*, which is a summary of who is eaten by whom within the community. In this saltwater marsh in midwinter, the arrows indicate that plants are eaten by grasshoppers, which are in turn eaten by small rodents and a variety of birds, which are in turn eaten by such larger animals as hawks and owls.

The grasshopper itself is a large grouping of objects. In Figure 1-9, the grasshopper has been taken apart to show its internal structure, which is organized into systems: the digestive and excretory systems that process food and remove waste products; the tracheal system that brings in and distributes oxygen to all parts of the body; the male and female reproductive systems that function in the formation of new grasshoppers; and the nervous system that controls and coordinates physiology and behavior.

The internal systems of the grasshopper can, in turn, be broken down into their components—the cells of the body—which are uniquely organized in each system. And these cells are made up of molecules, which can be broken down into the basic components of matter—the atoms—which are also uniquely organized in each type of cell in the body.

The biologist seeks to describe the forms of organization at each level and to determine how phenomena occurring at one level affect phenomena occurring at other levels (for example, how activity within the grasshopper's nervous system coordinates the grasshopper's behavior in the food web).

Cause-and-Effect Relationships

Everything in the universe is changing from one state to another. All living systems are born, grow to maturity, age, and die. Inanimate (nonliving) objects are also changing. For example, events occurring within the sun are continuously changing. The sun was formed some 4.6 billion years ago and since that time has

*Jacob Bronowski, *The Common Sense of Science* (Cambridge: Harvard University Press, 1953), p. 21.

Arrows connect consumer organisms to their food sources.

Figure 1-9
The Internal Organization of the Grasshopper

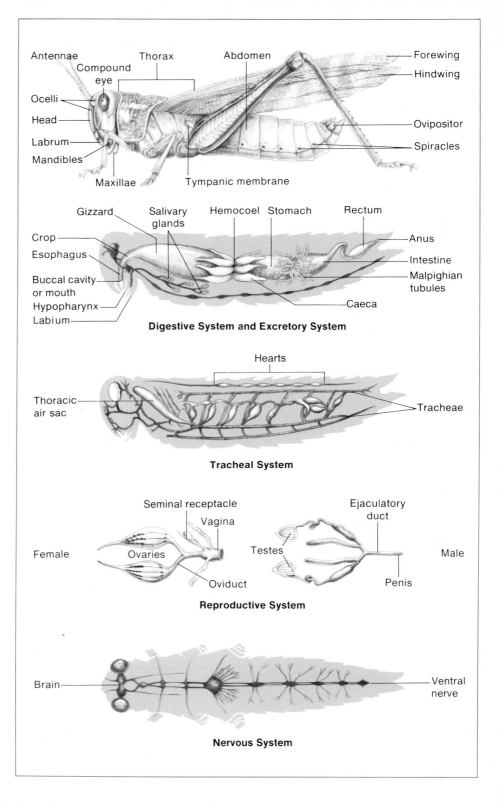

Digestive System and Excretory System

Tracheal System

Reproductive System

Nervous System

radiated energy in the form of heat and light onto the earth and other neighboring objects. This energy is gradually being dissipated, and scientists believe that the sun will cool and darken after several billion years. All of its planets will then be lifeless rocks whirling through space.

Scientists believe there is order in the changing universe—that it is possible to discover what is causing these changes. The relationship between a prior sequence of events (the cause) and the change (the effect) is the *cause-and-effect relationship.*

Cause-and-effect relationships between events occurring in the universe are rarely simple or easy to find. Consider, for example, the growth of a corn plant on a farm in Iowa. If the effect being studied is the final maturation of the plant to the corn-bearing stage, then it is necessary to consider a great number of events that preceded and could have influenced this maturation. Many different factors affect corn maturation, including the type of seed, the moisture content and chemistry of

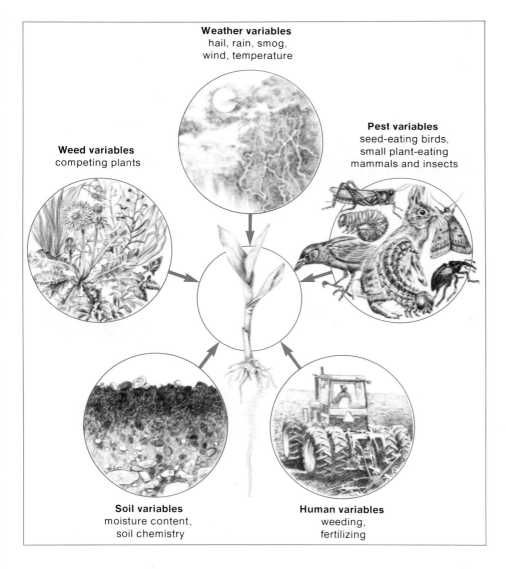

Weather variables
hail, rain, smog,
wind, temperature

Pest variables
seed-eating birds,
small plant-eating
mammals and insects

Weed variables
competing plants

Soil variables
moisture content,
soil chemistry

Human variables
weeding,
fertilizing

Figure 1-10
Variables in Growth
There are several variables that could influence the growth of a corn plant, from the time the seed is planted until the corn is harvested.

the soil, the weather (temperature, sunlight, and rain), and the farmer's activities including weeding and fertilizing. Each unique factor that could contribute to a cause-and-effect relationship is a *variable*. The different variables that must be considered in this growth example, from the time the seed is planted until the corn is harvested, are summarized in Figure 1-10.

Each variable in a cause-and-effect relationship may have a long, complex history. Thus, in our corn-growth example, it is important to know the history of the seed—its genetic origins, age, and physical condition. Rainfall and temperature are also important variables, and the daily weather conditions over the field are part of large-scale weather patterns that originated thousands of miles away.

It also should be noted that variables work together or *interact* to produce an effect. Rainfall, sunlight, and proper fertilization of the soil are all necessary for healthy plant growth. Without one of these or a number of other variable conditions, the plant will not grow to maturity.

So many variables can influence cause-and-effect relationships that determining their contributions and interactions is one of the more difficult problems in science. Later in this chapter, in the discussion devoted to multivariable cause-and-effect relationships, we will examine the procedures that scientists use to solve such problems.

Understanding the Past and Predicting the Future

Laws emerge from descriptions of specific events, and by the time these descriptions have been completed, the events are history. For example, if you witnessed a car accident just an instant ago, by the time you could describe the accident, the sights and sounds would be gone and you would be describing the past.

Scientists are able to predict certain aspects of the future from their understanding of the past. A common but dramatic example is the forecast of the exact times that the sun will rise and set in the next 24 hours. Such predictions are based on the laws of planetary motion, which were derived from descriptions of past movements of the earth and the other planets in relation to the sun. This form of prediction is easy to understand. Many sequences of events occur repeatedly, each time in essentially the same form. Based on descriptions of these sequences, future repetitions of past events can be predicted with a high degree of accuracy. However, even when recurring events have been observed and described many times, some element of uncertainty is still present in the prediction.

The most mysterious and significant form of prediction is the prediction of objects and events that have never been experienced or even imagined. This form of prediction requires scientists to extend their understanding from the known into the unknown by a combination of guesswork, imagination, and logical thought. The development of atomic bombs and nuclear reactors is an example of this process. From their knowledge of matter and energy, scientists predicted that a tremendous amount of energy would be released if the basic components of matter (the atoms) could be split apart. Prior to the first experimental attempts to split the atom, however, this prediction was just an educated guess.

A biological example of this mysterious form of prediction is the understanding we now have of the mechanisms of heredity, which transmit similar features from parent to offspring. The patterns of inheritance revealed by the carefully controlled matings of animals and plants have inspired biologists to predict that structures that could contain and transmit hereditary information would eventually be found in the cells of an organism. Now, with the aid of such sophisticated modern tools as the scanning electron microscope (Figure 1-3), these structures—the *chromosomes*—are routinely observed and studied.

The Scientific Revolution

About 300–400 years ago, *the scientific revolution* arose from the union of two kinds of human activities—applied technology and abstract thought. *Applied technology* was first employed by artisans, craftsmen, and engineers, who perfected tools, shelters, and agricultural methods through trial-and-error experimentation. *Abstract thought* was first practiced by philosophers and scholars, who were interested in such concepts as the meaning of life and the organization of nature. These early thinkers acquired knowledge by reasoning rather than by experimentation.

At the time of the scientific revolution, western civilization was in a state of rebirth, or renaissance. The human mind was breaking free from inhibiting beliefs and restrictive thinking in all areas of life. Business and commerce were flourishing, and practical businessmen were seeking the advice of scholars and philosophers about solving such problems as how to navigate the oceans to enhance exploration and trade. As barriers to communication were removed between the previously secluded intellectual community and the more practical artisans and craftsmen, all of the elements required for the development of modern science were joined together.

Renaissance intellectuals soon realized that the method of trial-and-error testing, or *experimentation,* which had been perfected over the last several thousand years by the men of technology could be of great use in understanding nature. An example of one of the earliest uses of this new insight is a classic experiment in biology performed by the Italian scientist Francesco Redi in 1668. For thousands of years prior to Redi, it had been commonly accepted that life could develop spontaneously from nonliving objects. Thus it was believed that mice suddenly emerged from decaying grain and that maggots arose spontaneously from rotting meat. But Redi questioned this concept of the spontaneous generation of life and proposed that life came from *prior life.* Redi observed that flies hovered around rotting meat and guessed that the appearance of maggots might somehow be caused by the flies.

Thus, the scientific community was presented with two competing ideas about the origins of maggots. In an earlier period, this competition would have been settled by determining which idea was more consistent with established and accepted beliefs. But this was the Renaissance, when all ideas—established or new—were being questioned and tested. To test these competing ideas, Redi performed one of the first biological experiments.

In an experiment, the scientist is not content simply to observe sequences of events as they occur in nature, but instead *intervenes* to change some aspect of the natural sequence to see what will happen. Redi placed rotting meat in three groups of jars: a group covered with tight lids, a group covered with fine gauze, and a group

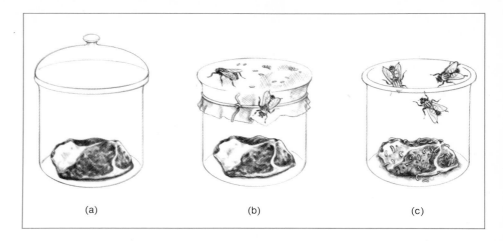

Redi placed rotting meat in three types of jars: (a) sealed, (b) covered with gauze, and (c) uncovered. Maggots appeared only on the meat in the uncovered jars. In the gauze-covered jars, however, where flies could see and smell but could not come in contact with the meat, fly eggs were found on the covers.

with no covers. Redi reasoned that if the maggots were spontaneously generated, they would appear on the meat in all three types of jars, but that if the flies caused the maggots, they would appear only on the meat in the uncovered jars. The results of Redi's experiment (Figure 1-11) provided evidence to support his idea and to refute the prevailing belief in the spontaneous generation of life.

The controversy over spontaneous generation did not end here, however. During the next 200 years, this important scientific concept was experimentally tested on different organisms and rejected. It is through this process of formulating, testing, and reformulating ideas, first employed by Redi and other Renaissance scientists, that we have achieved not only our modern understanding of how life forms originate and develop but all other current scientific knowledge.

Modern Science—A Process of Forming and Testing Hypotheses

The essence of modern science is not impressive machinery or abstract, symbolic language, but the process of forming and testing ideas. These ideas, which are essentially guesses about relationships in nature, are called *hypotheses*. We will examine how hypotheses are formed and tested by analyzing hypothesis testing in the Redi experiment. The sequence of steps in the Redi study is diagrammed in Figure 1-12.

What Is a Hypothesis?

Forming a hypothesis is the first step in the development of a law. Recall that laws account for the relationships between objects and events in the universe—common properties, levels of organization, cause-and-effect relationships—that have been tested in many ways and have yet to be proved wrong.

One characteristic of a hypothesis is therefore that it *describes a relationship between objects and events in the universe.* Redi developed his experiment to test two descriptions:

1. All maggots on rotting meat are produced by spontaneous generation (the accepted hypothesis).

2. All maggots on rotting meat are caused by direct contact between flies and the meat (Redi's hypothesis).

Both of these hypotheses describe cause-and-effect relationships. The accepted hypothesis alludes to a mysterious, unobservable process. Redi's hypothesis proposes a sequence of causal events that somehow begins with fly contact and ends with the appearance of maggots.

Another characteristic of a hypothesis is that it is *universal.* A hypothesis is a generalization about all members of a group or a population (all flies and all maggots) that holds true for all the times that an event occurs (maggots appear on rotting meat). The scientist's goal is to discover universal relationships that apply to all members of a group.

A third characteristic of a hypothesis is that, although it can seldom be proved true, it *can usually be disproved (proved false).* One reason a hypothesis cannot be fully proved is that it is a universal statement about all members of a group, and with rare exceptions, it is not possible to actually observe all members of a group. Redi *hypothesized* about all instances of maggots on rotting meat; certainly, he could never observe all instances of maggots appearing on meat throughout the world. However, just one instance of maggots appearing on meat without prior fly contact would disprove Redi's hypothesis. If maggots had appeared in the closed jars in Redi's experiment, the fly-contact hypothesis would have been proved false.

Another reason that a hypothesis cannot be fully proved is that hypotheses frequently concern phenomena that have never been directly observed and may

Figure 1-12
The Sequence of Steps in Francesco Redi's Study of Spontaneous Generation

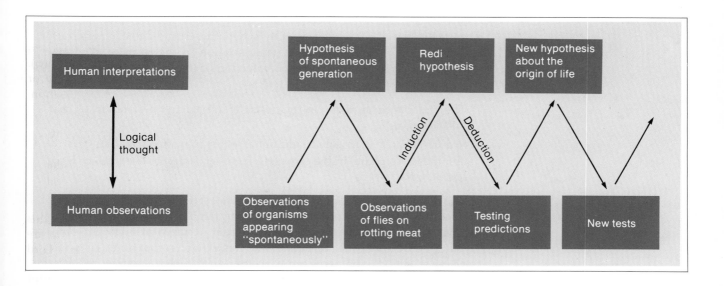

not even be observable. The laws of gravity describe forces between objects, but these forces cannot actually be seen. Genes—the mechanisms within cells that transmit hereditary information between generations—were hypothesized long before the technological capacity existed to observe these mechanisms.

If a hypothesis deals with observable or unobservable *universal* phenomena, then how can it be tested? Scientists must become detectives and examine all of the evidence for and against the hypothesis. This evidence includes all of the logical consequences of the hypothesis—all of the relationships that must also be true if the hypothesis is true. In the Redi example, if fly contact actually causes maggots, then the prevention of fly contact must logically prevent the appearance of maggots. Scientists examine the hypothesis or the universal description of a relationship to determine its logical consequences and to test the validity of these consequences. If the consequences are not true, then the hypothesis cannot be true. We will now discuss how logical thought processes are used to form and test hypotheses.

Observing Nature and Developing a Hypothesis

There are two forms of logical reasoning—*inductive* and *deductive.* Scientists use *inductive logic* to develop a hypothesis. This form of logic progresses from specific observations to a generalization. The scientist examines specific observations until a pattern is found and then makes an educated guess about the underlying relationships that cause the pattern. This guess is the hypothesis.

Scientists do not follow any rules in making this inductive jump from specific observations to the general relationship or hypothesis. Taking the most mysterious and creative step in the scientific process, which involves imagination, guesswork, and luck, the mind jumps from the known to the unknown.

Testing the Hypothesis to Determine Its Logical Consequences

Scientists employ the second form of logical reasoning—*deductive logic*—to examine the logical consequences of their hypotheses (inductive generalizations). Deductive logic progresses from a generalization in the form of a premise or hypothesis to a specific conclusion contained in or implied by the generalization.

An example of deductive logic might be:

The abundance of plants in the desert is limited by the amount of rainfall. *(hypothesis)*

If a desert area is irrigated (additional water is made available), *then* the abundance of plants in that area will increase. *(deductive conclusion or logical consequence)*

The deductive conclusion is always an *if*—*then* statement: *If* this is true, *then* this must be true. This type of reasoning will be familiar to those of you who are Sherlock Holmes fans. ("Watson, *if* the murder occurred at midnight, *then* Lord Harbuttle could not be the murderer, as he was with us at that time.")

Many scientific hypotheses are concerned with highly complex relationships and have many logical consequences. Each of these consequences must be tested and accepted before the hypothesis can be considered a law. If any of the logical consequences are proved false, then the hypothesis must be changed or completely rejected.

Testing the Hypothesis: Returning to Nature

The logical consequences of the hypothesis are determined by deductive reasoning, but it is necessary to return to nature to test these predictions.

Although all scientists form inductive hypotheses and use deductive logic to determine what these hypotheses predict, not all scientists can use experiments to test their predictions. In many cases, scientists cannot or do not wish to manipulate or change the phenomena being studied. Clearly, there is no way to manipulate and change events in the history of life that occurred millions of years ago. And scientists who study organisms in their natural environments usually want to keep human activity from interfering with sequences of events.

When scientists cannot or do not wish to conduct an experiment, they must employ other methods to test their predictions. A scientist might form the hypothesis that early humans migrated along a certain route to follow the big-game animals they were hunting. The logical consequence or prediction contained in this hypothesis is that evidence of early human existence (graves and skeletal remains, the fossilized embers of campfires from that time period, and so on) should be found along this route. The scientist can then test this prediction by searching for such evidence along the hypothesized route.

Multivariable Hypotheses

The hypotheses that scientists investigate are rarely simple one-to-one relationships. Many variables are usually involved in a cause-and-effect relationship, and determining what each variable contributes and how all of the variables act together (*interact*) is one of the more difficult problems scientists face. For example, we saw earlier that many variables must be considered in studying the growth of a corn plant (Figure 1-10). In this section, we will examine four techniques that scientists use to investigate such *multivariable problems:* control of variables, measurement, sampling, and statistics.

Control of Variables

Scientists employ the *control of variables technique* to study the contribution that each variable makes to the cause-and-effect relationship in a multivariable problem. In biology, the organism under study is taken out of nature and put into a laboratory where the variables in its environment can be manipulated (controlled). A series of experiments is then performed. Each time, only one variable is changed and all other variables are kept in some known constant state. For example, the effects of light on corn growth could be studied by subjecting young corn plants to differing amounts of light per day and keeping all other environmental variables constant. (All plants would be kept at the same temperature, receive the same amounts of water at the same intervals, be grown in the same type of soil, etc.)

The *controlled experiment* is a powerful tool for studying multivariable problems, because it permits the scientist to dissect the problem into its component elements and determine what each element contributes to the cause-and-effect relationship. However, complete control is an ideal that can rarely be achieved in biological experiments because living organisms are simultaneously influenced by many variables and laboratory studies provide a distorted picture of the behavior of organisms that are adapted to life in a complex, ever-changing environment. All of the factors in the natural environment can never be duplicated under the artificial and greatly simplified conditions created in a laboratory.

Therefore, to understand many biological problems completely, scientists must conduct controlled laboratory experiments *and* study organisms in their natural environments. In such *natural* or *field studies,* it is still possible to isolate the contributions made by individual variables. The effects of light on corn growth could be examined by planting the corn in parts of the same field that receive varying amounts of light (shaded or unshaded) or in different climates where the number of hours of sunlight per day varies during the growing season. Such studies would make use of natural controls of the light variable.

Measurement

Multivariable problems are rarely solved with a simple yes–no answer. In a study of the effect of light on plant growth, the scientist must be able to evaluate subtle differences in growth. Measurement is required to make such evaluations.

Measurement transforms a quality into a quantity (a number). Every member of your biology class can be described in terms of size, hair color, speed of movement, and other characteristics. Some of these qualities will be the same for everyone; other qualities will vary from person to person. Most people have a left hand and a right hand, but hand size varies over a wide range from large to small. Measurement can be used to show exactly how large each hand is relative to the hands in the rest of the biology class.

Until recently, the most commonly used units of measure in the United States followed the *English system,* which includes such units for length as inches, feet, yards, and miles. The English system is currently being replaced by the *metric system,* which is already used throughout most of the world and in all scientific communities. The basic unit of length in the metric system is the *meter,* which was originally defined as 1/10,000,000 of the earth's quadrant (a *quadrant* is one-quarter of the earth's circumference) and which is now defined in terms of the wavelengths of light emitted by moving atoms. Because you will encounter measurements from both the English and the metric systems in this book, the relationships between the basic units for length, mass, and volume in these two systems are presented in Figure 1-13.

Sampling

The values of hand size obtained for your biology class would be a small sample of the hand-size measurements for the entire human population. Most of the measurements in this book also pertain to small samples taken from large populations. Although biologists are primarily interested in establishing broad,

Figure 1-13
Relationships Between the Basic Units
in the English and Metric Systems of Measurement

Some of these numbers have a positive or negative (—) superscript to the right, such as 10^3 or 10^{-3}, which is called an *exponent*. Recall from algebra that an exponent tells you how many times a number is to be multiplied by itself. For example, $10^3 = 10 \times 10 \times 10$, or 1,000. Similarly, $10^6 = 10 \times 10 \times 10 \times 10 \times 10 \times 10 = 1,000,000$. The negative exponent indicates that the number should be converted to a fraction by dividing the number into 1, and then multiplying this fraction by itself the indicated number of times. Thus,

$$10^{-3} = \frac{1}{10} \times \frac{1}{10} \times \frac{1}{10} = \frac{1}{10 \times 10 \times 10} = \frac{1}{1,000}.$$

Numbers with positive or negative exponents are often used instead of very large or very small numbers, respectively.

Length

1 KILOMETER = 1,000 meters (also written 10^3 meters) = 0.62 mile

1 METER (abbreviated m) = 1/1,000 kilometer (also written 10^{-3} kilometer) = 39.37 inches

1 CENTIMETER (abbreviated cm) = 1/100 meter (also written 10^{-2} m)

1 MILLIMETER (abbreviated mm) = 1/1,000 meter (also written 10^{-3} m)

1 MICROMETER (denoted by μm) = 1/1,000 mm (also written 10^{-3} millimeter) = one-millionth of a meter (written 10^{-6} m)

1 NANOMETER (abbreviated nm; also called 1 millimicrometer and denoted by mμ) = 1/1,000 micrometer = one-billionth of a meter (written 10^{-9} m)

1 ANGSTROM (abbreviated Å) = 1/10 nanometer = 1/10,000 micrometer = one ten-billionth of a meter (written 10^{-10} m)

Mass

1 KILOGRAM = 1,000 grams (also written 10^3 grams) = 2.20 pounds

1 GRAM (abbreviated g) = 1/1,000 kilogram (also written 10^{-3} kilogram)

1 MILLIGRAM (abbreviated mg) = 1/1,000 gram (also written 10^{-3} gram)

1 MICROGRAM (denoted by μg) = 1/1,000 milligram (also written 10^{-3} milligram) = one-millionth of a gram (written 10^{-6} g)

1 NANOGRAM (abbreviated ng) = 1/1,000 microgram (also written 10^{-3} microgram) = one billionth of a gram (written 10^{-9} g)

Volume

1 LITER (abbreviated l) = 1,000 milliliters (also written 10^3 milliliters) = 1.06 liquid quarts

1 MILLILITER (abbreviated ml) = 1/1,000 liter (also written 10^{-3} l) = 1 cubic centimeter (abbreviated cc or cm^3); 1 ml (= 1 cc) of water at 4°C weighs 1 gram

1 MICROLITER (denoted by μl) = 1/1,000 milliliter (also written 10^{-3} milliliter) = one-millionth of a liter (also written 10^{-6} l); 1 microliter of water at 4°C weighs 1 milligram.

general relationships that describe large populations, it is usually impossible for them to measure more than a small sample from the population of interest.

However, it is essential that the largest possible sample be obtained, because it is too easy to construct a false hypothesis from only a few observations. The sample also should be truly representative of the population under study. Consider how representative your biology class is of the entire human population. Does it contain both sexes and all racial and ethnic groups at all age levels? If not, what subgroup of the human population does your class represent?

Statistics

If a sample is truly representative of a larger population, then measurements from the sample can be used to estimate characteristics of the population. These estimates are really hypotheses about an entire population based on a very small sampling from that population. To make these estimates, biologists employ procedures from a branch of mathematics called *statistics.*

One familiar population characteristic that is statistically estimated from a sample is the *mean* or *average value* of the population. Your biology class is a small sample from a large population, and although it probably contains some very tall and very short people, most of the students in your class are between these two extreme heights. The mean value describes the height of the "average" person in the class. It is calculated by summing the individual height measurements for all of the students in the class and dividing this total by the number of students in the class. If your class is truly representative of a large population, the mean height of your class will be an estimate of the mean height of the population.

Scientists employ statistical methods to reveal the differences between populations. To learn whether Americans are taller than Europeans, a sample of heights from each population could be taken and a mean height could be calculated for each population. Other statistical methods could then be applied to determine whether the observed differences between the two sample means did or did not indicate real differences between the heights of Americans and Europeans. Many other kinds of statistical techniques allow us to make accurate comparisons between populations. In general, scientists use statistical tools to summarize information about samples and to make decisions about whether the sampled populations do or do not differ from one another.

The control of variables and the use of such *quantitative techniques* as measurement, sampling, and statistics are particularly advantageous when multivariable problems are studied. These tools help the scientist to dissect such a problem into its component variables and then to determine the exact contribution each variable makes in the cause-and-effect sequence.

Hypotheses, Theories, and Laws

The world of life is a vast, interconnected web of objects and events. Biologists face the all-encompassing problem of understanding the total organization of this web. Individually, most biologists develop and test *hypotheses* about some limited aspect of a universal cause-and-effect relationship. However, some individuals undertake the enormous task of integrating all that is known in a particular field into general *theories.* For example, the *theory of evolution* developed by Charles Darwin currently dominates and integrates the science of biology. In Chapter 8, you will be introduced to the Darwinian theory and exposed to its influence in all areas of biology.

Theories are not final, perfect statements. Rather, they are stages in the process of developing and testing hypotheses. Theories are constructed by linking together many related hypotheses, and scientists examine theories regularly to test their component hypotheses. Unlike a simple hypothesis, however, a theory is so broad and can lead to so many logical consequences that it must be tested in a multitude of ways. It is often more appropriate to remove or to change component hypotheses than it is to reject the entire theory.

Simple hypotheses and broad aspects of integrated theories that have been repeatedly tested and not proved to be false are called *laws.* However, even laws may not be entirely correct. Some future scientist may devise a test that will prove that a law is inadequate. If this happens, the hypotheses and theories that led to the establishment of the law must be rejected or revised.

The Scientific Version of Reality

The body of scientific knowledge includes hypotheses, theories, and laws in various stages of development and testing that permit us to interpret the relationships that exist in nature. These interpretations of reality are based on observations of nature. However, reality lies somewhere in between our observations and our interpretation of these observations. The diagram in Figure 1-14 shows one way to view these concepts. There, the scientific process moves between observations and interpretations through logical thought patterns.

Hypotheses, theories, and laws are developed by inductive analyses and are examined deductively to determine their logical consequences—what they predict. Scientists then test these predictions by making additional observations. The test results and new observations may change previous hypotheses, theories, and laws

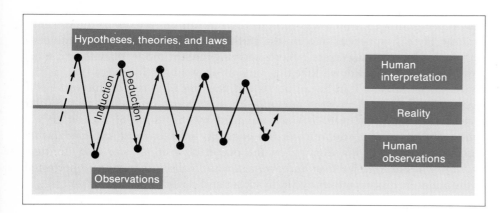

Figure 1-14
The Scientific Process

Figure 1-15
The Version of the Universe Accepted by Fifteenth-Century Scholars
This woodcut illustration was done by a fifteenth-century German artist.

or produce new ones, which will be tested in turn, and the cycle continues. The goal is to develop a scientific version of reality that is identical to what we observe, so that every prediction developed from our interpretation is always found to be true in nature.

To see how far this process has been developed, consider the version of reality, shown in Figure 1-15, which was accepted by scholars as recently as the fifteenth century. Here, a spherical, motionless earth lies at the center of the universe, surrounded by revolving crystal spheres carrying the sun, moon, planets, and stars. Now, of course, we "know" that this version of reality is incorrect. Yet, as the American astronomer Carl Sagan points out, the first *direct* test of this version of reality was not made until our unmanned space probes broke free of the solar system and this event was "not accompanied by the tinkle of broken crystal spheres." Our current scientific version of reality—the body of knowledge that describes a world of life existing on the planet of a minor star in a limitless universe—has been tested over and over again and will be changed and perfected as the scientific process moves cyclically closer and closer to reality.

Summary

Science remains an open frontier. There are many scientific discoveries still to be made. Scientists are motivated by *curiosity* and act on the belief that there is *order* in the changing universe. Their goal is to discover and describe this order. The scientist's goal is to determine the *cause-and-effect relationships*, *common properties*, and *levels of organization* among objects and events in nature. The laws that describe these relationships are human interpretations of the order in the universe.

Modern science arose from the union of two kinds of human activity—*applied technology* and *abstract thought.* The essence of modern science is a process of developing and testing *hypotheses*. Hypotheses—educated guesses about the relationships between objects and events in the universe—are developed by observing nature and employing a process of *inductive logic* to link the observations together into a general relationship. The hypotheses are examined by employing *deductive logic* to determine their logical consequences, or *what they predict.* Each hypothesis is then tested by making the additional observations specified by the predictions. If the predictions are found to be untrue, then the hypothesis must be modified or completely rejected.

A *theory* is a group of related hypotheses. When a theory or a simple hypothesis has been repeatedly tested and not disproved, it is referred to as a *law*. However, a law can still be altered. All aspects of our interpretation of nature are subject to change based on the consequences of new observations. The scientific process oscillates between human *observation* and human *interpretation.* Reality usually exists somewhere between. When continued observations and interpretations are made, the oscillations become less extreme and the human interpretation more and more closely approximates reality.

The Evolution
of Life

2

The Time Before Life

2

A t the most basic level of organization, all physical things—water, rocks, air, stars, plants, and animals—are composed of exceedingly small units called *atoms*. The atoms that form our bodies originated billions of years ago in stars. Before becoming a part of us, they were a part of the earth, its atmosphere, and millions of organisms. After being a part of us, they will be incorporated into other physical things. This recycling of the same atoms over billions of years unites us with all living and nonliving things.

The Structure of Atoms

Different kinds of physical things are formed from different structural combinations of atoms. Although the universe is immense and diverse, almost all of the things we encounter in daily life are made up of just a dozen different kinds of atoms. Only 92 types of atoms are known to occur naturally, and most of these are very rare.

  The atom is enormously complex and still not fully understood. The basic structure of the atom is comprised of a compact *nucleus* or center surrounded by a diffuse cloud of *electrons* (Figure 2-1). Although the electrons are much smaller than the nucleus, they occupy most of the space within the atom, because they are positioned far apart and each electron moves continuously within a specific path around the nucleus. The components of the nucleus, on the other hand, are packed together and are fairly stationary. You can visualize the relative volumes taken up by these structures if you imagine an atom magnified to 200 meters (about 220 yards) in diameter. The nucleus of this atom would be about $\frac{1}{2}$ centimeter ($\frac{1}{5}$ inch) across, and its electrons would appear as a large cloud of widely spaced pencil dots moving throughout the remainder of the volume.

The Atomic Nucleus

  Almost 200 types of particles have already been shown to occur within the nuclei of atoms, and additional particles are still being discovered. Because most of

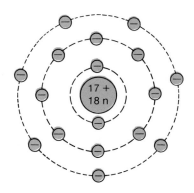

Figure 2-1
**The Basic Structure
of an Atom**

Each atom contains equal numbers of protons and electrons. The protons (+), which carry a positive charge, and the neutrons (n), which have no charge, are located within the nucleus. The electrons (−), which carry a negative electric charge, are continuously moving in patterns of concentric spheres, or shells, around the nucleus. The chlorine atom, diagrammed here, contains 17 protons, 17 electrons and 18 neutrons.

Table 2-1 Properties of the Major Components of an Atom		
Particle	Weight*	Electric charge
Electron	0.00055	−1
Proton	1.00732	+1
Neutron	1.00871	0

*Weight is measured in *atomic mass units* (AMU); 1 AMU = 1.66 × 10⁻²⁴ grams (g).

the properties of these nuclear components are not understood, we cannot describe the nucleus of an atom completely. However, in the study of biology, it is sufficient to consider only the two major components of the nucleus—protons and neutrons.

Protons and *neutrons* are relatively heavy particles and are closely packed together within the nucleus. Each of these particles is about 2,000 times heavier than an electron (see Table 2-1). Neutrons are electrically neutral; they do not have a positive or a negative charge. Protons have a positive charge, and their electrical properties cause them to repel one another. Although protons resist being close together, they exist in a clustered state due to other, little-understood particles in the nucleus that appear to hold the protons together.

Each type of atom has a characteristic number of protons. For example, the hydrogen atom contains a single proton, and the oxygen atom contains eight protons. However, each type of atom does not contain the same number of neutrons. When a type of atom occurs in several forms and each form contains a different number of neutrons, these forms are called *isotopes*. The isotopes of an atom have the same chemical properties, but different weights. For example, the carbon atom always contains six protons, but one type of carbon isotope contains six neutrons and another contains seven neutrons.

The Electron Cloud

Like protons, electrons exhibit electrical properties. All electrons have a negative charge and tend to repel one another. An electron (−) and a proton (+) have opposite electrical charges, which attract them to one another and hold them together. An atom is composed of equal numbers of electrons and protons, so that the number of negative particles is equal to the number of positive particles and the overall electrical charge of the atom is *neutral.* The *number* of electrons and protons characterizes different types of atoms.

Although the electrons appear to be scattered in a cloud around the nucleus, their spatial arrangement is not random. Electrons travel within *shells* or orbits that surround the nucleus. A shell is not an actual structure but an area located a particular distance from the nucleus through which a group of electrons moves in a path around the nucleus. As we can see in Figure 2-1, the first shell (the shell closest to the nucleus) of chlorine contains just two electrons, the middle shell of this atom contains eight electrons, and the outermost shell contains seven electrons. Because electrons always appear in regular patterns, scientists have developed specific rules

Table 2-2
Arrangement of Electrons into Shells

Only the 30 smallest atoms are shown. Shell 1 is closest to the nucleus, shell 2 occurs just after shell 1, and so on.

Atom	Total number of electrons in atom	Number of electrons in shell			
		1	2	3	4
Hydrogen	1	1	0	0	0
Helium	2	2	0	0	0
Lithium	3	2	1	0	0
Beryllium	4	2	2	0	0
Boron	5	2	3	0	0
Carbon	6	2	4	0	0
Nitrogen	7	2	5	0	0
Oxygen	8	2	6	0	0
Fluorine	9	2	7	0	0
Neon	10	2	8	0	0
Sodium	11	2	8	1	0
Magnesium	12	2	8	2	0
Aluminum	13	2	8	3	0
Silicon	14	2	8	4	0
Phosphorus	15	2	8	5	0
Sulfur	16	2	8	6	0
Chlorine	17	2	8	7	0
Argon	18	2	8	8	0
Potassium	19	2	8	8	1
Calcium	20	2	8	8	2
Scandium	21	2	8	9	2
Titanium	22	2	8	10	2
Vanadium	23	2	8	11	2
Chromium	24	2	8	12	2
Manganese	25	2	8	13	2
Iron	26	2	8	14	2
Cobalt	27	2	8	15	2
Nickel	28	2	8	16	2
Copper	29	2	8	18	1
Zinc	30	2	8	18	2

to describe how electrons are distributed in the shells surrounding the nucleus.

If we examine a series of atoms in which each atom contains one more electron than the previous atom, two general rules become clear: (1) each shell can hold only a certain number of electrons, and (2) the electrons occupy the innermost shells first (Table 2-2). The maximum number of electrons that can exist in any given shell n can be calculated from the expression

$$\text{Maximum number} = 2n^2$$

Thus, the first shell ($n = 1$) of an atom can hold only 2 electrons; the second shell ($n = 2$), up to 8 electrons; the third shell ($n = 3$), up to 18 electrons; and so on. The two general rules of electron distribution are demonstrated in Figure 2-1. The 17 electrons in the chlorine atom are arranged so that they completely fill the first and second shells and leave just 7 electrons in the third shell. The next largest atom, argon, contains 18 protons and 18 electrons. The additional electron occurs in the third shell, which contains a total of 8 electrons.

The rule governing the maximum number of electrons in a shell does not hold for the outermost shell of an atom, which can never contain more than 8 electrons. This shell is not filled to its maximum until it is no longer the outermost shell. The arrangements of electrons in the shells of some of the smaller atoms are provided in Table 2-2. In the argon atom, the third shell contains 8 electrons. The additional electrons in the two next larger atoms, potassium and calcium, are contained in the fourth shell. Once the fourth shell contains two electrons, as it does in calcium, the additional electrons found in still larger atoms (from scandium to copper) occur in the third shell until it holds its maximum of 18 electrons. The fourth shell then fills until it contains 8 electrons, and the pattern repeats itself for larger and larger atoms. In each type of atom, the number of electrons is equal to the number of protons, and this number determines both the number of electron shells and the number of electrons in the outermost shell.

The properties of an atom are determined by the number of electrons in the outermost shell. Atoms that have similar numbers of electrons in their outermost shell have similar chemical properties. For example, lithium and sodium interact with other atoms in a similar way because both atoms have a single electron in their outermost shell.

The Types of Atoms

The simplest atom is hydrogen, which contains one proton and one electron. The most complex atom, which has been constructed by nuclear chemists under unusual laboratory conditions, has 106 electrons and a nucleus containing 106 protons and 153 or 157 neutrons (two isotopes). Very large atoms like this unnatural one are unstable and tend to break down into smaller atoms.

All of the 92 naturally occurring atoms can be ordered in an unbroken sequence from hydrogen to uranium according to the number of electrons and protons they contain (see Table 2-3). Each atom in this sequence contains one more proton and one more electron than the preceding atom. Tables that list all of the atoms that occur naturally are called *tables of elements.* An *element* is a material that is constructed entirely of identical atoms. A piece of pure silver is composed entirely of silver atoms and is therefore an element.

The nuclei of smaller atoms are stable in that they do not come apart easily or spontaneously. These nuclei undergo structural changes only under the extraordinary degrees of temperature and pressure that exist in the interiors of stars or in atomic bombs, for example. However, atoms do readily gain or lose electrons, depending on the structural pattern of the electrons surrounding the nucleus, and this property is important in the study of biology.

The number of electrons in atoms with filled outer shells is stable. A filled outer shell contains eight electrons, except for the filled outer shell of helium, which

Table 2-3
Symbols and Atomic Numbers for the 92 Naturally Occurring Atoms

The *atomic number* is the number of protons or electrons in the atom. Asterisks indicate atoms that are essential to the formation and development of living organisms.

Atom	Symbol	Atomic number	Atom	Symbol	Atomic number
Hydrogen*	H	1	Silver	Ag	47
Helium	He	2	Cadmium	Cd	48
Lithium	Li	3	Indium	In	49
Beryllium	Be	4	Tin*	Sn	50
Boron	B	5	Antimony	Sb	51
Carbon*	C	6	Tellurium	Te	52
Nitrogen*	N	7	Iodine*	I	53
Oxygen*	O	8	Xenon	Xe	54
Fluorine*	F	9	Cesium	Cs	55
Neon	Ne	10	Barium	Ba	56
Sodium*	Na	11	Lanthanum	La	57
Magnesium*	Mg	12	Cerium	Ce	58
Aluminum	Al	13	Praseodymium	Pr	59
Silicon*	Si	14	Neodymium	Nd	60
Phosphorus*	P	15	Promethium	Pm	61
Sulfur*	S	16	Samarium	Sm	62
Chlorine*	Cl	17	Europium	Eu	63
Argon	Ar	18	Gadolinium	Gd	64
Potassium*	K	19	Terbium	Tb	65
Calcium*	Ca	20	Dysprosium	Dy	66
Scandium	Sc	21	Holmium	Ho	67
Titanium	Ti	22	Erbium	Er	68
Vanadium*	V	23	Thulium	Tm	69
Chromium*	Cr	24	Ytterbium	Yb	70
Manganese*	Mn	25	Lutetium	Lu	71
Iron*	Fe	26	Hafnium	Hf	72
Cobalt*	Co	27	Tantalum	Ta	73
Nickel	Ni	28	Tungsten	W	74
Copper*	Cu	29	Rhenium	Re	75
Zinc*	Zn	30	Osmium	Os	76
Gallium	Ga	31	Iridium	Ir	77
Germanium	Ge	32	Platinum	Pt	78
Arsenic	As	33	Gold	Au	79
Selenium*	Se	34	Mercury	Hg	80
Bromine	Br	35	Thallium	Tl	81
Krypton	Kr	36	Lead	Pb	82
Rubidium	Rb	37	Bismuth	Bi	83
Strontium	Sr	38	Polonium	Po	84
Yttrium	Y	39	Astatine	At	85
Zirconium	Zr	40	Radon	Rn	86
Niobium	Nb	41	Francium	Fr	87
Molybdenum*	Mo	42	Radium	Ra	88
Technetium	Tc	43	Actinium	Ac	89
Ruthenium	Ru	44	Thorium	Th	90
Rhodium	Rh	45	Protactinium	Pa	91
Palladium	Pd	46	Uranium	U	92

contains two electrons (see Table 2-2). Atoms with partially filled outer shells, however, commonly gain or lose electrons. When this happens, they are left with unequal numbers of electrons and protons and no longer have a neutral electrical charge. When electrons are lost, the atom acquires an overall positive charge; when electrons are gained, the atom becomes negative in charge.

Atoms with unequal numbers of electrons and protons are called *ions.* The chlorine atom shown in Figure 2-1 tends to gain a single electron to form a stable outer shell of 8 electrons. The chlorine ion contains 18 electrons and 17 protons and has an overall negative charge. Symbolically, the chlorine ion is represented by Cl^-, compared to chlorine in its neutral form, which is simply represented by Cl. Chlorine is transformed from a neutral to an ionized state whenever electrons in its environment are free to be added to its outer shell. A chlorine atom may acquire an electron if it is in the presence of another type of atom that has a single electron in its outer shell. Sodium (Na), for example, contains 11 electrons: two electrons in the first shell, eight in the second, and one in the outer shell (see Table 2-2). The sodium atom tends to lose this outermost electron in the presence of a chlorine atom. Consequently, both atoms become ions: chlorine acquires a negative charge (Cl^-), and sodium acquires a positive charge (Na^+). Both of these ions are common in biological systems.

The major effect of an electrical charge is to exert a force on other charged particles in its vicinity. The attractive and repulsive forces between charged particles determine the structure of atoms and are responsible for the physical and chemical interactions of all living and nonliving materials.

The Formation of Atoms

Hydrogen and small amounts of other atoms occur in vast clouds of gas between the stars. Within these clouds, hydrogen is present in the form of single atoms and in groups of two atoms (H_2). It also occurs in very small amounts in the form of hydrogen ions (H^+), which are single protons. When clouds of gas condense to form stars, larger atoms are formed from the hydrogen atoms. The earth's atoms were formed billions of years ago when stars developed from clouds of H_2 gas.

From Clouds of Gas to Stars

Atoms within a gas cloud usually tend to move away from one another, so that the cloud expands. Under certain conditions, however, the atoms move toward one another and the gas cloud becomes smaller. The force of mutual attraction that causes the atoms to move toward one another, called *gravity,* occurs between all bodies in the universe and is not the same as the attractive force between positive and negative charges.

Scientists still do not know why the forces of gravity are sometimes stronger than the tendency of atoms to move away from each other. But the atoms within a gas cloud are occasionally moved close enough to one another to be influenced by their mutual gravitational force. The coming together or *condensation* of atoms increases temperature and pressure within the cloud to such a degree that the electrons eventually become separated from the protons. The protons then aggregate in a reaction called *nuclear fusion,* a star is born from a cloud of gas, and the changes that occur in the protons cause larger atoms to form.

Stars occur in vast assemblages called *galaxies* that are held together by gravity. In addition to hundreds of billions of stars, a galaxy also contains gas clouds and "dust," or tiny groups of atoms. Our own galaxy (Figure 2-2) began to form many billions of years ago when stars created from condensing gas were held together by gravity. Stars within our galaxy continue to be born and eventually to die. Our galaxy is just one of many billions of galaxies in the universe.

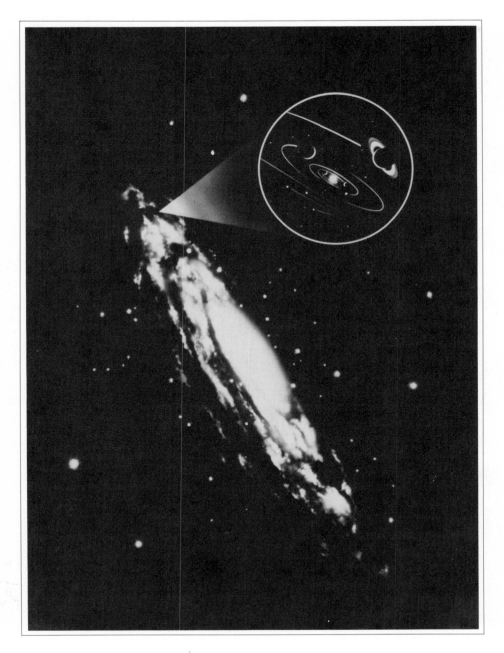

Figure 2-2
Our Galaxy
Our solar system (enlarged within the circle) is just one of the billions of solar systems in the Milky Way galaxy.

Figure 2-3
The Formation of Our Solar System

Our solar system is believed to have begun as a diffuse cloud of gas (a) that eventually condensed (b–d). The sun was formed from the dense center of the cloud, and the planets were formed from the cooler, less dense outer regions.

Nuclear Changes That Produce Larger Atoms

Many different types of atoms are formed from the original hydrogen atoms inside stars. Small atoms combine in a process of nuclear fusion to form larger atoms. The first step in the sequence of changes—the joining, or *fusion,* of hydrogen atoms to form helium atoms—occurs when the electrons and protons of the hydrogen atoms become separated as the hydrogen gas is condensing to form a star. The protons, moving at tremendous speeds, slam together and fuse to form the nuclei of helium atoms. The hydrogen atom consists of one electron and one proton, whereas the helium atom consists of two electrons, two protons, and two neutrons. In young stars, four protons from four hydrogen atoms come together to form a helium nucleus. During this process, two of the protons are restructured; some of the material becomes neutrons, and what remains of these protons is transformed into other particles that are not incorporated into the helium nucleus. At this stage, the newly formed nucleus contains two neutrons and two protons, and has a charge of +2. This formation attracts two electrons, becomes neutral in charge, and is then a helium atom.

A tremendous amount of energy in the form of heat and light is released during such changes in atomic structure. (The fusion of hydrogen atoms to produce helium atoms is the same reaction that occurs during the explosion of a hydrogen bomb.) Because hydrogen is so abundant and the amount of energy released in a hydrogen reaction is so great, scientists hope that someday nuclear fusion will be a safe energy source. The amount of energy that could be made available from controlled nuclear fusion is essentially unlimited.

In larger stars, after the hydrogen has been converted to helium, these atoms fuse with one another to form carbon atoms and again release energy in the process. Eventually, such an enormous amount of energy is released that the large stars explode. During the explosion, carbon is converted to even larger atoms by the same fusion reaction.

Iron, which has 26 protons and 26 electrons, is the largest atom formed by nuclear fusion. Atoms larger than iron are formed when neutrons produced during the explosion of a star are captured by the nuclei of iron atoms. These nuclei are heavy with neutrons and are so unstable that some of the neutrons are converted into electrons and protons. In this way, stable atoms with higher numbers of electrons and protons are formed. As the explosion continues, the atoms that have been formed are dispersed into space.

The earth and sun were formed about 4.6 billion years ago from a parent cloud of hydrogen gas mixed with small amounts of larger atoms produced in earlier stars (Figure 2-3). The dense, hot gas at the center of the cloud condensed to form our star, the sun. The cooler, less dense outer regions of the cloud formed the planets. The raw materials of primitive life are assumed to have been hydrogen, carbon, nitrogen, and oxygen. These same atoms are known to exist currently in the sun and other stars.

The Structure of Molecules

On the cooler planets like our earth, atoms rarely occur by themselves. Instead, they are joined in groups of varying numbers, called *molecules,* which are held

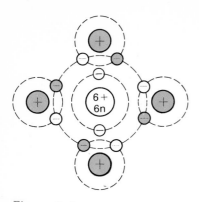

Figure 2-4
The Structure of Methane

The methane molecule is formed
from one carbon atom (unshaded)
and four hydrogen atoms (shaded).
The structure of methane is
diagrammed

```
        H
        |
  H  —  C  —  H
        |
        H
```

together by the mutual attraction of the protons and electrons in their atoms. A molecule may be made up of several identical atoms, such as a molecule of hydrogen gas in which two atoms of hydrogen are joined together, or a molecule may be made up of different types of atoms, such as a molecule of methane gas, which is formed from one carbon atom and four hydrogen atoms (Figure 2-4). To simplify communication, chemists have developed a shorthand for expressing the types and numbers of atoms in a molecule. A methane molecule is written CH_4, where C represents a single carbon atom and H_4 represents four hydrogen atoms.

The characteristics of most molecules differ greatly from the characteristics of their component atoms. For example, hydrogen and oxygen normally occur as gases composed of two-atom molecules, symbolized by H_2 and O_2, respectively. However, these two gases combine to form water (H_2O)—a substance with very different properties than either of its atoms. The reaction between hydrogen and oxygen to form water is written

$$2(H_2) + O_2 \longrightarrow 2(H_2O)$$

which tells us that two molecules of hydrogen gas plus one molecule of oxygen gas combine to form two molecules of water. The symbols to the left of the arrow represent the materials prior to the chemical reaction; the symbols to the right of the arrow represent the product materials after the reaction has taken place. This equation is *balanced,* because it accounts for all atoms involved.

If a few atoms in a molecular structure are added, removed, or rearranged, the properties of the molecule change greatly. The construction of many different forms of molecules from a few types of atoms can be compared to the formation of a language of millions of words from a few dozen letters. Just as the variety of words, not letters, contributes to the diversity of a language, the variety of molecules, not atoms, produces the great diversity of substances in the universe.

Chemical Bonds

When two atoms bump into each other, they may be held together by the electrical charges of their protons and electrons. Some atoms form a stable association because the electrons of each atom become attracted both to their own nucleus and to the nucleus of the other atom. Whether or not the attraction of the electrons to both nuclei is strong enough to hold the atoms together depends on how the electrons are arranged within the atoms. Attractive forces that hold atoms together are called *chemical bonds.*

Three types of chemical bonds occur in living systems—covalent, ionic, and hydrogen bonds. *Covalent bonds* form when the electrons of two atoms are shared in such a manner that both atoms acquire a stable outer shell. In the methane molecule, shown in Figure 2-4, the hydrogen and carbon atoms are held together by covalent bonds. Each electron in the outer shell of the hydrogen atoms and the carbon atom is influenced by the positive charge from both nuclei. The methane molecule is more stable than the individual atoms of carbon and hydrogen because the outer shells of all the atoms are completely filled. By *sharing* electrons with the four hydrogen atoms, the carbon atom achieves a stable configuration of eight electrons in its outer shell. Similarly, each hydrogen has one electron of its own and achieves a stable outer shell of two electrons in the presence of the carbon atom.

Multiple covalent bonds—two or three bonds which join a pair of atoms—may also be formed by electron sharing. Each bond involves one electron from each atom. For example, a double bond between carbon and oxygen occurs in each molecule of formaldehyde (Figure 2-5). An oxygen atom with six electrons in its outer shell shares two of its electrons with a carbon atom, thereby achieving a stable outer shell of eight electrons. Carbon obtains eight electrons in its outer shell by sharing two electrons with the oxygen atom and one electron with each of the two hydrogen atoms. Double bonds between carbon and oxygen are common in biological molecules.

The second type of chemical bond is the *ionic bond,* which is also called the *electrovalent bond.* Ionic bonds form when electrons are completely transferred from one atom to another. The chlorine atom (Cl) has an outer shell of seven electrons and tends to gain an electron to achieve a stable outer shell of eight electrons. The sodium atom (Na) has an outer shell of only one electron and tends to lose this electron to achieve a stable outer shell of eight electrons. If these two atoms collide, the sodium atom loses one electron to the chlorine atom. The resulting sodium ion (Na^+) has a positive charge and the resulting chlorine ion (Cl^-) has a negative charge, due to the transfer of the electron. Because these two ions have opposite electrical charges, they are attracted to each other and form sodium chloride (NaCl), or table salt. In an ionic bond, electrons are transferred and the oppositely charged ions are attracted to one another. In a covalent bond, the electrons are not displaced from their original shells and bonding is accomplished by sharing electrons.

The third type of chemical bond is the *hydrogen bond,* which is formed by an attraction between a hydrogen atom and two other atoms such that the hydrogen atom holds the two other atoms together. When a hydrogen atom is joined to another atom by a covalent bond, two electrons orbit its nucleus. These electrons spend more time on the side of the hydrogen atom where the covalent bond exists, so that much of the time the side opposite the bonding site consists of an exposed proton and has a slight positive charge. This *asymmetry* of hydrogen atoms is

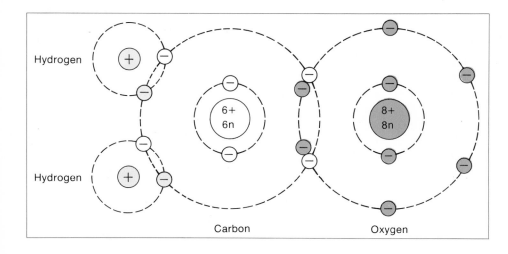

Figure 2-5
The Structure of Formaldehyde

The formaldehyde molecule is formed from one carbon atom (unshaded), one oxygen atom (dark brown), and two hydrogen atoms (light brown). The structure includes a double bond between the carbon and oxygen atoms. The formaldehyde molecule is diagrammed

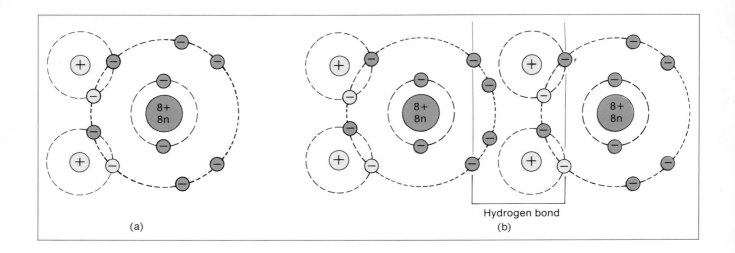

**Figure 2-6
The Hydrogen Bond**

(a) The water molecule has a slight positive charge on one side (left in the diagram) and a slight negative charge on the other side (right in the diagram), due to the manner in which the electrons travel within the molecule. The protons and electrons of the oxygen atom are dark brown in the diagram; the protons and electrons of the hydrogen atoms are light brown. (b) Two water molecules are held together by the attraction between the positive side of one molecule and the negative side of the other. The two molecules are bridged by a hydrogen bond.

illustrated by the water molecule, which is shown in Figure 2-6(a). The oxygen atom within the water molecule is also somewhat asymmetrical due to the covalent bonds it forms with hydrogen. The opposite side of the oxygen atom from its bonding point with hydrogen contains four exposed electrons and has a slight negative charge. A water molecule therefore exhibits a slight positive charge on one side and a slight negative charge on the other. When two water molecules lie near each other, the positively charged side of one molecule is attracted to the negatively charged side of the other, and the two molecules are weakly joined together, Figure 2-6(b). The hydrogen bond is much weaker than a covalent bond or an ionic bond.

This discussion would be incomplete if we failed to examine the energy changes involved in chemical bonding. *Energy* is not an easy concept to define. It is measured in terms of the capacity to do "work." Heat energy, for example, is measured in calories, and a calorie is the amount of heat energy required to do the work of raising the temperature of 1 gram of water 1 degree Celsius or centigrade (C). A rise in the temperature of the water reflects a gain in energy. The concept of energy will be explored further later in the chapter, but this greatly simplified definition will give you an idea of what is meant by a change in the amount of energy within a system.

Ordinarily, molecules contain less energy than their atoms contain. When a chemical bond forms between two atoms, some of the energy within the structure is released. The atoms then remain joined together until the same amount of energy is added to break the bonds and return the individual atoms to their previous energy levels. The forces of a chemical bond are similar to the forces that hold two magnets together. The magnets remain joined until energy is applied (in this case, the muscular movement of someone's arms) in a manner that pulls them apart. Similarly, the attraction between atoms can only be overcome by the addition of sufficient energy to overcome the energy of attraction.

Chemical bonds are important in biology not only because they permit a large variety of molecules to form but also because they produce changes in energy levels when they are formed or broken. For example, when altered, the chemical bonds in food molecules provide organisms with the essential energy to maintain life. We will explore molecules and chemical bonds further in Chapter 3.

The Earth's Early Atmosphere: Setting the Stage for Life

The earth's first atmosphere contained high concentrations of hydrogen and helium, but most of these atoms escaped during the formation of the planet before it was massive enough to hold such small, light materials by gravitational force. The loss of this first atmosphere explains the low concentrations of helium in our present atmosphere, despite the abundance of helium elsewhere in the universe.

The earth's second atmosphere formed as gases were released from the interior of the earth. These gases—predominantly methane (CH_4), ammonia (NH_3), water vapor (H_2O), hydrogen gas (H_2), nitrogen gas (N_2), and some carbon monoxide (CO) and carbon dioxide (CO_2)—provided the setting for the origin of life. Unlike our present atmosphere, the earth's second atmosphere contained no oxygen gas (O_2). Most of the water vapor condensed as the earth cooled, and oceans formed in the natural basins on the earth's surface.

In 1937, the Russian biochemist A.I. Oparin hypothesized that if suitable energy had been available to permit chemical reactions to occur, such a mixture of gases would have given rise to a variety of biologically important molecules. A great deal of energy was probably available on primitive earth. Some of the major sources of energy would have been radiation from the sun, electrical discharges from lightning, heat from volcanoes, and shock waves from the impact on the earth of rock particles from outer space. Oparin believed that some of this energy caused the atmospheric gases to join together to form larger molecules that eventually rained down into the oceans, where they accumulated to form a warm, dilute "soup" within which life began. The different types of energy vary in intensity, and the chemical reactions produced would have yielded different types of molecules.

Experimental Tests

Oparin's hypothesis about the origin of life has been indirectly tested in experiments designed to simulate the conditions that existed on primitive earth. In the first of these experiments, conducted in 1953 by Stanley Miller and Harold Urey at the University of Chicago, an electric current was passed through a mixture of methane, ammonia, hydrogen, and water. The apparatus used in the Miller–Urey experiment is shown in Figure 2-7. Under these simple conditions, many of the basic molecules found in all living organisms were formed.

Since this early experiment, many similar laboratory tests have produced a variety of biologically important molecules. The results of these tests support Oparin's hypothesis that the molecular components of primitive organisms could have been formed from the simple gases that are believed to have been present in the earth's early atmosphere. The requirements of life turn out to be remarkably simple. A few materials, treated appropriately, can produce the unique molecules that are the basic components of life.

Figure 2-7
The Apparatus Used in the First Experiment Designed to Simulate Conditions During the Origin of Life

Gaseous ammonia, methane, and hydrogen were placed in this apparatus. Boiling water in the small flask circulated the gases and water vapor through the large flask, where a spark was discharged to provide an energy source. Biological molecules were formed and accumulated in the water trap.

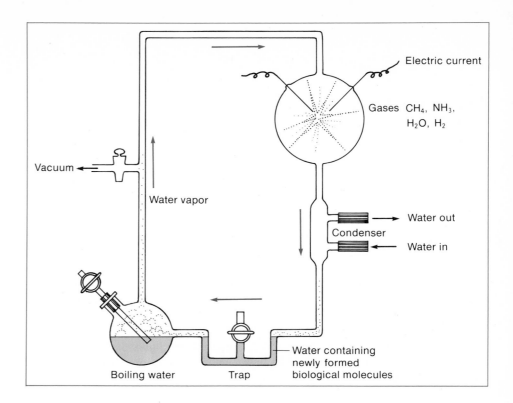

The Chemistry of Life

The composition of the earth and its atmosphere limited the number of atoms from which living organisms could form. Only 24 of the 92 types of atoms that occur naturally on earth have been shown to be essential to the formation and growth of living organisms. Most of the atoms that make up the human body are found in abundance in the earth's crust or its oceans (see Table 2-4).

In addition to the abundance of atoms, three other factors have played a major part in determining the chemistry of living forms. *Water* is the basic liquid within which materials are dissolved in living systems. A greater variety of substances will dissolve in water than in almost any other liquid. Many of these substances also tend to become ions when they are dissolved in water—a property that facilitates chemical reactions. In addition, water plays an important role in the development of hydrogen bonds. One side of the water molecule contains exposed protons, and the other side contains exposed electrons; these sides afford sites for the formation of weak hydrogen bonds that are integral to biological reactions. Many other materials that are essential to life interact with water. Moreover, water has unusual capacities for taking in and releasing heat energy without creating great temperature changes. This property, which promotes constancy of temperature, will be discussed in Chapter 17.

Carbon atoms, the central building blocks of large biological molecules, can

Table 2-4
The Atomic Composition of the Human Body Compared to the Atomic Composition of Seawater, the Earth's Crust, and the Universe

Type of atom	Human body (%)	Seawater (%)	Earth's crust (%)	Universe (%)
Aluminum	0	*	7.9	*
Calcium	0.31	0.006	3.5	*
Carbon	9.5	0.0014	0.19	0.021
Chlorine	0.03	0.33	*	*
Helium	0	*	*	9.1
Hydrogen	63.0	66.0	0.22	91.0
Iron	*	*	4.5	0.002
Magnesium	0.01	0.033	2.2	0.002
Nitrogen	1.4	*	*	0.042
Oxygen	25.5	33.0	47.0	0.057
Potassium	0.06	*	2.5	*
Silicon	*	*	28.0	0.003
Sodium	0.03	0.28	2.5	*
Sulfur	0.05	0.017	*	0.001

*Indicates trace amount.

bond together in long chains and rings to provide a framework for the construction of very large molecules. The carbon atom's ease and versatility of bonding is responsible for the millions of different biological molecules found on earth.

Biological materials are constructed from very *small, light atoms.* In smaller atoms, the positively charged nucleus of one atom is more exposed to the negatively charged electrons of another atom, and the resultant attraction readily leads to the formation of strong chemical bonds between the two atoms. Due to the greater tendency of small atoms to form bonds and to the greater strength of these bonds, small atoms are more frequently involved in the formation of the large molecules found in living systems. Only three of the 24 types of atoms known to be essential to life have more than 34 protons and electrons, and only very small amounts of these three atoms (molybdenum, tin, and iodine) are found in living organisms.

The Properties of Energy

Before we complete our discussion of the physical universe and proceed to an examination of the more specific biological systems, it is important to consider the general properties of energy. One of the more important biological characteristics is the way in which living materials utilize energy.

The study of energy is called *thermodynamics* (from the Greek words meaning "heat motion"), because the earliest studies of energy were concerned with the way in which heat flowed from one part of a system to another part. *Energy* can be defined as anything that makes it possible to do work—to bring about movement against resistance. To throw a ball into the air is to move an object (the ball) against the resistance of gravity. This motion requires energy, which is derived from muscular movements of the arm, which, in turn, are driven by chemical energy from within the molecules of the muscles.

Heat, light, electricity, magnetism, motion, sound, chemical bonds, and nuclear reactions are all forms of energy, and these energies can be converted from one form to another. The muscular movements required to throw a ball result from the conversion of *chemical energy* to *kinetic energy,* or motion. In later chapters, we will see that the chemical energy within muscles is ultimately derived from the light energy of the sun.

The Laws of Thermodynamics

Our basic understanding of energy can be expressed in the form of two laws. The most powerful and fundamental generalization that can be made about the universe is the *First Law of Thermodynamics,* which states that energy can be transferred from one place to another or transformed from one form to another but that it cannot be created or destroyed. From this generalization, it follows that the total amount of energy in the universe is constant. The First Law of Thermodynamics, often called the *law of conservation of energy,* is regarded as a law because all attempts to disprove it have failed, despite a large number of painstaking observations and ingenious experiments.

The *Second Law of Thermodynamics* cannot be as easily stated. The original expression of this law was that heat can never pass spontaneously from a colder to a hotter body and that a temperature difference can never appear spontaneously in a body that is originally at uniform temperature. A more general interpretation of the law is that any system, including the universe, naturally tends to increase its *entropy.*

Entropy

Entropy is the tendency of energy to become evenly distributed throughout a system. Wherever energy is more concentrated than usual, that concentration decreases; wherever energy is less concentrated than usual, that concentration rises.

Consider the distribution of heat—the form of energy with which we are most familiar. Heat is the random movement of the atoms or molecules that make up physical objects. As particles move more rapidly, heat energy increases and the object becomes warmer. Heat tends to flow from a hot place (an area of more concentrated energy) to a cold place (an area of less concentrated energy). An ice cube placed in a glass of water will melt as the heat energy in the water flows into the ice cube. As the water loses heat and the ice gains heat, the energy throughout the glass becomes more evenly distributed, or tends toward entropy. When the ice cube is first placed in the water, the water particles are moving more rapidly than the particles in the ice cube. When the particles collide and bounce apart, some of the kinetic energy *(momentum)* is transferred from one particle to the other. The most likely outcome of a collision is that the particle moving at a greater speed will lose some kinetic energy and slow down and that the particle moving at a lesser speed will gain some kinetic energy and speed up. Eventually, all of the particles within the glass will be moving at about the same speed and the energy will be evenly distributed throughout the glass. Entropy increases as a result of the collision of particles, which tends to reduce extreme concentrations of energy.

The increase in entropy that occurs spontaneously in the universe is often referred to as an *increase of disorder* or a *decrease in the degree of organization.* At first, the fast- and slow-moving molecules in the ice water are concentrated in different places in the glass; that is, the molecules are organized according to speed of movement. After a period of time, however, these molecules intermingle and the total energy is distributed among all of the molecules in the glass. Initially, the two systems—ice cube and water—are different, but later they become equivalent. Before mixing, work can be accomplished within the system because the two forms of water have different energy levels. Once the energy is uniformly distributed throughout the system, however, no further energy changes or *work* can take place.

It has been said that living systems are an exception to the Second Law of Thermodynamics because they represent a greater state of order or organization than that found elsewhere in the universe. However, the high levels of organization in living systems are maintained only by the regular input of energy, as we will see in later chapters. Without the continual addition of energy, all living systems would rapidly disintegrate into a disorderly array of atoms and would tend toward entropy.

Summary

The basic units of all physical things are called *atoms.* Each atom has a central *nucleus,* which contains *protons* and *neutrons* and is surrounded by a cloud of *electrons.* The protons have a positive charge, the electrons have a negative charge, and the neutrons have no charge. Electrons and protons are attracted to one another, but electrons repel one another, as do protons. These *attractive* and *repulsive* forces between the charged particles of atoms determine the physical and chemical states of all living and nonliving materials. The electrons of an atom travel around the nucleus in specific paths or *shells.* The number of electrons in the *outermost shell* of an atom determines how that atom will interact with other atoms.

Certain kinds of atoms interact with one another to form *chemical bonds*—attractive forces that hold atoms together in an assemblage called a *molecule.* Three kinds of chemical bonds are important in living systems:

1. The *covalent bond,* in which the electrons of two atoms are shared.
2. The *ionic bond,* in which electrons are transferred from one atom to another and the resulting ions are attracted to one another by their opposite charges.
3. The *hydrogen bond,* in which a hydrogen atom is linked to another atom by a covalent bond, so that its proton remains somewhat exposed and the area around the exposed proton has a slight positive charge that attracts the electrons of other atoms.

Chemical bonds between atoms are formed and broken by *changes in energy levels*, which are fundamental to the continuance of life.

Scientists have shown that the basic molecules of life could have formed readily from the materials they believe were present in the earth's atmosphere shortly after our planet was formed about 4.6 billion years ago. These materials were methane, ammonia, water, hydrogen gas, nitrogen gas, carbon monoxide, and carbon dioxide. The three most important factors that determine the chemistry of life are the unique properties of *water*, the bonding characteristics of *carbon atoms*, and the way in which *small atoms* readily form molecules. All living systems contain water, all biological molecules contain carbon, and most of the atoms known to be essential to life are relatively small.

Living systems are dynamic, and the changes that occur within living things are generated by the chemical energy within the bonds that join atoms together. Two fundamental laws describe the properties of energy. The *First Law of Thermodynamics* states that energy can be transferred from one place to another or transformed from one form to another but that it cannot be created or destroyed. According to the *Second Law of Thermodynamics*, all systems tend toward *entropy*—the even distribution of energy throughout a system.

The Molecules of Life: Structure and Origin

How do living things differ from nonliving things? This important and difficult question has puzzled philosophers and scientists for centuries. Even today, despite all that we know about the organization of atoms and molecules in the universe, we do not have a definitive answer to this question.

All of the objects on earth are constructed from different combinations of the same types of atoms. If these objects are arranged in order of increasing complexity of molecular organization, from simple structures containing a few atoms and chemical bonds to large structures composed of many atoms and chemical bonds, the simplest molecules are unquestionably not living and the most complex molecules are living. Between these extremes, however, it is not possible to draw a boundary line that absolutely divides all nonliving material from all living material.

Although we cannot always distinguish between living and nonliving materials, scientists know a great deal about the chemical nature of living structures and how they could have developed in the past. At least in part, the historical pathways from the simplest groupings of atoms to the elaborate organizations of living systems can now be reconstructed.

In Chapter 2, we examined how particles in the universe combine to form higher levels of organization within the atoms. Now we will consider a comparable process—how atoms join to become molecules and how molecules join to become more complex structures. This is the process that eventually led to the origin of life. These gradual changes from simple to complex levels of organization involve primarily carbon, hydrogen, oxygen, nitrogen, sulfur, and phosphorus atoms. The particular arrangement of these atoms within living organisms is the culmination of billions of years of gradual molecular changes and the persistence of certain forms of molecules at the expense of others.

The Formation of Biological Molecules
Prior to the Appearance of Cells

In Chapter 2, we studied the fundamental unit of molecules—the *atom*. Here, we will examine the fundamental unit of life—the *cell*. All things on earth that we can classify without hesitation as living—such as trees, insects, birds, and humans—are groupings of cells (Figure 3-1).

Cells are remarkably alike, regardless of the types of organisms they form. Because cells share so many common properties, scientists believe that all cells have descended from a single type of ancestral cell. Apparently, the characteristics of the first kinds of cells have been transmitted to all subsequent cells.

Cells are also enormously complex. They are constructed of intricate patterns of large molecules that function together to perform a variety of activities. Due to the complexity of cellular structure and function, it seems inconceivable that the first cell could have suddenly appeared on earth. A fairly long period of gradual molecular changes must have preceded the incorporation of molecules into the first complete cell.

Recent evidence indicates that the transition from large molecules to simple cells took place very early in the earth's history. Fossils found in South Africa are believed to represent cells that existed 3.5 billion years ago. These earliest known cells were already quite complex and apparently possessed much of the basic organization found in the cells of living organisms today.

Figure 3-1
Organisms Are Composed of Cells

(a) Nerve cell of a mouse ($\times$382), (b) human cheek cells ($\times$292); (c) cells in an onion bulb ($\times$125); (d) frog blood cells ($\times$389); (e) cells in a section of a cat trachea ($\times$146).

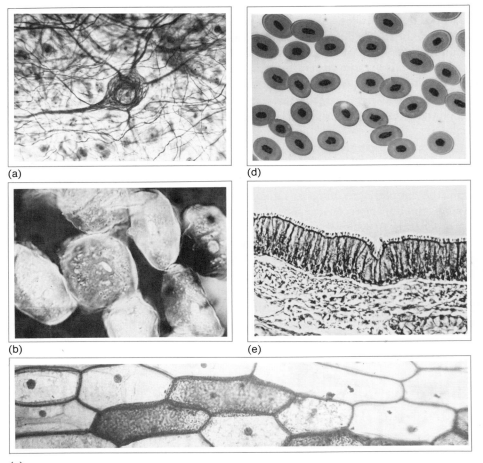

(a)

(d)

(b)

(e)

(c)

The very first cells must have originated from noncellular materials. Molecules of considerable variety and complexity probably accumulated slowly within the seas of primitive earth before the first cells appeared. Large molecules develop when smaller molecules interact to form chemical bonds. The frequency of these chemical interactions depends on how often molecules encounter one another. The development of large molecules would have occurred at a slow rate prior to the appearance of cells, because the molecules existing in the earth's oceans were widely separated. Once cells appeared, however, new types of molecules could have formed at a rapid rate as smaller molecules accumulated in high densities within the cell structure and, being close to one another, interacted frequently to form more complex molecules.

All that we know about the early molecules is derived from our observation of present forms of life. The primitive oceans probably contained a vast array of diverse molecules, many of which could have possessed the necessary characteristics to form the framework of life. However, only a few of these molecules existed for a sufficiently long period of time to eventually unite with one another and form cells. Modern life forms are a record of the few lines of descent that have managed to persist out of the many that have evolved in the past.

In this chapter, we will examine the structure of biological molecules in modern cells. Then we will consider how molecules could have formed in the oceans of primitive earth and how molecular structures could have been organized into the first cells.

The Framework of Life: Specific Large Molecules

In general, the molecules within nonliving (*inorganic*) objects are quite stable, whereas the molecules within living (*organic* or *biological*) systems are so fragile that their organization is relatively easy to destroy. Biological molecules are usually formed from a much greater number of atoms than inorganic molecules are. Greater stability can be achieved when only a few atoms join to form molecules. As atoms are added, the molecular structure becomes increasingly unstable.

An enormous variety of chemical reactions are required to maintain a living cell, and each of these reactions involves one or more biological molecules. Therefore, each cell must contain a large variety of molecules.

Large molecules have a greater potential for diversity than small molecules, simply because large molecules contain greater numbers of atoms in their structures and these atoms can be arranged in a greater variety of ways. Cells are formed from large molecules because variety of molecular function is essential to life. Because they are large, the molecules of life are also fragile. Living organisms tend to be made up of molecules that are large, fragile, and highly diverse in structure.

All biological molecules contain carbon atoms. Although the atoms of most elements combine with other atoms only in limited ways to form small molecules, carbon atoms join with each other and with different types of atoms to an almost unlimited extent. Very large molecules are formed from long chains and rings of carbon atoms and occur in a multitude of shapes, sizes, and varieties.

The Basic Biological Molecules

The four basic types of biological molecules are carbohydrates, lipids, proteins, and nucleic acids. All of these molecules occur as *polymers* (*poly* = many), which means they are constructed from multiple repetitions of a basic molecular unit, the *monomer* (*mono* = one). Monomers are joined together by covalent bonds to form polymers. When each of these bonds forms, a molecule of water is removed via a reaction called *dehydration synthesis.* Similarly, a polymer can be broken down into its component monomers by breaking each bond between adjacent monomers and adding a molecule of water. The addition of a water molecule during the breakdown of a polymer is called *hydrolysis.*

Carbohydrates

Carbohydrates are energy sources for all cells and also form important structural components, such as the wood of trees and the skeletons of insects. Carbohydrates are formed from carbon, hydrogen, and oxygen atoms, and their basic molecular pattern is the association of two hydrogen atoms and one oxygen atom with each carbon atom. The term carbohydrate means "watered carbon." The name originated when early chemists incorrectly concluded that carbohydrates were composed of water molecules (two hydrogen atoms joined with one oxygen atom) attached to carbon atoms. The group of molecules classified as carbohydrates actually turned out to have more complicated structures, but the original name has persisted.

The carbohydrates of greatest biological importance are the *sugars,* also known as *saccharides.* The simplest sugar, *monosaccharide* (a monomer), can contain from 2 to 10 carbon atoms. *Glucose,* a 6-carbon monosaccharide, is the primary energy source for most cells; 12 hydrogen atoms and 6 oxygen atoms are associated with the 6 carbon atoms of glucose, as shown in Figure 3-2(a) and (b). The chemical formula for glucose is written $C_6H_{12}O_6$.

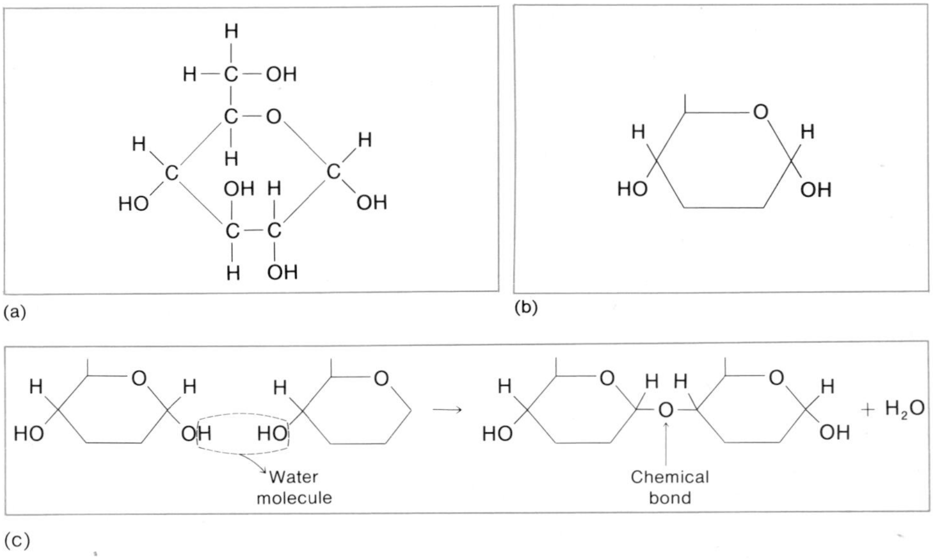

Figure 3-2
The Structure of Saccharides

(a) The arrangement of atoms within a glucose molecule. The basic structure or *frame* of this molecule is a circular ring formed of five carbon atoms and one oxygen atom. Other atoms, including the sixth carbon atom, are attached as side groups to the basic frame. When an oxygen atom and a hydrogen atom make up a side group, the group is usually written as OH (or HO) without drawing the bond between the two atoms. (b) The standard symbolized diagram of the structure in (a), omitting all of the carbon atoms and most of the side groups. The hydrogen and oxygen atoms on the left and right side are indicated because they play a role in the formation of chemical bonds between glucose molecules. (c) When monosaccharides join together to form a polymer, a molecule of water is removed at each bond. In the diagram, two glucose molecules join together to form a maltose molecule, which is a type of disaccharide. A water molecule is formed from the OH groups and released during the bond formation.

(a)

(b)

(c)

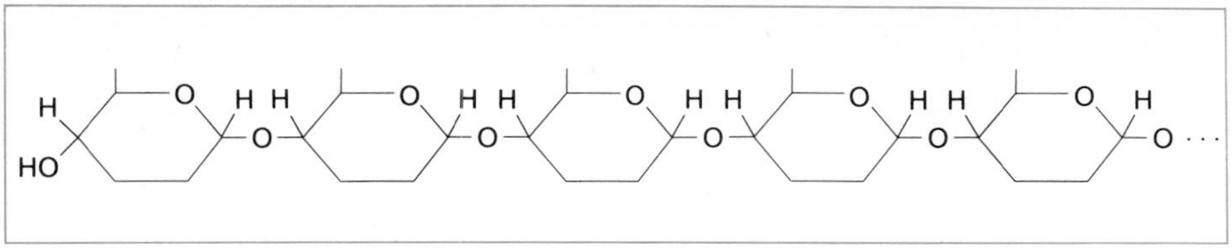

Figure 3-3
Part of a Polysaccharide
Molecule

Polysaccharides are polymers of simple sugar monomers. In this diagram, each glucose molecule contains the atoms shown in Figure 3-2(a) (except at the bonding site) but is shown in the standard symbolized form explained in Figure 3-2(b). One molecule of water is removed as each bond is formed between the glucose molecules.

Disaccharides are sugars formed by the union of two monosaccharide monomers, with the loss of one water molecule. The common disaccharide maltose is formed in this manner from two glucose molecules, as shown in Figure 3-2(c). The chemical formula of maltose is $C_{12}H_{22}O_{11}$. Maltose is formed during the breakdown of starches, which occurs when plant material is digested by animals. *Sucrose* (table sugar) and *lactose* (milk sugar) are two other important disaccharides in biological systems.

Polysaccharides are large polymers of the simple sugar monomers (Figure 3-3). The polysaccharides of greatest biological importance are *starch, glycogen,* and *cellulose.* Energy is stored in plants in the form of starch, and in animals in the form of glycogen. Both starch and glycogen must be broken down into their monomers (glucose molecules) before the energy they contain can be used by the cells. Cellulose is the main component of plant *cell walls*—the rigid structures that surround each cell within a plant. Materials made from plants, such as wood, cotton, and paper, consist almost entirely of cellulose molecules. Only a few organisms are able to break cellulose down into monomers of glucose to make use of the energy it contains. Many cellulose-digesting organisms are *unicellular* (composed of a single cell) and live within the digestive tracts of animals. These resident organisms permit cows and termites to obtain usable energy in the form of glucose from the cellulose molecules in grass and wood.

In addition to their importance as a source of energy and in plant structure, polysaccharides also form the support material in the outer skeletons of insects and crustaceans and are an important component of skin and other animal tissues.

Lipids

The distinctive characteristic of the broad category of biological molecules classified as *lipids* is that they tend to be insoluble in water. Like carbohydrates, lipids are composed of carbon, hydrogen, and oxygen atoms, but lipids contain fewer oxygen atoms than carbohydrates do.

The lipid group includes four types of molecules:

1. The *simple lipids,* which are the natural fats, oils, and waxes.
2. The *steroids,* which form vitamin D, cholesterol, a number of hormones, and many other diverse biological molecules.
3. The *compound lipids,* formed from the union of a simple lipid and another type of molecule, which make up much of the cell structure and form the cell membranes.
4. The *carotenoids,* which form many of the substances that give color to plant and animal materials.

Figure 3-4
Formation of a Fat Molecule
Each fat molecule is formed from an alcohol molecule (usually glycerol) and three fatty acid molecules. (a) The arrangement of atoms in a molecule of glycerol. (b) The arrangement of atoms in a molecule of fatty acid. (Only the portion of the molecule that is the same for all types of fatty acids is shown.) The Ⓡ indicates an attached side group of atoms that differs in the various types of fatty acid molecules. Note the double bond between the carbon and oxygen atoms. (c) A fat molecule is formed when chemical bonds develop between the three OH groups of a glycerol molecule and the OH groups of three fatty acid molecules. One molecule of water is eliminated from the structure at each of the three bonding sites.

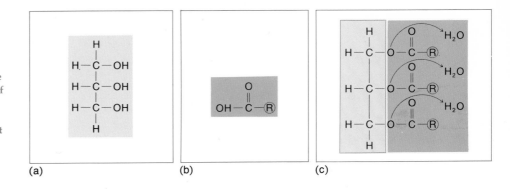

(a)   (b)   (c)

The most common type of lipid is *fat*, a polymer formed from an alcohol molecule (usually glycerol) and three fatty acid molecules (Figure 3-4). Fats differ in structure and function because they are constructed of different types of fatty acid molecules. Fat is the molecule in which most animals store energy on a long-term basis.

Proteins

The third component of living material, *protein*, is quite different from carbohydrates and lipids. In Greek, the word protein means "of first importance"—an appropriate term, since all of the basic biological processes are dependent on proteins.

Like carbohydrates and lipids, proteins are constructed from carbon, hydrogen, and oxygen atoms. In addition, proteins always contain nitrogen and often contain sulfur. Although they are composed of only a few types of atoms, proteins are large molecules, often containing thousands of atoms, that have the potential to form many different, complex structures.

Proteins are particularly fragile because they are large structures and can be destroyed by small changes in the physical and chemical environment. For example, only slight heating can cause irreversible changes in protein structure. Cooking an egg causes the egg white, which is a protein solution, to harden (coagulate). This change is irreversible; cooling the egg white cannot return it to its previous liquid form. For the same reason, a high fever is dangerous to human life. At temperatures exceeding about 110°F (43.5°C), the proteins within the human body lose their structures and, consequently, their functions. Like the change in the egg white, this loss of structure is irreversible.

Proteins make up a large part of every living organism and are crucial to all cellular activities. New proteins are assembled within a cell from simple monomers called *amino acids,* shown in Figure 3-5(a) and (b). The proteins of most organisms contain 20 different kinds of amino acids. However, all 20 of these amino acids do not appear in each protein molecule; each molecule contains many of the same type of amino acids. In a protein molecule, 50–1,000 or more amino acids unite by chemical bonding (with the removal of a water molecule) to form a long chain, as shown in Figure 3-5(c). Each bond between two amino acids is a *peptide bond* (a form of covalent bond). The particular sequence of amino acids in the chain determines the characteristics of the protein.

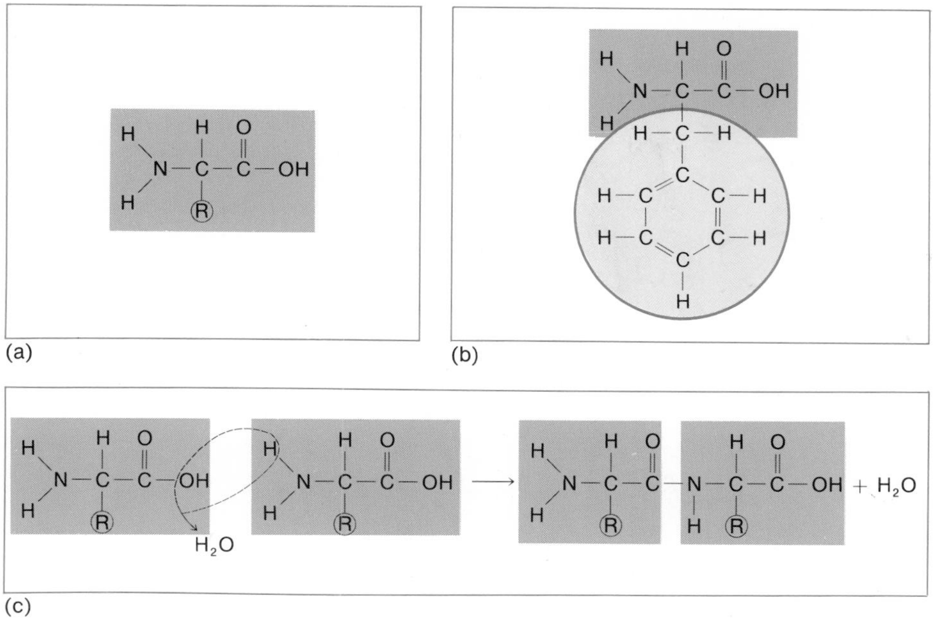

(a)

(b)

(c)

Figure 3-5
Amino Acids Join Together
to Form Proteins

(a) The general structure of all amino acids. The symbol Ⓡ represents a side group of atoms that is unique for each of the 20 basic types of amino acids. (b) A particular amino acid, phenylalanine, with its side group Ⓡ shown in detail. (c) When two amino acids join together, a water molecule is removed and a peptide bond is formed. A protein is made up of hundreds of amino acids linked together by peptide bonds.

Protein molecules are so diverse because a large variety of amino acid arrangements can occur in the chain. All of the words in the English language are formed from different combinations of 26 letters. Proteins have only 20 "letters" (the amino acids), but the number of "words" (proteins) they form can be very large; a single "word" may contain thousands of "letters."

The exact molecular structures or sequences of amino acids are known for only a few proteins. The first protein to be analyzed was insulin—an important molecule that controls the concentration of glucose in the blood. Insulin was found to have the chemical formula $C_{254}H_{377}N_{65}O_{75}S_6$, but it took ten years of research (1944–1954) for scientists to work out the exact sequence of amino acids in the insulin molecule. Recently developed techniques make this type of analysis somewhat less time-consuming, but the determination of the exact sequence of amino acids in a protein remains difficult.

Each protein has a particular shape, which is defined by the folding and bending of the long chain of amino acids. There are four levels of structural complexity in a protein molecule (Figure 3-6). The *primary structure* is the particular sequence of amino acids in the chain. The *secondary structure* is the twisting of the chain to produce a spiral, which is caused by the formation of hydrogen bonds between the different amino acids in the primary structure. Each bond joins an oxygen atom of one amino acid to a nitrogen atom of another amino acid in such a way that either a spiral or a pleated-sheet pattern is formed. Some proteins also form weak chemical bonds between parts of the secondary structure, which causes further folding. The resulting *tertiary structure* is usually globular or spherical in shape. Finally, some proteins have *quaternary structures* in which two or more protein molecules are held together by chemical bonds.

The final shape of a protein molecule is an important key to the determination of its properties. There are two basic types of protein molecules—fibrous proteins and globular proteins. *Fibrous proteins* are long molecules without the foldings of a

Figure 3-6
Levels of Structural Organization Within a Protein Molecule

The *primary structure* of a protein (not shown in detail) is the chain of amino acids linked together by peptide bonds. This chain (indicated by the black line) is twisted to form a spiral, which is the *secondary structure* of a protein. The spiral, in turn, is folded in on itself, so that the whole structure resembles a globule, which is the *tertiary structure* of a protein. The *quaternary structure* is the linking of this molecule with another protein molecule.

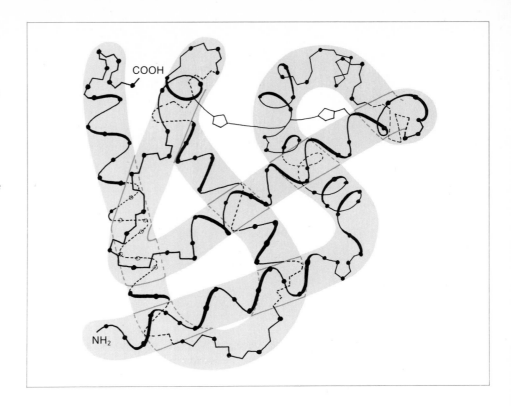

tertiary structure. Fibrous proteins maintain their structure in a variety of different types of solutions, and such cell structures as skin, bones, muscles, tendons, and ligaments are formed largely from cells packed with fibrous proteins. *Globular proteins* have tertiary and quaternary structures that can be altered by a variety of solutions, particularly salty, acidic, or alkaline solutions. These more fragile proteins form *enzymes* (the molecules that govern chemical reactions), hormones, and the proteins in the blood.

Nucleic Acids: RNA and DNA

Like the carbohydrates, lipids, and proteins, the nucleic acids are also polymers that are of key importance to life. The two forms of nucleic acids—*deoxyribonucleic acid (DNA)* and *ribonucleic acid (RNA)*—work together to direct the assembly of amino acids into proteins within the cell. The nucleic acids are composed of carbon, hydrogen, oxygen, nitrogen, and phosphorus atoms.

The monomers from which nucleic acid molecules are constructed are called *nucleotides*. Each nucleotide is formed from three molecules—a sugar, a phosphate, and a nitrogenous base (see Figures 3-7 and 3-8). An RNA nucleotide is formed from a ribose molecule (the sugar), and phosphate molecule, and one of four different types of nitrogenous-base molecules (adenine, guanine, cytosine, or uracil), as shown in Figure 3-9(a). In the formation of an RNA polymer, many nucleotides are joined together by chemical bonding between the ribose of one nucleotide and the phosphate of another, as shown in Figure 3-9(b). There are four

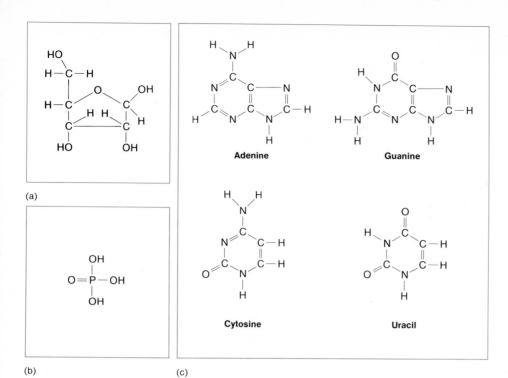

Figure 3-7
The Molecules That Form
the Nucleotides of RNA

(a) The sugar ribose. (b) The phosphate (P is the symbol for phosphorus). (c) The four nitrogenous bases.

(a)

(b)

(c)

Figure 3-8
The Molecules That Form
the Nucleotides of DNA

(a) The sugar deoxyribose. (b) The phosphate. (c) The four nitrogenous bases. Note that three of the nitrogenous bases (adenine, guanine, and cytosine) are the same as the nitrogenous bases in the RNA nucleotide, but that the fourth nitrogenous base (thymine) is different.

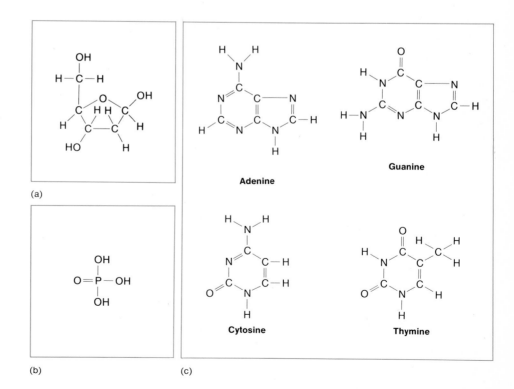

(a)

(b)

(c)

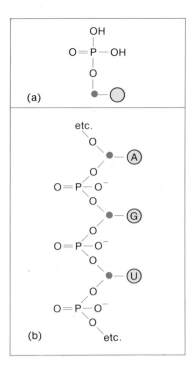

Figure 3-9
Part of an RNA Molecule
Each small circle represents a ribose
molecule; each large circle represents
a nitrogenous-base molecule. (a) A
nucleotide monomer is formed from a
sugar, a nitrogenous base, and a
phosphate molecule. (b) Many
nucleotides link together to
form an RNA molecule. Only three
nucleotides are linked here. The three
different bases in this particular
segment of RNA are adenine (A),
guanine (G), and uracil (U).

different types of RNA nucleotides, each containing one of the four nitrogenous bases.

The composition of the ribose–phosphate strand is the same in all RNA polymers, but the sequence in which the four types of nitrogenous bases are arranged varies from molecule to molecule. This sequence of bases gives each RNA molecule a unique structure and function. RNA is important in the synthesis of proteins, and several categories of RNA molecules are found within each cell.

The DNA molecule, or the "master molecule of life," is also a polymer formed from nucleotide monomers. Each DNA nucleotide contains a deoxyribose molecule (a sugar similar to ribose but containing one less oxygen atom), a phosphate molecule, and one of four different kinds of nitrogenous base molecules (adenine, guanine, cytosine, or thymine), as shown in Figure 3-8. The DNA nucleotide differs from the RNA nucleotide in that DNA nucleotides contain deoxyribose (instead of ribose) and their fourth possible nitrogenous base is thymine (rather than uracil).

As in the RNA molecule, the nucleotides in the DNA molecule are joined together by the chemical bonds that form between the sugar and the phosphate molecular units. Unlike the RNA molecule, which contains only one strand of nucleotides, the DNA molecule contains two strands of nucleotides joined together by bonds between their bases. The structure of the DNA molecule is similar to a ladder in which the "rungs" are the pairs of bases between the two sugar–phosphate strands (Figure 3-10). The nitrogenous bases pair up between the sugar–phosphate strands in a very specific manner. The adenine of one strand always pairs with the thymine of another strand and guanine always pairs with cytosine, so that the arrangement of nucleotides along one side of the molecule determines the arrangement of nucleotides along the other side. This fixed pattern of bond formation between the four types of nitrogenous bases is known as the *base-pairing rule.* In contrast, no fixed rules govern the way in which the four different types of nucleotides are arranged along the length of the molecule. The particular sequence of bases (the variable part of the nucleotide) within a DNA molecule designates its unique structure and function.

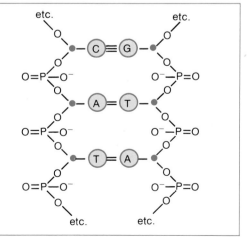

Figure 3-10
Part of a DNA Molecule
Each small circle represents a
deoxyribose molecule; each large
circle represents a nitrogenous-base
molecule (adenine, guanine, cytosine,
or thymine). Six nucleotides are
shown, joined together by bonds
between the deoxyribose and
phosphate parts of the molecules
as well as by bonds between the
nitrogenous bases. (Guanine and
cytosine are always joined by three
bonds; adenine and thymine, by two
bonds.)

To summarize the two forms of nucleic acids, the DNA molecule differs from the RNA molecule in three basic ways:

1. DNA is double-stranded, whereas RNA is single-stranded.
2. DNA contains the sugar deoxyribose, whereas RNA contains ribose.
3. One of the four nitrogenous bases is different in the two molecules. DNA contains thymine instead of uracil.

The first three-dimensional model of a DNA molecule was contructed by James Watson and Francis Crick in 1953 (Figure 3-11). Although the arrangement of atoms in the structure of the DNA molecule cannot actually be seen, all tests of the Watson–Crick model indicate that it is correct. A major achievement in the science of biology, the model DNA structure has revealed the chemical basis for heredity: the nitrogenous bases of the DNA structure form the genes—the basic units that transmit hereditary traits. The particular sequence of bases along the DNA molecule forms the blueprint, or genetic code, of an organism. We will examine how DNA and RNA molecules determine the characteristics of an organism in subsequent chapters.

The Formation of Polymers During Precellular Evolution

All of the complex molecules essential for life must have formed from the simple molecules present on primitive earth, but scientists do not yet know how the simple precursors of life joined together to form the large molecules that eventually became the framework of living systems. It is difficult to imagine the conditions required to effect the transition from monomers to polymers.

For two monomers to bond together, a molecule of water must be removed (Figures 3-2 to 3-5). Since the transition from monomers to polymers is envisioned as occurring in a watery environment, such as shallow oceans or tidepools, the removal of water molecules would be extremely difficult. Due to their own random movements and collisions with one another, molecules tend to move from areas of high concentration to areas of low concentration. This spontaneous movement, called *diffusion,* is exhibited by all molecules. Thus, a water molecule would tend *not* to move away from two monomers into the surrounding water. However, movement in a direction opposite to that of diffusion does occur under certain conditions and is a basic process in every living organism. Carbohydrates, lipids, proteins, and nucleic acids are all constructed from monomers by the water-removal process. This process occurs in the high water concentrations found inside cells. Living cells are equipped with special mechanisms that force this reaction to occur by concentrating the monomers within the cell and providing the energy required to move the water molecule from an area of low to an area of high concentration in the opposite direction of natural diffusion.

In the watery environment of the early oceans, some concentrative mechanism must have moved the monomers close enough to one another so that they would unite. Scientists have proposed two hypotheses to explain how polymers could

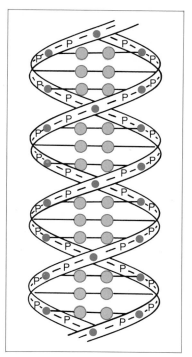

(a)

(b)

Figure 3-11
The Watson–Crick Model
of DNA Structure

(a) The components of this model of a DNA molecule are the same as those in Figure 3-10, but the entire structure is twisted to form a spiral or *helix.* The sugar–phosphate strands are drawn as bands in the diagram to emphasize the twistings. (b) James Watson (left) and Francis Crick (right) demonstrate their model of DNA structure.

have formed from monomers. One hypothesis assumes that such simple concentrative mechanisms as the reduction of water from tidepools by evaporation or the aggregation of monomers on a surface such as clay were sufficient to produce local areas of highly concentrated monomers. Energy sources in the form of electrical storms or solar radiation acting on these monomers would then have removed the water molecules, thereby joining the monomers into polymers.

The second hypothesis is that some type of precellular structures existed that allowed the monomers to separate themselves from their watery environment. Biologists base this hypothesis on the known tendency of molecules to travel by diffusion from areas of high to areas of low concentration. Proponents of this hypothesis believe that the removal of water molecules to form polymers took place inside a structure that did not otherwise contain water molecules. Laboratory demonstrations using dilute chemical solutions similar to those believed to have existed on primitive earth have shown that small, bubble-like structures develop spontaneously and that the interiors of these structures contain no water. Inside such a structure, any monomers that were present would readily join together to form polymers, and the water molecules formed during this reaction would quickly leave the molecule. The hypothesis further suggests that small changes in these precellular spherical structures increased their complexity and diversity and eventually led to the development of cells.

Based on what is currently known about polymer formation and diffusion, either of these hypotheses could be correct. In reconstructing the events that led to the origin of life, such details may never be known with certainty.

Summary

Life is organized within structural units called *cells*. Each cell contains a great variety of molecules that determine cellular activities. It is not always easy to distinguish between living and nonliving materials at the molecular level. In general, however, *biological molecules* tend to be large and fragile and contain carbon atoms.

There are four types of basic biological molecules: *carbohydrates*, *lipids*, *proteins*, and *nucleic acids*. All of these molecules are *polymers*—molecules constructed of multiple repetitions of one or a few types of smaller molecules called *monomers*. The monomers are joined together by *dehydration synthesis*, during which a molecule of water is removed as two monomers become united by a covalent bond. Similarly, a polymer is separated by breaking the covalent bond between each pair of monomers and adding a molecule of water. The process of breaking a polymer down into its component monomers is called *hydrolysis*.

Scientists presume that biological molecules formed on primitive earth long before the first cells appeared. How these large polymers formed from their basic monomers is not known. It is known that the process would have required a concentration of monomers, the removal of water molecules, and a source of energy.

The First Cells

4

A characteristic of living systems is the persistence of high concentrations of both energy and materials despite the natural tendency of energy to dissipate (*entropy*) and materials to move away from areas of high concentration (*diffusion*). Living systems can maintain this unusual state because the chemical reactions of life take place inside cells, where the environment is completely enclosed and controlled by a barrier—the *cell membrane.* The controlled movement of materials across the cell membrane permits certain types of molecules to accumulate and be converted into the biological molecules that the cell requires to perform its activities. A continuous input of energy is necessary to counteract the natural tendencies of entropy and diffusion.

Fossils of cells have been found in rocks 3.5 billion years old (Figure 4-1).

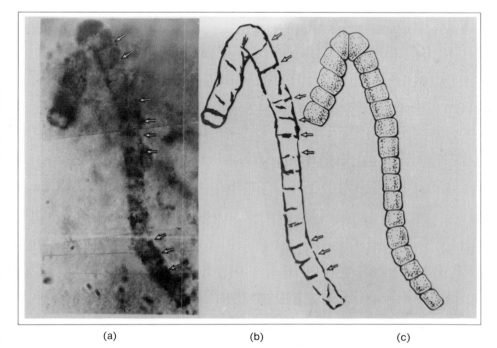

(a)   (b)   (c)

Figure 4-1
The Oldest Fossil Cells

(a) Traces of a group of simple cells found in rock that is 3.5 billion years old. (b) A rough sketch of what these fossils show. The arrows match points on the drawing with the fossil. (c) A more complete sketch of what these cells probably looked like. Although the cells appear in an aggregate, each cell is presumed to have formed a complete organism.

Although cells may have existed before this time, physical traces of these forms of life will probably never be found, because conditions that produce fossils did not exist much earlier in the history of the earth. The earth was formed about 4.6 billion years ago, but a solid crust was not present until about a billion years later. The oldest rocks on earth (found in Greenland) are approximately 3.8 billion years old, but have been too altered by heat and pressure to preserve traces of biological materials.

It is reasonable to assume that the first cells to appear on earth were simple in structure and function and that the more complex cells found in living systems today developed more recently. It is also reasonable to assume that each primitive cell was a complete organism.

From what we now know about living systems, the basic structure of the first cells must have contained:

1. An *outer membrane,* which controls movement of materials and promotes high concentrations of energy and molecules.
2. *Enzymes,* or molecules that regulate chemical reactions.
3. Special *molecules that store and transport energy.*
4. Special *structures capable of assembling protein molecules* (including enzymes) from amino acids.
5. *Nucleic acids,* which contain the instructions for building proteins from amino acids and can make identical copies of themselves.

Cells exist today that contain only these basic structures and never join together to form larger, more complex organisms. Each of these simple cells forms a unicellular organism. A bacterium—one type of organism formed entirely of a simple cell—is shown in Figure 4-2. The basic structures necessary for cellular life and the way in which these structures may have been organized into the first cells are the subjects of this chapter.

Figure 4-2
Diagram of the Interior of a Bacterium
This simple, unicellular organism contains only a few structures: a single circular (looped) DNA molecule, enzymes, ATP and RNA molecules (*not shown*), ribosomes, and a cell membrane. Most bacteria are also enclosed within a rigid outer cell wall.

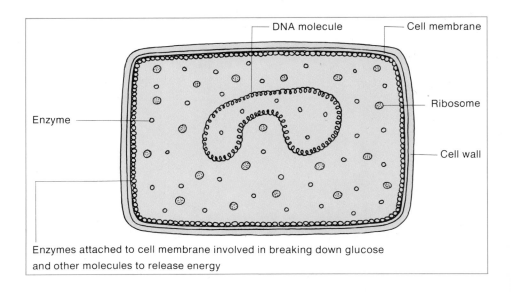

DNA molecule — Cell membrane

Enzyme

Ribosome

Cell wall

Enzymes attached to cell membrane involved in breaking down glucose and other molecules to release energy

Biological Membranes

All living cells are enclosed within a membrane, which not only contains the cell but also controls the movement of materials into and out of the cell. This outer structure of the cell plays a critically important role in maintaining an internal cellular environment that differs from the environment outside the cell.

Cell membranes can move materials in the opposite direction from their natural diffusion pattern. The cell membranes extract a substance in dilute solution on one side and transport it across the membrane structure to the other side, where the substance is much more highly concentrated. Cell membranes are also selectively permeable, allowing only certain materials to cross their boundaries. The outer cell membranes control the intake of nutrients and the removal of waste products. The permeability of a cell membrane to a given material is not fixed but can be altered in response to changes in the environment of the cell.

Structure

All of the membranes found in biological systems are essentially alike in structure. Cell membranes are composed almost entirely of two classes of molecules—lipids and proteins.

The gross structure of the cell membrane is made up of *lipids*. These lipids are primarily *phospholipids*, which are formed from one glycerol molecule, two fatty acid molecules, and one phosphate molecule (a molecule composed of one phosphorus atom joined with four oxygen and three hydrogen atoms). Other structures are often attached to the phosphate (Figure 4-3).

One end of each phospholipid molecule (the glycerol and phosphate *heads*) attracts water and the other end of the molecule (the two fatty acid *tails*) repels water. Two sheets of these molecules, lying end to end, form a double layer. The water-attracting ends form the two outer layers, like the two pieces of bread that form a sandwich, and the water-repelling ends are arranged in between (Figure 4-4). The fundamental structure of a biological membrane—a double layer of lipid molecules—forms a barrier to the free movement of material between the cell and its external environment.

Cell membranes also contain proteins and a few other kinds of molecules. Unlike the phospholipids, the protein molecules do not occur in orderly patterns. Apparently, the proteins have important functional rather than structural properties, which will be discussed in the following section.

Function

Although most membranes are quite similar in *structure*, the *properties* of membranes vary considerably. The kinds of materials that enter and leave a kidney cell, for example, differ greatly from the types of materials that pass through the outer membrane of a muscle cell. Movement of materials across membranes occurs in one of three ways:

1. *Passive transport*, in which certain types of molecules are permitted to move spontaneously across the membrane from areas of high to areas of low concentration. (This process of *diffusion* requires no input of energy.)

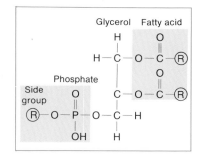

**Figure 4-3
The Structure of a
Phospholipid**

The structure is similar to that of a fat molecule (Figure 3-4), except that one of the fatty acids in the fat molecule is replaced by a phosphate group in the phospholipid. The left side of this molecule (glycerol and phosphate) attracts water and is often referred to as the *head* of the molecule. The right side of the molecule consists of the fatty acid *tails* and repels water. ⓡ represents an attached side group, which differs for the various types of phospholipids.

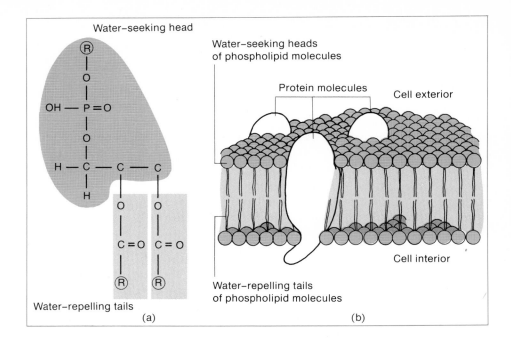

Figure 4-4
The Basic Structure of Biological Membranes

(a) Symbolized diagram of the parts of a phospholipid molecule. (b) The arrangement of molecules in a membrane. The phospholipids form the structure; the proteins provide specific functions.

Water-seeking head

Water-seeking heads of phospholipid molecules

Protein molecules

Cell exterior

Water-repelling tails

Water-repelling tails of phospholipid molecules

Cell interior

(a)

(b)

2. *Facilitated transport,* in which special transport mechanisms within the membrane carry molecules across the membrane from areas of high to areas of low concentration. (Carrier molecules permit the diffusion of materials to which the cell membrane is not normally permeable; this process requires no input of energy.)

3. *Active transport,* in which certain materials are transferred from areas of low to areas of high concentration (requires the input of energy).

Many types of molecules can freely cross the cell membrane by diffusion or passive transport. Water is the most important material moved in this way. When water is permitted to diffuse across a cell membrane but dissolved materials, such as salts and sugars, are not, the process is called *osmosis.* The interior of a cell is a water solution that contains a number of dissolved substances. The concentration of water molecules inside the cell is therefore less than the concentration of molecules in pure water. (Pure water, a liquid comprised entirely of water molecules, is a 100% concentration.) Under normal conditions, a cell is in *osmotic balance,* which means that the concentration of water molecules inside and outside the cell membrane is the same. However, if a cell is placed in pure water and the cell membrane is permeable only to water, then only water molecules will move into the cell (from an area of high to an area of low concentration). As the amount of water inside the cell increases, the cell membrane expands and eventually bursts from the pressure. In contrast, a cell placed in a highly concentrated salt solution will undergo osmosis in the opposite direction: water molecules will tend to leave the cell, and in time, the cell will shrink and die from lack of water. The passive transport of water across cell membranes can cause critical problems for living systems. In Part 3, we will examine the survival mechanisms that various organisms have evolved in response to these problems.

The capacity of a cellular membrane to allow the passive transport of certain materials and not others is called *selective permeability*. Therefore, even if a concentration difference exists between the inside and the outside of a cell, diffusion will not take place unless the cell membrane allows the material to pass. Each membrane has its own permeability characteristics, which are subject to change depending on the activities of the cell and the conditions in its external environment.

In both facilitated and active transport, some sort of carrier is required to transport molecules across the lipid barrier of the membrane. Each membrane contains a variety of proteins of different sizes and structures (see Figure 4-4). These proteins are specific to a cell type, and their production is controlled by the DNA molecules in the cell. Unlike lipids, the proteins in cellular membranes do not form orderly arrays but appear to float freely within the lipid structure. Biologists believe that these proteins (called *permeases*) carry materials across the membrane and that they are similar in structure and function to enzymes, which we will discuss later in the chapter. If these proteins do actually carry materials, then the fact that different membranes contain different types of proteins is not surprising: the proteins could provide each membrane with its specific permeability.

Evidence suggests that these proteins function (1) as specific sites that enable certain materials inside and outside the cell to attach themselves to the cell membrane, thereby altering its permeability characteristics, and (2) as carriers of materials from one side of the membrane to the other. One model that describes how carrier proteins may function can be compared to a revolving door (Figure 4-5). A carrier protein picks up a so-called *passenger molecule* by forming temporary chemical bonds. The carrier then rotates 180° to position the passenger molecule on the other side of the membrane. The temporary bonds are broken, the passenger molecule moves away from the membrane, and the carrier protein subsequently revolves back to its original position.

In facilitated transport, carrier proteins help to diffuse materials to which the cell membrane is relatively impermeable. The carrier improves the membrane's permeability to a substance without altering the direction in which the material naturally tends to move. In contrast, in active transport, carrier proteins move materials in the opposite direction from the direction in which they would naturally tend to move. Active transport therefore requires energy.

Membrane Potentials

In addition to controlling materials within the cell, the selective permeability of membranes also affects the electrical charge of a cell. Most biological membranes are impermeable to *ions*—the atoms or molecules with an electrical charge. Ions form when an atom gains or loses an electron and is left with an unequal number of protons and electrons. Molecules that contain such atoms have an overall positive or negative charge.

Although the interior of the cell contains a variety of both types of electrically charged biological molecules, there is a net excess of negatively charged ions inside

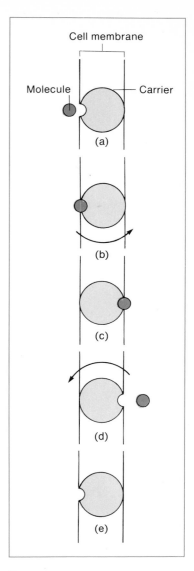

Figure 4-5
How Carrier Proteins Function

(a) The carrier protein within a membrane has specific sites on which it binds with a particular kind of molecule. (b) A temporary bond forms between the molecule and its carrier. (c) The carrier then rotates 180°, carrying the molecule to the other side of the membrane. (d) The temporary bond is broken, and the molecule is released. (e) The carrier molecule rotates 180° back to its original position.

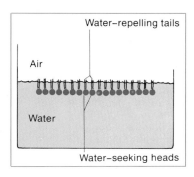

Figure 4-6
Formation of Artificial Membranes

When lipid molecules are dropped onto the surface of water, they spontaneously form continuous sheets of molecules or membranes.

the cell and a net excess of positively charged ions outside the cell. If they were free to move, the negative ions would travel from the inside to the outside of the cell (toward the positive charge), and this movement would be electrical energy. Because cell membranes are normally impermeable to ions, this energy of movement remains *potential energy.* The difference in the number of electrons between two points, a measure of the potential energy of the electrons, is expressed in *volts.* (For example, a flashlight battery usually carries 1.5 volts.) The magnitude of the difference between the inside and outside of a living cell is between 20 and 100 *millivolts* (1 millivolt = 0.001 volt). This electrical property of cells is called the *membrane potential.*

The difference in electrical charge between the inside and the outside of a cell can be neutralized or reversed only if the cell membrane changes its permeability. The controlled movement of ions across cell membranes is vital to the functions of nerve and muscle cells.

The Formation of Membranes

The first membranes must have appeared before the first cells were formed, but the exact origins of cell membranes are unknown. Artificial membranes form readily from solutions of lipids (Figure 4-6). Sometimes the lipid molecules virtually assemble themselves into small, saclike structures.

Proteins added to the lipid molecules in these simple experiments also become incorporated into the artificial membranes. This integrated structure behaves like a living membrane, even exhibiting selective permeability. If lipid molecules were present in the primitive oceans, simple membranes could easily have formed in this manner.

Enzymes—Regulators of Life's Chemistry

Almost all of the chemical processes that occur within a living organism are controlled by *enzymes,* which increase the speed of chemical reactions. Enzymes are globular proteins, and each enzyme controls a specific chemical reaction. This specificity of function is due to what are called *active sites*—small areas on the surface of the enzyme molecule that attract and hold together specific molecules. Each enzyme allows only certain molecules to fit onto its active sites, just as only one type of key fits a lock or two pieces of a jigsaw puzzle fit together (Figure 4-7). Chemical bonds are more readily formed or broken in this close association. The structure of the enzyme itself is not changed during the reaction, and therefore an enzyme can facilitate subsequent reactions.

Before a chemical reaction can occur, a certain amount of energy, called the *energy of activation,* must be added to the molecules to break up the existing molecular structure (Figure 4-8). In the absence of enzymes, energy in the form of heat or pressure is required to increase the rate at which chemical reactions occur. In the presence of an enzyme, the molecules involved in the reaction are bound to the enzyme molecule in such a way that their structure is distorted and the existing arrangement of atoms can be broken up more easily. Less energy is therefore required to activate the reaction, and processes that would take a long time to complete or that would require the input of large amounts of energy in the absence

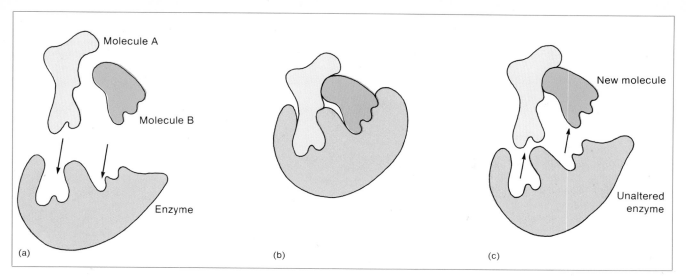

Figure 4-7
Model of Enzyme Action

(a) Enzymes have active sites that attract and hold particular molecules. (b) While held by the enzyme, the molecules undergo structural rearrangements that allow a chemical reaction to occur. The enzyme itself does not participate in the reaction; it simply holds the molecules in a position that promotes the reaction. Here, molecules A and B are held to the enzyme in such a way that a chemical bond is formed between them. (c) The enzyme is unaltered after the chemical reaction is complete and can facilitate further reactions of this type.

of enzymes can be carried out rapidly under the conditions of low pressure and temperature found within a living cell.

Because enzymes are highly specific, they rarely work independently. Enzymes are not randomly distributed throughout the cell, but occur in orderly arrays within particular parts of the cell. In most cases, a series of chemical reactions—a *biochemical pathway*—is controlled by a variety of enzymes working in sequence such that the product of one enzyme reaction is used by another enzyme to make yet another substance.

Although the basic structure of an enzyme is always a globular protein, some enzymes cannot function unless other structures are attached to the molecule. Many enzymes require a *cofactor,* which is an atom or a molecule in ionic form. The ions of both calcium and copper atoms are known to be cofactors of certain enzymes. Other enzymes may require the attachment of a *coenzyme*—a type of molecule usually classified as a *vitamin.*

Enzymes are often named according to the materials on which they act. For example, if an enzyme acts on phosphate molecules, it is called a *phosphatase.* The ending "ase" is simply added to the name of the molecule being acted on in the reaction.

Enzymes are vitally important to all living systems. Cellular structure and function are determined by the chemical reactions that take place within a cell, and almost all of these reactions are governed by enzymes. The kinds of enzymes that a cell is able to construct therefore determine what the cell is able to do. Instructions for the building of these important molecules are contained within the DNA molecules of the cell.

Figure 4-8
Changes in Energy Levels During a Chemical Reaction—The Effect of Enzymes

(a) In the absence of enzymes, much energy must be added to initiate the reaction between glucose and oxygen to form carbon dioxide and water. Once the reaction occurs, energy is released. The end products of the reaction (carbon dioxide and water) contain less energy than the original molecules (glucose and oxygen). (b) In the presence of specific enzymes, which are found inside a living cell, less energy is required to activate the reaction due to the manner in which glucose and oxygen are held to enzymes. The difference in energy levels between the starting materials (glucose and oxygen) and the end products (carbon dioxide and water) is the same whether or not enzymes are present.

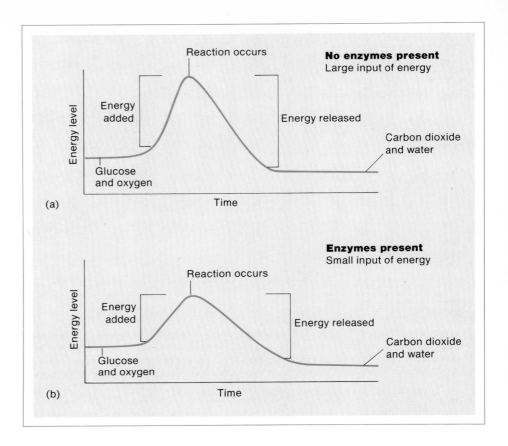

Conserving and Storing Energy: ATP

Most cells obtain energy from the breakdown of sugar (usually glucose) molecules into smaller units. In the presence of oxygen, glucose is converted into carbon dioxide, water, and energy (see Figure 4-8). The biological term for this process is *cellular respiration,* and this topic will be discussed in Chapter 5.

When sugar molecules are burned outside living systems, energy in the form of heat and light is released in one sudden burst. Because heat of this intensity would destroy the molecular organization within the cells and light energy cannot be used by most cells, sugar molecules are not broken down this rapidly within cells. Instead, about 20 enzymes gradually dismantle a sugar molecule, releasing energy for use by the cells in a slower and more controlled way. This entire sequence of reactions is accomplished under the normal temperatures and pressures of living systems.

Although about one-half of the energy released during the cellular breakdown of molecules is in the form of heat, the slow rate of the reaction prevents the development of high temperatures. The remaining released energy is transferred to the chemical bonds within molecules of *adenosine triphosphate* or ATP. This single type of molecule is involved in almost all of the energy transactions within cells.

Although some organisms obtain their energy from the sun or from eating plants or animals, most of this energy is transferred to ATP molecules before it is released in any reaction. Examples of reactions that require energy are the construction of molecules, active transport, and muscle contractions.

The ATP molecule is made up of a nitrogenous base (adenine), a sugar (ribose), and a series of three phosphate groups, as shown in Figure 4-9(a). Energy released from the breakdown of molecules is transferred to the chemical bonds that join the phosphate groups of ATP together. The bonds that attach the second and terminal phosphate groups (indicated by $\sim$ in the figure) are known as *high-energy bonds,* because their formation requires more than twice the amount of energy required to form other chemical bonds and this large amount of energy is released when these bonds are broken.

Energy is released to do biological work when the terminal phosphate group of the ATP molecule is removed, as shown in Figure 4-9(b). The transformed molecule, in which a phosphate is replaced by a water molecule, is *adenosine diphosphate* or ADP (*di* indicates the presence of the two remaining phosphate groups). Occasionally, the second phosphate group (the middle group in the ATP molecule) is also removed to release energy, but usually only the terminal phosphate group is removed.

When energy is released during the breakdown of sugar (or other) molecules, ADP is converted to ATP by the addition of a third phosphate group and the removal of the water molecule. The process of formation and breakdown of ATP is summarized in Figure 4-10.

The ATP molecule allows cells to accumulate a large amount of energy in a small space. Because the cells of all forms of living organisms use ATP, it probably appeared early in the evolution of life.

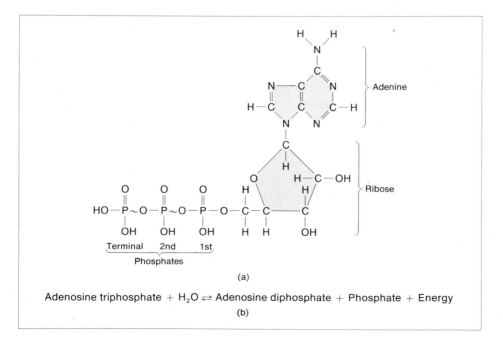

(a)

Adenosine triphosphate + H_2O ⇌ Adenosine diphosphate + Phosphate + Energy

(b)

Figure 4-9
Energy Storage and Transport
(a) The arrangement of atoms in a molecule of adenosine triphosphate (ATP). The wavy lines between the phosphorus (P) and oxygen (O) atoms indicate high-energy chemical bonds. When one of these bonds is broken (usually the one holding the outermost or *terminal* phosphate group to the molecule), a great deal of energy is released. An equivalent amount of energy is required to form the bond. (b) The reactions that occur in energy release and storage. The removal of a phosphate group from the ATP molecule (left to right in the equation) causes the release of energy. The addition of a phosphate group to the ADP molecule (right to left in the equation) requires the input of energy, which is then stored in the ATP molecule.

When a phosphate group is removed
from the ATP molecule, energy is
released, and a water molecule
replaces the phosphate group in the
ADP molecule. The addition of a
phosphate group and energy to an
ADP molecule converts it into the
energy-storage molecule ATP; the
phosphate group replaces a water
molecule.

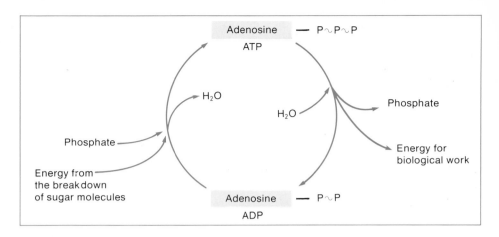

Protein Synthesis

Proteins form enzymes and some of the structural components of cells. The particular types of enzymes that a cell manufactures control the activities of the cell. Proteins are constructed (*synthesized*) from amino acids according to the genetic information coded in the DNA molecules (*genes*) of each cell. The genes of an individual are a "list" of all the proteins its cells can manufacture along with precise instructions for the assembly of each protein.

The *genetic code* is provided by a sequence of nitrogenous bases along each sugar–phosphate strand of the DNA molecule. This sequence of bases directs the assembly of a sequence of amino acids into a protein. Each amino acid is "spelled out" by a particular set of three adjacent nitrogenous bases. In a long segment of the DNA molecule, instructions are provided for the exact sequence of many amino acids in a protein. A series of 300 bases along one strand of a DNA molecule might provide the directions for the construction of a protein containing 100 amino acids. A *gene* can be defined as a sequence of bases that provides the code for the assembly of a specific protein.

The events that occur during the synthesis of proteins can be separated into two primary phases: (1) the *transcription* of the genetic code, in which the sequence of bases within the DNA molecule is transferred to a sequence of bases within an RNA molecule, and (2) the *translation* of the genetic code, in which the sequence of bases within the RNA molecule directs the assembly of amino acids into proteins.

Transcription

The actual construction of proteins does not occur directly on the DNA molecule. Instead, the genetic code is incorporated into the structure of an RNA molecule, which carries the information to other parts of the cell where proteins are actually synthesized. To transfer the genetic instructions, the two strands of the DNA molecule separate, as shown in Figure 4-11(a) and (b), so that the bases along one strand are exposed. Individual RNA nucleotides are then attracted to the exposed bases, and temporary chemical bonds form between the bases of the DNA molecule and the bases of the RNA nucleotides. These bonds develop according to the base-pairing rules (adenine always pairs with thymine; cytosine always pairs

Figure 4-11
Transcription—The Transfer of the Genetic Code in a DNA Molecule to an RNA Molecule

(a) Part of a DNA molecule before transcription. The unshaded triangles represent deoxyribose, the circles represent the nitrogenous bases, and P indicates a phosphate group. (b) The part of the DNA strand that is to be transcribed separates, breaking the bonds between the nitrogenous-base pairs. (c) RNA nucleotides are attracted to and pair with the DNA nucleotides. The brown triangles represent ribose. Guanine pairs with cytosine, and adenine pairs with uracil. As the RNA nucleotides become arranged along the DNA molecule, bonds form between the ribose and the phosphate units of adjacent RNA nucleotides. (d) After transcription, the newly formed RNA molecule moves away and the DNA molecule re-forms.

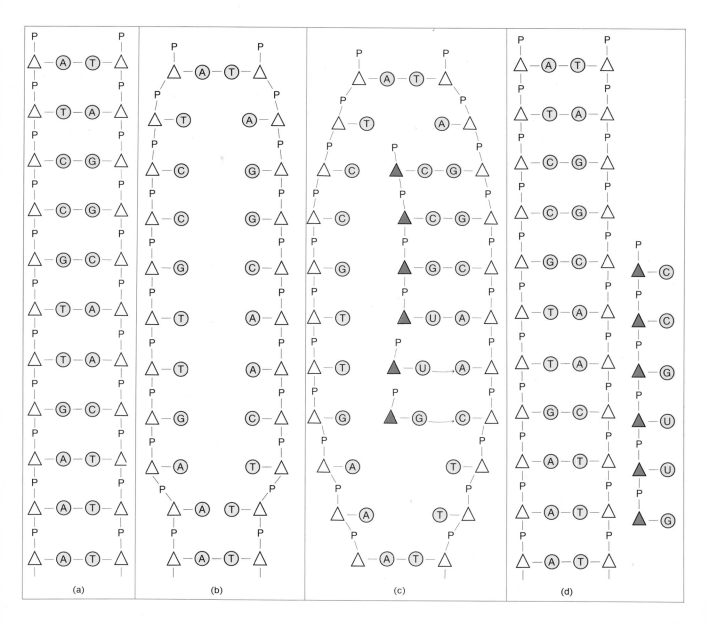

with guanine), as shown in Figure 4-11(c). Since RNA nucleotides contain uracil instead of thymine, wherever the base-pairing rules call for an RNA nucleotide with thymine, one carrying uracil is substituted. As the RNA nucleotides become joined to the DNA molecule, chemical bonds also form between the sugar and the phosphate units of the adjacent RNA nucleotides. Once this occurs and the RNA nucleotides are joined together to form a strand, the temporary chemical bonds holding the RNA molecule to the DNA molecule are broken. The assembly of an RNA molecule in accordance with the base sequences of a DNA molecule is called *transcription.*

Once the complete code for a protein has been incorporated into the newly assembled RNA molecule, as shown in Figure 4-11(d), this molecule leaves the DNA structure and carries the information to the *ribosomes*—special structures on which amino acids are assembled into proteins (Figure 4-2). The molecules of RNA act as intermediaries, carrying the genetic information originally coded within the DNA structure to the site of protein synthesis. These molecules are called *messenger RNA (mRNA) molecules.*

Although DNA is a large molecule, it is held together by a great many chemical bonds, making it a very stable molecule. Stability is essential to the DNA molecule; the genetic code would be scrambled if the structure were disrupted. When the two sugar–phosphate strands of DNA separate in the process of RNA construction (transcription), no structural changes occur within either strand or in the sequence of attached bases. After the genetic code has been transcribed, the two strands of DNA reassociate to form the original structure, which is then preserved for future transcription.

All known living systems use nucleic acids and even the same *three-base code* (a sequence of three bases that provides the code for an amino acid) to direct the synthesis of proteins from amino acids. Because the code is universal, biologists believe that it may have originated in the earliest cells.

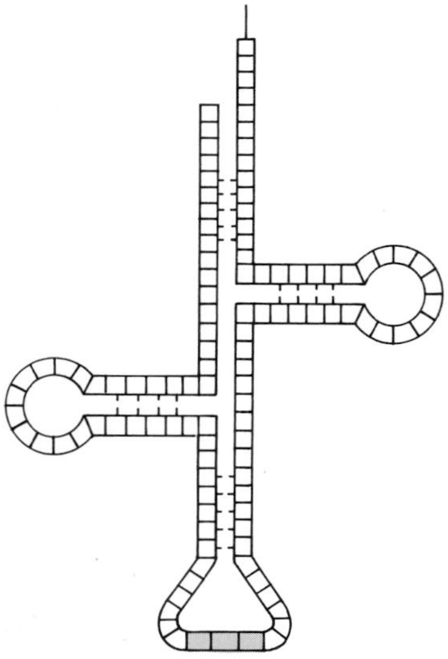

Translation

Proteins are synthesized on the ribosomes, which are small, spherical particles suspended throughout a simple cell (see Figure 4-2). Ribosomes are constructed of protein and a special type of RNA molecule called *ribosomal RNA.* In addition to providing a site on which amino acids can assemble into proteins, ribosomes clearly play a role in coordinating the assembly process. However, the details of their role are not well understood.

Once the entire sequence of bases that code for a particular protein molecule has been transcribed, the newly constructed messenger RNA (mRNA) molecule moves toward and attaches itself to the outer surface of a ribosome. Amino acids and another form of RNA molecule called *transfer RNA* (abbreviated tRNA) are dispersed throughout the cell and act together to construct a protein molecule, according to the genetic code in the mRNA molecule now positioned on the surface of a ribosome.

All of the amino acids used in the synthesis of proteins are present in the area surrounding the ribosomes. The tRNA molecules act as vehicles, picking up the amino acids and transporting them to the mRNA molecule on the ribosome. The tRNA molecule is a small structure with four active sites (Figure 4-12). At one end

Site of amino acid attachment

Three–base code

Figure 4-12
Transfer RNA Molecule
This small RNA molecule is constructed of nucleotides (represented by blocks in the model). Some of the bases that form part of the nucleotides are united by bonds (dashed lines) to give the molecule its cloverleaf shape. The lower end of the molecule contains the three bases that will eventually attach to the mRNA molecule according to the base-pairing rules. The appropriate amino acid (defined by the lower three-base code) will attach itself to the upper end of the molecule. An active site that recognizes its amino acid activating enzyme appears on the left. An active site that recognizes the ribosome appears on the right.

of the molecule is a sequence of three unpaired nitrogenous bases (adenine, guanine, cytosine, or uracil), which provide the code for a particular amino acid. At the opposite end of the molecule is the site where the particular amino acid will become attached. The two remaining active sites are located on each side of the molecule. One site recognizes the enzyme that facilitates attachment of the amino acid (Figure 4-13); the other site recognizes the ribosome. There is a specific tRNA molecule and attachment enzyme for each type of amino acid. Once the enzyme has coupled the amino acid to its appropriate tRNA molecule, the tRNA molecule carries the amino acid to the ribosome.

The mRNA molecule is positioned on the ribosome in such a manner that the first three bases of the genetic code are on the center of the ribosome, as shown in Figure 4-14(a). These three bases attract and pair with a tRNA molecule carrying the three matching bases, according to the base-pairing rules. For example, if the mRNA sequence is adenine, guanine, and cytosine, then the tRNA complex carrying the bases uracil, cytosine, and guanine will approach and attach to the mRNA molecule at this particular site. Once pairing between the bases occurs, the mRNA molecule will shift its position so that the next three-base code in its structure is on the center of the ribosome. This particular sequence of three bases attracts and pairs with a tRNA molecule carrying a matching sequence of three bases, as shown in Figure 4-14(b). A bond then forms between the amino acid carried by this second tRNA molecule and the amino acid carried by the first tRNA molecule. Once the two amino acids are joined, the first tRNA molecule is released and the mRNA molecule again shifts across the surface of the ribosome in such a way that the third sequence of three bases is available to attach itself to the appropriate tRNA molecule, as shown in Figure 4-14(c).

This process continues until a complete protein is constructed as a linear sequence of amino acids joined according to the linear sequence of bases on the

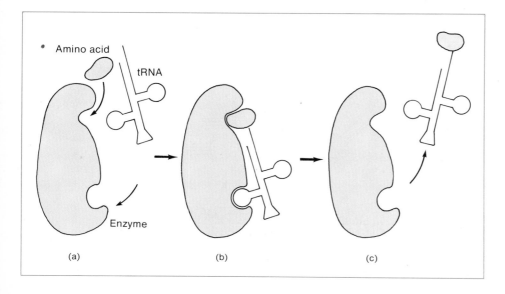

(a)   (b)   (c)

Figure 4-13
Joining an Amino Acid to the Transfer RNA Molecule
(a) An enzyme attracts the appropriate amino acid and tRNA molecule, (b) binds them both to its surface, and (c) facilitates the formation of a chemical bond between the amino acid and the tRNA molecule.

Figure 4-14
The Assembly of Amino Acids into a Protein Molecule

The code for assembling proteins from amino acids is contained within the mRNA molecule, which moves across the surface of a ribosome (left to right in the figure). (a) The first three-base code on the mRNA molecule attracts and unites with the matching three-base code on a tRNA molecule. (b) The second tRNA molecule attaches to the next three-base code on the mRNA molecule, and the amino acid it is carrying joins with the first amino acid. (c) The first tRNA molecule leaves without its amino acid as the fourth tRNA molecule approaches the ribosome. This process continues until all of the amino acids coded within the mRNA molecule have been joined and form a long protein molecule.

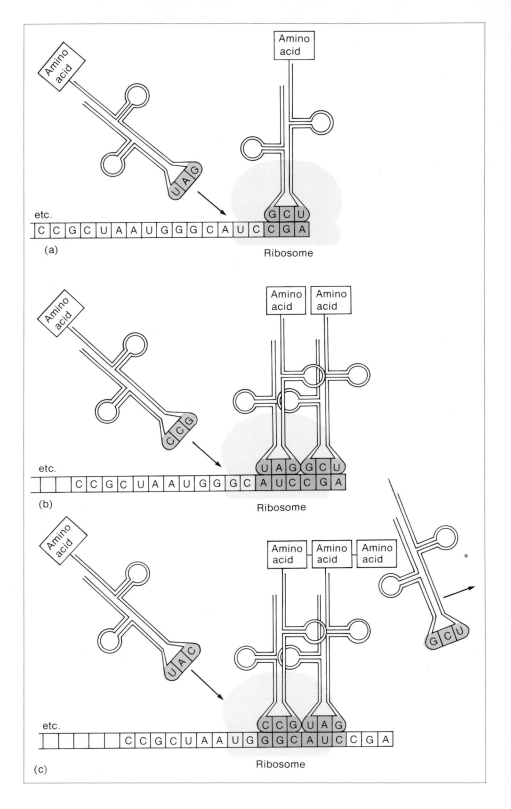

mRNA molecule. Thus, the genetic code incorporated in the structure of the mRNA molecule is *translated* into the amino acid sequence of a protein molecule. Once the complete code of the mRNA molecule has been translated, the molecule breaks apart into its component RNA nucleotides, which return to the DNA structure for reuse. The newly formed protein molecule moves away from the ribosome to the place in the cell where it is to be used.

A Comparison of Modern Simple Cells and the First Cells

We have just reviewed the basic characteristics of the simplest cells: (1) a selectively permeable cell membrane; (2) enzymes; (3) ATP molecules; (4) ribosomes and RNA molecules, which assemble protein molecules from amino acids; and (5) DNA molecules, which contain the instructions for building proteins.

All three forms of the simple cells that exist today—bacteria, methanogens, and blue-greens—are unicellular organisms and are considered to be primitive life forms. These organisms, called *prokaryotic* (prenuclear) *cells,* have no nuclear membrane to isolate the genetic material (usually in the form of a single, circular DNA molecule) from the rest of the cell. Many simple cells also have a rigid wall surrounding the cell membrane, which provides additional support and protection.

The three forms of simple cells differ in the way they acquire energy from the environment. Most *bacteria* obtain energy by breaking apart biological molecules, which they obtain by feeding on living or dead organisms. *Methanogens* split hydrogen molecules (H_2) to obtain energy to convert carbon dioxide and hydrogen into methane (CH_4), a biological molecule. These unusual prokaryotes live within the digestive tracts of larger organisms and in the mud beneath stagnant water. *Blue-greens,* like plants, contain special pigments, located on membranes within the cell, which capture light energy from the sun.

Since the oldest cells known resemble bacteria in structure, we can speculate about their functions by comparing them with modern forms of bacteria. Some modern bacteria can capture and use light energy from the sun, but most bacteria can obtain energy only from the chemical bonds that have already been formed in biological molecules, such as carbohydrates, lipids, and proteins. Organisms that obtain energy in this way are called *heterotrophs.* The oldest fossil cells were probably heterotrophic, capable of moving biological molecules from their surrounding environment into the cell interior and breaking them down into smaller units to release the energy and building materials necessary to maintain life.

Were the earliest cells similar in structure to the oldest known fossil cells and therefore to the modern forms of heterotrophic bacteria that require biological molecules for food? Not according to recent evidence, which suggests that the oldest known fossils represent a later stage in cellular evolution. Comparisons between the biochemical pathways and nucleic acid structures of bacteria and methanogens show that these two forms of simple cells are quite different. The implication of these findings is that bacteria and methanogens had a common ancestor, which existed prior to the appearance of either of these simple cells. The possible evolution of simple unicellular organisms is diagrammed in Figure 4-15.

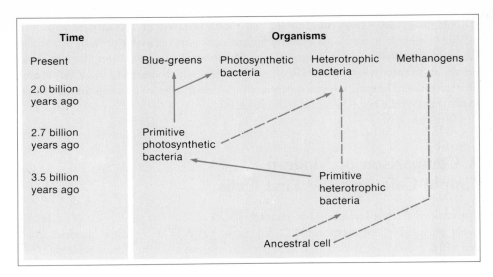

**Figure 4-15
The Possible Evolution of
Simple, Unicellular
Organisms**

The dashed arrows indicate
evolutionary sequences that have
not been verified; the solid arrows
represent established lines of descent.

Biochemical Pathways—
Life's Solution to the Early Food Shortage

The first cells on earth lived in an environment rich in large biological molecules
that had gradually formed from smaller monomers in the warm oceans. Because
these molecules were similar in structure to the biological molecules synthesized
and used by the first cells, only a few enzymes were necessary to convert these food
molecules into the molecules needed to maintain early cellular activity. In contrast,
many modern organisms must synthesize all of their complex molecules from a
few very simple starting materials—a process that requires many enzymes.

The sequences of the reactions by which the cells of living organisms convert
simple starting materials into complex molecules are referred to as *biochemical
pathways.* In the hypothetical pathway shown in Figure 4-16, each step is controlled
by a specific enzyme which, in turn, is produced according to the instructions of a
particular gene in the cell. The entire set of enzymes in a biochemical pathway
probably appeared gradually, one enzyme at a time, as large food molecules in the
early environment became less abundant. The end product (molecule E in Figure
4-16) may have been essential to the activities of early cells. If it had been available
in the environment, cells could have incorporated and used this molecule without
altering its structure. When molecule E became scarce in the environment, cells
could have developed the ability to take in molecule D and convert it to molecule E
through the use of a specific enzyme. Similarly, as D became scarce, an enzyme
would have been required to convert C to D. This course of events probably
continued until long biochemical pathways, each containing the appropriate series
of enzymes to effect molecular conversion, were constructed.

**Figure 4-16
Diagram of a
Biochemical Pathway**

All of these steps occur within the
cell. Each step is controlled by a
specific enzyme, which is synthesized
according to the code of a particular
gene. A is a molecule obtained from
the food eaten by an organism; E is
the product molecule required to
perform some specific cellular
function.

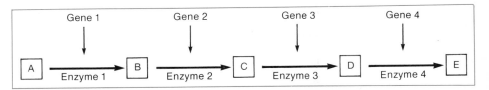

A New Source of Energy—The Primitive Autotroph

The next phase of cellular evolution probably began when the simplest biological molecules in the oceanic environment became scarce. The only potential nutrients available then would have been inorganic materials, and any organism capable of converting these simple materials into biological molecules would have had an advantage at this time.

Autotrophs are organisms that can convert carbon dioxide into biological molecules. A few organisms, including methanogens, can derive energy for this conversion from simple inorganic materials in the environment. However, most autotrophs utilize light energy from the sun to convert carbon dioxide into sugar molecules—a process called *photosynthesis*. The hydrogen atoms required for the reaction are usually provided by water molecules. The most common photosynthetic reaction, which occurs in all green plants, is

$$6CO_2 + 12H_2O \xrightarrow{light} C_6H_{12}O_6 + 6O_2 + 6H_2O$$

(Carbon dioxide)   (Water)   (Glucose)   (Oxygen gas)   (Water)

To accomplish photosynthesis, an organism must be able to construct several specific enzymes and to capture and hold light energy from the sun. Molecules that readily absorb light are called *pigments*. The color of the pigment depends on the wavelength of light that it absorbs. The pigment that produces the green color of most plants is *chlorophyll*. In photosynthesis, the energy of the absorbed light is transferred to the pigment molecule and is then used by the cell to convert carbon dioxide and water (or some other hydrogen source) into glucose molecules, which can then be converted into other forms of biological molecules.

Fossil cells with pigments have been found in rocks known to be 2.7 billion years old. These early photosynthetic cells, which resemble bacteria in structure, probably evolved from the heterotrophic "bacteria" and later diverged into the photosynthetic bacteria and blue-greens found on earth today. (The oldest fossils of blue-greens are 2 billion years old.) This evolutionary sequence is shown in Figure 4-15.

Photosynthesis represented a great evolutionary advance because it was no longer necessary for cells to feed on the slow-forming biological molecules in their noncellular environment. Biological molecules could now be formed much more rapidly and efficiently within the cell itself from glucose—the product of photosynthesis.

An important consequence of early photosynthesis was the release of the end product, molecular oxygen (O_2), which accumulated in the environment and caused a dramatic change in the composition of the waters and the atmosphere. At first, this change in the environment must have interfered with normal heterotrophic activity, which had evolved without molecular oxygen under *anaerobic conditions*.

The presence of molecular oxygen altered the subsequent pace and direction of evolution by providing cells with the opportunity to use this gas in energy-yielding and other biochemical reactions. During *cellular respiration*, simple molecules (such as the sugars formed by autotrophs) are broken down to yield smaller molecules and energy. In the absence of oxygen during *anaerobic respiration* (which occurred in the earliest cells), sugar molecules are only partially broken down to form ethyl alcohol (C_2H_5OH) or lactic acid ($C_3H_6O_3$) and considerable energy is retained within the alcohol or lactic acid molecules. A much more complete breakdown of sugar molecules occurs in the presence of oxygen during *aerobic respiration,* which results in end products of carbon dioxide (CO_2) and water. More energy is therefore made available to cells when an aerobic reaction occurs.

Today, most organisms utilize oxygen and die if they are deprived of it. The gradual accumulation of oxygen due to early photosynthesis also caused a layer of ozone (O_3) to form in the upper atmosphere. Later, this ozone layer played a great role in reducing the amount of ultraviolet radiation that reached the earth's surface. This type of sunlight can be very damaging to biological materials, and its reduction was critical in the development of life on land.

Summary

Biologists assume that the first cells to appear on earth contained the minimum structures necessary to sustain life—a cell membrane, enzymes, ATP molecules, ribosomes, RNA molecules, and DNA molecules.

The *cell membrane,* which is constructed of phospholipids and proteins, serves as a barrier, separating the interior of a cell from its external environment. The membrane is *selectively permeable* in that it controls the materials that enter and leave the cell, thereby permitting the cell to accumulate structural materials and energy and to dispose of wastes. The membrane also produces an *electrical potential* between the interior of the cell and its external environment by allowing a net excess of negatively charged *ions* to accumulate inside and a net excess of positively charged ions to accumulate outside the cell.

Chemical reactions within a cell are regulated by *enzymes*, which speed up reactions and also reduce the amount of energy required to initiate reactions. Enzymes are proteins, and the instructions for building them are located within the DNA molecules of the cell.

Energy released from the breakdown of molecules inside a cell is transferred to the chemical bonds of *adenosine triphosphate* (ATP). These small molecules transport the energy where it is needed and release it by breaking the chemical bond between the second and terminal phosphates of their structure.

Proteins are particular sequences of *amino acids* that are linked together according to the instructions contained within DNA molecules. These instructions form the *genetic code:* each sequence of three *nitrogenous bases* along one strand of the DNA molecule is the symbol for a particular amino acid. A sequence of hundreds of bases comprises the instructions for building one protein. In the construction of a protein molecule, the code within the DNA molecule is *transcribed* into a messenger *RNA, or mRNA, molecule,* which carries the code to a *ribosome*—the site of protein synthesis. Amino acids are brought to the ribosome and linked together according to the instructions within the mRNA molecule by special molecules called *transfer RNA,* or *tRNA, molecules.* Linking specific amino acids to one another according to the arrangement of bases in the mRNA molecule is called *translation.*

The oldest fossil cells resemble bacteria and probably acquired energy and structural materials by engulfing the large biological molecules that were abundant in the warm waters of the early oceans. These fossils probably do not represent the very first cells to appear on earth, but it is unlikely that fossils of these first cells will ever be found.

As the bacteria-like primitive cells became more abundant in the warm oceans, the large food molecules in their environment became scarce, forcing the cells to rely on smaller molecules that were less similar to the molecules they needed to maintain their activities. Enzymes were required to convert these smaller molecules into materials that could be used by the cells. As time passed, primitive cells acquired the ability to manufacture a large variety of enzymes and long *biochemical pathways* developed.

At some point in cellular history, certain types of primitive cells acquired the ability to construct all of their own molecules by using the energy of sunlight, carbon dioxide, and water. This process—called *photosynthesis*—which appeared at least 2.7 billion years ago, has greatly altered the environment. An end product of the photosynthetic reaction— oxygen gas (O_2)—is now essential to the lives of most organisms. It is used in chemical reactions and also forms a shield (the *ozone layer*) in the atmosphere, which protects living forms against damaging ultraviolet radiation from the sun.

Modern Cells: The Basic Units of Life

5

The cell was first recognized as the basic structural and functional unit of life in the nineteenth century. The central role of the cell in living systems is expressed in the cell theory, which states that:

1. All living things are organized into cells.
2. Each individual cell exhibits the basic properties of life, which include the ability to obtain energy from the environment and to use this energy to construct particular types of molecules.
3. All cells arise from previously existing cells. They do not develop directly from noncellular material.

Types of Cells

There are many different types of cells. Each cell has the ability to synthesize a unique array of enzymes that determines all of its chemical activities. Despite their chemical differences, cells are remarkably alike in terms of their basic structure.

On the basis of their internal organization, all cells can be categorized into two basic types: the *simple cells* described in Chapter 4, which have relatively little internal structural organization, and the more *complex cells,* in which different aspects of chemical activities occur within a variety of distinct structural subunits.

In Chapter 4, we learned that the simple forms of unicellular organisms are bacteria, methanogens, and blue-greens. These cells are considered to be the most primitive forms of life. They are called *prokaryotic* or prenuclear cells because they have no nucleus to isolate the genetic material from the rest of the cell. We know that prokaryotic cells contain a cell membrane, genetic material (usually DNA in the form of a single, circular molecule), RNA molecules, ribosomes, enzymes, and ATP molecules. In addition, many of these simple cells have a rigid cell wall surrounding the cell membrane, and the photosynthetic forms (the blue-greens and some of the bacteria) have light-absorbing pigments arranged on membranes within the cell.

All organisms other than bacteria, methanogens, and blue-greens are composed of *eukaryotic* or truly nuclear cells. In these complex cells, genetic material is confined within a membrane-enclosed structure (the nucleus) which separates it from the rest of the cellular material (the cytoplasm). All multicellular forms of life (plants, animals, and fungi) and many unicellular forms of life (the *protists*) are composed of eukaryotic cells. In addition to a nucleus, eukaryotic cells contain a number of other well-defined structural subunits, called *organelles*. Each type of organelle has a distinctive morphology and performs a specific cellular function.

The fossil record indicates that prokaryotic cells arose about 3.5 billion years ago and were the only life forms for approximately 2 billion years. Evidence suggests that the first eukaryotic cells originated about 1.4 billion years ago. These two major classes of cells appear to have had a common ancestry; the modern forms of each class display close similarities in biochemical pathways and structures. The eukaryotes probably arose from the early prokaryotes. Because these two types of cells are so distinct and no fossil evidence of intermediate types of cells exists, it is difficult to reconstruct the evolution of eukaryotes from prokaryotes.

Table 5-1
A Comparison of Structures Within Prokaryotic and Eukaryotic Cells

Structure	Function	Prokaryote	Eukaryote
Cell membrane	Encloses materials; controls movement of materials into and out of cell	Present	Present
DNA molecules	Contain genetic instructions	Naked form	In chromosomes
RNA molecules	Synthesize proteins	Present	Present
ATP molecules	Store and transport energy	Present	Present
Enzymes	Speed up chemical reactions; decrease energy of activation	Present	Present
Ribosomes	Provide sites for protein synthesis	Present	Present
Nuclear membrane	Isolates genetic material from rest of cell	Absent	Present
Chromosomes	Contain DNA and control its delegation to daughter cells during mitosis	Absent	Present
Nucleolus	Synthesizes ribosomes	Absent	Present
Endoplasmic reticulum	Transports materials; synthesizes lipids and certain proteins	Absent	Present
Golgi complex	Stores materials and transports them out of cell; synthesizes carbohydrates	Absent	Present
Lysosomes	Break down unwanted molecules	Absent	Present in some forms
Mitochondria	Perform cellular respiration	Absent	Present
Plastids	Perform photosynthesis	Absent	Present in some forms
Cell wall	Protects and supports cell	Present in some forms	Present in some forms

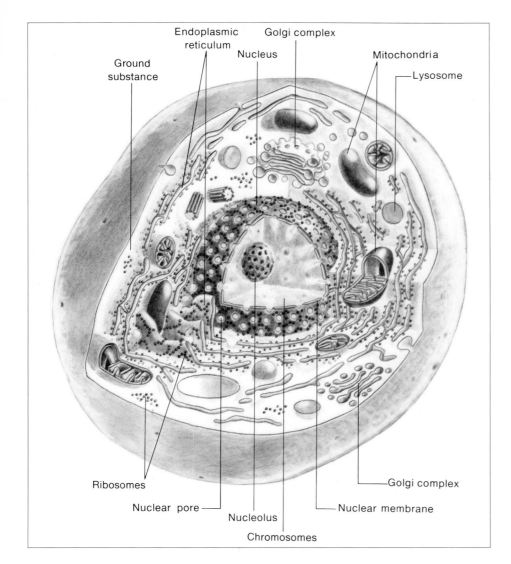

Figure 5-1
The Structure
of a Eukaryotic Cell
This complex cell contains many
components. All eukaryotic cells are
similar in structure, whether they
form unicellular organisms or are part
of multicellular organisms.

The basic structure of a eukaryotic cell is the same, whether it forms an entire organism or part of a multicellular individual. More than 10,000 different enzymes may be working simultaneously within each eukaryotic cell. The localization of all of the enzymes involved in each particular chemical process within distinct *organelles* in a eukaryotic cell allows cellular chemical activities to be compartmentalized, thereby promoting a greater degree of specialization and efficiency in these more complex cells than is found in simpler prokaryotic cells (see Table 5-1).

In Chapter 5, we will describe the structure and function of each type of organelle in a eukaryotic cell (Figure 5-1). Then we will examine the alternative hypotheses that biologists have proposed to explain the origin of eukaryotic cells and their organelles.

The Nucleus: Isolation of the Genetic Code

The most prominent feature within a eukaryotic cell is the *nucleus*—a relatively large, spherical structure that is enclosed within two distinct nuclear membranes (Figure 5-2). Each of these membranes is similar in structure and function to the cell membrane. The small breaks or *pores* occurring at various intervals along the pair of nuclear membranes permit certain large molecules to pass into and out of the nucleus. These membranes isolate the genetic material from the chemical activities occurring in the rest of the cell, or *cytoplasm*.

Chromosomes

As in the prokaryotes, cellular activities in the eukaryotes are ultimately controlled by the DNA molecules. However, the molecular organization of the genetic material differs greatly in these two basic cellular forms. In the prokaryotes, DNA exists as a simple molecule, often referred to as "naked DNA," that does not associate with other kinds of molecules. In the eukaryotes, DNA forms part of a larger, elongated structure—the *chromosome*—and many chromosomes are contained within each eukaryotic cell.

The chromosome may be the least understood cellular organelle. The primary molecules in a chromosome are DNA, two kinds of proteins, small amounts of

Figure 5-2
The Nucleus of a Cell

The large, circular structure in the center of this cross section of a cell (× 5,500) is the *nucleus*. The light, cloudlike area within the nucleus is the *nucleolus*. The *chromosomes* appear here only as dark areas. Note the double *nuclear membranes* separated at intervals by *pores*. The network of parallel membranes outside the nucleus is the *endoplasmic reticulum*.

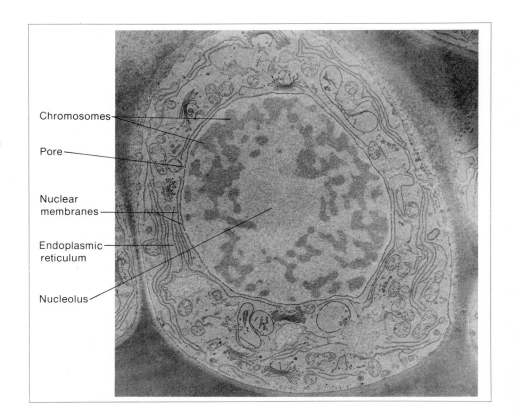

Chromosomes

Pore

Nuclear membranes

Endoplasmic reticulum

Nucleolus

RNA, and lipids. Although biologists have learned a great deal about the molecular structure of DNA, they still know very little about how DNA and these other components are integrated into the chromosomal structure. For example, it is not yet clear whether a DNA molecule extends continuously from one end of a chromosome to the other or whether it occurs as a sequence of short segments within the chromosome.

At some point along the length of each chromosome, a special area of constriction called the *kinetochore* (kinetics = pertaining to motion; chore = body) appears (Figure 5-3). The kinetochore is involved in the orientation and movement of chromosomes during cell division. The position of the kinetochore and the length of the chromosome provide each type of chromosome in a cell with a unique, distinctive shape.

The cells of an organism and the cells of all individual organisms in a species have the same number of chromosomes. With the exception of the reproductive cells, the genetic material of all human cells is distributed among 46 distinct chromosomes, whether the cell forms part of the skin, nerves, heart, or liver.

The Nucleolus

The *nucleolus* is a visible body within the nucleus (Figure 5-2), but has no membrane to separate it from the rest of the nuclear material. The nucleolus is packed with granules of RNA, which are the precursors of ribosomes. A nucleus may contain several nucleoli, depending on the cell's ribosome requirements.

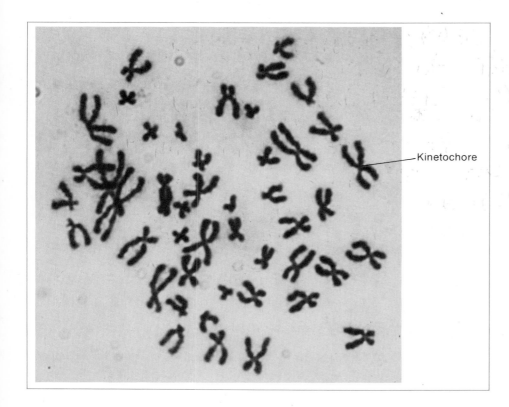

———Kinetochore

Figure 5-3
The Chromosomes of a Human Cell
This photograph was taken during cell division when the chromosomes are short and thick and the nuclear membranes have disintegrated. Each chromosome has already been duplicated and is ready to be distributed to two daughter cells. Each pair of duplicate chromosomes is held together by a kinetochore, which appears as a constriction.

Ribosomes

Ribosomes are small, spherical organelles that are suspended throughout the cytoplasm. They are constructed, within the nucleolus, of proteins and a special form of RNA molecule called *ribosomal RNA.* The role of the ribosome as a site for the important cell function of protein synthesis has already been described in Chapter 4. Protein synthesis takes place in the same way in prokaryotic and eukaryotic cells.

A *polyribosome* is a cluster of from 5 to 10 ribosomes that occurs within the cytoplasm of eukaryotic cells. A single messenger RNA (mRNA) molecule becomes associated with a ribosomal cluster in such a way that each ribosome in the cluster can translate the genetic code into the protein molecule. Many copies of the protein molecule for which the mRNA molecule codes can therefore be synthesized at the same time.

Not all of the ribosomes in a eukaryotic cell are suspended within the cytoplasm. Some ribosomes are attached to membranes, and these organelles will be described in the next section.

Endoplasmic Reticulum

An elaborate network of membranes is located throughout the cytoplasm of eukaryotic cells (see Figures 5-1 and 5-2). These membranes, called the *endoplasmic reticulum* (endoplasmic = internal cytoplasm; reticulate = network), provide a surface, or location, for chemical reactions and form small canals for the transport of newly synthesized products. Biologists have identified two types of endoplasmic reticulum: a rough form with ribosomes attached (Figure 5-4), and a smooth form without ribosomes.

The *rough endoplasmic reticulum* is involved in the synthesis of proteins that are to be stored within the membrane network or, in the case of multicellular organisms, that are to leave the cell and be used in other parts of the body. The ribosomes, which are similar to the ribosomes found elsewhere in the cytoplasm, are attached to the outer surface of the membranes. After proteins are constructed on these ribosomes, they are stored or transported through the membrane canals to other parts of the cell.

The *smooth endoplasmic reticulum* (without ribosomes) aids in the synthesis of a number of different types of molecules, particularly the lipids, and also breaks apart damaging molecules that are taken in by the cell (such as drugs) or that are produced by the cell as waste products. In addition, the smooth endoplasmic reticulum plays an important part in the transport of materials from one part of the cell to another.

The Golgi Complex and Lysosomes

The *Golgi complex* (called *dictyosomes* in plants) is similar in structure to the smooth endoplasmic reticulum. Each complex consists of several large cavities or vesicles

Figure 5-4
The Rough Endoplasmic Reticulum

A close-up of the endoplasmic reticulum, showing the membranes covered with ribosomes. Some of the cell's proteins are synthesized on these ribosomes; others are synthesized on ribosomes scattered throughout the cytoplasm.

surrounded by a series of parallel membranes arranged to form flattened sacs (Figures 5-1 and 5-5). The size, pattern, and number of Golgi vary greatly, depending on cell type.

The primary function of the Golgi complex is to accumulate and remove certain large molecules from the cell, especially the molecules that form on the endoplasmic reticulum. In multicellular organisms, these molecules include the digestive enzymes and hormones, which are produced in some cells and used elsewhere in the body. As the vesicles become filled, they move to the outer part of the cell and fuse with the cell membrane. The contents of the vesicles are then dumped outside.

Evidence suggests that the Golgi also serve as sites of carbohydrate synthesis and as points where some carbohydrates and proteins become linked together before being transported out of the cell. In addition, the Golgi may be sites of membrane synthesis.

The Golgi complex also constructs *lysosomes*—special membrane-bound vesicles found only in animal cells (Figure 5-1). These structures confine potentially destructive enzymes that are not needed by the cell inside membrane-bound sacs to prevent them from destroying other components of the cell structure, including nucleic acids, polysaccharides, and proteins.

Occasionally, however, lysosomes do break apart the organelles within a cell. Usually these structures are no longer functioning properly. During food deprivation, this activity may enable a cell to use parts of its own structure as a source of energy and raw materials. Under certain conditions, the membrane of a lysosome ruptures and the enzymes are released to destroy the entire cell. An injured or dead cell is decomposed in this manner. The destruction of certain living cells by their own lysosomes is also a normal part of the development of multicellular organisms. For example, this process causes the formation of fingers from a clublike structure during the early embryonic development of the human hand.

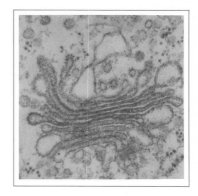

Figure 5-5
The Golgi Complex
This organelle ($\times 25,800$) is comprised of a number of parallel membranes and several vesicles. Certain molecules formed by the cell, including digestive enzymes and hormones, are concentrated within the membrane-enclosed vesicles. These molecules are released from the cell when the vesicles move to the outer part of the cell and fuse with the cell membrane.

Mitochondria: Energy Release and Storage

Mitochondria are large, sausage-shaped organelles that are distributed throughout the cytoplasm of all eukaryotic cells (Figures 5-1 and 5-6). Each mitochondrion is constructed of an outer, enclosing membrane and an inner membrane with elaborate folds or *cristae* that extend into the interior of the organelle. Both membranes are covered with very small projections.

Mitochondria play a part in cellular respiration by transferring the energy within biological molecules to the chemical bonds of ATP. The most common source of molecular energy is the carbohydrate monomer glucose. A large battery of enzymes, which are synthesized on the ribosomes of the cell, work within the mitochondria in an integrated fashion. These enzymes slowly release energy from the glucose molecule by dismantling its structure one carbon atom at a time. Biologists believe that the small projections found on the membranes of the mitochondria contain assemblages of enzymes that are involved in particular aspects of this breakdown process.

Figure 5-6
The Structure of a
Mitochondrion

(a) Photograph of a cross section of a mitochondrion ($\times 40,000$), showing its outer enclosing membrane and its inner membrane with folds (*cristae*) extending into the interior.
(b) Diagram of a mitochondrion with a side cut away to expose the elaborate structure of the cristae. (c) Highly magnified photograph of the small projections that cover the membranes and are believed to be sites for the localization of groups of enzymes.

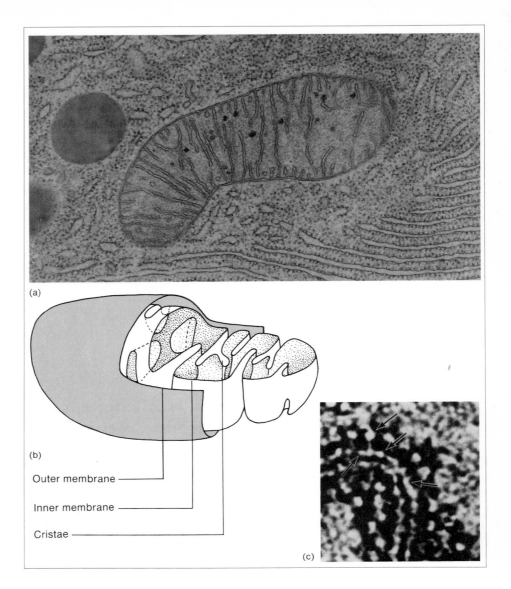

(a)

(b)

Outer membrane ———

Inner membrane ———

Cristae ———

(c)

Cellular Respiration

Prior to entering a mitochondrion, the glucose molecule is divided in half to form two molecules of pyruvic acid. This reaction, called *glycolysis,* is represented by the equation

$$C_6H_{12}O_6 \longrightarrow 2C_3H_4O_3 + 4H + \text{Energy}$$

(1 glucose molecule)   (2 pyruvic acid molecules)   (4 hydrogen atoms)

Two chemical bonds between carbon atoms are broken in the conversion of the 6-carbon structure (glucose) into two 3-carbon structures (pyruvic acid). Many enzymes are required to break these bonds, which release energy and transfer it to the chemical bonds of ATP. This reaction occurs in all cells, and biologists believe that an identical reaction occurred in the first cells to appear on earth.

The next step in cellular respiration varies, depending on the type of cell (Figure 5-7). Cells process pyruvic acid in three major ways: alcoholic fermentation, lactic acid fermentation, and aerobic respiration. The first two reactions occur in the absence of oxygen and are referred to as *anaerobic* (without oxygen) *cellular respiration.*

In primitive cells and in some eukaryotic cells, such as yeast cells, glucose is broken down by *alcoholic fermentation* through the conversion of pyruvic acid into alcohol (C_2H_5OH) and carbon dioxide. This reaction is represented by the equation

$$2(C_3H_4O_3) + 4H \longrightarrow 2(C_2H_5OH) + 2CO_2$$

The four hydrogen atoms that enter the reaction were released during glycolysis.

Both alcohol and carbon dioxide are waste products of the reaction. When bread is made, the alcohol in the yeast evaporates during baking and the remaining carbon dioxide causes the bread to rise. When wine is fermented, the alcohol in the yeast remains in the liquid and the carbon dioxide diffuses away.

In *lactic acid fermentation,* each pyruvic acid molecule is converted into lactic acid ($C_3H_6O_3$) in the absence of oxygen gas. During the reaction, each molecule of pyruvic acid is rearranged and hydrogen atoms are added to the structure:

$$2(C_3H_4O_3) + 4H \longrightarrow 2(C_3H_6O_3)$$

Again, the four hydrogen atoms that enter the reaction are released during glycolysis. This form of anaerobic respiration occurs in the muscle cells of humans and other animals during extreme physical activity, such as sprinting, when oxygen cannot be transported to the cells as rapidly as it is needed.

Both alcoholic and lactic acid fermentation yield relatively small amounts of energy from glucose molecules. Only about 2% of the energy present within the chemical bonds of glucose is converted into adenosine triphosphate (ATP). Most of the energy remains in the waste products of alcohol or lactic acid. The more complete breakdown of the glucose molecule, in which the chemical bonds between all of the carbon atoms are broken, can occur only in the presence of oxygen gas and is called *aerobic respiration*

Aerobic respiration occurs in most eukaryotic cells and in many prokaryotic cells. Instead of being converted into alcohol or lactic acid, the pyruvic acid molecules, together with the four hydrogen atoms, are taken into the mitochondria of the eukaryotic cells. During the final stages of respiration, which occur in the presence of oxygen gas within these organelles, each pyruvic acid molecule passes through a series of more than 100 enzyme-facilitated chemical reactions. A complete description of this remarkable process is beyond the scope of this book. Here, we will examine only the general nature of the three major reactions that occur within the mitochondria during *aerobic respiration:* (1) the conversion of pyruvic acid to acetyl coenzyme A, (2) the Krebs cycle, and (3) oxidative phosphorylation.

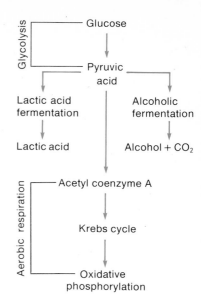

Figure 5-7
The Major Steps
in Cellular Respiration

Glycolysis—the conversion of glucose to pyruvic acid—occurs in all cells. Pyruvic acid can then be converted into lactic acid or into alcohol and carbon dioxide. Both of these reactions occur in the absence of oxygen and are forms of anaerobic respiration. Pyruvic acid can also be converted into acetyl coenzyme A and pass through the Krebs cycle and oxidative phosphorylation. This reaction is aerobic respiration.

In the first phase, each pyruvic acid molecule is transformed from a 3-carbon molecule to a 2-carbon molecule, giving off carbon dioxide in the process. The 2-carbon molecule immediately joins with a molecule called *coenzyme A*. The newly formed molecule is *acetyl coenzyme A*.

In the Krebs cycle—a series of gradual steps within the interior of the mitochondria—about 10 enzymes work together to remove energy from the acetyl portion of acetyl coenzyme A. The remaining chemical bonds are broken, and the released carbon atoms join with oxygen to form carbon dioxide. Some energy is trapped within ATP molecules, but most of the energy is captured by the electrons within the hydrogen atoms. These hydrogen atoms, together with those formed during glycolysis, break apart into their component protons and energy-rich electrons.

In the third phase of aerobic respiration—the *oxidative phosphorylation reactions*—the energy-rich electrons are passed along a chain of specialized protein molecules in such a way that the electrons slowly release energy, which is then used to form ATP molecules. This sequence, or *electron transfer*, occurs on the cristae of the mitochondria. After passing along this chain, the low-energy electrons join with the protons formed earlier and the electrons and protons combine with oxygen atoms to form water. (Two electrons and two protons form two hydrogen atoms, and two hydrogen atoms and an oxygen atom form water, H_2O.)

Energy is released during all phases of aerobic respiration: glycolysis, the conversion of pyruvic acid to acetyl coenzyme A, the Krebs cycle, and oxidative phosphorylation. But most of the energy is released from the high-energy electrons during the oxidative phosphorylation reactions, which produce 24 of the 38 ATP molecules formed during aerobic respiration. About 40% of the chemical energy originally contained in the glucose molecule is made available to the cell in the form of ATP during aerobic respiration; the remainder escapes as heat energy. The complete aerobic cellular respiration reaction is

$$C_6H_{12}O_6 \;+\; 6O_2 \longrightarrow 6CO_2 \;+\; 6H_2O \;+\; Energy$$

| (1 glucose molecule) | (6 oxygen molecules) | (6 carbon dioxide molecules) | (6 water molecules) |

Chloroplasts—The Synthesizers of Glucose

The cells of all photosynthetic organisms, such as plants, contain pigments, usually in the form of chlorophyll and carotenoid molecules, which absorb light energy. The pigments in eukaryotic organisms are located within special organelles called *plastids*, which are distributed throughout the cytoplasm of the cell. The most common plastids—the *chloroplasts*—contain green chlorophyll pigments. The chloroplasts also contain *carotenoids*—yellow, orange, and red pigments that play a relatively minor role in photosynthesis. The green chlorophyll usually masks the colors of these other pigments except in autumn, when the amount of chlorophyll in the cells of many plants decreases and the carotenoids become apparent.

Each chloroplast is enclosed by an outer membrane similar to the cell membrane. Inside, the chloroplast is composed of an elaborate series of parallel membranes—some spanning the entire organelle, others occurring as stacks of

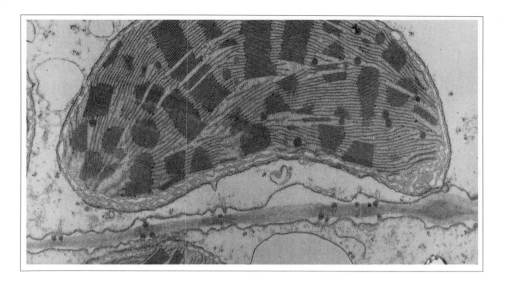

Figure 5-8
The Chloroplast
The *light reaction* of photosynthesis occurs on the membranes inside the organelle. Pigments are incorporated in the structure of these membranes. The *dark reaction* of photosynthesis occurs in the areas between the membranes. The entire organelle (×22,900) is enclosed within a membrane similar to the cell membrane.

smaller membranes (Figure 5-8). The pigment molecules are incorporated in the structure of these internal membranes.

In *photosynthesis,* light energy from the sun is converted into chemical energy in such a way that covalent (electron-sharing) bonds form between the carbon atoms of glucose. After this first step, the more than 70 separate chemical reactions involved in the photosynthetic process are similar to the reactions that occur in the synthesis of any biological molecule. However, the first step—the *light reaction,* during which light energy is initially converted into chemical energy—is remarkable and unique.

The Light Reaction

Light energy occurs in the form of almost weightless particles (*photons*) that travel at the incredible speed of 186,000 miles per second. Photons carry an immense amount of energy in the form of motion. When they collide with pigments, some of this energy is transferred to the pigment molecules.

The structure of pigments is unusual in that the carbon atoms that form part of the molecule are joined together in such a way that the electrons vibrate back and forth, changing position from one carbon atom to another. Compared to the electrons of other chemical bonds, these electrons are not as closely bound to the pigment structure. When a photon of light strikes a pigment molecule, the energy it contains is transferred to one of the vibrating electrons. After such an electron receives the energy from two photons of light, it moves about so rapidly that it breaks away from its original position in the pigment molecule and is attracted to other molecules, with which it shares its energy.

Eventually, the energy within an *excited electron* is transferred to the high-energy phosphate bonds within an ATP molecule or a similar structure—a nicotinamide-adenine dinucleotide phosphate (NADP) molecule. When the energy is transferred to an ATP molecule, the process is called *cyclic photophosphorylation,* shown in Figure 5-9(a). ("Cyclic" indicates that once the energy of the excited electrons is transferred, these same electrons are recycled back into the original pigment molecule;

Figure 5-9
The Two Pathways Taken by Electrons That Have Captured Light Energy

(a) *Cyclic photophosphorylation.* Two excited electrons (e!) release their energy, which is used to form ATP (Ⓟ symbolizes phosphate). After releasing their energy, the two electrons return to the chlorophyll molecule. (b) *Noncyclic photophosphorylation.* Two excited electrons join with NADP and two protons from a water molecule to form $NADPH_2$. The electrons from the water molecule replace the two electrons that have left the chlorophyll molecule. The single oxygen atom joins with another oxygen atom from an identical noncyclic photophosphorylation reaction to form oxygen gas (O_2).

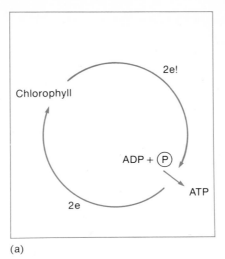

(a)

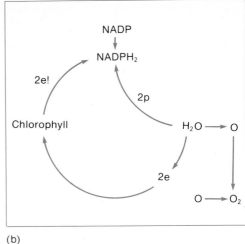

(b)

"photophosphorylation" indicates that light energy is transferred to the phosphate bonds.) When the energy is transferred to NADP molecules, the process is called *noncyclic photophosphorylation,* shown in Figure 5-9(b), because the same electrons are not recycled back into the pigment molecule. In the noncyclic reaction, two excited electrons leave the pigment molecule and eventually join an NADP molecule. Two protons from a water molecule are also added to the new structure, which becomes $NADPH_2$ (each proton and electron forming a hydrogen atom). $NADPH_2$ is then used as a source of energy and as a source of the hydrogen atoms needed to construct glucose molecules in the series of steps called the *dark reaction* (to be discussed in the next section). Only two electrons and an oxygen atom remain in the original water molecule. The two electrons are passed through a complex route back to the original pigment molecule to replace the excited electrons that are now carried within the $NADPH_2$ molecule. The remaining oxygen atom combines with another oxygen atom to form oxygen gas (O_2), which is released from the chloroplast. Both the cyclic and noncyclic forms of photophosphorylation occur during the light reaction in all photosynthesizing cells.

The Dark Reaction

The energy captured within the ATP or $NADPH_2$ molecules during the light reaction is used to construct glucose molecules from carbon dioxide during the dark reaction. The dark reaction of photosynthesis can occur in the light, but it does not require light.

The energy that is obtained from the sun in the light reaction and transferred to the ATP and $NADPH_2$ molecules is used primarily to bind hydrogen atoms from $NADPH_2$ to molecules of carbon dioxide and to join the carbon atoms together to form glucose ($C_6H_{12}O_6$). The dark reaction, also called carbon *fixation,* is a series of straightforward reactions that are basically similar to the reactions involved in the synthesis of other molecules within a cell although they differ in detail.

The complete equation for the light and dark reactions of photosynthesis is

$$6CO_2 \ + \ 12H_2O \ + \ Energy \longrightarrow \ C_6H_{12}O_6 \ + \ 6O_2 \ + \ 6H_2O$$

(6 carbon dioxide molecules)   (12 water molecules)   (1 glucose molecule)   (6 oxygen molecules)   (6 water molecules)

The newly formed glucose molecules are then stored in the form of large starch molecules, converted into other types of molecules (such as amino acids), or broken apart again so that the released energy can be transferred to ATP molecules for immediate use within the cell.

Most organisms ultimately depend on photosynthetic reactions for energy. Animals that eat plants obtain their energy directly from the products of photosynthesis. Some of the ingested plant materials are converted into animal tissues which, in turn, may be a source of energy for other animals (predators and parasites). Most living organisms depend on the conversion of light energy into chemical energy—a process performed only by photosynthetic cells.

The Ground Substance of the Cytoplasm

We know that the cytoplasm—the area of the cell outside the nucleus—contains a variety of special organelles. The medium in which these organelles are suspended is called the *ground substance.*

The ground substance is not simply a structureless fluid. It contains an elaborate network of fibrous protein strands that perform a variety of important functions, including holding organelles in place (Figure 5-10), maintaining cell shape, directing the movement of materials within the cell, and directing the movement of the entire cell. The network may also hold enzymes in place, thereby coordinating the activities of certain biochemical pathways.

This network of protein strands and organelles is immersed in a fluid that is about 50% water and is rich in small molecules, such as glucose, amino acids, carbon dioxide, and oxygen gas. The ground substance positions these small molecules so that they are available to participate in chemical activities and also coordinates the organelles into a single functional unit.

The Cell Wall: A Support for Plant Tissues

Most plant cells are encased in a *cell wall*—a rigid structure that completely surrounds the cell membrane (Figure 5-11) and provides the cell with both mechanical support and protection. Because the cell wall is a porous structure that permits the passage of virtually all types of molecules, it plays little or no role in determining the movement of substances into or out of the cell.

The cell interior contains high concentrations of molecules and ions, so that when a cell is placed in fresh water, it takes in the water and begins to swell. This process is *osmosis*—a special form of diffusion we have already discussed in Chapter

Figure 5-10
The Fibrous Proteins of the
Ground Substance

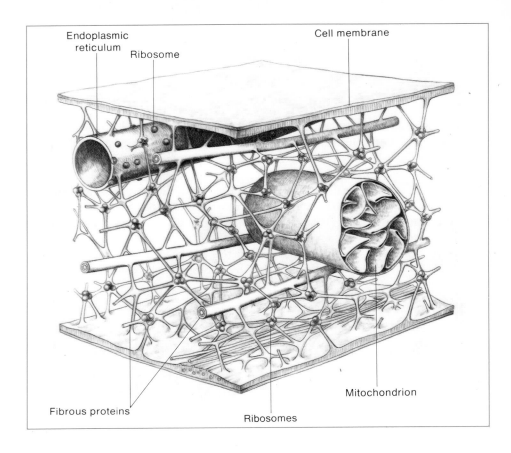

Endoplasmic reticulum
Ribosome
Cell membrane
Mitochondrion
Ribosomes
Fibrous proteins

Figure 5-11
Cells from the Leaf
of a Sunflower

Each cell is enclosed within a cell
wall, which surrounds the cell
membrane. In the central cell
($\times$10,000), the nucleus (N), several
chloroplasts (Chl), and mitochondria
(M) are visible.

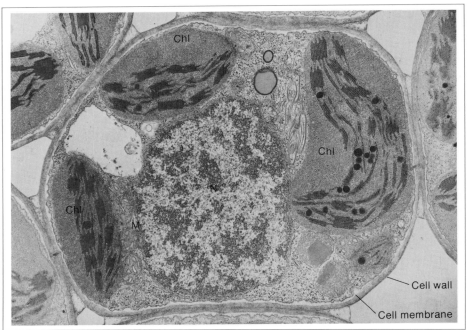

Chl
N
Chl
Chl
M
Cell wall
Cell membrane

4. Osmosis also occurs within the plant cell, where water accumulates and the elastic cell membrane expands. When the cell membrane presses against the rigid cell wall and can expand no further, the intake of water is stopped, thereby protecting the membrane from rupturing. The increased pressure caused by the accumulation of water within the cell is called *turgor pressure.* By making the cells rigid, turgor pressure helps to hold plant tissues upright despite the force of gravity. Deprived of fresh water, a plant loses its turgor pressure and wilts.

The cell walls of most plants are composed primarily of cellulose—a polysaccharide that the cell produces from glucose monomers. Cellulose molecules are organized into bundles; each bundle contains about 40 long cellulose molecules (Figure 5-12). Polysaccharides, waxes, and other types of materials may also be incorporated into the structure of the cell wall.

The cell walls of a plant often remain long after the plant itself has died, thereby preserving the plant's structure. Because the cell walls in plants are constructed of cellulose molecules arranged in bundles, each molecule is relatively inaccessible to the digestive enzymes of the small organisms that feed on dead materials. Plant tissues therefore take much longer to decompose than animal tissues do.

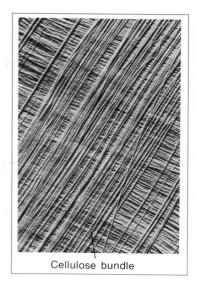
Cellulose bundle

Figure 5-12
Close-up of the Structure of a Cell Wall
The cellulose molecules are arranged in bundles ($\times$ 24,200), which give the cell wall its strength.

Possible Origins of Eukaryotic Cells

Fossils of cells of an intermediate structure between prokaryotes and eukaryotes have yet to be discovered. Biologists therefore know nothing about how the cellular organelles characteristic of eukaryotic cells originated. Many hypotheses have been developed to explain the transition from prokaryotic to eukaryotic cells. Here, we will consider two of the more widely accepted hypotheses: (1) the *progressive differentiation hypothesis,* which views the evolution of eukaryotic cells as a gradual increase in the complexity of prokaryotic cells, and (2) the *symbiotic hypothesis,* which views the origin of eukaryotes as a more abrupt process involving the union of several prokaryotic cells.

The Progressive Differentiation Hypothesis

According to this more traditional hypothesis, as prokaryotic heterotrophs became increasingly abundant and depleted the supply of large biological molecules in their environment, the photosynthetic prokaryotic autotrophs—cells that could construct their own molecules from inorganic materials—began to dominate the early environments of the earth. The subsequent evolution of the autotrophs eventually produced two major groups of eukaryotes: (1) *plants,* which retained their photosynthetic abilities, and (2) *animals,* which lost their photosynthetic abilities as they developed methods of eating plants. Complex structures within eukaryotic cells, such as the mitochondria and plastids, evolved gradually as indentations and modifications of the cell membrane. Through successive generations of cells, the organelles became more specialized in structure and function. The main strength of this hypothesis is that it does not depart from the prevailing scientific view that the evolutionary process has been a series of small, gradual changes.

The Symbiotic Hypothesis

This more radical hypothesis suggests that structures within the eukaryotic cells arose more suddenly from the union of different kinds of prokaryotic cells. According to this view, primitive heterotrophic cells varied considerably in size. Some of the larger cells acquired the ability to obtain energy and structural materials by ingesting smaller cells and developed into predators. Occasionally, the ingested cell continued to survive intact within the predator cell without being disassembled into its component nutrients. The two cells formed what is called a *symbiotic relationship,* in which two different kinds of organisms live together in close association (Figure 5-13). Over time, the cells in this symbiotic association became increasingly dependent on one another until each cell lost some of its original functions and became more specialized in structure and function.

When the smaller cell living inside the larger, predatory cell was a heterotroph capable of utilizing oxygen to break down sugar molecules, the smaller cell benefited by gaining a sheltered environment and the larger cell benefited by gaining a membrane-enclosed structure that contained the materials required for aerobic respiration. According to the symbiotic hypothesis, what biologists now identify as mitochondria were once smaller, free-living, prokaryotic heterotrophs that have subsequently undergone millions of years of evolution within the larger cell.

Many biologists also believe that chloroplasts were once free-living, photosynthetic prokaryotes—probably in the form of blue-greens (Figure 5-13), which they resemble in both structure and function. Many other structures within eukaryotic cells may also have been formed by the union of prokaryotes.

The discovery that both mitochondria and chloroplasts contain circular DNA molecules, which are found in the prokaryotes, as well as RNA molecules and ribosomes, and the finding that these organelles can synthesize some of their own proteins provide strong evidence to support the symbiotic hypothesis. These

**Figure 5-13
Symbiosis—Two Different Organisms Living Together in Close Association**

Here, blue-green prokaryotes live inside a protozoan—a unicellular eukaryote (×13,000). The numerous parallel membranes within the blue-green cells carry the light-absorbing pigments involved in photosynthesis. There is a striking resemblance between the structures of a blue-green and a chloroplast.

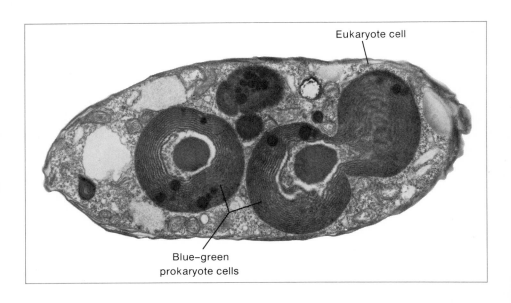

Eukaryote cell

Blue–green prokaryote cells

organelles are similar in size to prokaryotic cells, suggesting that the organelles within eukaryotic cells are remnants of prokaryotic cells that once existed as independent organisms.

Until cells or fossils of cells representing the intermediate stages of evolution between the prokaryotes and the eukaryotes are found, both the progressive differentiation hypothesis and the symbiotic hypothesis provide valid explanations of the origin of cellular organelles.

Summary

The *cell* is the basic unit of life. All living things are organized into cells, and each cell exhibits the properties of life. Cells are formed only from the division of previously existing cells.

Cells can be classified according to the complexity of their internal structure: (1) *prokaryote* or *simple cells*, which have relatively little internal structural organization, and (2) *eukaryote* or *complex cells*, which have a number of internal structures, including a nucleus, chromosomes, endoplasmic reticulum, Golgi complexes, lysosomes, mitochondria, and chloroplasts.

The *nucleus* is the part of the eukaryotic cell that contains the *chromosomes* and *nucleolus*; its interior is isolated from the rest of the cell by a nuclear *membrane*. Chromosomes contain DNA molecules and other kinds of materials.

The *cytoplasm*—the area of a eukaryotic cell outside the nucleus—contains many organelles. The *endoplasmic reticulum* occurs throughout the cytoplasm. It is composed of networks of membranes, which form canals through which materials are transported from place to place within the cell. Ribosomes attached to the outer surface of the *rough endoplasmic reticulum* provide sites for the synthesis of proteins destined to be stored within the cell or to be transported out of the cell. Other ribosomes that are not attached to the membranes of the endoplasmic reticulum are involved in the synthesis of proteins to be used immediately within the cell. The *smooth endoplasmic reticulum*, with no ribosomes attached, may provide sites for lipid synthesis.

The main functions of a *Golgi complex* are to store materials within the cell and to transport materials out of the cell. This organelle may also be involved in the synthesis of large carbohydrate molecules. *Lysosomes*—special sacs constructed by a Golgi complex— are filled with enzymes that can be used to break apart unwanted molecules within the cell.

Energy for cellular activities is supplied primarily by the *mitochondria*. Within these organelles, pyruvic acid (formed from glucose) is broken down and the energy released in the reaction is taken up by ATP molecules. This process is referred to as *cellular respiration*. There are two forms of cellular respiration: *aerobic respiration* (with oxygen) releases about 20 times more energy from each glucose molecule than *anaerobic respiration* (without oxygen) releases.

Plastids are organelles within eukaryotic cells that undergo *photosynthesis*—the reaction in which light energy is used to build glucose molecules. The most common type of plastid is the *chloroplast*, which contains the green pigment *chlorophyll*. *Pigments* located inside the plastids capture the light energy from the sun, which electrons then carry to ATP or $NADPH_2$ molecules. Cells use these high-energy molecules to build glucose molecules from carbon dioxide and water. Most organisms are ultimately dependent on photosynthesis for life, since it is the only process by which light energy is converted into the chemical energy within biological molecules.

The organelles of a cell are suspended within a medium called the *ground substance*, which contains both a network of fibrous protein strands and a fluid composed of water and small molecules. The ground substance plays an important role in coordinating all the activities of a cell.

The structure of a plant cell is characterized not only by the presence of plastids but also by a *cell wall.* This rigid, porous structure is made of cellulose and surrounds the cell membrane, protecting the cell and giving it support.

Two possible explanations of how eukaryotic cells may have originated from prokaryotic cells are the *progressive differentiation hypothesis* (gradual changes increased the complexity of cell structure) and the *symbiotic hypothesis* (eukaryotic cells formed from the union of several prokaryotic cells).

The Reproduction of Cells

6

No cell continues to grow indefinitely. After it reaches a certain size, a cell divides to form two cells. Cells limit their growth to offset the disadvantages they would encounter if they were larger. A major disadvantage of an unregulated increase in cell size would be an increase in the total amount of chemical activity required to maintain normal cell functions, which would place a greater demand on the cell membrane's ability to transport materials into and out of the cell. Beyond a certain cell size, the cell membrane is unable to exchange materials at an adequate rate and cellular activities are curtailed.

When a cell divides, the two new *daughter cells* are exact copies of the original cell. This process of cell division is called *cellular reproduction.* In order to function, each daughter cell must contain a complete copy of the DNA molecules that direct protein synthesis. A cell reproduces itself by making two exact copies of the DNA molecules and then splitting in two, so that each daughter cell receives a complete set of genetic instructions.

In Chapter 6, we will examine the process of *DNA replication,* how prokaryotic and eukaryotic cells divide, and the role cell division plays in the development of a multicellular organism.

The Replication of DNA Molecules

A critical stage in biological evolution was the development of the ability of DNA molecules to make exact copies of themselves, or to *replicate.* The replication of DNA molecules is similar to the assembly of RNA molecules from a DNA molecule during transcription (shown in Figure 4-11 on page 71). In replication, however, both strands of the DNA molecule are copied and the DNA nucleotides, which contain the bases adenine, guanine, cytosine, and thymine, are assembled.

Replication is initiated by the separation of the two strands that make up a

DNA molecule (Figure 6-1). The bonds between the paired bases are broken, and the DNA nucleotides—each containing a sugar, a phosphate, and a base—join together with the exposed bases on the sugar–phosphate strands of the original DNA molecule, according to the base-pairing rules (adenine pairs with thymine; guanine pairs with cytosine). As the nucleotides become joined to the intact strand, they also form bonds between the sugar of one nucleotide and the phosphate of another so that a new strand is assembled. Each new strand is joined to one of the previous strands to form a double-strand DNA molecule. In this way, two new DNA molecules are formed, each having a sequence of base pairs identical to that of the original DNA molecule.

It is important that neither sugar–phosphate strand of the original molecule breaks apart during the replication process. The integrity of these strands preserves the genetic code.

Figure 6-1
DNA Replication

The DNA molecule separates to form two new strands. Bonds between base pairs connect each new strand to one of the original strands, thereby forming two DNA molecules that are identical to each other and to the original DNA molecule. Note that adenine pairs only with thymine and that cytosine pairs only with guanine. The small circles represent the sugar deoxyribose; the large circles represent the nitrogenous bases.

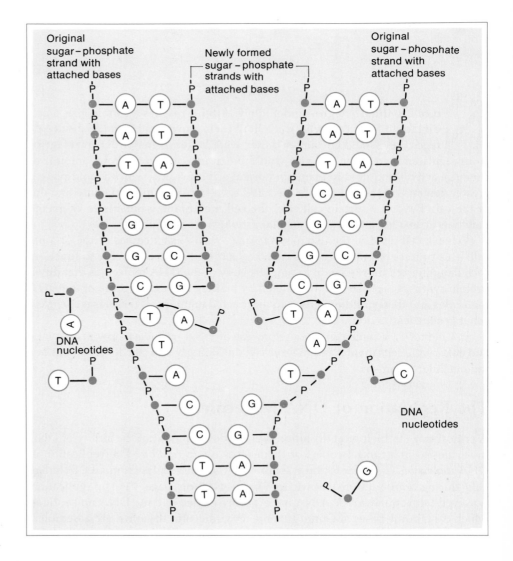

The two newly formed DNA molecules separate as the original DNA molecule replicates itself. Each new DNA molecule contains one sugar–phosphate strand with attached bases from the original DNA molecule and one newly assembled sugar–phosphate strand with attached bases.

Cell Division

The replication of DNA molecules occurs in the same way in all cells. However, the process of cellular reproduction—the way in which a cell divides into two cells—differs in prokaryotic and eukaryotic cells.

Binary Fission in Prokaryotic Cells

In a prokaryotic cell, each of the two newly formed DNA molecules moves to a different side of the cell. The cell then simply pinches in two so that each daughter cell receives one of the DNA molecules and about half of the cellular material from the parent cell. This simple method of cell division is called *binary fission* (Figure 6-2).

The earliest cells probably reproduced by binary fission, just as prokaryotic cells do today. After the cell divides, each daughter cell grows until it reaches a certain size and then undergoes DNA replication and cell division. In this way, prokaryotic organisms multiply in number. Under ideal environmental conditions, a prokaryotic cell, such as a bacterium, can divide as frequently as every 20 minutes.

Mitosis in Eukaryotic Cells

Despite the enormous diversity of life forms on earth, all eukaryotic cells multiply by forming two cells from one cell in the same manner. Cell division in eukaryotic cells is called *mitosis.* Mitosis occurs during the asexual reproduction of cells, during the development of multicellular organisms from a single fertilized egg, and during the renewal of cells for the purpose of tissue maintenance in adult multicellular organisms.

Mitosis is a sequence of highly coordinated chemical and structural cellular changes. During this process, the genetic information contained in a single cell is replicated and distributed equally between two daughter cells. The equal distribution of genetic material is more complex during cell division in eukaryotes than it is during cell division in prokaryotes, because DNA occurs within a number of separate chromosomes in eukaryotic cells. Each daughter cell must receive one copy of each chromosome, which requires a high degree of coordination within the eukaryotic cell.

There are five phases of mitosis: interphase, prophase, metaphase, anaphase, and telophase (Figure 6-3). However, mitosis is a continuous process, and each of these phases flows into the next.

Interphase The interval between successive divisions of a cell is called the *interphase.* This stage is sometimes referred to as the *resting phase,* since it is not characterized by dramatic changes in the cell structure. However, the cell is not

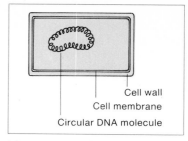
(a)

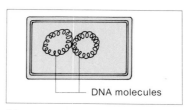
(b)

(c)

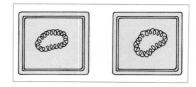
(d)

Figure 6-2
Binary Fission in a Bacterium
(a) The cell prior to division. (b) DNA replication produces two identical circular molecules. (c) As the cell pinches itself in two, one DNA molecule is incorporated in each half of the cell. (d) After fission, each of the two daughter cells contains a DNA molecule and about one-half of the cellular material.

Figure 6-3
Chromosomes During the Different Phases of Mitosis

The cell shown in these drawings and photographs contains six chromosomes. (a) During *interphase*, the chromosomes are so long, thin, and intertwined that individual structures cannot be recognized. (b) The chromosomes shorten during *prophase*. At this stage, two *chromatids* joined together at the kinetochore are visible within each chromosome. (c) During *metaphase*, the chromosomes are very short and oriented toward the center of the cell. *Spindle fibers* form between the kinetochores and each of the two cell *poles*. The kinetochores divide at this stage, and each chromatid becomes a chromosome. (d) The two chromosomes (former chromatids) in each pair move toward opposite poles of the cell during *anaphase*. Each kinetochore is pulled by a shortening *spindle fiber*, so that the chromosome appears to be doubled as its two segments move together after the kinetochore. (e) Two identical groups of chromosomes, each enclosed within a nuclear membrane, form clusters at the two poles of the cell during *telophase*. In the latter part of this phase (not shown), the cell membrane pinches apart between the two nuclei. The end result of these mitotic activities is the formation of two identical cells from a single parent cell.

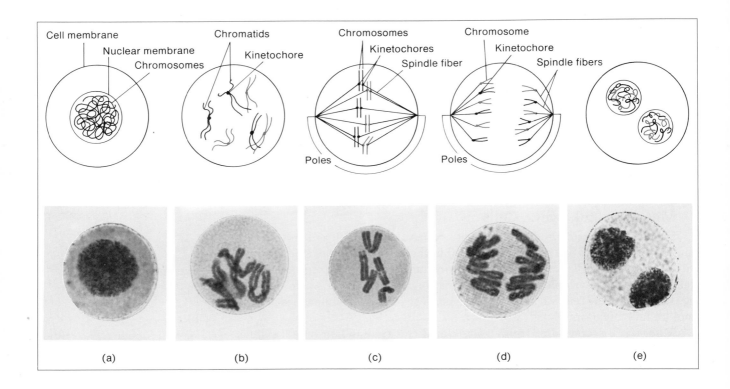

actually resting during interphase. Normal cellular activities, such as protein synthesis and cellular respiration, which are disrupted during the other phases of mitosis, occur during interphase.

When a cell first enters interphase, it has just been formed from a previously existing cell. The new cell is smaller than a normal cell because it has received only one-half of the cytoplasm of the parent cell. In the early part of interphase, the cell grows and forms new structures, such as ribosomes and mitochondria. About midway through interphase, each chromosome doubles. At this stage, the chromosomes are so long and thin that they are not visible as separate bodies, as shown in

Figure 6-3(a). The DNA molecules within the chromosomes replicate to make exact copies of themselves.

Other components of the chromosomes are also synthesized at this time. As a result of these activities, each chromosome becomes two identical, elongated structures—the *chromatids*—which are joined together in the region of the kinetochore. This constricted part of the chromosome is not duplicated until later in the mitotic sequence. The paired chromatids intertwine around each other to form a coil and are not distinguishable at this stage.

Prophase   During the next stage of mitosis, *prophase,* the paired chromatids contract, becoming short and thick. The chromatids, which are now visible through a microscope, can be distinguished from each other by their length and by the position of their kinetochores, as shown in Figure 6-3(b). As the chromatids of each pair contract, they gradually unwind and the coil disappears. *Spindle fibers,* which are long hollow tubes composed primarily of fibrous protein molecules, begin to form and eventually attach to the kinetochores (Figure 6-4). These fibers play an important role in moving the chromosomes to the daughter cells in a later phase.

Near the end of prophase, the nucleoli disappear and the nuclear membrane surrounding the chromosomes disintegrates.

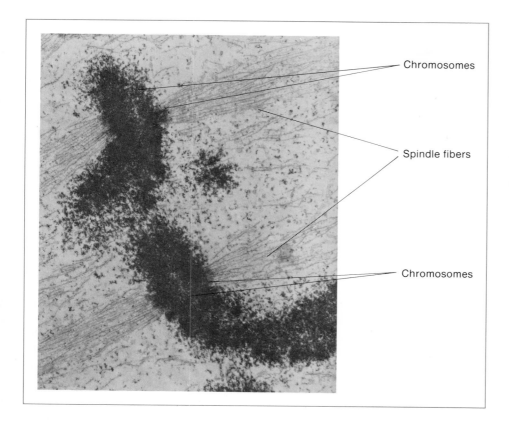

Chromosomes

Spindle fibers

Chromosomes

Figure 6-4
The Attachment of Spindle Fibers to Kinetochores
Here, many spindle fibers are attached to each of the kinetochores on two pairs of chromosomes. In this photograph, the kinetochores have already duplicated themselves and are beginning to separate as the spindle fibers shorten.

Metaphase In *metaphase,* which begins after the nuclear membrane disappears, the paired chromatids (each pair still joined together by a kinetochore) move toward the center of the cell, where they arrange themselves into a single line. Once this formation has been achieved, some of the spindle fibers position themselves between the kinetochores and the two ends, or *poles,* of the cell at right angles to the line of paired chromatids. Each of these spindle fibers extends from a kinetochore to a pole of the cell. In some cells, the spindle fibers attach themselves to special structures at each pole. A set of spindle fibers is now attached to either side of each kinetochore, so that spindle fibers connect each kinetochore to both poles of the cell. The spindle fibers that are not attached to kinetochores extend from pole to pole of the cell.

The kinetochores then divide in two and the paired chromatids are completely separated from one another, as shown in Figure 6-3(c). Each chromatid has its own kinetochore and is now a chromosome.

Anaphase During *anaphase,* the spindle fibers attached to the kinetochores shorten and the paired chromosomes are pulled away from one another. This movement continues until one complete set of chromosomes is positioned at each pole of the cell, as shown in Figure 6-3(d).

Telophase During *telophase*—the final stage of mitosis—one set of chromosomes is grouped compactly at each cell pole, as shown in Figure 6-3(e). Nuclear membranes re-form from the endoplasmic reticulum and enclose each group of chromosomes, creating two new nuclei. Nucleoli reappear within the newly formed nuclei. The cell membrane then pinches apart between the two nuclei to form two separate cells.

At this time in plant cells, a new cell wall develops between the two newly formed cells. This cell wall begins as a string of disconnected cellulose plates, which eventually join together to form a partition between the two cells (Figure 6-5). The growing partition spreads until it eventually joins the mature walls of the original cell.

At the end of telophase, each daughter cell contains the same number of chromosomes that the parent cell originally contained. The chromosomes elongate once again, and the mitotic cycle is completed as the daughter cells reenter interphase.

Specialization of Functions Between Cells: Multicellular Organisms

For more than 2 billion years after the first appearance of cells on earth, single cells were the only forms of living organisms. Fossil evidence indicates that the first multicellular organisms appeared about 1 billion years ago. Because all multicellular organisms are composed of eukaryotic cells, these organisms probably originated from unicellular eukaryotes. After their appearance, multicellular organisms rapidly became abundant and diverse, although these larger and more complex organizations of life did not replace unicellular organisms.

We have seen how unicellular organisms grow to a certain size and then divide

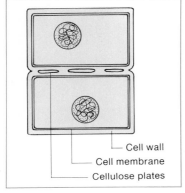

(a)

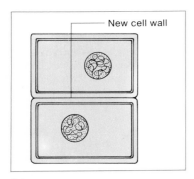

(b)

Figure 6-5
The Formation of a New Cell Wall During Late Telophase in a Plant Cell

(a) Disconnected cellulose plates appear between the two newly formed daughter cells. (b) The plates join together and eventually unite with the walls of the original parent cell.

into two cells and how each of these daughter cells follows its own independent course of subsequent growth and division. Cellular reproduction is the same in multicellular organisms, except that the daughter cells remain together instead of separating. Each multicellular organism begins as a single cell, which divides repeatedly by mitosis until a particular number of cells have formed. The cells within the aggregate then differentiate in both structure and function to perform the diverse tasks required to maintain a multicellular organism. Each type of cell performs a specific function that contributes to the survival and reproduction of the entire organism. In this way, the number of chemical activities that occur within each cell can be restricted and greater cellular efficiency can be achieved.

All multicellular organisms are composed of eukaryotic cells, which carry genetic information within chromosomes inside the nucleus. Eukaryotic cells reproduce by mitosis, so that all of the cells in a multicellular organism contain exact copies of the same set of chromosomes carrying the same genetic information. But if all of these cells are genetically identical, how do they become structurally and functionally different from one another during the development of the organism? This may be the most fundamental unanswered question in biology.

Cells that perform different functions must manufacture different enzymes, since these factors control the chemical activities of a cell. The fact that enzymes are produced according to information stored within the DNA molecules implies that each cell of a multicellular organism expresses only a limited amount of its full genetic potential in the form of its enzymes. Genetic information expressed in one type of cell may therefore remain dormant in another cell.

Expression of genes within a cell occurs when DNA directs the synthesis of mRNA (*transcription*) and mRNA, in turn, directs the synthesis of proteins (*translation*). But what mechanism does the cell employ to select the particular genes to be expressed within each cell of a multicellular organism? Evidence suggests that some of the genes of a cell direct the expression of other genes and that these so-called *regulatory genes* determine which DNA segments will be made available for transcription. It is probable that much of the genetic information within the cells of more complex organisms like human beings is involved in the control of gene expression rather than in the synthesis of proteins. Although gene expression is precisely controlled during the development of multicellular organisms, biologists still do not know what mechanisms are involved in this process. A more detailed discussion of cell specialization during development will be presented in Chapter 22.

Summary

After a cell has grown to a certain size, it divides to form two *daughter cells*, which are identical to each other and to the original parent cell. In order to function, these daughter cells must receive a complete set of the DNA molecules that were contained in the original cell.

DNA molecules have the capability to make identical copies of themselves. This process, called *DNA replication*, always precedes cell division. In a simple, prokaryotic cell, after the DNA molecule replicates, the two identical molecules move apart and the cell pinches in two. This method of cell division, or *cellular reproduction*, is called *binary fission*. Each of the resulting cells contains a complete copy of the genetic instructions and about half of the cellular material from the parent cell.

Eukaryotic cells divide by a special sequence of chemical and structural changes called *mitosis*. During the five phases of mitosis—interphase, prophase, metaphase, anaphase, and telophase—the chromosomes are duplicated and each chromosome is separated from its identical copy in such a way that each daughter cell contains a copy of all of the chromosomes in the parent cell. The two daughter cells are therefore genetically identical to the original cell and to each other. Each daughter cell also receives about half of the cellular material from the parent cell, including the organelles.

Multicellular organisms are composed of eukaryotic cells. Each multicellular organism develops from a single eukaryotic cell by repeated cell division (mitosis). Since all daughter cells formed by mitosis receive the same set of genes, all of the cells of a multicellular organism contain identical sets of genetic instructions. Yet the cells of a multicellular organism acquire different structures and functions during development. How this happens remains one of the more basic unsolved biological mysteries.

Genetics: The Physical Basis of Heredity

7

Generational patterns of heredity have been recognized for centuries. The expectation that offspring will resemble their parents has been a guideline in the selective breeding of domestic plants and animals for more than 10,000 years.

However, the physical basis of hereditary patterns was not discovered until the twentieth century. We now know that the characteristics of an individual are largely determined by the particular genes (DNA) present within the cells of the body. The genetic information contained within the structure of DNA molecules transmits many of the same characteristics from parent to offspring.

Although offspring usually resemble their parents, they seldom have completely identical features. The resemblance between parents and offspring is not exact for two reasons: (1) the characteristics of an individual are partially determined by the environment, and (2) most organisms inherit genes from two parents and therefore do not receive a set of genes identical to that of either parent.

Environmental factors interact with an individual's genes to influence the development of anatomical, physiological, and behavioral characteristics. Identical twins inherit exactly the same sets of genes, but if they are raised separately and experience different social and physical conditions as children, they will not be identical adults. Both the genes and the environment play a role in the expression of many hereditary features. The particular set of genes an organism inherits is its *genotype;* the characteristics the organism exhibits comprise its *phenotype.* Genes interact with environmental factors to produce the phenotype. For example, insufficient nutrition can permanently stunt body growth and brain development. Other environmental factors, such as seasonal changes, may temporarily alter the phenotype. The hair of some mammals and the feathers of some birds change from white in the winter to brown in the summer (Figure 7-1). The genotype remains the same, but the phenotype changes with the seasons to enable an animal to blend into

Figure 7-1
Interactions Between Genotype and Environment Determine the Phenotype
The plumage of a ptarmigan is brownish in summer and white in winter. Although the phenotype changes in response to different weather conditions, the genotype of each bird remains the same throughout its lifetime.

107

Figure 7-2
Phenotypic Changes in a
Flounder

Figure 7-2
Phenotypic Changes in a
Flounder

This bottom-dwelling fish has
adjusted its color and pattern to match
its background. When the flounder
moves to a different area, it will again
make phenotypic adjustments so that
it will be less visible to predators.

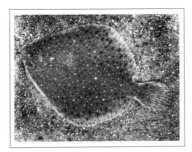

Figure 7-3
The 46 Chromosomes
(23 Homologous Pairs) That
Occur Within Each Diploid
Cell of a Human

(a) In a male, the sex chromosomes
(pair number 23) include an X and
a Y chromosome, shown in the lower
right-hand corner. (b) In a female,
both sex chromosomes are type X,
shown in the lower right-hand corner.
The chromosomes of a cell are not
normally arranged in such an orderly
pattern. Here, the individual
chromosomes appearing in a
photograph of a cell have been cut
out, matched, arranged in pairs, and
rephotographed. These chromosomes
are from dividing cells, so that each
chromosome has already been
replicated and the two identical
chromatids are held together by a
single kinetochore.

its environment and avoid detection by predators. Other animals, like the flounder shown in Figure 7-2, change color more frequently to camouflage themselves.

In Chapter 7, we will deal primarily with the most common method of reproduction among plants and animals—*sexual reproduction*—during which each individual receives genetic information from two parents and acquires some characteristics from each parent. During the other form of reproduction—*asexual reproduction*—each individual produces genetically identical offspring. Asexual reproduction is common in many unicellular organisms and in some multicellular organisms. Inheritance is a straightforward process in asexual reproduction, but a somewhat complex process in sexual reproduction.

Life Cycles

Before we begin our analysis of how genes are transmitted during sexual reproduction, it is important that you understand the *life cycles* of various organisms. The human life cycle is fertilized egg → adult → sperm or egg → fertilized egg. During this sequence of events, the number of chromosomes in a human cell changes from 46 in the cells of an adult to 23 in the sperm or egg to 46 again in the fertilized egg. This is only one of several modes of life cycle.

In sexual reproduction, a new individual is formed when two cells, one from each parent, unite to form a single cell. In most organisms, the cells that eventually unite (the *gametes*) are highly specialized in anatomy, physiology, and chromosome number. The union of the gametes is called *fertilization,* and the single offspring cell is called the *zygote* or *fertilized egg.* Each gamete carries a single set of chromosomes, so that the zygote receives a double set of chromosomes—one from each parent. A cell in which each type of chromosome is present in duplicate is called a *diploid cell.*

Chromosomes carry the genetic information for the construction of a large number of different proteins, which, in turn, determine many characteristics of the individual. In a diploid cell, each pair of similar chromosomes (one from each parent) is called a *homologous pair.* Both chromosomes in a homologous pair carry information for the same genetic traits and are similar in length and in the position of the kinetochore (Figure 7-3).

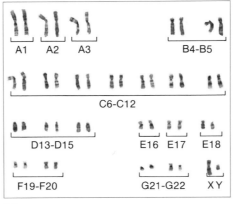

(a)

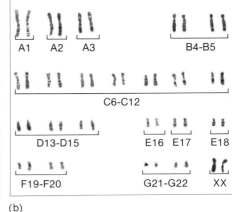

(b)

Prior to fertilization, a diploid cell in each parent undergoes a special form of cell division, called *meiosis,* which produces the gametes. During meiosis the number of chromosomes within the cell is reduced by half. If this did not occur, the number of chromosomes would double each time fertilization occurred and the amount of chromosomal material within a cell would become unmanageable after several generations. If the 46 chromosomes in human diploid cells were not halved during the production of gametes, then the zygote would contain 92 chromosomes (46 from each parental gamete) after fertilization. Each cell in the next generation would contain 184 chromosomes, and each cell in the tenth generation would contain 23,332 chromosomes. The reduction of the number of chromosomes during meiosis compensates for the doubling of chromosomes at fertilization.

The Alternation of Chromosome Numbers

The two sets of chromosomes within a diploid cell separate during meiosis so that each daughter cell receives one chromosome of each type (one chromosome from each homologous pair). Cells that contain only one set of chromosomes are called *haploid cells.* At various points in the life cycle of all sexually reproducing individuals, cell types alternate between the diploid and the haploid state. A diploid cell is reduced to a haploid cell by meiosis; two haploid cells (the gametes) then unite at fertilization to produce a diploid cell.

The pattern of this alternation between diploid and haploid cells varies, depending on the organism. Immediately after fertilization in all animals and most plants, the diploid zygote undergoes a series of *mitotic divisions* (cell division without change in chromosome number) to produce an adult multicellular organism composed of diploid cells. This life cycle is diagrammed in Figure 7-4(a). Prior to reproduction by the adult form, the number of chromosomes within the reproductive cells is halved during meiosis to yield haploid cells—the gametes (sperm and eggs). In animals, haploid cells are produced in the *ovaries* of females and in the *testes* of males; in flowering plants, haploid cells are produced by the *ovaries* and the *anthers,* which form part of the flower.

The adult forms of some types of algae and fungi are composed of haploid cells and undergo the opposite life cycle, diagrammed in Figure 7-4(b). The adult organism produces gametes by mitosis, and two gametes unite at fertilization to form a diploid zygote. Immediately after fertilization, the zygote undergoes meiosis to reduce the number of chromosomes to the haploid condition. Each haploid cell then divides by mitosis to produce the multicellular adult form. Diploid cells occur very briefly in the life cycle of these organisms.

In some plants, both types of multicellular individuals occur: one form consists entirely of haploid cells; the other, entirely of diploid cells. The life cycle of these organisms is diagrammed in Figure 7-4(c). Ferns, for example, are comprised of two types of multicellular individuals—the large, conspicuous diploid individual and the small, inconspicuous haploid individual. Meiosis occurs within certain cells in the large, diploid plant, producing haploid cells, which drop to the ground and develop into smaller haploid plants after a series of mitotic divisions. These haploid individuals grow to only 1 or 2 centimeters in length in the soil beneath the larger diploid form of the fern. Each of the smaller multicellular

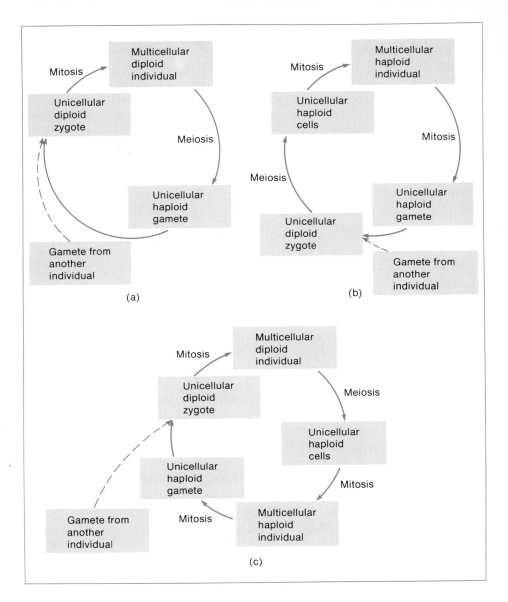

Figure 7-4
Three Forms of Life Cycles
(a) In all animals and most plants, the multicellular individual is composed of diploid cells. (b) In some algae and fungi, the multicellular individual is composed of haploid cells. (c) Ferns and mosses occur in two multicellular forms—one diploid, the other haploid.

plants produces sperm and/or eggs by mitosis and releases these haploid gametes into the moist soil. A sperm from one plant fertilizes an egg from another plant, and the resulting zygote divides and redivides by mitosis to produce another adult diploid form of the fern.

Mosses also occur in both haploid and diploid multicellular forms. In contrast with the fern, the more conspicuous part of the life cycle of a moss is a plant formed of haploid cells.

Characteristics of Diploid Cells

The genetic code for the construction of a particular type of protein is located within the same segment of DNA in both homologous chromosomes and in all chromosomes of that type within the population. Each diploid cell therefore contains two genes for each trait. Different forms of the same gene, called *alleles,* have DNA structures with slightly different base sequences located in the same place on the same type of chromosome. The base sequences that characterize alleles yield proteins that may function differently due to the variation of one or more amino acids. A gene may exist in a variety of alleles in the population, although a single individual can have a maximum of only two alleles—one in each homologous chromosome.

When the two genes for a trait are identical, they contain identical codes for the formation of protein structures. An individual carrying two such genes is said to be a *homozygote* (homo = "same"; zygote = "fertilized egg"), or *homozygous* for that particular trait. When the two genes carrying the code for a trait differ slightly (are alleles), the individual is said to be a *heterozygote* (hetero = "different") or *heterozygous* for that trait.

The two alleles of a heterozygote may act together to determine the precise nature of the trait, or the effect of one allele may prevent the expression of the other allele. When only one of the two alleles is expressed in the phenotype, that allele is said to be *dominant* over the unexpressed allele. For example, several alleles for hair color exist in the human population. An individual carrying the alleles that code for brown and for red hair will have brown hair, because the allele for brown hair will prevent the expression of the allele for red hair. An individual who is homozygous for the allele that codes for red hair will actually have red hair, because no brown allele will be present to prevent the expression of the allele for red hair.

Adult organisms have several advantages if they are composed of diploid rather than haploid cells. In many cases, one gene is sufficient for normal cell function, and the other gene may be held in reserve. If the instructions contained within one gene produce a faulty enzyme that does not function, the same gene on the homologous chromosome may be normal and may provide enough of the enzyme to permit the cell to function properly.

Another advantage of diploidy is that having two different forms of the same gene (two alleles) sometimes appears to be more beneficial than having only one gene in duplicate. The superiority of heterozygous individuals has been demonstrated in several domestic plant forms, including corn and wheat. This advantage may be due in part to the fact that each of the two enzymes produced by the two alleles functions better under different environmental conditions. For example, slightly different forms of an enzyme often function optimally at different temperatures, so that a particular chemical reaction could be facilitated by one of the enzymes when environmental temperatures are cold and by the other enzyme when temperatures are warm. An organism that is heterozygous for the enzyme is able to carry out this chemical reaction rapidly over a broader range of temperatures than an organism that is homozygous and contains only one of these enzymes.

Although there are clearly advantages to having a diploid cell structure, many organisms are composed of haploid cells in their adult forms. Their continued existence demonstrates that the diploid condition is not always superior to the haploid condition.

Meiosis—Morphological Changes

In the process of cell division called *meiosis,* the homologous pairs of chromosomes in a diploid cell separate and one chromosome in each pair is inherited by each daughter cell. After meiosis, a haploid daughter cell contains one chromosome of each type. In most animals and plants, the haploid cells then undergo structural and physiological changes to become sperm (or pollen grains) and eggs.

In many ways, meiosis is similar to mitosis—the process by which one cell divides into two daughter cells, each containing the same number of chromosomes that were present in the parent cell (see Figure 6-3 on page 102). In meiosis, however, the number of chromosomes is reduced by half during a sequence of two cellular divisions called *meiosis I* and *meiosis II.* First, we will consider the morphological changes that reduce the number of chromosomes. Then we will examine several important genetic events that result from meiosis.

Meiosis I

In *interphase,* prior to the first meiotic division, DNA replication occurs while the chromosomes are highly elongated and contained within the nuclear membrane of the diploid cell, as shown in Figure 7-5(a). As in mitosis, each chromosome then becomes short and thick during *prophase,* shown in Figure 7-5(b), and visible through a microscope as two chromatids held together by a single kinetochore. Near the end of prophase, the nuclear membrane and the nucleoli disappear.

During the next stage of the sequence, *metaphase,* the pairs of homologous chromosomes line up in the center of the cell, as shown in Figure 7-5(c). The two separate chromosomes of a homologous pair (four chromatids) lie close to each other but are not physically joined. All of the chromosomes in the cell therefore occur in pairs, and each chromosome is composed of two chromatids. Each pair of chromosomes has two kinetochores, and each kinetochore joins the two chromatids of a single chromosome together. The two kinetochores of a homologous pair of chromosomes are connected by spindle fibers to different poles of the cell.

During *anaphase,* the spindle fibers shorten to pull the chromosomes of each homologous pair apart and toward opposite poles, as shown in Figure 7-5(d). It is important to note that the *homologous chromosomes* are separated during anaphase in meiosis I, whereas the *chromatids* are separated during anaphase in mitosis. During the final stage of meiosis I, *telophase,* the chromosomes are separated into two distinct groups, one at each pole of the cell, as shown in Figure 7-5(e). Each group contains one chromosome from each of the homologous pairs. The cell membrane then pinches in two to produce two separate cells, as shown in Figure 7-5(f). A nuclear membrane does not normally form at the end of meiosis I.

Meiosis II

After telophase in meiosis I, the two daughter cells usually bypass interphase and enter directly into prophase of meiosis II. At this point, each cell contains

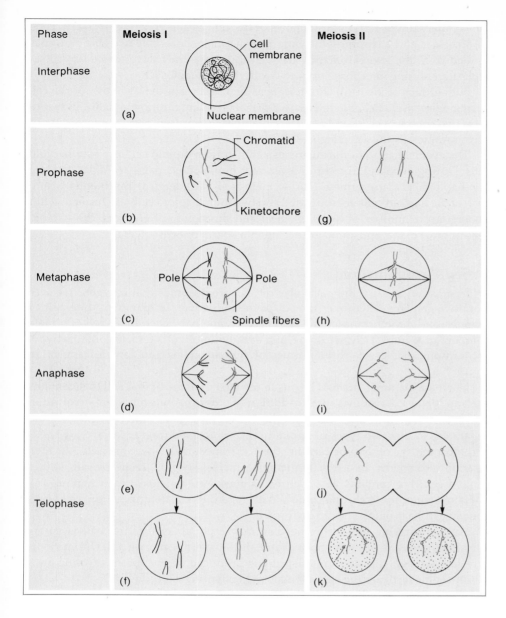

Phase	Meiosis I	Meiosis II

Interphase

(a) *Cell membrane, Nuclear membrane*

Prophase

(b) *Chromatid, Kinetochore*

(g)

Metaphase

(c) *Pole, Pole, Spindle fibers*

(h)

Anaphase

(d)

(i)

Telophase

(e)

(f)

(j)

(k)

Figure 7-5
Meiosis: Cell Division with Reduction of Chromosome Number

The diploid cell shown here contains six chromosomes, or three homologous pairs. After meiosis is complete, each of the haploid cells contains three chromosomes, one from each homologous pair. (a) Meiosis I begins with *interphase*, during which the chromosomes are elongated and enclosed within a nuclear membrane. Each chromosome replicates itself during interphase to form two chromatids that are joined together by a single kinetochore. (b) During *prophase* in meiosis I, the chromosomes become short and thick. Each chromosome is composed of two chromatids and a single kinetochore, and the nuclear membrane has disintegrated. (c) During *metaphase* in meiosis I, homologous chromosomes lie next to each other and are attached to opposite poles of the cell by spindle fibers. (d) During *anaphase* in meiosis I, each pair of chromosomes is pulled apart as the spindle fibers shorten. (e), (f) The cell membrane pinches in two during *telophase* to form two separate cells, each containing one chromosome from each homologous pair. Both of these daughter cells then undergo meiosis II, although this process is diagrammed in only one of the daughter cells here. (g) Meiosis II begins with *prophase*. (h) During *metaphase* in meiosis II, the chromosomes line up individually in the center of the cell and spindle fibers become attached to both sides of each kinetochore. (i) The kinetochores divide, and the two chromatids of each chromosome are pulled apart during *anaphase*. (j) During *telophase*, the cell membrane pinches in two to form two new cells. (k) A nuclear membrane forms within each cell, which then enters *interphase*. The single diploid cell shown in (a) has formed four haploid cells.

one-half of the original number of chromosomes and each chromosome is composed of two chromatids held together by a kinetochore, as shown in Figure 7-5(g).

During metaphase in meiosis II, the chromosomes line up singly (rather than in pairs, as in meiosis I) and spindle fibers connect each kinetochore to both poles of the cell, as shown in Figure 7-5(h).

The kinetochore joining the two chromatids of each chromosome divides at the beginning of anaphase. The spindle fibers shorten, and the two chromatids are pulled apart and toward opposite poles of the cell, as shown in Figure 7-5(i). Each chromatid is now a chromosome with its own kinetochore. The two chromosomes formed from the two chromatids are identical, since they were formed by DNA replication during interphase prior to cell division in meiosis I.

Both groups of identical chromosomes then localize at the two poles of the cell during telophase, as shown in Figure 7-5(j). The cell membrane pinches in two to form two complete cells, and a nuclear membrane develops around each group of chromosomes, as shown in Figure 7-5(k).

The events that occur during meiosis transform a diploid cell into four haploid cells. During meiosis I, the chromosomes of homologous pairs are separated and two haploid cells are formed. During meiosis II, the identical chromatids within each of the haploid cells are separated and two more haploid cells are formed. After undergoing a number of morphological and physiological changes, the haploid cells of most plants and animals develop into sperm or eggs.

Sex Determination

In most organisms, the particular set of chromosomes that a zygote receives from two gametes determines the sex of the individual. In mammals, one pair of homologous chromosomes is the sex chromosome pair. Each body cell of a male contains an X and a Y chromosome, and each body cell of a female contains two X chromosomes. The sex chromosomes of human males and females appear in Figure 7-3.

In a male mammal, meiosis begins in the testes with a diploid cell that contains both an X and a Y chromosome in addition to other chromosomes not involved in sex determination. During meiosis, the two homologous sex chromosomes are separated, so that two types of haploid cells are formed (one type carrying the X chromosome, the other type carrying the Y chromosome). These haploid cells then develop into sperm. In a female mammal, meiosis begins in the ovaries with a diploid cell that contains two X chromosomes. After meiosis, each haploid cell contains one of the X chromosomes. With regard to sex determination, only eggs carrying an X chromosome can be formed.

An egg carrying the X chromosome may unite at fertilization with a sperm carrying either a Y or an X chromosome. If the sperm carrying a Y chromosome fertilizes the egg, a male (XY) offspring is produced. If the sperm carrying an X chromosome fertilizes the egg, a female (XX) offspring is produced.

The sex of the offspring is therefore determined by the gametes produced during meiosis in the male. Normal males produce equal numbers of sperm carrying X and Y chromosomes, so that it is equally likely for the offspring to be male or female.

The Independent Assortment of Chromosomes

Because each parent forms genetically different gametes during meiosis, a particular male and female are able to reproduce a large variety of offspring that are rarely, if ever, identical. One of the factors contributing to the differences between gametes is the independent assortment of chromosomes that results when homologous pairs of chromosomes are separated during meiosis I. Prior to meiosis, each

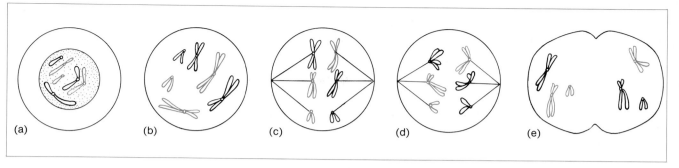

Figure 7-6
Independent Assortment of Chromosomes During Meiosis I
Prior to meiosis, the diploid cell shown here contains six chromosomes, or three homologous pairs. One chromosome in each homologous pair is black; the other is brown. (a) During *interphase*, the chromosomes double. Here, DNA replication has not yet occurred, and the chromosomes are shorter and thicker than they would be at this stage in meiosis. (b) During *prophase* in meiosis I, each chromosome appears as two chromatids connected by a single kinetochore. These chromosomes are distributed throughout the cell, since the nuclear membrane has disintegrated. (c) During *metaphase*, the homologous chromosomes lie next to each other and spindle fibers become attached to their kinetochores. (d) The direction in which the two homologous chromosomes are pulled toward the poles of the cell during *anaphase* depends how they lined up during metaphase. In this example, the black chromosome from one homologous pair and the brown chromosomes from the other two homologous pairs are positioned on the left during metaphase and are therefore pulled toward the pole on the left. Likewise, the brown chromosome from one homologous pair and the black chromosomes from the other two homologous pairs are positioned on the right during metaphase and are therefore pulled toward the pole on the right. The important point here is that the way in which the black and brown chromosomes of one homologous pair are arranged during metaphase has no influence on the way in which the black and brown chromosomes of the other two homologous pairs arrange themselves. The separation of the two chromosomes of one homologous pair is *independent* of the separation of the chromosomes of the other pairs. (e) During *telophase* in meiosis I, the cell membrane pinches in two to form two haploid cells, each containing some black and some brown chromosomes.

diploid cell contains two sets of chromosomes, which appear in Figure 7-6 as a black and a brown set. During metaphase in meiosis I, each chromosome from one set pairs with its homologous chromosome from the other set. Whether a particular chromosome in each pair is positioned on the left or on the right when the chromosomes line up in the center of the cell is entirely a matter of chance, as shown in Figure 7-6(c), and in no way influences how the chromosomes of other homologous pairs arrange themselves. During telophase in meiosis I, the cell divides so that all of the chromosomes on the left side of the parent cell are incorporated in one cell and all of the chromosomes on the right side of the parent cell are incorporated in the other cell, as shown in Figure 7-6(e). Therefore, the arrangement of the homologous chromosomes during metaphase in meiosis I determines which chromosome from each homologous pair is distributed to a particular gamete. As a result of the independent way in which the chromosomes of each homologous pair become separated (assorted) during anaphase, some of the chromosomes in each cell are from each side of the parent cell.

Variability Due to Independent Assortment

Many different genetic types of gametes can be formed. If we assume that at least some of the genes in a chromosome differ from some of the genes in its homologous chromosome, we can calculate the number of different types of gametes that can be formed from the independent assortment. For example, the

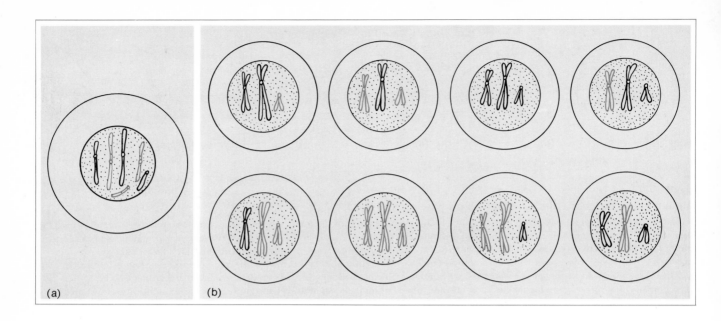

Figure 7-7
The Different Types of Haploid Cells That Can Form from the Independent Assortment of Chromosomes During Meiosis I

(a) A diploid cell during interphase, prior to DNA replication. (b) Eight types of gametes, each containing different groupings of chromosomes from each homologous pair, that can be produced from this cell. Only two of these cells form during each cycle of meiosis I.

eight different types of gametes that can be formed during meiosis I from the diploid cell in Figure 7-6 are shown in Figure 7-7.

For cells that have higher numbers of chromosomes, the simple calculation 2^n, where n is the number of chromosome pairs in the cell prior to meiosis, can be used to predict the number of different possible types of gametes that a diploid cell can produce. This number increases rapidly as the number of chromosome pairs increases. A simple cell containing six chromosomes (three pairs) would produce eight (2^3) gamete types, whereas the number of possible gamete types produced by a human is 2^{23}, or 8,388,608. Given such a large variety of possible types of sperm and eggs, it is impossible to predict the exact genetic composition of an offspring from a particular set of parents.

Mendel's Experimental Work

Two fundamental laws of heredity were discovered by Gregor Mendel (1822–1884), an Austrian monk whose experimental work preceded the discovery of chromosomes or meiosis (Figure 7-8). Mendel conducted several controlled breeding experiments with pea plants from which he drew inferences about the existence of chromosomes and the manner in which they are distributed to daughter cells during the formation of gametes. Although his discoveries were presented to the scientific community as early as 1865, the importance of Mendel's work was not recognized until 1900, after chromosomes had been discovered and postulated to be the physical basis of heredity. Mendel's work laid the foundation for modern genetic theory. However, later genetic research has shown that some subtle patterns of heredity were either overlooked by Mendel or were not uncovered by his experiments.

Mendel's remarkable success in revealing the basic patterns of inheritance was largely due to his precise experimental designs. Other scientists had performed similar breeding experiments but had tested only a few organisms and had attempted to monitor the simultaneous inheritance of a great number of traits. Mendel performed the same experiments on thousands of individual plants but investigated only one or two traits at a time. Mendel classified and counted the characteristics of all offspring, thereby isolating clear patterns in the transmission of traits from parent to offspring.

Many of Mendel's experiments were designed to test seven different traits of garden peas. Each trait was present in only two alternative forms. One of the traits studied, plant height, occurred in either tall or dwarf form.

The first step in Mendel's experimental series was to develop pure lines of individuals by mating only individuals with the same trait. Tall plants were mated with tall plants, and dwarf forms were removed from the population whenever they appeared as offspring. After several generations of these matings, individuals that deviated from the single trait (tall plant) no longer appeared in the offspring. At this point, Mendel had produced a pure line with regard to the particular trait under study.

Next, Mendel mated individual plants from two different, pure strains by removing pollen (sperm) from one plant and placing it directly on the female parts of another plant. For example, he took pollen from a plant of the tall variety and placed it on the female parts of a plant of the dwarf variety. Offspring from these matings, called the F_1 (*first filial*) generation, were examined and their characteristics were recorded. Individuals of the F_1 generation were then mated with themselves by taking the pollen from a plant and placing it on the female parts of the same plant. This method of reproduction, called *selfing,* often occurs naturally in plants. The offspring of the individuals mated with themselves comprised the F_2 (*second filial*) generation.

Before we examine some of Mendel's classic experiments in detail, we need to know how to predict the genotypes of the offspring from matings between two parents with known genotypes. When the inheritance of a single gene is to be predicted, it is relatively easy to determine the possible genotypes of the offspring. When two or more genes are considered simultaneously, however, the process is more complicated. A simple way to solve such genetic problems is to use the *Punnett square* (Figure 7-9), which lists the different types of sperm that could be produced in a row along the top of the square and the different types of eggs that could be produced in a column along the left side of the square. By combining each type of sperm with each type of egg (filling in the empty boxes), we obtain the expected genotypes of the zygotes (offspring). You should master this simple technique before reading further.

Figure 7-8
Gregor Mendel, Discoverer of the Laws of Heredity

The Law of Segregation

In one set of experiments, Mendel examined the inheritance of plant height (Figure 7-10). Using two pure strains (one of only tall plants and the other of only dwarf plants), Mendel obtained F_1 individuals by mating plants from one strain with plants from the other strain. The offspring from Mendel's matings between tall and dwarf parents always resembled the tall parent.

Figure 7-9
The Punnett Square

Using the Punnett square, shown here, it is relatively easy to predict the genotypes of offspring if we know the different types of gametes that can be produced with regard to the particular genes under study. In this example, each parent is heterozygous for the particular gene. *A* denotes one allele and *a* denotes the other allele, indicating that there is an *A* gene on one chromosome and that there is an *a* gene at the same place on the homologous chromosome. The genotype of each parent with regard to this trait is therefore written *Aa*. After the homologous chromosomes are separated during meiosis, each parent produces two types of gametes: those carrying the *A* allele, and those carrying the *a* allele. Placing the two types of sperm along the top of the Punnett square and the two types of eggs along the left side of the square and writing the various combinations of eggs and sperm in the boxes, we obtain the different genotypes of the zygotes (offspring) that are possible. In this example, one-half of the offspring will be of genotype *Aa*, one-fourth will be of genotype *AA*, and one-fourth will be of genotype *aa*. These fractions are referred to as *relative frequencies*.

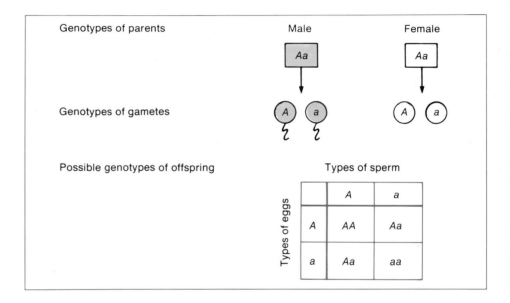

Figure 7-10
Matings Between Two Pure Lines of Plants

Mendel took sperm (pollen) from a tall plant and placed it on the female parts of a dwarf plant. All of the offspring from this mating were tall. Mendel's hypothesis about the inheritance of plant height appears on the right. He postulated that each plant carried two "factors" for the trait: plants from the pure strain of tall plants carry two factors for tall (*TT*); plants from the dwarf strain carry two factors for dwarf (*tt*). Each gamete carries only one factor: each sperm carries one tall factor, and each egg carries one dwarf factor. At fertilization, the two factors combine and the zygote inherits one of each factor (genotype *Tt*). Mendel further postulated that the *T* factor prevents expression of the *t* factor, so that the individual offspring is tall when both the *T* and *t* factors are present.

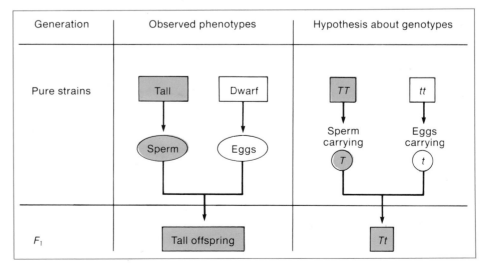

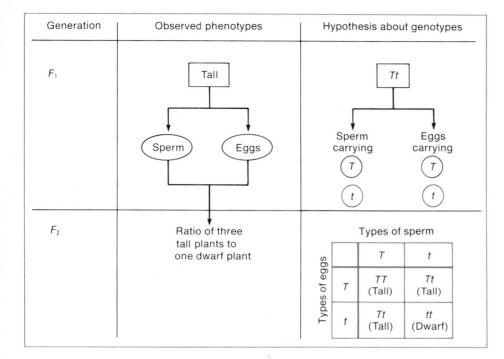

Generation	Observed phenotypes	Hypothesis about genotypes
F_1	Tall → Sperm, Eggs	Tt → Sperm carrying T, t; Eggs carrying T, t
F_2	Ratio of three tall plants to one dwarf plant	Types of sperm / Types of eggs

Types of sperm / Types of eggs:

	T	t
T	TT (Tall)	Tt (Tall)
t	Tt (Tall)	tt (Dwarf)

Figure 7-11
Selfing Experiments: Mating F_1 Individuals with Themselves

The F_1 plants in this experiment have been produced from the matings shown in Figure 7-10. Mendel postulated that all of these plants would be of genotype Tt. When the F_1 plants were *selfed* (sperm from a plant was placed on the female parts of the same plant), some of the offspring in the F_2 generation were tall and some were dwarf. For every four offspring, three were tall and one was dwarf (a 3:1 ratio). Mendel's hypothesis about the inheritance of plant height appears on the right. If each F_1 individual actually carried the Tt genotype, then one-half of the sperm and one-half of the eggs would carry the T factor and the other half of both types of gametes would carry the t factor. Constructing a Punnett square to predict the genotypes of the offspring when the F_1 plants are selfed, we find that one-fourth of the offspring will be of genotype TT, one-half will be of genotype Tt, and one-fourth will be of genotype tt. If we assume that the Tt individuals will always be tall, then the predicted phenotypes are three tall plants to one dwarf plant. Mendel's hypothesis fits the observed phenotypes.

Mendel then mated these F_1 offspring with themselves to obtain F_2 offspring (Figure 7-11). In this generation, tall and dwarf individuals occurred in a 3:1 ratio: three out of every four offspring were tall, and one out of every four offspring was dwarf. Thus, some individuals in the F_2 generation (the dwarf types) were unlike their parents but resembled one of their grandparents.

This simple set of experiments clearly demonstrated that the factor causing dwarf plants was not lost in the F_1 generation, because it reappeared intact in the F_2 generation. Mendel concluded that hereditary material occurs in discrete "factors," which we now call genes, and that these factors are not blended when two types (such as tall and dwarf) are present in the same individual. The factors affecting plant height are *segregated,* rather than blended, during reproduction.

From this experiment, Mendel developed the hypothesis that each pea plant carries two factors for height. In the pure strains, both factors are identical: each tall plant carries two factors for the tall characteristic, and each dwarf plant carries two factors for the dwarf characteristic (Figure 7-10). Each gamete formed during reproduction carries only one factor. In matings between the two strains, each parent plant contributes one of these factors (in its gamete) to the offspring, and each F_1 individual therefore inherits one factor for tall and one factor for dwarf.

To explain the fact that all of the F_1 individuals were tall, Mendel postulated that when two different types of factors occur within the same individual, one factor may be dominant over (prevent the expression of) the other factor. Mendel called the tall factor *dominant* and the dwarf factor *recessive.* The characteristics determined by a recessive factor develop only when the dominant factor is absent.

Each F_1 plant produces two types of gametes—one carrying the tall factor and the other carrying the dwarf factor (see the Punnett square in Figure 7-11). These gametes can unite to form three different types of offspring carrying (1) two tall factors, (2) a tall and a dwarf factor, or (3) two dwarf factors. If the gametes are united at random, these three types of individuals should occur in a 1:2:1 ratio. If the tall factor is dominant over the dwarf factor, all of the offspring that carry both types of factors (one-half of the offspring) should resemble the offspring that carry two tall factors (one-fourth of the offspring), so that three tall plants will occur for every dwarf plant in the F_2 generation.

To support his hypothesis, Mendel needed to demonstrate that each F_1 individual carried both a tall factor and a dwarf factor. He did this by mating the F_1 individuals with individuals from the dwarf line of plants to reveal any recessive factors. If Mendel's hypothesis was correct, one-half of the offspring of these matings would be tall and the other half would be dwarf. The experimental results confirmed Mendel's predictions (Figure 7-12).

Mendel repeated his experiments with pea plants, testing several other traits, including color of flower and shape of peas. All of these experiments led Mendel to the same conclusion. With no knowledge of chromosomes, Mendel demonstrated that the cells of pea plants are diploid, carrying two factors (genes) for each trait, and that in the formation of gametes during meiosis, the two genes (on two homologous chromosomes) *segregate* so that one gene is distributed to each gamete. At fertilization, or the union of two gametes, the offspring receive one gene carrying the trait from each parent. This process is referred to as Mendel's *Law of Segregation.*

Figure 7-12
Experiment Confirming Mendel's Hypothesis About the Inheritance of Height in Pea Plants
If the F_1 individuals from the experiment shown in Figure 7-10 were heterozygous, then matings between F_1 individuals and plants from pure dwarf strains should produce offspring of either the *Tt* or the *tt* genotype in a 1:1 ratio. The predicted phenotypes of the offspring are therefore: one-half tall and one-half dwarf. Mendels's prediction was found to hold true.

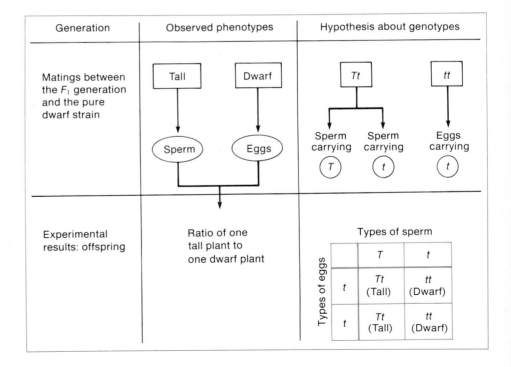

More than Two Alleles

In all of Mendel's experiments, only two alleles (two forms of the gene determining a certain trait) occurred in the population (for example, only tall and dwarf forms of the gene for height were considered). Subsequent experimentation has shown that some genes have a number of different alleles that exist in the same population. One gene may have a whole series of alternative forms, each of which affects the same trait in a different way. A single individual with only two chromosomes of each type can have only two alleles, but more than two forms of these alleles can exist in the entire population.

An example of a gene that occurs in more than two forms is the gene that determines human blood groups. An individual human carries two genes that code for proteins on the cell membranes of red blood cells, which transport oxygen throughout the body. The three different forms (alleles) of this gene that occur in human populations are represented by the symbols A, B, and O. The A and B forms cause the production of proteins A and B, whereas the O form does not produce any recognizable protein. Every human has two of these genes, so that there are six possible combinations: AA, AB, BB, AO, BO, and OO. An individual in blood group AB, for example, has the A allele on one chromosome and the B allele on the homologous chromosome. An individual in blood group AO carries one allele for the A protein and another allele that does not produce a protein.

An AB individual produces two types of gametes with regard to the blood-group alleles: those carrying the A allele and those carrying the B allele. Similarly, an individual of the AO type produces A-carrying gametes and O-carrying gametes. If these two individuals mate, they can produce offspring in blood groups AA, AO, AB, and BO (Figure 7-13).

In the inheritance traits studied by Mendel, one of the two alleles was

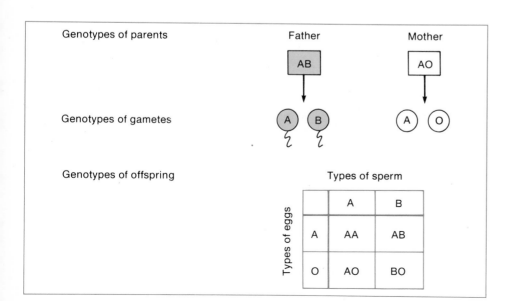

Figure 7-13
Inheritance of Blood Groups in Humans

In this example, the genotypes of offspring are estimated from knowledge of the parental genotypes with regard to blood group. If the father is type AB, he will produce sperm carrying the A gene and sperm carrying the B gene. If the mother is type AO, she will produce eggs carrying the A gene and eggs carrying the O gene. After the random union of eggs and sperm, four genotypes of offspring are possible in equal frequencies: AA, AB, AO, and BO.

dominant; when both alleles were present, only one allele determined the trait. The blood-group alleles provide another example of dominance. The A allele is dominant over the O allele, and both AA and AO individuals have the same type of protein on their red blood cells. Similarly, the B allele is dominant over the O allele, and BB and BO individuals cannot be distinguished by their red blood cell proteins. However, neither the A nor the B allele is dominant over the other. When both alleles are present, both A and B proteins appear on the red blood cells. Genes acting in this manner are said to be *codominant*.

The Law of Independent Assortment

In another set of experiments with pea plants, Mendel examined the simultaneous inheritance of two different traits. Designating one trait as the height of the plant and the other trait as the shape of the peas, Mendel developed two pure strains of plants: one strain of tall plants with smooth peas and another strain of dwarf plants with wrinkled peas.

Mendel then mated individuals from the two pure strains to obtain an F_1 generation of plants (Figure 7-14). All of these offspring were tall with smooth peas, resembling only one parent. When the F_1 individuals were mated with themselves (Figure 7-15), four types of offspring occurred in the F_2 generation in a 9:3:3:1 ratio; nine tall plants with smooth peas, three tall plants with wrinkled peas, three dwarf plants with smooth peas, and one dwarf plant with wrinkled peas.

Mendel explained the characteristics of the F_1 generation in precisely the same manner that he had explained the F_1 individuals in his earlier experiments designed to test single traits. Each F_1 plant received a tall factor and a smooth factor from one parent and a dwarf factor and a wrinkled factor from the other parent (Figure 7-14). Because tall is dominant over dwarf and smooth is dominant over

Figure 7-14
The Simultaneous Inheritance of Two Traits: Matings Between Pure Strains

(T = tall; t = dwarf; S = smooth; s = wrinkled) Mendel believed that T was dominant over t and that S was dominant over s. In the first of a series of experiments testing the simultaneous inheritance of two traits, Mendel took the sperm from a pure strain of tall plants with smooth peas and placed it on the female parts of a pure strain of dwarf plants with wrinkled peas. All of the offspring from these matings were tall plants with smooth peas. Each gamete carried one factor for each trait. The predicted genotypes appear on the right.

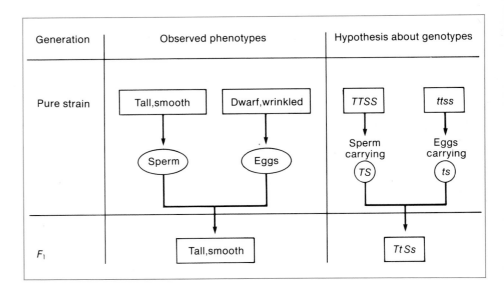

Figure 7-15
Simultaneous Inheritance of Two Traits: Selfing of the F_1 Plants

The F_1 plants in this experiment were produced from the matings shown in Figure 7-14. Mendel postulated that all of these plants would be of the *TtSs* genotype. When the F_1 plants were mated with themselves (selfed), a variety of different phenotypes appeared. For every 16 offspring in the F_2 generation, nine were tall with smooth peas, three were tall with wrinkled peas, three were dwarf with smooth peas, and one was dwarf with wrinkled peas. This is the exact ratio of phenotypes that Mendel predicted in his hypothesis about genotypes, shown on the right here. Because the genes for these two traits (plant height and pea shape) are positioned on different chromosomes and the assortment of chromosomes during the formation of gametes is independent, a gamete that receives the *T* (tall) gene can also receive either the *S* (smooth) or the *s* (wrinkled) gene. Similarly, a gamete that receives the *t* (dwarf) gene can also receive either the *S* or the *s* gene.

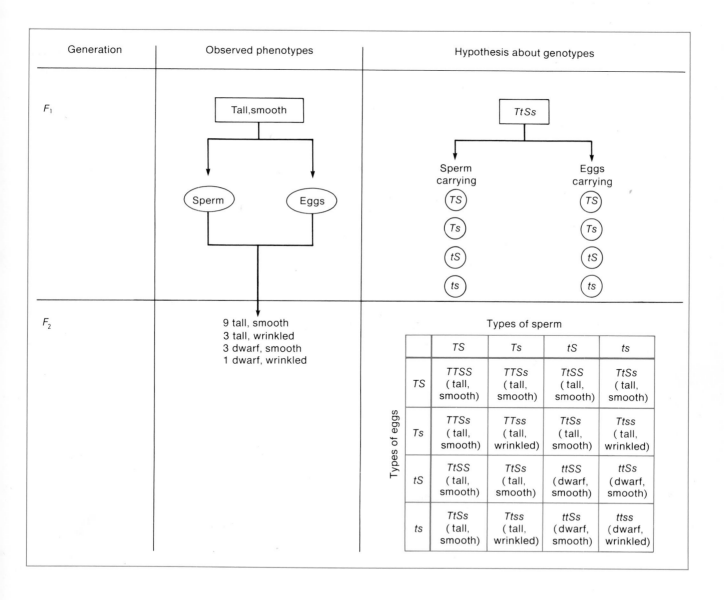

wrinkled, all of these plants exhibited only tall and smooth characteristics. The dwarf and wrinkled characteristics were not expressed in the phenotype.

Whether a particular gamete produced by an F_1 individual (Figure 7-15) receives the smooth or the wrinkled factor is in no way related to whether that gamete receives the tall or the dwarf factor. Four types of sperm and four types of eggs are therefore possible—gametes carrying genes for tall and smooth, tall and wrinkled, dwarf and smooth, and dwarf and wrinkled—so that there are 16 genotypes in the zygotes. Because tall is dominant over dwarf and smooth is dominant over wrinkled, there are just four kinds of phenotypes (Figure 7-15), which should occur in a 9:3:3:1 ratio.

As in his earlier experiments, Mendel tested this hypothesis by mating the F_1 individuals with the pure strains of dwarf plants with wrinkled peas (Figure 7-16). If the F_1 individuals were heterozygous for both traits (*TtSs*), then they would produce four different kinds of gametes. The pure strain of dwarf plants with wrinkled peas (genotype *ttss*), however, should produce just one type of gamete carrying a factor for dwarf (*t*) and a factor for wrinkled (*s*). Combining these types of gametes would then produce four kinds of genotypes in the zygotes (Figure 7-16) in equal numbers (a 1:1:1:1 ratio). Mendel's predictions proved to be correct, and his hypothesis about the genotypes was supported. The way in which the factors for one trait are segregated (assorted) during reproduction is *independent* of

Figure 7-16
Confirmation of Mendel's Independent Assortment Hypothesis

If the genes for two traits are carried by different sets of homologous chromosomes and if these chromosomes are moved to gametes independently, then a cross between an individual who is heterozygous for both traits and an individual who is homozygous for the recessive forms of both traits should yield four phenotypes in equal frequencies. Mendel found this to be true.

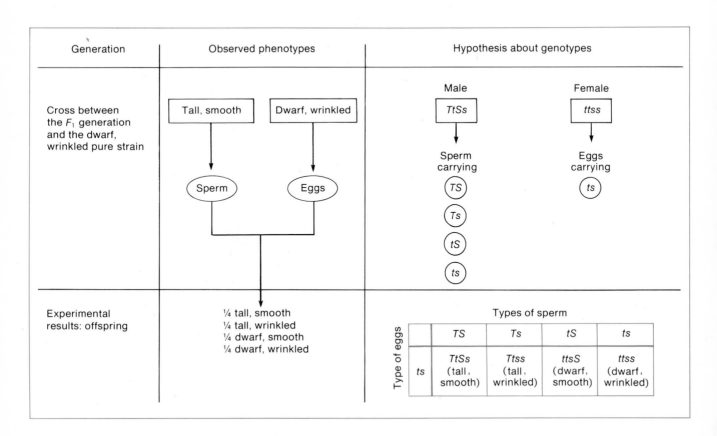

the way in which the factors for another trait are segregated. This process is referred to as Mendel's *Law of Independent Assortment.*

Subsequently, it has been shown that the genes (Mendel's "factors") for plant height and pea shape are located on two different pairs of homologous chromosomes. We now know that the assortment of chromosomes during meiosis is independent—that the way in which the chromosomes are distributed from one homologous pair to the gametes in no way influences the movement of the other homologous pairs. Mendel's studies correctly predicted that genes located on different sets of homologous chromosomes are distributed independently during the formation of gametes. Mendel did not observe the inheritance of genes linked on the same chromosome or the phenomenon of crossovers (to be discussed in the following sections), because all of the traits that he studied were carried by genes located on different sets of homologous chromosomes.

Genes That Are Linked in Inheritance

There are several thousand different genes but rarely more than 50 chromosomes in any individual cell of an organism. Each chromosome therefore contains many genes. All of the genes that make up a single chromosome (a *linkage group*) are normally inherited together, because they move as a single unit into the same gamete during meiosis.

One of the first demonstrations of linkage was between flower color and shape of pollen grains in certain varieties of sweet peas. The flowers of these varieties can be purple or red, and the pollen grains can be long or round. The allele for purple is dominant over the allele for red, and the allele for long pollen is dominant over the allele for round pollen.

To begin these experiments, two pure stains were established—a strain of plants with purple flowers and long pollen grains and a strain of plants with red flowers and round pollen grains. Following Mendel's example, individuals from these two pure strains were mated to produce an F_1 generation (Figure 7-17). As predicted by Mendel's hypothesis, all of the F_1 plants had purple flowers (dominant over red) and long pollen grains (dominant over round).

However, when the F_1 individuals were mated with themselves, the offspring did not occur in the 9:3:3:1 ratio predicted by Mendel's hypothesis (nine individuals with purple flowers and long pollen grains, three individuals with purple flowers and round pollen grains, three individuals with red flowers and long pollen grains, and one individual with red flowers and round pollen grains). Instead, a 3:1 ratio was found—three individuals with purple flowers and long pollen grains to one individual with red flowers and round pollen grains (Figure 7-17).

Surprisingly, a 3:1 ratio is also observed when only one trait at a time is considered. In fact, these two traits—flower color and pollen grain shape—are inherited as a unit. The genes determining the two traits are located in the same chromosome and are coupled during meiosis. The F_1 individuals produce only two types of gametes with regard to these traits—one type carrying the genetic information for purple flowers and long pollen grains, and the other type carrying the genetic information for red flowers and round pollen grains.

Figure 7-17
Effects of Having the Genes for Two Traits on the Same Chromosome

When genes are linked together on the same chromosome, they remain together during meiosis. The traits carried by these genes are usually inherited together. In this experiment, the gene determining purple flower color P is always positioned on the same chromosome as the gene determining long pollen grains L, and the gene determining red flower color p is always positioned on the same chromosome as the gene determining round pollen grains l. Thus, although the F_1 individuals have two alleles for both traits ($PpLl$), they can only form two types of sperm and two types of eggs (PL and pl). The other two combinations (Pl and pL) are not possible, because P and L are positioned on the same chromosome, as are p and l, and cannot be separated. The F_2 generation therefore exhibits a 3:1 ratio of phenotypes, as shown in the Punnett square here, rather than the 9:3:3:1 ratio that would be predicted if the genes for the two traits were carried on different chromosomes (see Figure 7-15).

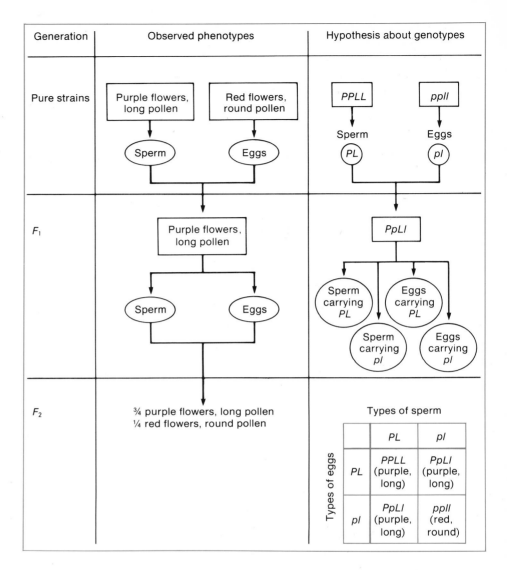

Crossovers

Occasionally, the continuity of a chromosome is disrupted, and two linked genes that were to be inherited as a unit become separated. The process that results in this discontinuity is called a *crossover.*

A crossover is the exchange of genetic material between the two chromosomes of a homologous pair during late prophase in meiosis I. At this point in meiosis, each chromosome occurs as two identical chromatids held together by a single kinetochore. When the two homologous chromosomes lie close to each other, blocks of genetic material are sometimes exchanged between the chromatid of one chromosome and the chromatid of the other, as shown in Figure 7-18(b).

The exchange of genetic material between homologous chromosomes is apparently caused by the breakage and subsequent rejoining of chromatids. When two chromatids (one from each of two homologous chromosomes) break at the same point, they may exchange places and rejoin the "wrong" chromatid rather than the chromatid from which they originated. Because the breaks occur in corresponding places on the two chromatids, genetic material is not lost but simply exchanged.

The new combinations of characteristics that result from this rearrangement of genetic material within the chromatids are now linked in inheritance in the offspring. In sweet peas, for example, the genes carrying the traits for purple color and long pollen grains usually occur on the same chromosome, as do the genes

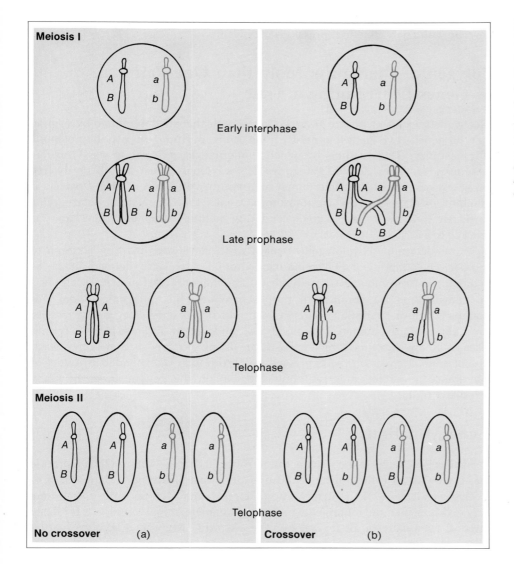

Figure 7-18
Inheritance of Genes With and Without Crossover During Meiosis

Only one pair of homologous chromosomes is shown, and only two of the many genes that occur on these chromosomes are identified. (a) Two kinds of gametes (*AB*, *ab*) are produced when no crossover occurs. (b) Four kinds of gametes (*AB*, *Ab*, *aB*, *ab*) are produced when a crossover does occur.

carrying the traits for red color and round pollen grains. In some experiments with pea plants, however, a few individuals in the F_2 generation inherited purple flowers and round pollen grains and a few individuals inherited red flowers and long pollen grains. During the formation of gametes prior to the development of these F_2 plants, in some of the plants a crossover occurred between the genes determining flower color and the genes determining pollen shape to produce two new linkage groups—a chromosome that carried the genetic information for both purple flower color and round pollen grains and a chromosome that carried the genetic information for red flower color and long pollen grains.

In addition to influencing patterns of inheritance, chromosomal rearrangement by crossing over may lead to new expressions of genes. The activity of some genes is influenced by interactions with neighboring genes. When one or more blocks of genetic material from one chromosome in a homologous pair exchange places with corresponding blocks of genetic material from the other chromosome, new neighbors and new genetic expressions may result.

Polygenic Inheritance: More than One Pair of Genes Determining a Trait

Most genetic experiments are designed to test traits that are determined by a single pair of genes—two genes located in the same position on two homologous chromosomes. This method of genetic inheritance, referred to as *Mendelian inheritance,* produces definite categories of individuals (such as tall or dwarf). However, many traits are determined by a number of different genes located at different positions on the chromosomes. Genes that interact to determine a particular trait are called *polygenes,* and this method of genetic inheritance is referred to as *polygenic inheritance.*

A trait determined by polygenes usually varies continually from one extreme to another. The offspring of a given mating do not fall into clearly defined categories that occur in the same relative frequencies. The inheritance of practically all traits in animals bred for commercial purposes follows the polygenic model. For example, continual variation occurs in the production of chicken eggs, hog fat, and cow milk. Similarly, human skin color, body build, height, and resistance to certain diseases all appear to have genetic bases consisting of many separate genes, each having a small effect on the expressed trait. Although the Mendelian model of inheritance is correct, it is too simple to explain the genetic bases of many hereditary traits.

Gene Mutations

A *mutation* is a change in the sequence of nitrogenous bases within a DNA molecule that occurs as an error during replication prior to cell division. Daughter cells receive the altered DNA structure, copy the error, and then transmit it each time they divide. When the mutation is present in a gamete, it may occur in a fertilized egg and eventually in all the cells of the offspring. Mutations account for the existence of two or more forms of a gene (alleles) in a population.

Three types of errors can occur during DNA replication: the addition of an extra base, the deletion of a base, or the substitution of one base for another base. A change in the total number of bases, which occurs if a base is added or deleted, is a serious error that alters the meaning of the entire genetic code. In Chapter 4, we learned that a sequence of three bases in a row along the DNA molecule is the code for a particular amino acid. A particular segment of a DNA strand may have the sequence of bases

AGCTCCAAAGCTTTTAGA . . .

in which each three-base unit is the code for an amino acid. The first amino acid in this protein molecule is coded by the bases AGC (adenine, guanine, and cytosine); the second, by TCC (thymine, cytosine, and cytosine); the third, by AAA; and so on. If the second base, guanine, is accidentally deleted, then the code for the first amino acid in this protein molecule would be the bases ACT; the second, CCA; the third, AAG; and so on. In this case, the deletion of a single base would alter each of the subsequent three-base units so that it codes for a different amino acid, and the entire protein constructed from this DNA segment would be changed. Similarly, the addition of an extra base in the molecule would alter all subsequent three-base codes.

However, the most frequent type of mutation is the substitution of a different base for an existing base in the DNA molecule. This type of error affects only one of the three-base codes, causing one amino acid to be substituted for another amino acid in the protein molecule. This substitution may have no effect on or may completely alter the function of the protein, depending on which amino acid in the structure is affected.

Sickle-Cell Anemia

One mutation caused by the substitution of one nitrogenous base for another in a DNA molecule is *sickle-cell anemia*—a human "disease" of the red blood cells (Figure 7-19). The protein molecule that allows red blood cells to transport oxygen is *hemoglobin.* In individuals who have sickle-cell anemia, one of the 287 amino acids

Figure 7-19
Red Blood Cells from an Individual with Sickle-Cell Anemia
(a) Blood of an individual with the defect, showing both abnormal and normal red blood cells (×500). As long as the oxygen content of the blood remains high, sickled cells maintain a normal shape. (b) Higher magnification of a normal blood cell (×6,770). (c) Higher magnification of a sickled red blood cell (×6,750).

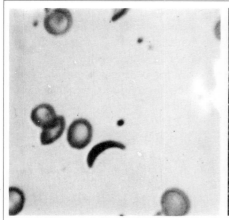
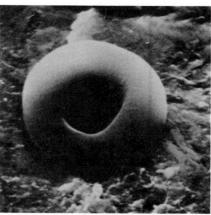

(a)   (b)   (c)

that make up the structure of the hemoglobin molecule is altered and the red blood cells are unable to transport oxygen fast enough to meet the needs of the body's cellular activities. Individuals who carry both genes in the mutated form are seriously debilitated and usually do not live long. Individuals who carry one mutated gene and one normal gene (heterozygotes) are debilitated to some degree, because one-half of their red blood cells are sickle cells. Oxygen is not transported as readily in these individuals as it is in individuals who carry two normal forms of this gene.

Curiously, individuals who are heterozygous for the sickle-cell gene are more resistant to malaria than individuals who carry two normal genes. Malaria is a disease caused by parasites that live within red blood cells. These parasites complete their life cycles less successfully in individuals who are heterozygous for the sickle-cell gene. The mutated form of this gene therefore occurs more commonly in areas where malaria is a serious threat to human survival. The heterozygous individuals survive in greater numbers and produce more offspring than normal individuals (who succumb more frequently to malaria) and individuals carrying two mutated forms of the sickle-cell gene (who rarely survive). This is one of the best documented examples of the advantages of heterozygosity—of having two different forms of the same gene.

Mutation Frequency

Alterations in the structure of the DNA molecule occur only in extremely rare instances under normal circumstances (about once in every 100,000,000 replications). The normal rate at which mutations occur can be increased by a variety of environmental factors. All forms of *radiant energy*—energy transmitted in the form of waves, including ultraviolet light, X-rays, gamma rays, and beta rays—can disrupt the chemical structure of DNA. *Radioactive materials*—atoms composed of unstable combinations of protons and neutrons—emit electrons, protons, neutrons, and other *subatomic particles* that often result in the formation of *ions* (atoms with unequal number of electrons and protons) within DNA molecules. The molecular disturbances created by these particles cause the nitrogenous bases to pair in unnatural ways. The subatomic particles given off during *nuclear reactions* have a similar effect on the structure of DNA molecules.

The number of mutations induced by radiant energy and radioactive materials is directly proportional to the amount of these materials that actually reaches the body cells. A small concentration of radiant energy or subatomic particles in a great many people causes just as many total mutations in the cells as a large concentration of these materials causes in the cells of only a few individuals, which accounts for the widespread concern over the release of nuclear materials into the atmosphere. The concentration of radiation received by any one person may be quite small, but many mutations will probably occur when a large population is exposed to small amounts of radioactive materials.

In addition to radiant energy and radioactive materials, many chemicals in our environment have been found to cause genetic mutations. Food additives, drugs, tobacco smoke, charcoal-broiled meats, pesticides, and a number of other substances have been implicated. In many cases, compounds that resemble the normal bases in the DNA molecule are mistakenly substituted into the DNA structure during replication. This has been found to be true of the caffeine molecule, although a fairly large amount of coffee would have to be consumed each day to produce a mutation. Other compounds actually cause the chromosomes to break into fragments, some of which have no kinetochore and cannot be appropriately segregated to daughter cells during cell division.

Effects of Genetic Mutations on the Individual

Mutations are random events that can occur at any time in any gene in any cell. Changes in the nonreproductive cells of adult tissues have little or no effect on total body function and are usually obscured by the activities of normal cells. A mutation that causes a cancer cell to develop, however, produces profound and serious changes in the adult organism. A cell becomes cancerous when it loses the ability to control mitosis and continually divides to form more and more cells of its type. Some forms of cancer appear to begin as a genetic mutation within a cell. Evidence to support this conjecture is provided by the increased incidence of leukemia in the survivors of the atomic explosion in Hiroshima during World War II. In leukemia, the white blood cells are cancerous and continue to divide by mitosis, apparently having lost their control mechanism.

Mutations that exist in gametes (sperm and eggs) are much more serious than mutations that occur in body cells. When a mutation is incorporated into the set of genes carried by a fertilized egg, the error is perpetuated each time the cell divides, so that every cell of the adult organism carries the altered form of DNA. Mutations that exist in all the cells of an organism are usually detrimental to some degree. The normal form of the organism is an efficient system that has undergone a long history of gradual improvements. Any random change in such a system is apt to be more disruptive than helpful.

Genetic Diseases

One advantage of diploidy is that a cell is provided with two copies of each gene, so that as long as a mutated gene does not interfere with cellular activities, the normal gene may cause the synthesis of sufficient quantities of its protein to allow the individual organism to lead a healthy life. In such cases, the mutated gene is recessive to the normal gene in that the heterozygote cannot be distinguished from the normal homozygote.

A large number of human diseases are known to be caused by abnormal genes. Many of these diseases occur only when an individual carries two abnormal genes. In some cases, scientists have identified the particular protein that is faulty and have established an effective treatment for the disease.

For example, the affected protein has been identified in galactosemia—a disease caused by the failure of the abnormal gene to produce one of the enzymes required to break down galactose, a simple sugar formed during the digestion of milk. Individuals who lack this enzyme are unable to convert galactose into glucose, so that galactose accumulates in the tissues and interferes with many body processes.*

Babies who inherit one of these abnormal genes from each of their parents are unable to process galactose. Build-up of this molecule causes a number of symptoms, including severe mental retardation. Unless they are treated, these babies usually die at an early age. However, the treatment of galactosemia is simple now that the cause is known: infants with galactosemia are simply fed synthetic "milk" made from soybean extract.

Parents can take a relatively simple test to determine whether they carry the defective gene and, if so, the probability that their children will be born with galactosemia. Each parent drinks water containing galactose and the galactose level in the blood is analyzed a few hours later. If the galactose level is higher than normal, the individual is a carrier of the recessive gene. A person with two normal genes will produce enough enzymes to convert all of the galactose into glucose rapidly, but a person with only one normal gene will convert galactose more slowly and considerable amounts of it will still be in the blood several hours after ingestion.

Many other diseases are known to be caused by gene mutations. Genetic diseases in which defective enzymes interfere with essential biochemical pathways are called *inborn errors of metabolism.* As scientists determine which enzyme is defective and what the enzyme would do if it were present, techniques can usually be developed to prevent the symptoms of a genetic disease.

Population Variability

In addition to protecting an individual from mutation, diploidy also allows the mutated form of a gene to remain in the population. Haploid individuals that carry an inferior form of a gene usually fail to reproduce and the allele disappears from the population as rapidly as it occurs. However, diploid individuals carrying the same inferior allele may produce just as many offspring (some of which will inherit the allele) as individuals carrying two normal forms of a gene.

The coexistence of different alleles in a population may be of no benefit to a particular individual but can be of enormous importance in determining whether a population is able to adjust to environmental changes. Altered forms of genes that are not advantageous in one environment may be advantageous in another environment. For example, mutations in the genes that determine the hair color of mammals have produced a variety of colors. In a forest environment where brown provides camouflage from predators, a recessive allele coding for white coat color may have no effect on heterozygous individuals but may be lethal for individuals that are homozygous for this allele and have white coats. As long as the heterozygote develops the normal brown coat, however, the mutated form of the gene will remain in the population. At another time or place, a white coat may provide camouflage against a background of snow. In this case, having a relatively high

*The development of diarrhea after the consumption of milk—a relatively common symptom among Asians and Africans—is in no way related to the more severe disease of galactosemia.

frequency of the allele for white coat color will allow the population to survive, because some individuals will always show the white coat color. If the allele for white coat color were not present in the population, no white individuals would occur and the population might be completely eliminated by predators in a winter environment.

The Recombination of Genes

Although mutations are the original sources of all genetic variation, differences between the offspring of each generation result from the *recombination* of genes that have previously accumulated in the population through mutations. A wide range of different offspring can be produced from matings between two parents of different genotypes.

Genes are recombined during sexual reproduction through three processes — the independent assortment of chromosomes, the crossovers between homologous chromosomes, and fertilization. During the first two processes, each parent produces a large number of genetically different gametes. At fertilization, however, only two gametes — one from each parent — unite. The two gametes that eventually form the zygote are determined solely by chance. Each offspring of a mated pair therefore receives a unique set of genes.

Gene Exchanges Among Populations

Another source of genetic variability in a population results from the movement of individuals among the populations of a species. These individuals carry unique alleles that originated as mutations in their former populations. As the immigrants mate with resident individuals, new alleles become incorporated into the population.

Genetic variability resulting from matings among populations has played a particularly important role in determining the characteristics of human populations. Prior to the development of rapid forms of transportation, most human populations were quite isolated from one another. A mutation that occurred in one population and, by reproduction, became common in that population might therefore have been entirely absent from another population. Over the past few centuries, however, the "gene flow" among human populations has accelerated rapidly. Human mobility has increased to the point that individuals frequently move to different geographical areas and find mates in the local population. In Hawaii, for example, the native population has incorporated immigrants from a diverse group of human populations, which has resulted in a wide variety of human types.

Genetic Engineering with Recombinant DNA

Genetic engineering is the use of the knowledge of genetics to make unnatural changes in populations. For centuries, genetic engineering has been employed in the selective breeding of domestic plants and animals. An example of the effect of selective breeding of animals with a large amount of genetic variability appears in Figure 7-20. Over the past decade, however, a revolutionary technique of genetic

**Figure 7-20
Selective Breeding Reveals
the Extent of Genetic
Variability in a Species**
Breeding programs led to the
development of these different types
of dogs. All of these breeds were
derived from an original wolflike
ancestral population.

engineering—called *DNA recombination,* or *gene splicing*—has been developed in which DNA is transferred from the cells of one organism to the cells of another organism.

DNA recombination allows DNA from any organism to be introduced into a bacterial cell. The most common method of DNA recombination involves the bacteria *Escherichia coli* (abbreviated *E. coli*), which have tiny closed loops of DNA in addition to the main strand of genetic material (Figure 7-21). These small loops, called *plasmids,* normally contain only a few genes, which direct the synthesis of "optional" enzymes useful, but not essential, to the cell's survival and reproduction. Plasmids move from bacterium to bacterium, occasionally picking up genes from one host cell and depositing them in another.

Geneticists have found ways to insert segments of DNA into plasmids. Enzymes are used to cleave a DNA segment from the genetic material of any type of organism and attach this segment to the DNA loop that forms a plasmid. The recombinant plasmid is then allowed to enter a bacterial cell, where it replicates both its own DNA and the "foreign" DNA that has been incorporated into its structure. The transferred gene instructs the RNA and ribosomes of the host cell to synthesize its protein.

Scientists have developed many uses for this genetic engineering technique. A gene incorporated into the structure of a plasmid can form an indefinite number of copies of itself by replication. The genes made available in this manner could be introduced into other cells, where their activities might improve the organism. For example, biologists have proposed that the genetic instructions to produce the enzymes that convert atmospheric nitrogen (N_2) into molecules for plant use (enzymes now found only in some forms of prokaryotes) could be incorporated into the genetic material of eukaryotic plant cells. If this genetic transfer gave plant cells the ability to utilize atmospheric nitrogen, it would eliminate the need for nitrogen fertilizer.

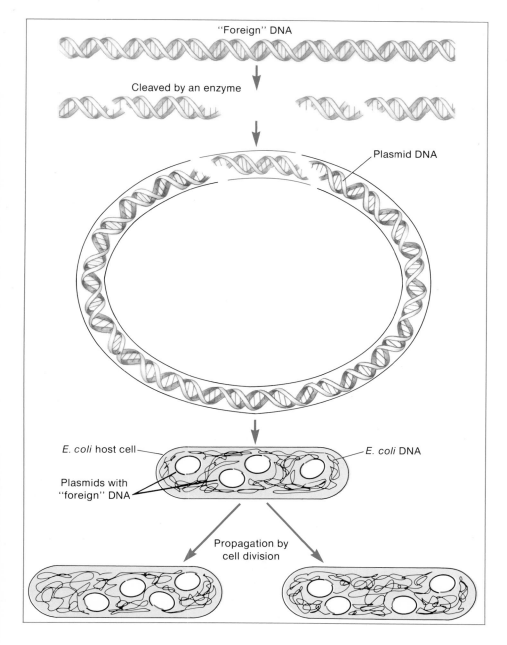

Figure 7-21
The Technique of DNA Recombination
A segment of DNA from the "foreign" organism is broken off and allowed to combine with the circular DNA molecule of a plasmid. The recombinant DNA then invades a bacteria cell (in this case, *E. coli*), where it replicates. The inserted genes can produce an indefinite number of identical copies in this manner and may direct the synthesis of their product proteins.

Another use of the recombinant DNA technique is the preparation of unlimited quantities of a gene product. The genes responsible for the synthesis of a particular hormone, enzyme, or antibody can be removed from the cell of a mammal, introduced into a bacterium, and allowed to direct the synthesis of their product in large quantities. Proteins produced in this way are used in the medical treatment of human disorders.

The recombinant DNA technique of gene transfer has become a controversial issue. Many people are opposed to the use of scientific knowledge to manipulate life. Others fear that scientists could develop new and uncontrollable strains of viruses and bacteria that would cause epidemics in human populations and that the possibility of such an epidemic is increased by the use of *E. coli*—a type of bacteria that normally inhabits the intestinal tract of vertebrate animals, including humans. However, the particular strain of *E. coli* used in recombinant DNA experiments has been purposely infused with genetic deficiencies that make its survival outside of laboratory conditions essentially impossible. The use of deficient strains of bacteria in genetic research is referred to as *biological containment*. Other questions of public concern include the possibility that the recombinant DNA technique could be used to manipulate human genes for harmful rather than beneficial purposes.

Summary

The characteristics of an individual are largely determined by the particular *genes* (DNA) contained in the body cells. An individual organism inherits genes from two parents during *sexual reproduction* and from a single parent during *asexual reproduction*. The particular characteristics exhibited by the offspring (its *phenotype*) are determined by interactions between the set of genes the individual carries (its *genotype*) and the environment.

The number of chromosomes in the cells of a sexually reproducing organism alternates between the *diploid* (two complete sets) and the *haploid* (one complete set) state during the *life cycle*. Various types of organisms undergo different life cycles. In animals and most plants, the body cells are diploid and the *gametes* (eggs and sperm) are haploid. A cell is reduced from diploid to haploid by *meiosis* and increased from haploid to diploid by *fertilization* (the union of an egg and a sperm). A diploid cell contains two sets of chromosomes, originally obtained from the set within a sperm and the set within an egg. Each type of chromosome in the sperm matches a chromosome in the egg, and these two chromosomes form a *homologous pair*. A diploid cell usually contains several pairs of homologous chromosomes.

Because a diploid cell carries two complete sets of chromosomes, it contains two copies of the gene determining each hereditary trait—one from each parent. If the two copies of a gene are identical, the individual is said to be *homozygous* for that trait; if the two copies of a gene are different, the individual is *heterozygous* for that trait. Different forms of the same gene, called *alleles*, originate from accidental changes in the DNA structure (*mutations*). A population may have many alleles, but a diploid individual can have a maximum of two alleles for each trait.

Meiosis is a special form of cell division during which a single diploid cell is transformed into four haploid cells. There are two phases of this form of cell division—*meiosis I* and *meiosis II*. Just prior to meiosis I, each chromosome of the diploid cell is replicated to form two *chromatids*, which are held together by a *kinetochore*. During meiosis I, the diploid cell divides to form two haploid cells, both of which receive one chromosome from each homologous pair. During meiosis II, both of these haploid cells divide again, and each new cell receives one of the chromatids from each of the chromosomes. The four haploid cells formed during meiosis then develop into gametes.

Several important genetic events occur during meiosis. In the mid-nineteenth century, before chromosomes were known to exist, Gregor Mendel performed a series of breeding experiments with plants and analyzed patterns of inheritance. Mendel discovered two basic laws of heredity.

Mendel's *Law of Segregation* states that there are two "factors" (genes) for each trait in a diploid individual and that during the formation of gametes, these two factors are segregated so that each gamete receives one factor for each trait. At fertilization, an egg and a sperm unite to form a diploid cell (the *zygote* or *fertilized egg*), which contains two factors (genes) for each trait. The Law of Segregation was particularly important because it predicted the fact that genetic material is not mixed or blended within the cells of an individual: each gene remains intact from one generation to the next.

Mendel's *Law of Independent Assortment* states that the manner in which two factors for one trait are delegated to gametes during meiosis in no way influences the manner in which two factors for another trait are delegated. We now know that this law holds only when the genes determining the two traits are positioned on different chromosomes; genes that appear on the same chromosome are usually inherited together and are called a *linkage group*. Mendel was actually describing the pattern of separation of *homologous chromosomes* during meiosis I.

According to Mendel's Law of Independent Assortment, the manner in which one pair of homologous chromosomes segregates in no way influences the manner in which another pair in the cell segregates. Thus, if one daughter cell receives a chromosome originally carried in the sperm prior to fertilization, the chromosomes that the daughter cell receives from the other homologous pairs may be from the original set of chromosomes in the sperm or in the egg. This process of chromosomal inheritance enables each individual to form a large variety of different gametes and is therefore an important source of genetic variability among the offspring of two individuals.

Another important event that takes place during meiosis is the *crossover* between homologous chromosomes—a source of genetic variability overlooked by Mendel. Usually, the genes on a particular chromosome are linked in inheritance and the chromosome is transferred as a unit. However, homologous chromosomes lie close together prior to their segregation during meiosis and sometimes exchange blocks of genetic material, allowing units of heredity to change over time.

Many traits are determined by *polygenic inheritance*, in which many genes (*polygenes*) interact to effect a particular trait in an individual. In such cases, the inherited trait exhibits a continuous range of variation within the population, because these genes can recombine within each offspring in a number of ways.

A *mutation* is a change in the sequence of nitrogenous bases within the DNA molecule. Such a change may alter the form of the protein for which a DNA segment codes. Mutations are usually rare events, although their rate of occurrence may be increased by a number of environmental factors. A population normally contains many mutations.

However, the effects of most mutations are masked because they occur in the heterozygous form, so that the normal form can compensate for the faulty protein. When both parents carry the same mutation, some of their offspring will probably inherit two faulty genes and will exhibit some defect. Effective treatments are now available for many genetic diseases.

Genetic differences between individuals allow populations to adjust to new environments. Sources of genetic variation are mutations, the independent assortment of chromosomes, crossovers between homologous chromosomes, fertilization, and migrations of individuals among populations. Future techniques of *genetic engineering* may enable geneticists to transfer pieces of DNA from one individual to another or to construct specific genes in the laboratory to be added to the human genetic repertoire.

Natural Selection: Matching Individuals with Environments

8

Until this point, we have been examining the physical basis of life—how atoms are organized into the unique molecules of life and how instructions for building these molecules are passed from generation to generation. In the remaining chapters of Part 2, we will explore quite a different concept—*evolution by natural selection.* Originally proposed by Charles Darwin in 1859, the *theory of evolution* attempts to explain how the first few types of living cells gave rise to the millions of different kinds of organisms found on our planet today.

Darwin's Theory—A New Era of Scientific Thought

In 1831, a British naval vessel, H.M.S. *Beagle,* began a five-year cruise to chart the oceans of the world, particularly the coastal waters of South America. It was customary at that time to include a naturalist on such expeditions to record the different forms of life and to collect museum specimens. The naturalist on the *Beagle* was Charles Darwin (Figure 8-1)—a young man of genius, who, at 22, had already tried and rejected a number of professions, including medicine and the ministry. The captain of the *Beagle* was a devout fundamentalist, who believed in the exact interpretation of the Bible and who viewed Darwin's assignment as an opportunity to gather information about God's creation—the earth and its inhabitants.

Instead of adding support to prevailing religious concepts, however, Darwin's work led to one of the most fundamental intellectual revolutions in human history. Charles Darwin found evidence to support the theory that organisms have been present on earth in an uninterrupted succession of generations over vast periods of time and that the life forms found in any one period differ to some degree from their ancestors in earlier periods. These findings led Darwin to conclude that the characteristics of plants and animals are not fixed, but gradually change from generation to generation.

Figure 8-1
Charles Darwin
as a Young Man

Although the mechanisms of heredity had yet to be discovered, Darwin was familiar with the methods of breeding animals and plants for domestic purposes and knew that selective characteristics could be passed from parents to offspring. Darwin reasoned that if hereditary instructions were transmitted from one generation to the next, then the individuals within a family line would always retain the same characteristics; they would not change over time. He was therefore puzzled to discover that the characteristics of closely related populations change over the course of time when they occupy new environments.

Darwin found that a striking relationship existed between the characteristics of individuals and their environments when he conducted a study on the Galapagos Islands, located in the Pacific Ocean near Central America. Darwin was particularly impressed by the 14 different types of finches that live on the islands. These birds differ primarily in body size and feeding structures, especially in the form of their beaks. Otherwise, the finches are remarkably similar (Figure 8-2), indicating that they are closely related birds. It occurred to Darwin that of all the different types of finches on the islands might have descended from a single ancestral form of finch that came to the islands from the South American mainland and survived long enough to produce offspring there. The descendants of these first colonists could have migrated to all of the islands within the Galapagos archipelago. In time, each population of finches would then have developed unique structures in accordance with their new environments.

From this and other studies, Darwin concluded that organisms gradually change from generation to generation in response to the demands of new environments and proposed that such changes accounted for the origin of all of the different *species* (kinds of organisms) present on earth. Darwin's conclusion contradicted the established belief that each species was created in its present form by a Creator.

After he returned to England in 1836, Darwin developed a theory outlining the mechanisms of the interactions between organisms and their environments that could cause species to change in form. Two years later, Darwin completed his theory of evolution, although it was not presented to the public until 1859, when his book, *On the Origin of Species by Means of Natural Selection,* was published. Even then, Darwin felt the publication of his theory was premature and submitted the manuscript only because he learned that another English naturalist, Alfred Russell Wallace, was developing a similar theory. The long delay in publication was largely due to Darwin's determination not to present his ideas until he had accumulated an overwhelming amount of evidence to support them. Both Darwin and Wallace have been recognized for the development of the theory of evolution, because both naturalists drew the same conclusions about the mechanisms that cause species to change form over time. However, Darwin is given more credit for the development of the theory of evolution because his documentation of the hypotheses from which the theory is derived was so massive that this revolutionary idea was plausible to the scientific community.

Darwin amassed enormous quantities of information from his own observations and experiments and from correspondence with other naturalists. From the factual information he collected, which pertained to only a small percentage of all living organisms, Darwin made generalizations about the characteristics of all forms of life. These hypotheses were educated guesses about general patterns of nature.

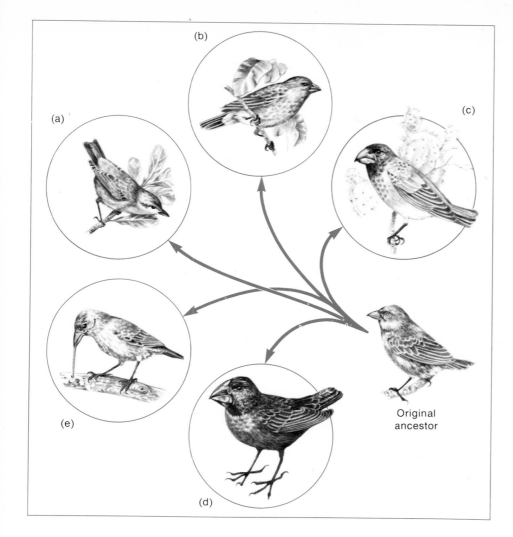

Figure 8-2
Different Types of Galapagos Finches
Many different types of finches evolved from the ancestral form that colonized the Galapagos Islands. (a) Some finches adapted to perch on trees, where they foraged for insects by removing the bark with their sharp beaks; (b), (c) others subsisted on small seeds, becoming ground-feeders with moderate-size beaks similar to our continental sparrow. (d) Other finches evolved large, powerful beaks capable of crushing large or thickly coated seeds. (e) The most remarkable of these birds, the cactus finch, uses the spines of a cactus plant to probe for hidden insects.

Original ancestor

Darwin then evaluated what he could predict if his hypotheses were true and determined by deductive reasoning what the mechanisms of natural selection were. Darwin's *theory of evolution by natural selection* is based on only four hypotheses and two deductions (predictions) derived from the hypotheses:

Hypothesis I: All organisms tend to leave more offspring than are needed to replace themselves when they die.

Hypothesis II: The number of individuals within each species remains approximately constant over time.

First Deduction: If hypotheses I and II are correct, then not all individuals survive and/or not all individuals who survive leave offspring. There is a struggle for existence.

Hypothesis III: Not all individuals of a species are alike.

Hypothesis IV: Some of the characteristics of individuals are passed from parent to offspring.

Second Deduction: If individuals within a species are different (hypothesis III), then certain kinds of individuals are more successful in the struggle for existence and leave a greater number of offspring. If the characteristics that contribute to superior survival and reproduction are passed from parent to offspring (hypothesis IV), then these same characteristics will increase in frequency from generation to generation. Whether a particular characteristic is superior or inferior depends on local environmental conditions.

Is the theory of evolution correct? In Chapter 1, we pointed out that a *theory* is a human interpretation of reality that must be repeatedly tested before it is accepted by the scientific community. A theory is tested by examining its logical consequences; predictions arising from the theory must hold true if the theory is correct.

The predictions from Darwin's theory have been repeatedly tested and thus far have been found to be correct. We will devote the remainder of this chapter to a discussion of the evidence related to the theory of evolution that has been accumulated since Darwin's work. This evidence has refined but not altered the basic structure of Darwin's theory. Darwin's conclusions form the theoretical basis of modern biology and are the currently accepted scientific explanation of how living systems have changed since the first cells appeared on earth more than 3.5 billion years ago.

The Neo-Darwinian Theory

The basic theory developed by Darwin and all of the information acquired from repeatedly testing the theory have been incorporated into a modified or *Neo-Darwinian theory of evolution by natural selection*. Major changes reflected in the Neo-Darwinian theory include a redefined unit of evolution, a knowledge of genetics, and the appreciation that not all kinds of mortality produce evolutionary change.

The unit of evolution is now recognized to be the population. Although Darwin's theory deals with changes that occur in species, we now know that most species are subdivided into local populations of individuals. A *species* is a group of similar organisms that can mate and reproduce successfully with one another. A *population* is a smaller group of individuals within a species. The individuals of a population live in close proximity and have a high probability of actually mating and producing offspring. Individuals of the same species are said to be in different populations when they are so isolated from one another that it is highly unlikely they will meet (mating could occur, but does not). Evolution is now defined as a change in *gene frequency* within a population in response to local environmental conditions.

Another important addition to Darwin's theory is a knowledge of the physical basis of heredity—the discovery of genes and how they are transmitted from parent to offspring, which we examined in Chapter 7. We now know that differences among individuals result from an accumulation of mutations (alleles) in the population and from the recombination of genes (independent assortment, crossovers, and fertilization) during sexual reproduction. Each offspring of a mated pair inherits a unique array of genes, and interactions between genotypes and environmental factors determine the phenotypes. These expressed traits influence which individuals will reproduce more successfully in a particular environment and therefore which genes will increase in frequency within the population.

A number of complicated genetic factors obscure the results of natural selection. One gene may affect many characteristics, or many genes may determine one characteristic. Therefore, it is difficult to discover the exact effects of selection on the population. A particular gene is combined with thousands of other genes in an individual, so that a beneficial and a detrimental gene may occur in the same individual or a gene that is beneficial at one stage of life may be detrimental at a later stage of life. Natural selection affects the *whole* individual. The individual either lives or dies, reproduces or fails to reproduce. Whether a particular gene is perpetuated depends on its effects and on the effects of all other genes in the organism. It is therefore not always possible to predict the precise rate at which a particular gene will increase or decrease within the population.

In his second deduction, Darwin predicts that individuals with certain traits will be more successful in the struggle for existence and leave more offspring than other individuals in the species. We are now aware that this is not always true. Many of the mortalities in populations are *nonselective;* particular genotypes do not make individuals more or less vulnerable to certain causes of mortality. Catastrophies such as floods and fires can remove the majority of the individuals in a population, regardless of their genotypes. Thus, the degrees of mortality or reproductive failure are not always reliable measurements of how rapidly certain genes increase or decrease within the population. If natural selection is to cause changes in gene frequencies, reproductive success must depend on the particular genes carried by the individuals.

The Environmental Factors That Cause Selection

Environmental factors, such as predators, weather and seasonal changes, chemicals in the soil or food, parasites, competition with other species, and competition among individuals of the same population for resources and for mates, can selectively remove certain genotypes from the population. Because populations respond to these environmental factors by *differential reproduction* (some individuals leave more offspring than others), certain genes occur more frequently and the entire population is eventually comprised of organisms that can meet the challenges of a particular environment.

Because environmental factors continually remove maladapted genotypes from the population, it seems that a perfect fit between genotypes and their environments should be achieved at some point. This is not true, however, because environments are also changing. Not only is the physical environment altered as a result of climatic changes, but other species in the environment also change, either as a result of their own evolution or the extinction and replacement of other species. A predator population, for example, causes evolutionary changes in its

(a)

(b)

(c)

Figure 8-3
Each Mated Pair Has the Potential to Leave More Offspring Than the Number Required to Replace Themselves

(a) Mother cat and kittens. (b) A snail lays 30–70 eggs each season. (c) A dandelion blossom yields over 200 seeds. Each seed is attached to a "parachute" to facilitate dispersal over long distances.

prey population by selectively removing the genotypes that are easiest to capture. If only certain individuals in the predator population can capture food, then changes in the prey population will, in turn, cause evolutionary changes in the predator population. Similarly when resources are scarce and individuals must compete for a limited food supply, the best competitors will leave the most offspring. Successive generations of a population are always composed of individuals whose parents performed best in the past, making the competition more and more keen.

The factor in the environment that causes the largest number of selective deaths or reproductive failures in a population is the most influential factor contributing to evolutionary change. This factor operates to prevent further population growth. To understand the role that the environment plays in evolution, we must first consider how populations increase in number and then examine the environmental limits to population growth in more detail.

Population Growth

All organisms appear to have the potential to leave more offspring than necessary to replace themselves when they die (Figure 8-3). This general characteristic formed Darwin's first hypothesis. The potential is seldom realized, as Darwin's second hypothesis suggests, but when it does occur, a population grows *exponentially* (increases in number very rapidly). The mathematical explanation of exponents is provided in Figure 1-13 on page 23. The use of exponents to describe population growth is demonstrated in the following example.

A human population would grow exponentially if each couple had four surviving children, because the population would double (increase by a factor of 2) during each generation. Two (2^1) adults in the first generation would produce four (2^2, or 2×2) adults in the second generation, who would produce eight (2^3, or $2 \times 2 \times 2$) adults in the third generation, and so on. There would be 1,024 (or 2^{10}) adults in the population in the tenth generation, and the population would consist of 1,048,576 (or 2^{20}) adults in the twentieth generation. In this example, the exponential increase in population size over a series of generations would be 2, 4, 8, 16, 32, 64, 128, ... In contrast, an arithmetic increase of two adults in each generation would be 2, 4, 6, 8, 10, 12, 14, ...

Exponential growth is shown as a function of time in Figure 8-4 for a hypothetical population. This type of growth can cause extremely rapid increases in the numbers of individuals in a population. Exponential growth is currently occurring in the human population, as shown in Figure 8-5. If the human population continues to grow at the present rate, all of the inhabitable areas of the earth will soon become overcrowded.

Most populations have a much higher rate of reproduction than the human population. A female Pacific salmon, for example, lays about 28,000,000 eggs before she dies. If each egg were fertilized and permitted to develop to maturity, there would be about 14,000,000 adult female salmon in the next generation (approximately one-half the offspring are male and one-half are female). If each of these females gave birth to the same number of offspring, tremendously rapid population growth would result.

Even more impressive is the potential growth of bacterial populations. Under ideal conditions, a single bacterium can divide (reproduce) every 20 minutes. At this rate, single bacterium could leave enough progeny after 36 hours to cover the surface of the earth with a layer of bacteria one foot deep. After 37 hours, the bacterial population would be over our heads. After a few thousand years, the bacteria would weigh as much as the visible universe and the population would be expanding outward from the earth at the speed of light.

Obviously, if any one population were permitted to grow in this explosive fashion for very long, its expansion would completely disrupt the activities of all other organisms. The rapid growth of the human population is beginning to cause such a disruption. Fortunately, populations rarely exhibit this extreme type of growth, but it is important to recognize that all populations have the potential for exponential growth.

Conditions That Foster Rapid Population Growth

Occasionally, the number of individuals increases greatly when a population immigrates or is introduced to a new area. If the population encounters abundant resources and few predators in the new environment, it may expand rapidly. Eventually, however, environmental limits will check this expansion.

One example of exponential population growth was the environmental transplantation of the striped bass from Atlantic to Pacific coastal waters. In 1879 and 1881, a number of one-year-old bass were caught off the New Jersey coast, transported across the continent by train, and released into San Francisco Bay. Only 435 fish survived the train journey, but by 1899, the size of this bass population had increased to such an extent that commercial fishermen caught 1,234,000 pounds in one season. The striped bass population had increased more than a million-fold in less than 20 years.

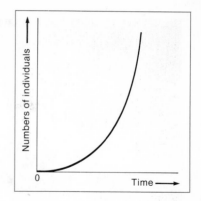

Figure 8-4
The Exponential Growth of a Population

A rapid increase in the number of individuals can occur when the factors that limit population growth are removed. All populations have the potential to grow in this manner.

Figure 8-5
Human Population Growth Has Exhibited Exponential Increases in Recent Years

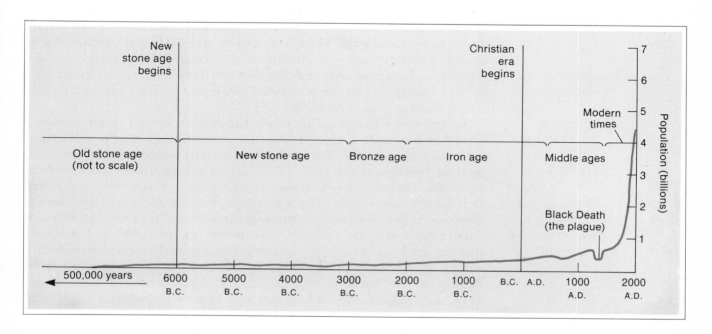

(a)

(b)

Figure 8-6
Growth of the Prickly Pear
Population in Australia
(a) A dense stand prior to 1926. (b) The
same stand in 1929, three years after
introduction of the cactus moth.

Another well-documented example of exponential population growth concerns the prickly pear cactus—a plant native to North and South America. In 1839, a single cactus was transported to Australia and planted in a garden there. Due to favorable environmental conditions, the plant reproduced with great success and soon expanded from urban to rural areas. By 1900, the prickly pear population occupied approximately 10,000,000 acres; by 1925, it covered 60,000,000 acres and was so dense in many places that a person could not walk between the plants, as shown in Figure 8-6(a). A number of methods of control proved unsuccessful or too costly to apply to such a large area. Homesteaders were forced to abandon their properties, because it became impossible to raise cattle or sheep in the area. During the 1920s, the cactus moth, which feeds only on the prickly pear cactus during its larval (caterpillar) stage, was introduced into the area. This insect eventually reduced the cactus population and controlled its density, as shown in Figure 8-6(b). By 1940, the prickly pear cactus occurred only in small, widely scattered populations throughout parts of Australia.

Environmental factors do not play a large role in natural selection when a population is expanding rapidly because almost all individuals are reproducing at their maximum potential. This maximum rate varies according to differences in genotype, however, so that some individuals are able to produce more offspring than others and the relative frequency of their genes increases in the population. The type of natural selection that occurs under these conditions in populations is referred to as *r-selection,* where *r* represents the maximum population growth rate. Characteristics that are *r*-selected (that permit an individual to leave more offspring under conditions of rapid population growth) include reproductive maturity at an earlier age, larger litters of offspring, and smaller sized offspring. This type of selection occurs in organisms that regularly colonize new environments, such as an area that has been burned or a temporary pond.

Environmental Limits to Population Density

The actual *rate of population growth* is the difference between the birth and the death rates within a population. If the number of births is exactly equal to the number of deaths over the same time period, then the population is not increasing or decreasing in number. Such a balance between birth and death rates is referred to as *zero population growth.*

Darwin's second hypothesis—that the number of individuals within each species remains approximately constant over time—appears to hold true for most populations. Although they may experience short periods of exponential growth, most populations reach a point beyond which they do not increase in number. An upper limit to population size is achieved when a population cannot expand its range due to the lack of appropriate new environments and cannot increase its density in the present environment due to the effects of predation, parasitism, weather conditions, or limited resources.

When a population is stablized at zero growth, either the birth rate or the survival rate is depressed below the maximum rate attained during exponential growth. When the reproductive output or the probability of survival is not depressed to the same extent in all individuals, natural selection causes evolutionary changes to occur. Interactions between the genotypes of individuals and the factors maintaining constant population density determine which individuals will leave the greatest number of surviving offspring and, therefore, the type of evolutionary changes that will occur.

Population Control by High Death Rates

Many populations sustain high birth and death rates to maintain a zero growth rate. For example, a pair of Pacific salmon mate only once, producing about 28,000,000 fertilized eggs, and then die. Because salmon populations are fairly constant in size, this indicates that 27,999,998 of the offspring from a single mating probably die before they reach reproductive age. On the average, each mated pair of Pacific salmon produces only two surviving offspring for the next generation.

Similarly, an individual pine tree may live for several hundred years and produce thousands of seeds each year. To maintain a constant population size, however, each pine tree would be required to produce only one surviving offspring (a single pine tree has both male and female parts) in its entire lifetime. Of course, such precise reproductive control does not occur in natural populations. Due to natural selection, some individuals may leave many surviving offspring and others may leave none.

Predation

What happens to all the offspring that never reach maturity? Most of them are eaten—seeds, seedlings, and young animals are common foods for other organisms. Because predators locate more food when prey are abundant and less food when prey are scarce, they maintain fairly constant prey populations of intermediate densities. The size of a predator population is, in turn, often limited by the

Figure 8-7
The Number of Lynx
(Predator) and Snowshoe
Hare (Prey) Pelts Collected
Annually by the Hudson Bay
Company (1845–1935)
The numbers of predators and prey
often oscillate over time.

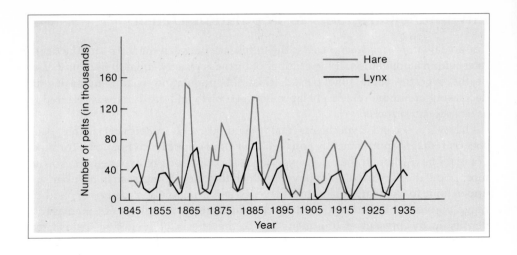

density of available prey populations. The predator population can expand as prey populations expand, because predators reproduce more successfully when food is abundant. The prey populations then decline due to heavy predation, and the predator population also declines due to starvation. Biologists believe that this interaction is a major cause of the regular oscillations sometimes observed in predator and prey populations (Figure 8-7). Although these populations fluctuate considerably around their mean densities, they are relatively constant in size over long periods of time. Both predator and prey populations continue to evolve more effective survival methods, and as long as their rates of evolution remain approximately the same, the two populations are able to coexist without causing the extinction of each other. In the complete absence of natural predators, prey populations often increase to incredibly high densities (Figure 8-8).

A dramatic example of evolutionary change caused by predation has occurred in the moth populations in industrial areas of Great Britain. Within a century, the color of most individuals in these populations has changed from white to black. Prior to the industrialization, the black forms occurred infrequently in the moth populations and were quickly removed by predatory birds. As Figure 8-9(a) shows, black moths were highly visible against the light-colored lichens (organisms composed of both fungi and algae) that covered the tree trunks. The genes that determined dark color in the moths were present in the population, but black moths were strongly selected against. As industrial fumes became more concentrated, the lichens on the trees were killed by pollutants in the air, exposing the dark tree trunks. This environmental change made the white moths more visible to predators and the dark moths harder to detect, as shown in Figure 8-9(b). The black moths therefore became more abundant in the populations, and as a result of the difference between the survival of the two forms, the moth populations changed from predominantly white to predominantly black. Due to the current reduction of air pollution in Britain, the trees are becoming re-covered with lichens and the environment is again favoring the light form of moth.

Figure 8-8
High Densities of Rabbits in Australia due to the Absence of Predators
This unnatural situation was caused by the introduction of European rabbits into Australia, where no natural predator population for rabbits exists.

(a)   (b)

Figure 8-9
The Two Forms of Moths on Different Backgrounds
Light and dark varieties (a) on a lichen-covered tree truck and (b) on a dark, lichen-free tree trunk.

Disease

Population size is also regulated by death from diseases caused by parasites and *pathogens* (infectious agents, such as viruses, bacteria, protozoa, and fungi). Disease epidemics are most effective when the *host* (diseased) *population* is crowded, because the pathogens can spread rapidly from individual to individual. This is one reason why specialized agricultural crops are so vulnerable to disease. Because single species of crops are maintained in very dense stands, once a pathogen successfully infects one plant, it can easily spread throughout the entire field.

Crowded populations usually cannot obtain adequate nutrients, so that their susceptibility to disease is increased. Animals seldom die of starvation or malnutrition; the cause of death is usually predation or disease. But susceptibility to predators and disease is greatly increased by an insufficient food supply. Undernourished individuals display a lower resistance to all types of environmental stress. Epidemics among humans are especially common in starving populations, in which resistance to infectious diseases has been lowered.

Plant populations are equally vulnerable when they are undernourished. The massive destruction of Ponderosa pine forests in western America by the bark beetle resulted from overcrowding—an unnatural condition caused by human intervention in natural regulatory processes. Under natural conditions, these forests would burn at regular intervals and their numbers would be reduced by a natural mechanism that has always been a normal part of their environment. Since humans have lived in these areas, attempts have been made to control forest fires to protect the trees and the structures that humans have built within the forests. A consequence of this fire protection is that the trees now grow in stands of unusually high densities. Under crowded conditions, the resistance of a tree is lowered and it is much more vulnerable to the effects of the bark beetle. A healthy tree that does not have to compete with its neighbors for moisture and nutrients is relatively invulnerable to this beetle.

Emigration

Many animal populations maintain appropriate densities by forcing excess individuals to leave the area, or to *emigrate.* These outcasts are forced to search for new areas in which to establish themselves and reproduce. Emigrants suffer high mortality rates due to predation and starvation; the survivors rarely contribute to population growth, because they are seldom able to find an area in which they can successfully reproduce. This form of population regulation is common among such mammals as muskrats, squirrels, mice, and, most notably, lemmings (Figure 8-10), whose spectacular population explosions and subsequent emigrations have been observed at regular intervals in arctic environments. Scandinavian lemmings do not commit suicide by marching into the sea, as many believe; instead, they are dispersed from areas with high-density populations. Because lemmings are small and do not have good eyesight, many of the dispersing individuals fall into fiords, which are narrow inlets of water along the seacoast. The factors that determine which individuals emigrate and which remain to contribute offspring to the population have yet to be established. If genetic differences between the two groups are found to exist, then emigration is a form of natural selection.

Population Control by Low Birth Rates

The typical response when a population is introduced into a new and favorable environment is depicted graphically in Figure 8-11. After an initial period of rapid growth, resource levels are insufficient to support a higher population density and the number of individuals in the population becomes fairly constant. The upper limit to population density is referred to as the environment's *carrying capacity* for that population. When a population is at its carrying capacity, the number of individual births is approximately equal to the number of individual deaths and there is zero population growth.

Competition for resources, such as food and nesting sites, is the main selective force in populations that have achieved their carrying capacity, and the evolutionary trend is toward improved competitive ability. Resources are often secured by individuals within the population who either monopolize resource areas or gain

Figure 8-11
The Population Growth of Sheep Introduced into Tasmania

Each point on the graph represents the average number of sheep over a five-year period. During the first few decades after the sheep were introduced to their new environment, their food source (grass) was plentiful and their population grew at an exponential rate. After 1854, however, no further increases in population size occurred. The number of sheep in the population after 1854 is considered to be the *carrying capacity* of the environment for this breed of sheep. At this population density, the sheep were eating the grass at about the same rate as it grew.

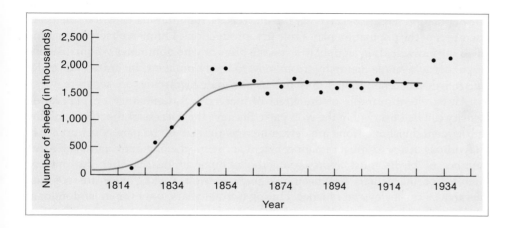

first access to available resources by intimidating other individuals in the population. Because larger individuals can usually monopolize resources more successfully, the characteristic of larger body size is often selected within a population. This type of natural selection is referred to as *K*-selection (in contrast to *r*-selection), where *K* denotes carrying capacity. When individuals compete for resources, the selected characteristics often include larger adult size, fewer (but larger) offspring in each litter, longer adult life spans, more than one litter per female per life span, and a prolonged developmental period prior to reaching adult status.

Limiting the Number of Reproducing Individuals

One of the more dramatic revelations in recent decades is that the social behavior of some animals controls the rate at which births occur in the population. Many species of birds and mammals do not continue to increase in number until they exhaust their food supplies. A remarkable repertoire of built-in behavioral and physiological mechanisms controls the birth rate in such a way that the population remains at or below the carrying capacity of the environment.

Many animals control population growth by limiting the number of females that are permitted to mate. In such a social structure, only a limited number of individuals have access to the resources necessary for successful reproduction. A contest of some sort usually determines which individuals will occupy the valuable spaces or *territories* that contain resources. The size of each territory, or the number of individuals that can occupy territories in a given area, is determined by the amount of available resources, which may vary from year to year. When the population is too large to be maintained by the available resources, some individuals are not permitted to occupy territories and, consequently, cannot reproduce. Many of these individuals starve or die from predation, but some of them do survive and are provided with another opportunity to reproduce. During their year of exile, some of these individuals may become larger or more experienced than others in the reproductive population and may perform better in the contest for territories when given a second chance. Resources may also be more abundant during the following year, so that each member of the population may occupy a smaller territory, making additional spaces available.

The number of reproducing individuals in a population is also limited by *hierarchical dominance,* which is determined through contests between individuals in the population (Figure 8-12). Only the individuals that dominate most of the other members of the population play a role in reproduction. A human observer can rank all of the individuals in a population on the basis of who dominates whom. In some populations, a single hierarchy (from most to least dominant) includes both males and females. In other populations, female and male hierarchies are separate.

An excellent example of the effect of hierarchical dominance on population density can be observed in the wolf pack. Because the success of the hunts and the subsequent division of food are determined by the size of the pack, the number of individuals in a pack must remain constant at the most efficient or *optimal* size (the number of births must approximate the number of deaths after the optimal number of individuals has been reached). In wolf packs, there are separate hierarchies for males and females. During normal years, only the most dominant male and female mate, and they prevent other members of the group from mating.

(a)

Figure 8-12
Dominance Contests for the
Privilege of Reproduction
(a) Bighorn sheep. (b) Elephants.

(b)

As a result, there is only one litter in a pack each year. However, this system of limiting reproduction is flexible, allowing the pack to compensate for unpredictable mortality. In regions of Alaska where the wolf is heavily hunted by humans, packs often suffer sudden reductions in size. When this occurs, more individuals within the pack are permitted to mate and more than one litter is produced each year. In this manner, the pack rapidly returns to its optimal size.

Most animal contests to determine reproductive rights are highly ritualized. Two animals fight only until it is clear which contestant will win; they seldom fight to the death. Ritualized contests lead to the selection of characteristics that *suggest* the animal will be a winner, such as large antlers, long tail feathers, large body size, good coordination, and superior stamina.

Limiting the Number of Offspring per Female

In some populations, most females reproduce, but the number of offspring that each female produces is reduced when the size of the population reaches its carrying capacity. In high-density populations, social interactions among individuals increase and may produce emotional stress and physiological imbalances. Such changes interfere with normal reproductive behavior and physiology. Overcrowding may cause a reduction in the number of eggs released from the ovary, an increase in the occurrence of premature or defective births, and homosexuality (no offspring from such matings), and may disrupt normal courtship behavior.

This type of reproductive regulation has been observed in populations of small mammals, many of whom undergo regular cycles of high and low population densities every three or four years (Table 8-1). Biologists do not yet know how natural selection has produced this type of response to population density, since it is unnatural for selection to favor a decrease in reproduction. However, when individuals of a population are crowded together and resources are scarce,

Table 8-1
Attributes of House Mice at Different Levels of Population Growth and Crowding
(The animals were given all the food and water required for normal growth and maintenance. The population was confined to an area of 150 square feet, so that the only variable was the degree of crowding.)

Attribute	Stage 1 Early growth (uncrowded)	Stage 2 Midgrowth (moderate crowding)	Stage 3 Terminal growth (definite crowding)
Population size	8–40	40–100	100–125
Percent of pregnant adult females	33%	15%	8%
Number of live births	51	157	118
Percent of surviving infants	72%	15%	12%
Aggression score	0.05	0.43	1.17
Number of adults per nest box with litters	1.60	3.2	0.29
Condition of nest	Good	Fair but polluted	Very poor; heavily polluted
Number of wounded adults	1–3	4–18	26–45

Source: Southwick, C., *Ecology and the Quality of Our Environment* (New York: Van Nostrand Reinhold Co., 1972), p. 220. Used by courtesy of Willard Grant Press, Boston, MA.

offspring have a very low probability of surviving to adulthood. Individuals that are sensitive to crowding and respond by not reproducing may be selectively favored, because they will be more likely to survive when resources are inadequate and to reproduce in some later season when the population is less crowded.

Delaying the Age of First Reproduction

Population growth is also curbed when the age of the first reproduction in females is delayed, so that fewer generations are alive at any given time. Because early reproduction tends to shorten an organism's life span, one selective advantage to delaying reproduction when the population is crowded is that the individual will live longer and may reproduce if environmental conditions become more favorable or if its position in the population improves at a later time. When larger body size or experience in life provide a competitive advantage, the individuals in a population that continue to grow and develop, rather than reproduce at an early age, will ultimately leave more surviving offspring.

This subtle method of curbing population growth may offer at least a temporary solution to the expansion of the human population and may currently be the most acceptable reproductive limitation in terms of human morality. The long-term solution, however, can only occur when the number of births is decreased or the number of deaths is increased. If such a system were implemented, each woman could have as many children as she desired as long as she did not begin to reproduce until an established time after the age at which reproduction normally begins in the population. To illustrate the effectiveness of this method, we will consider a hypothetical example.

Two women each want to have six children. Female A begins to reproduce at the age of 18 and has one child every year until she is 24. Her daughters follow the same reproductive pattern. Female B begins to reproduce at the age of 30 and has one child every year until she is 36. Her daughters do the same. (It is customary to consider only the females in population studies, because female parenthood is easier to trace.) We will assume that each female leaves three daughters and three sons. The total number of progeny left by females A and B at various ages during their lifetime would therefore be

Female A

Age		Number of female progeny
24		3 daughters
48		9 granddaughters (3 from each daughter)
72		27 great-granddaughters
96		81 great-great-granddaughters
	Total	120 living female progeny

Female B

Age		Number of female progeny
36		3 daughters
72		9 granddaughters (3 from each daughter)
108		27 great granddaughters
	Total	39 living female progeny

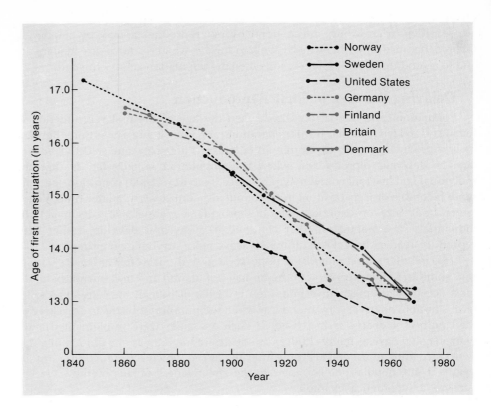

Figure 8-13
The Age at Which
Menstruation First Occurs
Has Decreased in the United
States and Europe
On the average, girls begin to
menstruate 2.5–3.3 years earlier than
they did a century ago.

If Female A, who reproduces between the ages of 18 and 24, lives to be about 100 years old, she will produce 120 living female progeny within her lifetime. Female B, who reproduces between the ages of 30 and 36, will produce only 39 living progeny within her lifetime.

To decrease the growth rate of the present human population, it would be advantageous for women to reach reproductive maturity at a later age. However, the opposite trend is actually occurring. The age at which females have their first menstrual period (an indication of reproductive ability) has decreased since the mid-1800s (Figure 8-13), probably due primarily to improved nutritional habits—especially the increase in protein in the human diet. Obviously, unless mating is prevented or birth-control methods are employed during the early reproductive years, the ability to reproduce earlier will continue to contribute to the rapidly increasing density of human populations.

Selection in Human Populations

Natural selection occurs continuously in human populations, although its effects have been greatly reduced by the use of medicines, improved sanitation, and our technological abilities to maintain a more constant and favorable environment.

In human populations today, the primary form of natural selection may be the regulation of family size. Some individuals are leaving several offspring, who will survive to reproduce many duplicates of their genes in the next generation; other individuals are leaving few or no offspring. If the genes of individuals who are leaving more offspring differ to a marked degree from the genes of individuals who are leaving fewer offspring, then evolution is taking place. But if the individuals who are leaving more offspring are only a random sample of the genotypes in the population, then no evolutionary change will occur. Because culture exerts an enormous influence on human characteristics, it is difficult to determine whether the individuals who leave many offspring represent a random sample or a selective sample of genotypes from the population.

In human populations, natural selection is also achieved through *differential mortality.* Selective deaths occur among humans, especially prior to birth. The frequency of human miscarriages may be as high as 50% of all conceptions. These deaths usually occur so early in a pregnancy that they are not detected by the mothers.

A greater number of genetically defective individuals survive today than ever before due to medical advancements. A major criticism of the treatment of genetic diseases is that the affected individuals are likely to produce offspring who will inherit the same defect, thereby perpetuating these diseases in human populations. As long as the medical treatment is available, however, the accumulation of such genes may produce no serious consequences.

There has been considerable controversy over the prospect of a genetically deteriorating population. *Eugenics* is the general term used to describe programs designed to improve human populations by permitting genetic deaths and/or limiting reproduction in individuals with genetic defects. A sophisticated solution to the problem of genetic defects may ultimately be provided by the new technology of genetic engineering, discussed in Chapter 7.

Summary

The *theory of evolution by natural selection* was developed by Charles Darwin and Alfred Russell Wallace in the mid-nineteenth century. An outline of the theory follows:

1. All organisms tend to leave more offspring than required to replace themselves, but *populations* tend to remain relatively constant in size. Therefore, not all of the offspring survive and/or not all of the individuals leave offspring.
2. Individuals carry different genes, which they pass on to their offspring during reproduction. An individual who produces several offspring passes on more genes to the next generation than an individual who leaves only a few offspring. The number of offspring that an individual produces depends on how well the individual meets local *environmental challenges*, including predators, pollution, seasonal variations, and inadequate food supplies. To some extent this ability to survive is determined by the genes that the individual carries. Thus, some genes increase in frequency and others decrease in frequency in the population.

Evolution is defined as a change in *gene frequency* within a population.

All populations have the potential to grow at an *exponential rate* because most individuals have the potential to leave more than one offspring. This potential is seldom realized, however, because one or more environmental factors limit further increases in population density, either by increasing the death rate or by decreasing the birth rate (or both). When a population reaches its *carrying capacity*, it contains the precise number of individuals that the environment can support in terms of such resources as food and nesting sites. At this point, the death rate is equal to the birth rate and *zero population growth* is achieved.

An increase in the death rate occurs when predators, parasites, or pathogens are able to attack a greater number of individuals because the density of their prey or host population has increased. An increase in the death rate also occurs when food supplies are insufficient to support the existing number of individuals in a population. In some animal populations, excess individuals are forced to *emigrate*, or leave the area, when resources are insufficient to support all members of the population. *Differential mortality* occurs when certain genotypes die or emigrate, so that they are completely removed from the population. Differential mortality causes changes in gene frequency, or evolution, because genotypes that are decreasing in the population leave fewer copies of their genes in the next generation.

The birth rate may decline when the population has reached its carrying capacity due to a decrease in the number of reproducing individuals (usually determined by *dominance* or *territorial rights*), a decrease in the number of offspring born to each female, or an increase in the age at which reproduction begins in an individual. *Differential reproduction* occurs when only certain genotypes in the population are permitted to reproduce. Like differential mortality, differential reproduction is a basis of natural selection and causes evolutionary changes in the population.

Natural selection has been reduced in human populations due to medical and technological advancements and improvements in sanitation, which have tended to equalize the ability of all humans to reproduce. *Eugenics* is the term used to describe proposed programs that would prevent genetically defective individuals from reproducing, thereby removing deleterious genes from the human population.

Ecology: Relationships Between Populations and Their Environments

9

E*cology* is the study of the environmental factors that govern the distribution and abundance of populations of organisms. The *distribution* of a population is a kind of map that indicates the geographical location of the population. The *abundance* of a population is the *density,* or the number of individuals per unit area. Both the distribution and the abundance of a population reflect how well the individuals that occupy an environment are suited to it.

Organisms usually function in highly complex environments. An *environment* includes:

1. Physical conditions, such as climate and type of soil.
2. The availability and abundance of resources, such as food and nesting sites.
3. All of the other kinds of organisms present in the environment that could affect the survival and reproductive success of the population.

The physical conditions and resources in an environment are relatively easy to quantify. The third factor—the other kinds of organisms present—usually determines the complexity of the environment of a population. To make accurate predictions about the distribution and abundance of a particular kind of organism, it is necessary to understand how that organism interacts with all other kinds of organisms in the area.

The relationships between populations in an area are seldom obvious and may be apparent only after some component of the environment changes. For example, in a mosquito-control program carried out in Borneo by the World Health

Organization, the number of malaria-carrying mosquitos was reduced by spraying the area with the insecticide DDT. However, a certain type of predatory wasp, which feeds on caterpillars in the area, was also killed by the poison. Removal of this wasp population caused an exponential increase in the caterpillar population. Since the roofs of houses in Borneo were made of plants, the caterpillars, which feed on plants, severely damaged the houses. The program also resulted in the accumulation of the DDT in houseflies, which were eaten by small lizards. The sick and dying lizards were, in turn, eaten by house cats, which died in large numbers as the insecticide became concentrated in their bodies. The cats normally kept the rat population at a low density. As the cats died, the rats rapidly increased in number and invaded the houses. The fleas carried by the rats were transferred to the human residents. Since the fleas were the host population for the bacteria that cause a serious human disease—the plague—the program to control malaria actually contributed to a potential plague epidemic and damaged houses as well. We could cite countless examples like this one in which well-intended tampering with an environment has produced unexpected and unfortunate repercussions. The ecologist's goal is to understand the relationships between organisms and all aspects of their environments sufficiently to make accurate predictions about the consequences of altering an environment.

An *ecosystem* is the term used to describe an assemblage of populations and their nonliving environment in a particular area. A *community* is a group of populations that have a high probability of interacting and, consequently, of influencing one another's distribution and abundance. Neither an ecosystem nor a community remains constant over time. The physical environment is altered by the populations that exist in an area, and the living environment is altered by changes in the density and evolution of populations. A change in the physical environment or in one of the populations may greatly affect other residents in the area. Because the consequences of human behavior play an important role in all ecosystems, it is important for us to understand our environment so that we can make accurate predictions about the future.

Environmental Resources—Energy

Because individuals have the ability to leave more offspring than are required to replace themselves, a population will increase in size until something in the environment imposes a limit on further population growth. A number of factors, including predators and unfavorable weather, may prevent a population from reaching its *carrying capacity* (the maximum number of individuals in the population that the available resources in the environment can sustain). If these controlling factors are not operative or effective, the carrying capacity will eventually be reached when an essential resource is being used at the same rate that it is being made available in the environment.

Food resources are always available in limited amounts and will ultimately set an upper limit on the density of a population. All organisms rely on food to provide the energy and structural materials needed for their maintenance, growth, and reproduction. First we will examine the relationship between energy and the community of organisms.

Every living organism must be able to obtain energy from its environment—not only to produce new tissues during growth and reproduction, but also to maintain a high degree of organization within the tissues of an organism. Living material is subject to the Second Law of Thermodynamics: the large molecules that characterize life tend to disintegrate, creating an increase in entropy, unless they are maintained by a regular input of energy. Much of the energy in food is used just to maintain the existing level of organization within the body of the organism.

Practically all energy enters the biological world through *photosynthesis*—the remarkable process during which plants and other autotrophs convert light energy from the sun into the chemical energy contained in molecules. This molecular energy is then available to be used by the autotrophs as well as by other types of organisms. Autotrophs convert only 0.1–5% of the light energy that reaches the earth's surface into chemical energy, largely because the pigments autotrophs use during photosynthesis can only capture certain wavelengths of light. The rate of photosynthesis is further limited by both the physical conditions of the environment and the form of the autotrophs existing in the community. These limits to photosynthesis include such variables as light intensity, temperature, moisture, atmospheric gases, soil nutrients, total leaf area, and the geometry of the vegetational canopy.

Once this small percentage of light energy is converted into chemical energy in autotrophs, it becomes a food source for other organisms. Autotrophs therefore determine the total amount of energy available to all other organisms in the *community*. The transfer of energy throughout a community—by the process of eating and being eaten—is referred to as *energy flow*.

Ecologists group organisms according to the types of food they eat, placing all organisms that obtain their energy from a similar source on the same *trophic level* ("trophic" = feeding). The trophic levels in a community are: *autotrophs* (unicellular and multicellular), *herbivores* (organisms that eat plants), *primary carnivores* (organisms that eat herbivores), *secondary carnivores* (organisms that eat primary carnivores), *tertiary carnivores,* and so on. *Decomposers*—organisms that consume dead plant and animal materials—may be classified as a special group or defined according to the trophic level in which their particular food is found.

Loss of Energy in a Food Chain or Web

A *food chain* is a particular sequence of energy flow among organisms. A simple food chain might be grass ⟶ mice ⟶ coyotes. Other examples of food chains are provided in Figure 9-1. However, communities are seldom organized into such simple food chains. Each organism usually eats more than one type of food and, in turn, is usually consumed by several different types of animals. These more complex trophic relationships are called *food webs.* The organization of a particular community—a freshwater pond—into its food web is diagrammed in Figure 9-2.

The food molecules that animals consume are broken down, and the chemical energy within the molecules is released. Some of this energy is used to construct

Figure 9-1
Some Food Chains
A particular sequence of energy flow among organisms is a *food chain*. (Arrows indicate the direction of flow.)

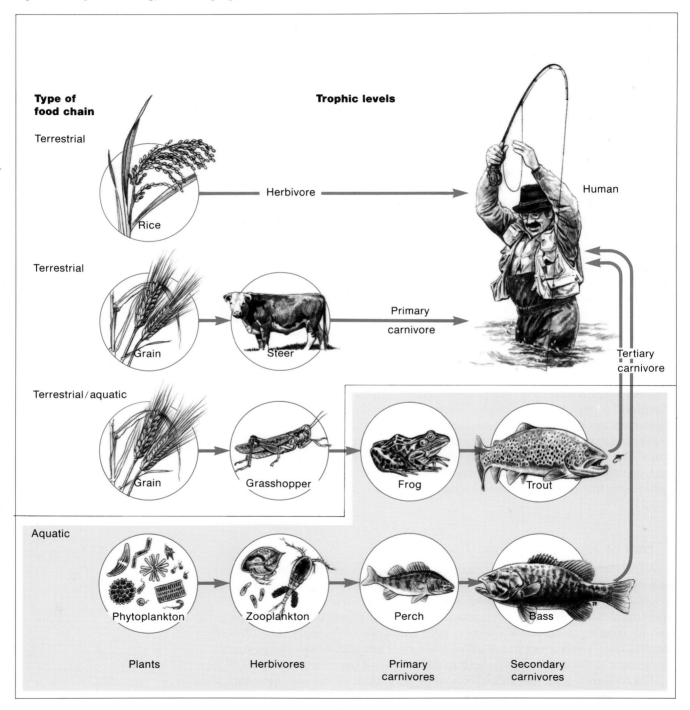

Type of food chain

Terrestrial

Terrestrial

Terrestrial / aquatic

Aquatic

Trophic levels

Rice

Herbivore

Human

Grain

Steer

Primary carnivore

Grain

Grasshopper

Frog

Trout

Tertiary carnivore

Phytoplankton

Zooplankton

Perch

Bass

Plants

Herbivores

Primary carnivores

Secondary carnivores

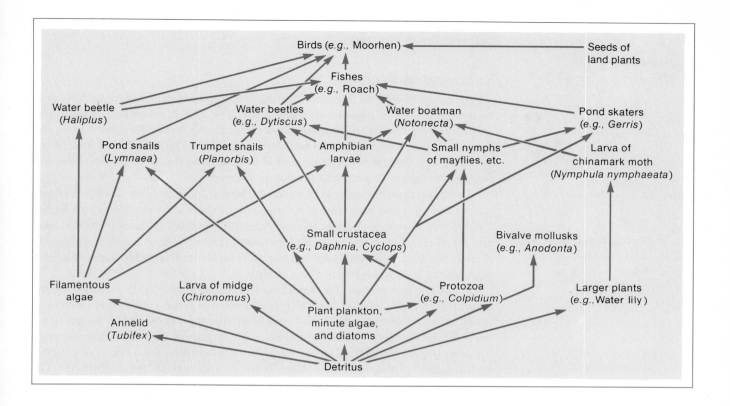

Figure 9-2
The Food Web of
a Freshwater Pond

In many communities, such as the one
shown here, trophic relationships are
highly complex. (Arrows indicate the
direction of energy flow from food to
consumer.)

and maintain biological molecules within the consumer's body. In this way,
chemical energy is transferred from the molecules of one organism (food) to the
molecules of another organism (the consumer). Whenever one type of molecule is
broken down and the energy it releases is used to construct another type of
molecule, some of the chemical energy is converted into heat energy. This
conversion is in accordance with the Second Law of Thermodynamics, which
predicts that whenever energy is transferred, a portion of it is dispersed as heat
energy to the surrounding environment. Although heat is a form of energy, it is less
useful to biological systems than chemical energy.

In addition to the loss of usable energy in the form of heat, the consumer is
unable to convert some of the energy contained in food to a usable form, because
the organism's system cannot break the chemical bonds that hold the molecules
together. These materials pass unchanged through the digestive system and out of
the animal in the form of *feces*. For example, humans cannot convert much of the
chemical energy contained in salad greens to a usable form because we do not have
the enzymes required to break apart the cellulose molecules that form the cell walls
in this food material.

Only about 5–15% of the energy in food material can be utilized by the
consumer to form new tissues. For example, if chickens were fed 1,000 pounds of
grain, only about 100 pounds of this food would be incorporated in their bodies. If
humans ate 100 pounds of chicken, they would produce only about 10 pounds of

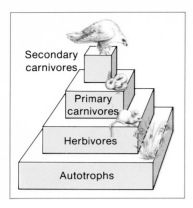

Figure 9-3
An Ecological Pyramid

The size of each trophic layer represents the *biomass,* or total chemical energy, at that trophic level.

Figure 9-4
The Effect of Toxic Substances in a Food Chain

Due to the accumulation of high levels of DDT in the brown pelican's food chain, two of these pelican eggs had such thin shells that they were crushed by the weight of the mother's body.

human tissue. As chemical energy flows through a food chain, a great amount of usable energy is lost at each stage of the chain.

Ecological Pyramids

An *ecological pyramid* (Figure 9-3) diagrams the amount of tissue or *biomass* at each trophic level of a particular community. The total biomass of all autotrophs (plants and photosynthesizing unicellular organisms) forms the lowest level, the biomass of herbivores forms the next level, and so on. The size of each level of the pyramid is proportional to the amount of biomass and therefore to the amount of chemical energy at each trophic level in a community.

The amount of biomass at the next highest trophic level is smaller, because usable energy is lost each time molecules are transferred from a lower to a higher trophic level. Because energy can only enter the pyramid at the autotrophic level and is lost at each subsequent step of the food chain, a pyramid is usually limited to three to five trophic levels. Sufficient energy is rarely transferred to support animal populations at higher levels in a community.

In most communities, the amount of biomass is a reliable estimate of the amount of chemical energy at each trophic level. The largest amount of biomass is found at the autotrophic level, and each subsequent level exhibits a proportional decrease in biomass. A very different pattern occurs in a few communities, however. At any given time in some aquatic communities, the amount of biomass in the autotrophs is less than the amount of biomass in the herbivores. The explanation for this unusual pattern is that aquatic autotrophs are small, rapidly growing organisms that are consumed by large, slowly growing herbivores. Although more energy flows through the autotrophic level than the herbivore level, it is not allowed to accumulate in the form of tissue because the autotrophs are eaten almost as rapidly as they are formed. The biomass is therefore not an accurate measure of the actual amount of energy available at each trophic level in these kinds of aquatic communities.

Biological Magnification

An unfortunate consequence of the reduction of biomass at each stage of a food chain is that some toxic substances become increasingly concentrated at each trophic level. This effect is known as *biological magnification.* Materials such as DDT, mercury, and radioactive elements may be ingested with food and stored in the body. If, for example, there are 100 molecules of DDT in 1,000 pounds of grain, the same 100 molecules of DDT will be present in 100 pounds of chicken tissue formed by consuming the grain and in 10 pounds of human tissue formed by eating the chickens. Some of the animals at the end of a long food chain may contain concentrations of toxic substances up to 1 million times higher than the concentrations found in their environments (Table 9-1).

Toxic substances have adverse effects on biological systems and, as they become more concentrated in body tissues, these effects become more destructive (Figure 9-4). The presence of toxic materials in the environment is a serious threat to such carnivores as eagles, tuna, and humans.

Table 9-1
Biological Magnification

In each of these communities, the first organisms listed are the lowest members of the food chain and the last organisms are the highest members. In the Long Island Estuary, plankton are eaten by minnows, minnows are eaten by predatory fish, and predatory fish are eaten by birds. The numbers indicate the relative concentrations of materials compared to the concentrations found in the environment. The environment of the Long Island Estuary is water, and the concentration of DDT in this environment is 1. The concentration of DDT in plankton in the estuary is 800 times greater than the concentration of DDT in the water. The concentrations of toxic materials are always greater at the next higher trophic level in a community.

DDT in the Long Island Estuary (1967)		Radioactive phosphorus (^{32}P) in the Columbia River (1956)		Radioactive strontium (^{90}Sr) in a Canadian lake receiving atomic wastes (1963)		Radioactive iodine (^{131}I) in a Washington desert following stack release (1956)	
Water	1	Water	1	Water	1	Vegetation	1
Plankton	800	Insects	3	Sediments	200	Thyroids of jackrabbits	500
Minnows	11,600	Swallows	75,000	Aquatic plants	300		
Predatory fish	34,600	Duck eggs	200,000	Minnows	1000		
Fish-eating birds	92,000			Perch bone	3000		
				Muskrat bone	3900		

Environmental Resources—Elements

In addition to energy, all organisms require *structural materials* to survive and grow. More than 30 elements (see Table 2-3, page 35) are essential in the construction of biological molecules. Unlike energy, structural elements are not altered by use; atoms can be used and reused and may be cycled repeatedly between nonliving and living systems.

Biogeochemical cycles describe the movement of elements through the various components of the ecosystem. The two main forms of these cycles are the gaseous cycles and the sedimentary (mineral deposits) cycles. *Gaseous cycles* include the carbon, oxygen, and nitrogen cycles; the primary reservoirs of these elements are the atmosphere and the ocean. Most of the other elements have *sedimentary cycles;* the primary reservoirs of these elements are the soil and rocks of the earth's crust.

All biogeochemical cycles are important to the maintenance of life forms on earth, because they permit elements to be used repeatedly to construct biological molecules. We will consider two biogeochemical cycles in detail here—the nitrogen (gaseous) cycle and the phosphorus (sedimentary) cycle.

The Nitrogen Cycle

One of the most important elements required by all organisms is *nitrogen*—an essential component of proteins and genes. The earth's atmosphere contains 78%

nitrogen in gaseous form (N_2). Plants introduce nitrogen into the biological community, but not in its gaseous form. Plants can only process nitrogen when it is dissolved in water, usually in the form of *nitrate* (NO_3) or *ammonia* (NH_3).

Nitrogen gas from the air is converted into water-soluble forms by the process of *nitrogen fixation*, in which the nitrogen molecule is split into two nitrogen atoms and combined with hydrogen to form ammonia. Nitrogen fixation requires energy and usually occurs when (1) energy from lightning or volcanic activity acts on the nitrogen molecules in the air, or (2) bacteria and blue-greens (two forms of prokaryotes) are active in the soil or water.

The tissues of dead plants and animals, as well as animal urine and feces, provide another source of nitrogen. The nitrogen in dead material must be released and recycled if new organisms are to grow and life is to continue. Certain bacteria and fungi, the decomposers of the community, break down the molecules containing nitrogen in dead biological materials to obtain energy and structural materials and then release ammonia as a waste product.

The ammonia produced from nitrogen fixation and from the activities of decomposers is then used directly by some plants or further acted on by bacteria to produce *nitrite* (NO_2) and nitrate (NO_3). The conversions of ammonia to nitrite and of nitrite to nitrate both yield energy that is then utilized by the bacteria. Nitrate can be used directly by plants. Nitrate may also be utilized by a second group of bacteria and fungi; the by-product of their activities, *nitrogen gas* (N_2), returns to the atmosphere.

Ammonia and nitrate are absorbed through plant roots and incorporated into biological molecules. The nitrogen is then transferred through food chains, primarily in the form of *amino acids*, and ultimately distributed throughout the community. The nitrogen cycle is diagrammed in Figure 9-5.

Human activities interfere with this natural nitrogen cycle in many ways. Because humans remove plants and animals from their natural environment to eat them or to use them for other purposes, the nitrogen from these organisms is not returned naturally to agricultural lands and forests. Some of the molecules that contain nitrogen are eaten by humans, incorporated in body tissues, and eventually concentrated in cemeteries. The remaining molecules containing nitrogen and consumed by humans are converted into urine and feces and are not returned to the soil in which the nitrogen originated. The isolation of human remains and waste materials from agricultural areas prevents the natural recycling of biologically important elements.

Farmers are aware of the importance of nitrogen in the environment and return it to the soil in the form of fertilizers. However, nitrogen is added occasionally in massive doses, whereas the natural system regularly recycles small quantities of nitrogen. During periodic overfertilizations, rain or irrigation water carries much of the nitrogen beyond the roots of plants to the groundwater below and eventually into streams.

The practice of keeping cattle in feedlots, where large amounts of urine and feces accumulate and flood the soil, also concentrates nitrogen in groundwater and may eventually penetrate drinking-water sources. Although nitrogen is important to life, certain molecular forms of it are toxic. A number of human deaths, particularly infant deaths, have been attributed to the accumulation of large amounts of nitrogen compounds in drinking water.

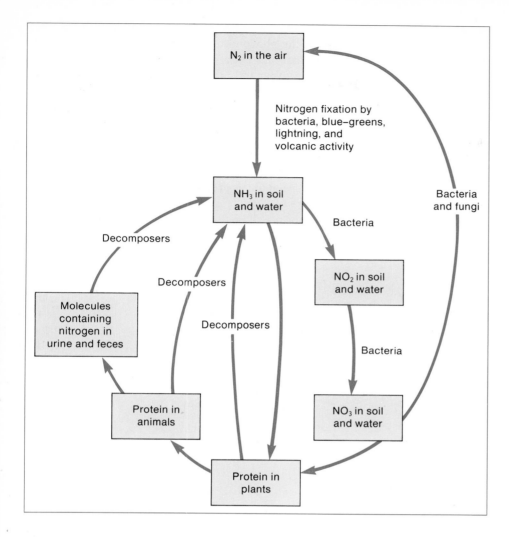

Figure 9-5
The Nitrogen Cycle (Arrows indicate the direction of movement.)

Most of the nitrogen in plants and animals is incorporated in proteins, although nitrogen also plays an important role in the formation of other biological molecules.

The Phosphorus Cycle

Phosphorus, another important element in all biological systems, is a structural component of cell membranes, genes, and the energy-carrying molecule ATP. Phosphorus comprises about 1% of animal body weight. It is abundant in rocks and slowly becomes available to life forms when it is dissolved by flowing water. The dissolved phosphorus is absorbed by plants, transferred to animals, and eventually recycled by decomposers acting on dead plants, dead animals, and animal waste products. In aquatic systems, the phosphorus in the waste materials of very small animals (*zooplankton*) can be used directly by plants. However, dissolved compounds containing phosphorus (*phosphates*) may not be available to living systems, because these compounds may combine with other elements and become immobilized as sediment. Streams and rivers carry much of the phosphorus in the soil and water out to the ocean, from which it cannot return because phosphorus does not exist in a gaseous form. The phosphorus cycle is diagrammed in Figure 9-6.

Humans have interfered with the phosphorus cycle by depositing sewage and other human waste products containing large amounts of phosphorus, such as detergents, in many freshwater streams and lakes. In the presence of abundant phosphorus, populations of blue-greens increase in density and eventually form a thick layer on the surface of the water (Figure 9-7). The organisms near the bottom of the layer die because they do not receive enough light to complete photosynthesis and are decomposed by bacteria—a process that requires oxygen. The tremendous increase in the rate of decomposition depletes the oxygen supply in the water and causes the deaths of larger aquatic animals, including fish. This sequence of changes, called *eutrophication,* is progressing rapidly in many freshwater lakes and streams.

Ecologists predict that phosphorus will be the first element required to sustain life to be depleted from terrestrial ecosystems. Without phosphorus, no life forms

Figure 9-6
The Phosphorus Cycle
(Arrows indicate the direction of movement.)

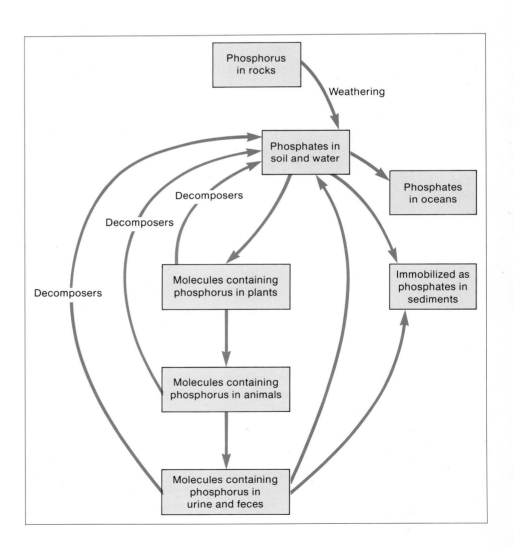

will be able to exist on land. Increased human activities have eroded soils to the point that much of the earth's phosphorus is being washed out to sea and cannot be recycled to the land. Phosphorus is currently obtained primarily from mining rocks and is also recovered from animal bones and feces. An important source of this element is the droppings (*guano*) of sea birds that have fed on oceanic fish and returned some phosphorus to seashores in the form of feces. All of these extraction processes are costly.

If current phosphorus sources are depleted, all terrestrial organisms will again be dependent on the rate at which flowing water naturally dissolves phosphorus out of rocks and the total number of organisms living on earth will be greatly reduced. The maximum human population size that could be sustained under these conditions is estimated to be 1–2 billion people—less than half the present number.

Other Important Elements

All biological molecules contain carbon and hydrogen atoms, and most biological molecules also contain oxygen. None of these elements are likely to become scarce. Carbon enters the biological community in the form of carbon dioxide from the air, which is taken in through the leaves of plants. Hydrogen and oxygen atoms are introduced into the community in water, which is absorbed through the roots of plants. Within plants, carbon, hydrogen, and oxygen atoms are incorporated in sugar molecules during the process of photosynthesis. These elements may then become components of other types of molecules when sugar is altered to form various biological materials. These materials are later broken down by plants and animals, and the carbon, hydrogen, and oxygen atoms are returned to the environment in the form of carbon dioxide and water.

Scientists know little about the natural cycles of most of the other elements required for life. Small quantities of calcium, potassium, sulfur, sodium, iodine, molybdenum, zinc, iron, magnesium, chromium, copper, boron, cobalt, and vanadium are known to be essential in the construction of some biological molecules. As we continue to remove organic materials from our agricultural lands, we also remove these essential atoms. Many of these elements are further removed from biological use when they are incorporated in such products as automobiles and metal containers. These materials must be returned to the soil in a form that living systems can use if they are to be naturally recycled.

Community Structure Within Trophic Levels

In most ecological communities, more than one type of organism—more than one species of plant, herbivore, or carnivore—exists at each trophic level, because the environment is highly variable from place to place and from time to time. No single type of organism can be well adapted to all environmental conditions.

Ecologists use the term *niche* to indicate all of the requirements that must be met for a particular species to exist in an environment. A niche has many dimensions, including temperature, humidity, food types, foraging place and method, and nesting sites. Most of the niche requirements of a species are determined by genetic factors and cannot be altered without evolutionary changes.

Figure 9-7
The Effect of Too Much Phosphorus
When human waste products containing large amounts of phosphorus are discharged into freshwater lakes and streams, the populations of blue-greens become extremely dense. Many of these blue-greens die because they do not receive enough light and are decomposed by bacteria—a process that requires oxygen. The greatly increased rate of decomposition in such waters depletes the oxygen supply, resulting in the deaths of larger aquatic animals, including fish.

Figure 9-8
Herbivores in the Grassland Community of the Serengeti Plain
These species can coexist because their food sources differ. Each species eats different types of plants or different parts of the same type of plant.

Because every environment supports a variety of physical conditions and resources, many different species (with different niche requirements) can coexist in the same area (Figure 9-8). The number of species at each trophic level of a particular community is determined by the outcome of three processes: colonization, competition, and coevolution.

Colonization

All organisms have some method of *dispersal*—of moving from one place to another. In many species, this mobility occurs only when the organism is young. For example, the seeds of plants are capable of moving from the parent to other areas, but adult plants are immobile. *Colonization* occurs when one or more individuals move to a new area to establish a population.

Some individuals of a species are always dispersing into new areas. Whether or not they successfully establish a population depends on the physical conditions and the resources in the new environment. If all of their niche requirements are satisfied and available (not monopolized by another species), the colonizers will be able to exist in the new area.

Competition Between Species

Competition occurs when two species require a resource that is in short supply. The use of a resource by the individuals of one species reduces its availability to the individuals of another species. This form of competition is called **inter**specific *competition*. Competition among individuals of the same species is called **intra**specific *competition*.

When a species colonizes an area, it often must compete with one or more resident species. If the niche requirements of the competitors and the colonizers are very similar, intense interspecific competition will take place to determine which species will use the resources to survive and reproduce. An early indication of interspecific competition for resources is a reduction in the reproductive rate of one or both species. A study of the effect of field mice on the reproductive rate of house mice in central California grasslands demonstrates this principle. In the absence of field mice, each adult house mouse produced an average of three offspring. When field mice occurred in densities above 90 mice per acre, however, each house mouse produced only one offspring. A reduction of the food resources available to the resident population due to the presence of a competitor caused a reduction in the reproductive rate.

Long-term competition between species can result either in the *extinction* of one population or in *evolutionary changes* in the niches of the competing species. The extinction or competitive exclusion of one species from an area by another species is a frequent outcome of interspecific competition. When both species require the same resources, one species will usually become extinct before different niche requirements can evolve.

A familiar example of competitive exclusion can be found in an untended garden. Most domestic varieties of plants are not as well adapted to their environments as native plants. Unless they are protected from direct interspecific competition by the removal of native plants (weeds), domestic plants will usually be excluded from a garden within a short time.

Figure 9-9
Evolutionary Changes in the Niches of Two Competing Species

(a) Before natural selection, both species feed on prey of approximately the same size. (b) After natural selection, both species have become specialized to feed on different sizes of prey. This specialization results from the greater reproductive success of individuals from both species that feed on prey not found in the area of overlap.

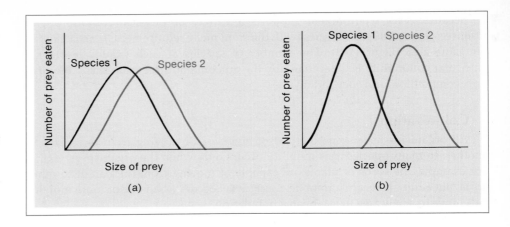

Coevolution Within Trophic Levels

Coevolution occurs when the evolution of one species is strongly influenced by the evolution of another species, and vice versa. Such interactions between species have played an important role in determining the characteristics and distributions of living organisms. Interspecific competition for resources within a trophic level often results in the coevolution of the competing species through the process of natural selection.

An example of evolutionary changes in the niches of two competing species is diagrammed in Figure 9-9. Initially, the two species compete for the same food resources (intermediate-sized prey), and in the area of overlap neither species can obtain enough resources to survive and reproduce. Through natural selection, this niche requirement is modified in both species, so that mutual use of the same resource is reduced.

The actual niches of five coexisting species of warblers in the eastern United States have been studied. All five species feed on the insects found on spruce trees, but careful observation has revealed that each species feeds on a somewhat different part of the tree (Figure 9-10). The niche of each species has been altered through competition with other species in the area.

As competing species develop divergent uses of the same resources, the physical characteristics of the species may also change. Figure 9-11 illustrates the evolution of different beak shapes, each adapted to procuring a different food source, in three competing species of birds.

Anatomical, physiological, or behavioral adaptations that have evolved among similar, competing species and that facilitate different uses of the same resources are called *character divergences.* Genetically determined characteristics, such as beak shape, temperature tolerance, or times during which foraging occurs, are altered as a result of competition among coexisting species.

Intense competition between species appears to be a transitory phenomenon in natural communities. In time, competitors are excluded or the niches of competing species are molded by natural selection so that little interspecific competition remains in the community.

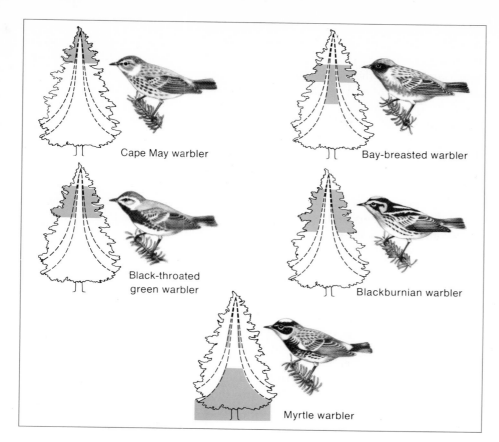

Cape May warbler

Bay-breasted warbler

Black-throated green warbler

Blackburnian warbler

Myrtle warbler

Figure 9-10
Coevolution Between Competitors Produces Unique Niches
These five species of warblers in the eastern United States all feed on the insects found on spruce trees. As a result of competition and coevolution, each species feeds on a somewhat different part of the tree. This division of available resources permits all five species to coexist in the same environment.

Coevolution Between Trophic Levels

Coevolution also occurs between the consumer and the consumed. By removing the individuals that are easiest to locate and capture, predators exert a strong selective force on their prey populations, causing more effective predator-avoidance methods to evolve. Similarly, the genes of predators that are unable to find and capture prey are removed from the predator population and, in time, more adept predators evolve. Many animal characteristics, such as keen sense of smell, good vision, mobility, and coloration, are the result of coevolution.

Coevolution Between Plants and Herbivores

Each herbivore population can eat only some types of plants in its environment. Many plants are inedible because they contain substances that are toxic to certain herbivores. Although each type of herbivore has evolved methods of storing or detoxifying some of these substances, no single herbivore population can process all of the diverse toxic materials produced by the plants in its environment. In addition to chemical defenses, many plants also have thorns or hooks that discourage herbivores, and some plants exude sticky substances that can trap insect herbivores.

Many fruit and pine trees have evolved a mechanism called the *feast-and-famine*

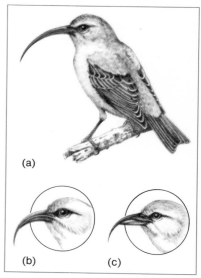

(a)

(b)   (c)

Figure 9-11
The Effect of Resource Specialization on Beak Shape
The three species of Hawaiian honey creepers have beaks that are adapted to their different feeding habits.
(a) *Hemignathus obscurus* picks insects from crevices in tree trunks and branches; (b) *H. lucidus* uses its shorter bottom bill to chip away loose bark to expose insects; (c) *H. wilsoni* uses its extremely stubby bottom bill like a woodpecker to pound into soft wood to obtain insects.

strategy that prevents herbivores from consuming all of their reproductive products—their fruits and cones. These trees reproduce only in alternate years, so that herbivores find an abundance of food one year (feast) but no food the following year (famine) and the herbivore population is greatly reduced. When fruits or cones are abundant again in the next year, the much smaller herbivore population is unable to consume the entire crop. In response to the feast-and-famine strategy, herbivore populations have evolved such strategies as storing of nuts and cones, attempting to reproduce only in alternate years, and migrating long distances to locate other food sources during off years.

The acacia tree exhibits one of the most elaborate defense mechanisms found in plants. This tree produces regular amounts of nectar, which serves as food for an aggressive type of tree-nesting ant. The ant colony vigorously defends its food supply from intruders and, in doing so, also defends the tree from a variety of herbivores. The acacia expends some energy to produce nectar for the ants, but it gains protection from herbivores in return.

Coevolution Between Herbivores and Carnivores

Just as each type of herbivore cannot consume all types of plants, each type of carnivore cannot consume all types of herbivores. Each species of herbivore employs a unique predator-avoidance method, and no single species of carnivore is capable of evolving the many adaptations necessary to permit it to catch and consume all of the different types of prey in its environment. For example, wild dogs are able to catch deer by running them to exhaustion but can rarely catch prairie dogs, which sprint to safety in nearby burrows. Wild dogs are long-distance runners, not sprinters. Cats, on the other hand, are adept sprinters that wait until an animal is vulnerable and then quickly run down their prey. But cats have little endurance for long-distance running. No single species of carnivore can excel at both sprinting and long-distance running. Each carnivore has evolved highly specialized adaptations to the characteristics of certain species of herbivores.

Mimicry and Camouflage

Like plants, many animals produce toxic substances. These animals are often brightly colored, so that predators will remember that they taste bad. Some nontoxic species have evolved the same coloration as toxic species. This process is

Figure 9-12
Mimicry in the Butterfly

The monarch butterfly (left) is toxic to most predatory birds. Its color patterns are closely mimicked by the nontoxic viceroy butterfly (right). Both butterflies are vulnerable to the same bird predators.

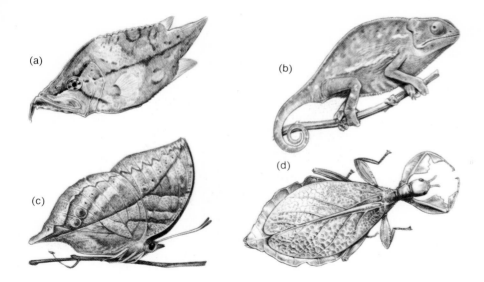

Figure 9-13
Four Types of Animals That
Resemble Leaves
(a) Fish; (b) lizard; (c) butterfly;
(d) leaf insect.

called *mimicry* (Figure 9-12). A predator that encounters a toxic species will avoid all individuals of that particular coloration, including nontoxic species.

As a means of protection from predators, many animals have evolved *camouflage*. They essentially hide in their environment by resembling such objects as leaves, twigs, or the bark of trees (Figure 9-13). Some animals camouflage themselves by carrying parts of the environment—dead insects, stones, or pieces of bark—around on their backs.

Community Changes

Each trophic level in a community contains several different species, because no one species can make the necessary adaptations to utilize all of the resources found at that level. Resident members of communities also differ from place to place and over time because each species cannot adapt to all possible environments.

Changes in Environment from Place to Place

Changes in the environment are reflected in the communities of organisms that live there. Familiar environmental changes result as elevation increases; from the bottom to the top of a mountain, gradual changes occur in temperature, moisture, soil texture, length of growing season, and other environmental factors. Each set of environmental conditions at a particular elevation supports a different community of organisms. If you walked up the side of a mountain, you would observe a gradual change in the types of plants and animals in the environment as you moved from the streamside communities at the bottom to the tundra communities at the top.

In addition to changing physical conditions in the environment, competition between species contributes to the differences in species composition from place to place. Two species that require almost identical resources will compete until one species is excluded from that particular environment. There will be little or no

Figure 9-14
Geographical Replacements That Do Not Coexist

Swamp rabbits and marsh rabbits are geographical replacements that have adjacent but nonoverlapping distributions in similar environments.

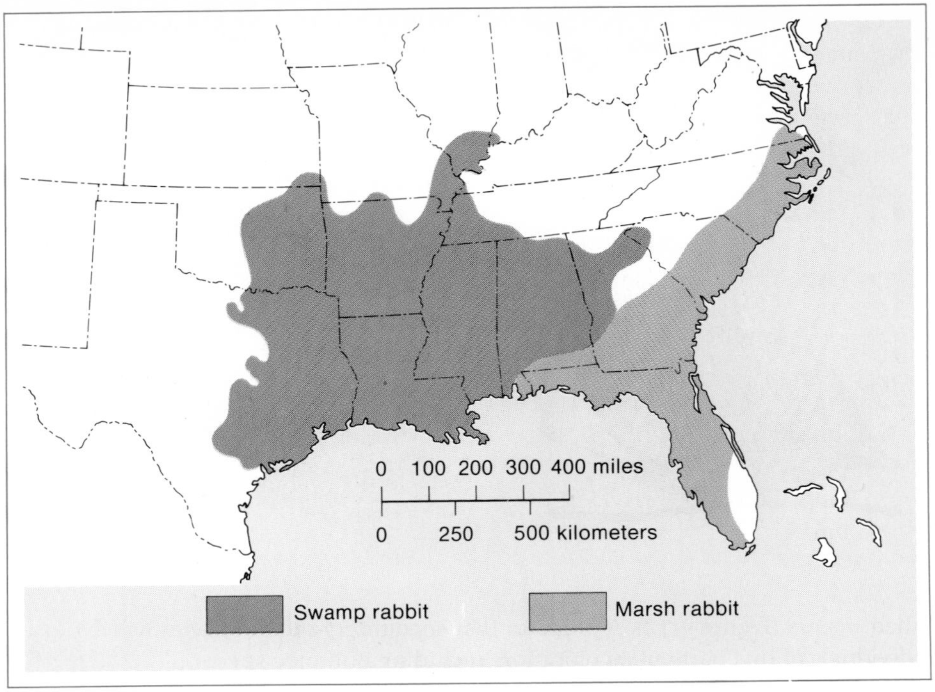

Figure 9-15
Trees with Slightly Overlapping Distributions

(a) Lodgepole pine; (b) jackpine. (Dark areas indicate overlapping distributions of the two species.)

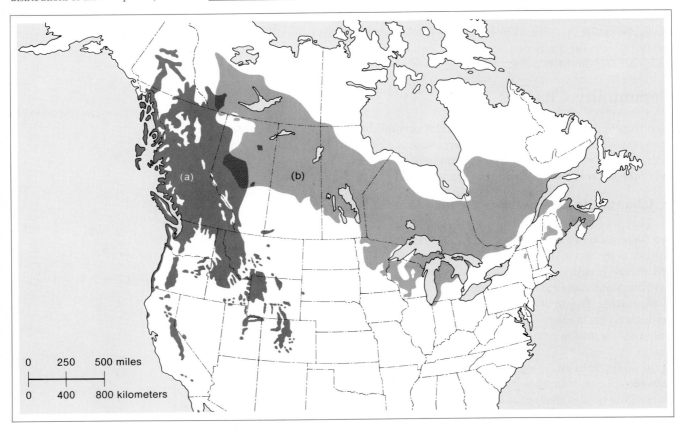

overlap in the geographical distributions of these species. The particular set of environmental conditions, which varies from place to place, will determine which species will populate an area. Ecologically similar species that establish themselves in separate communities are called *geographical replacements.* Two almost identical geographical replacements are swamp rabbits and marsh rabbits, which have adjacent but nonoverlapping geographical distributions (Figure 9-14). Two other geographical replacements are the lodgepole pine and the jackpine (Figure 9-15). Except for two small overlapping areas, similarities in these two species of pine prevent coexistence. One species is predominant in certain areas; the other, elsewhere.

When geographical replacements differ more markedly, their distribution patterns overlap to a greater degree. Some geographical replacements live together in the same area when the outcome of interspecific competition is ambiguous and in geographically separate areas when one species clearly dominates the other. The overlapping distributions of geographical replacements can be illustrated by the North American meadowlarks (Figure 9-16). One species has adapted to conditions found in the eastern half of the continent; the other, to conditions found in the western half. These distributions overlap in the center of the continent, where environments are intermediate and competitive exclusion does not occur.

Changes in the Environment Over Time

Species occupy a particular area because the physical environment and the available resources meet their niche requirements. If the environment does not change over time, a species may occupy an area indefinitely. If the environment changes, however, the resident species may be replaced by other species.

The most rapid environmental changes occur after an area has been severely disturbed, as it would be after farming. Once a farm is abandoned and the soil is no longer worked, a series of different communities will occupy the area. In the eastern United States, the first types of plants to invade the plowed land are primarily annual plants, which grow and reproduce within a year. The area is then occupied by grasses, shrubs, and trees, in that order. This process of change in communities is an *ecological succession* that may take decades to accomplish.

Each community in the line of succession will contain species that are adapted to the present environment (Figure 9-17). However, resident species alter the environment by changing the texture and chemical composition of the soil, the amount of sunlight that reaches the ground, the availability of resources, and other conditions. As these changes occur, the environment becomes more appropriate for a new set of species, which colonizes and replaces the earlier community for which the environment has now become less appropriate. A sequence of different communities colonizes the land until the environment is returned to its condition prior to the disturbance (for example, prior to being farmed or prior to a fire). At this point, the area will contain a group of species similar or identical to the species present before the disturbance, and no further environmental changes will occur, except evolutionary changes over time. This final stage is referred to as the *climax community.* Thus, a disturbed area that once supported a pine forest will undergo a sequence of community changes that eventually reestablish pine trees; desert environments will undergo community changes that eventually reestablish desert

Eastern meadowlark
Western meadowlark

Figure 9-16
Geographical Replacements with Broadly Overlapping Distributions
The eastern and western meadowlarks of North America are geographical replacements that have adjacent but overlapping distributions in similar environments.

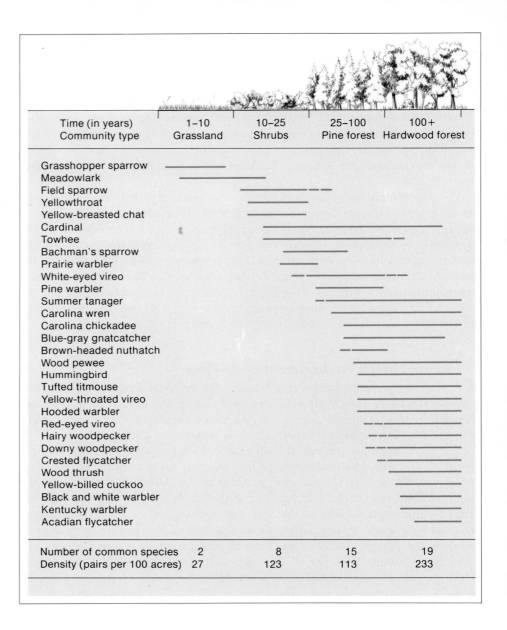

Figure 9-17
Succession: Changes in Bird
Species on an Abandoned
Farm in the Southeastern
United States

Time (in years) Community type	1–10 Grassland	10–25 Shrubs	25–100 Pine forest	100+ Hardwood forest
Number of common species	2	8	15	19
Density (pairs per 100 acres)	27	123	113	233

plants; and so on. The animals suited to each plant community will also appear and disappear during each successive change in the environment.

Species Diversity

A community can only support a limited number of species. Each new species uses some of the resources required by the resident species, which then suffer a reduction in their population sizes. Because the available resources in any area are limited, any additional species that successfully establishes itself in the community after a certain number of species are present will cause a resident population to

Table 9-2		
Species Diversity and Latitude		
The number of bird species increases as latitude decreases.		
Region	Latitude (degrees North)	Breeding birds (number of species)
Greenland	80–60	56
Labrador	60–52	81
Newfoundland	47–52	118
New York	41–45	195
Florida	25–30	143
Guatemala	14–18	469
Panama	6–9	1,100

Source: Dobzhansky, T., "Evolution in the Tropics," *American Scientist,* Vol. 38 (1950), p. 209.

leave or die. When this happens, a community has reached its limit of *species diversity.*

Some communities contain more species than other communities. In Table 9-2, a considerable increase in the number of bird species is revealed as latitude decreases from Greenland to Panama. Why do the tropics support so many more bird species than the more northerly latitudes? Ecologists have offered many suggestions. One hypothesis is that weather conditions in tropical areas can be predicted with greater reliability from year to year, so that each species can become more specialized on a reliable food resource. A great number of specialized species with small populations may coexist in a tropical community because smaller fluctuations in population sizes occur when resources are dependable, thereby reducing the possibility of extinction. A second hypothesis is that tropical areas have provided bird species with more time to evolve and adapt to the environment and to one another. In the past, communities have been totally destroyed in the northern latitudes when glaciers covered the land. Northern bird species are therefore still in the process of colonizing these areas and evolving specializations through interspecific competition. This hypothesis suggests that if there are no new glacial disturbances, the northern latitudes will eventually support as many bird species as the tropics.

Many other hypotheses have been developed to explain the differences in species diversity that occur with changes in latitude. Local environments also support varying numbers of species, so that different numbers of species coexist in communities within each latitude. Desert communities, for example, contain about the same number of species wherever they occur, and the number of species characteristically found in a desert community differs from the number of species found in a pine forest community.

Ecological Equivalents

The environment determines both the number of species that exist in a community and the evolution of species within a community. Similar environments should therefore contain similar communities, and this premise seems to hold

**Figure 9-18
Ecological Equivalents
Among Rain Forest Mammals**
The animals on the left occur in
Africa; the animals on the right occur
in South America. Each pair of
animals is drawn to the same scale.
The pairs are virtually unrelated, but
have similar niches. (a) Pigmy
hippopotamus and capybara;
(b) African chevrotain and paca;
(c) royal antelope and agouti;
(d) yellowback duiker and brocket
deer; (e) terrestrial pangolin and giant
armadillo.

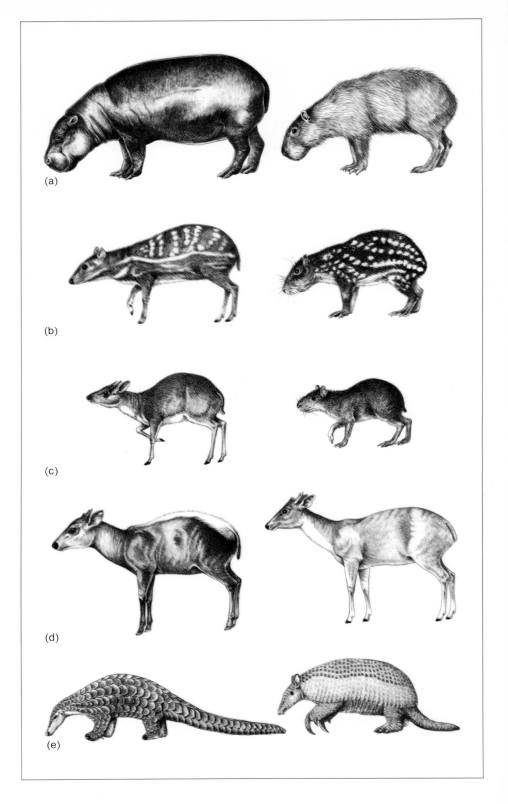

true. Similar environments on different continents contain about the same number of species per community, and resources in these environments are partitioned into similar niches. Even when the species within similar communities are not closely related, they are molded by natural selection to have equivalent niches.

Ecologically similar species in nonadjacent communities (communities on two different continents, for example) are *ecological equivalents.* Not only do they have similar niches, but the species have often evolved similar physical structures in response to the same selection pressures. Ecological equivalents demonstrate that evolution can be predicted to some extent. Comparable environments cause comparable species to evolve. Some examples of ecological equivalents are shown in Figure 9-18.

Speciation

A new species does not arise whenever the space and resources it requires become available in a community. Species are formed under unusual circumstances that are quite unrelated to the development of communities.

Although species are usually organized into local populations, all members of a species are potentially capable of mating and reproducing with one another. The genetic information carried by two individuals of the opposite sex in a species combines to produce offspring that contain genes from both parents. Individuals from different species cannot reproduce successfully, and the genetic information from a member of one species therefore cannot be transferred to a member of another species.

Forming Two Species from One

Two different species can be created when a *geographical barrier*, such as a mountain or a glacier, forms within the distribution of a species and separates the individuals into two populations (Figure 9-19). Because the barrier prevents any

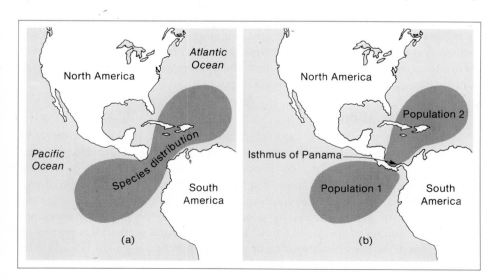

Figure 9-19
The Formation of a New Species by the Development of a Geographical Barrier
(a) Distribution of a marine species before the formation of the Isthmus of Panama. (b) The species was isolated into two separate populations when the Isthmus of Panama formed about 2 million years ago.

Figure 9-20
Separated Populations Evolve Unique Features

Each pair of these mollusks once belonged to a single species. Differences arose after the formation of the Isthmus of Panama. In each group, the Pacific forms appear on the top; the Atlantic forms, on the bottom.

movement between the two populations, no genes are exchanged and each population evolves independently (Figure 9-20). The two populations can become so different that even if the barrier is removed later, the individuals from one population can no longer reproduce successfully with the individuals from the other population. The two populations have become two different species. Recently evolved differences in mating structures, reproductive seasons, or courtship behaviors could prevent reproduction between two such species.

Speciation by Allopolyploidy

New plant species can form through a process called *allopolyploidy* ("allo" = different; "polyploid" = more than one set of chromosomes). When the gametes from two species of plants unite, the offspring (the F_1 generation) are usually sterile, because the chromosomes of the two parent species are unable to pair properly (are not homologous) and therefore do not undergo meiosis. Such a mating is diagrammed in Figure 9-21(a). When all of the chromosomes in the F_1 generation appear in duplicate, however, each chromosome can pair with its identical form and meiosis can be carried out to produce gametes, as diagrammed in Figure 9-21(b). The doubling of all chromosomes in a plant cell often results from a "mistake" in mitosis: the chromosomes divide but the cell does not. If the gametes from individuals in the F_1 generation unite, the offspring in the F_2 generation belong to a new species. These offspring will receive a different number of chromosomes and will be unable to mate with individuals from either parent species. It is estimated that approximately one-half of the 235,000 existing species of flowering plants have originated in this manner.

Species Numbers:
The Balance Between Speciation and Extinction Rates

More than 2 million different species live on earth today. This great diversity of organisms is maintained by a balance between *rates of speciation* and *rates of extinction*. Both of these processes have occurred regularly in the past and continue to occur today. New species are formed arbitrarily, and a species survives only if it has the adaptations required for existence in its environment. If a species is not adapted to its environment or is not successful in competing for resources, it will become extinct. A species may also become extinct after adapting to an environment that then undergoes such rapid changes that the species cannot make evolutionary adjustments quickly enough to maintain its population.

Linear Speciation

A species evolves over time, gradually adapting to a changing environment. At some point during its lengthy evolution, the species becomes so different from its original ancestors that it is classified as a different species. This process of forming a new species, called *linear speciation,* is quite different from the formation of two species from one species by geographical separation or the formation of a new species from two parent species by allopolyploidy. Linear speciation does not increase the number of species that exist at any given time. Rather, it represents a gradual change in the characteristics of organisms within a single lineage. The three types of speciation are diagrammed in Figure 9-22.

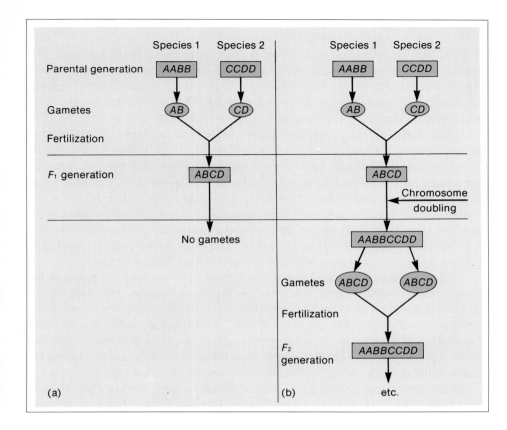

Figure 9-21
Formation of a New Species by Allopolyploidy
Each letter (*A, B, C, D*) represents a type of chromosome. Two identical letters (*AA, BB, CC, DD*) denote a homologous pair of chromosomes. (a) Although two species of diploid plants (each containing two homologous pairs of chromosomes in this example) may mate, their offspring will be sterile, because only a single chromosome of each type will be present. The pairing of homologous chromosomes during meiosis therefore cannot take place, and gametes cannot be formed. (b) When all of the chromosomes in the cells of the F_1 generation double in number, identical chromosomes can pair during meiosis and gametes can be formed. These gametes can unite with each other to form an F_2 generation and continue the hereditary line. Here, the individuals receiving eight chromosomes belong to a new species. They can mate and reproduce with one another, but not with either parent species.

Figure 9-22
Three Ways in Which New Species May Form
(a) New species form when one species splits into two species. The ancestral species does not exist after this division. This type of speciation commonly occurs when populations within a species become *geographically separated* for long periods of time. (b) In plants, a new species may be formed from two parent species through *allopolyploidy*. In this case, all three species—the new species and both parent species—continue to exist. (c) New species form over time by gradual evolutionary changes. At some point, a species may become so changed that it is classified as a different species from its ancestors. In this type of speciation, called *linear speciation*, only one species exists at any given time.

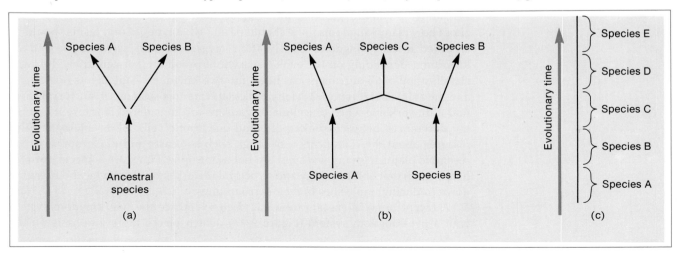

Classification

All organisms are related to one another; at some distant point in the history of life, there was a common ancestral stock. However, some species are more closely related or genetically similar than others due to recent speciation events. Humans are more closely related to monkeys and apes than they are to dogs or cats; birds and lizards are more closely related to each other than either is to insects.

A system of classifying organisms is used to indicate the degree of relationship among species. The exact evolutionary history of an organism is not usually known, so that it is difficult to determine precisely when each type of organism first became a species and from which species it originated. In the absence of historical evidence, the *criterion of similarity* is used to measure the degree of the relationship between two species. Many primarily anatomical characteristics of the organism are examined. The more similar the characteristics are between individuals of different species, the more genes the organisms are considered to have in common and the more closely related they are therefore believed to be.

Similar species are grouped into *genera* (singular, *genus*), and the name of the genus is then combined with the name of the species. Since, for example, the domestic dog and the wolf are anatomically similar, both species are placed in the same genus and given the genus name of *Canis.* The dog is given the species name of *familiaris* and the wolf is given the species name of *lupus,* so that their respective names are *Canis familiaris* and *Canis lupus.* Both the genus and the species names are italicized; the genus name is capitalized, but the species name is not.

Similar genera are grouped into *families.* Dogs, wolves, coyotes, jackals, and foxes belong to the family Canidae. Similar families are further grouped into *orders.* Such animals as dogs, bears, raccoons, weasels, skunks, hyenas, and cats all belong to the order Carnivora. Similar orders are grouped into *classes.* Dogs belong to the class Mammalia, which includes humans, bats, whales, rabbits, elephants, horses, cattle, mice, anteaters, shrews, and many others. Similar classes are grouped into *phyla* (singular, *phylum*). Animals in the class Mammalia are grouped with fishes, snakes, lizards, frogs, and birds into the phylum Chordata. Successively larger classifications include organisms that share a decreasing number of characteristics.

Similar phyla are grouped into *kingdoms.* Biologists have different opinions about how many kingdoms there should be and what types of organisms should be classified in each kingdom. Classification systems contain between two and six kingdoms, depending on the evolutionary interpretation. For centuries, the basic classification of all organisms has been into two kingdoms—the plants (*autotrophs*) and the animals (*heterotrophs*). Many biologists find this system satisfactory; others find it unworkable when unicellular organisms are considered, because it ignores the differences between prokaryotic and eukaryotic cells. Some biologists also disagree about the classification of fungi, such as yeasts and mushrooms, which resemble plants in many ways but are not autotrophs. Fungi are a special form of heterotroph that obtain energy and structural materials by absorbing and breaking down (reducing) molecules in their surroundings.

A recent trend in classification has been to replace the two-kingdom system with a five-kingdom system (Figure 9-23), which places more emphasis on the

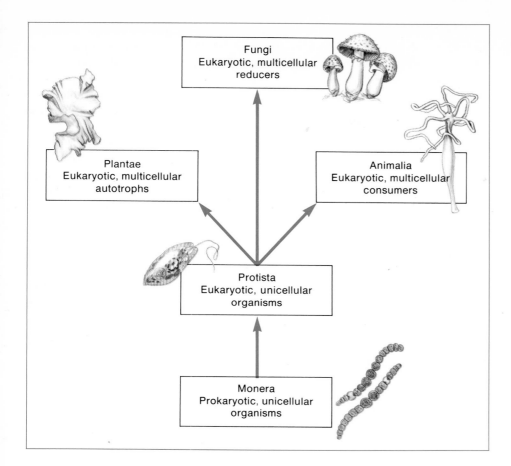

Figure 9-23
Outline of the Five-Kingdom System of Classification
This outline is based on the symbiotic theory of eukaryotic evolution. Beginning with simple, prokaryotic cells, a number of symbiotic events lead to the existence of unicellular eukaryotic organisms. The eukaryotic cells eventually produce the three forms of multicellular life—plants, fungi, and animals.

differences between prokaryotic and eukaryotic cells. The five-kingdom system is also based to a large extent on the symbiotic theory of eukaryotic evolution discussed in Chapter 5. The five kingdoms proposed are the prokaryotic unicellular organisms (Monera), the eukaryotic unicellular organisms (Protista), the eukaryotic multicellular autotrophs (Plantae), the eukaryotic multicellular reducers (Fungi), and the eukaryotic multicellular consumers (Animalia). We will refer to the five-kingdom system throughout this book.

Summary

Ecology is the study of the environmental factors that govern the distribution and abundance of populations of organisms, including physical conditions, available resources, and the existence of other kinds of organisms in the area. An *ecosystem* describes all of these factors; a *community* describes an assemblage of interacting populations.

Every living organism must be able to acquire energy and structural materials from its environment. *Energy* enters an ecosystem primarily in the form of light and is converted into the chemical energy found in biological molecules by photosynthesis. Organisms capable of acquiring energy in this way are called *autotrophs.* These organisms obtain *structural materials* from the air, water, and soil. *Herbivores* acquire their energy and structural materials from eating plants and other autotrophs; *carnivores*, from eating other animals. A *food chain* is a particular sequence of energy transfer from autotroph to herbivore to carnivore. A *food web* describes the transfer of energy when each organism eats more than one kind of food.

Organisms that obtain energy in the same way are said to be at the same *trophic* (feeding) *level*. Trophic levels of a community can be arranged in a sequence of food transfer to form an *ecological pyramid*, with autotrophs on the bottom and carnivores on the top. Most of the energy that passes through the various trophic levels of the pyramid is lost from the system as heat energy. Lesser amounts of organic *biomass* (organisms) can therefore be supported at higher trophic levels.

Biological magnification is the accumulation of toxic materials in various organisms as energy passes to higher and higher trophic levels. Materials such as DDT, mercury, and radioactive elements become more concentrated in the tissues of organisms as they are passed from autotrophs to herbivores to carnivores. The bodies of top carnivores often contain lethal concentrations of these materials due to the indiscriminate disposal of wastes into the environment by humans.

Biogeochemical cycles describe the movements of elements, such as nitrogen, phosphorus, and carbon, through an ecosystem. Unlike energy, these materials remain unchanged and can be recycled through the components of an ecosystem again and again. Any alteration of these cycles due to human interference can have a tremendous effect on the composition of the community.

A *niche* is all of the environmental requirements of an organism or a population. Niches are determined by evolutionary responses to the physical environment and to other organisms in the community.

No organism lives in isolation. Each type of organism is a component of a large community of plants, animals, fungi, and microorganisms. A community develops through the processes of *colonization, interspecific competition*, and *coevolution*. Each community is a dynamic system of interrelated activities among many species. A change in any population in the community, especially the addition or removal of a species, can have far-reaching effects on the other resident species.

Similar environments in different places support about the same number and types of species. Ecologically similar species that occur in separate communities instead of coexisting in the same place are called *geographical replacements.*

Ecological succession is the orderly sequence of communities that develops in an area after an environmental disturbance. Each community in a stage of succession contains species that are adapted to the environment at that stage. The environment is undergoing continuous changes due to the activities of the resident species. As the environment is slowly altered, it becomes more appropriate for a new set of species, which colonizes and replaces the earlier community. A sequence of different communities occurs until the environment becomes similar to its condition prior to the disturbance. This stage is referred to as the *climax community*. During this period, the environment remains fairly stable.

Some communities contain more species than others. One conspicuous pattern is an increase in the number of species as latitude decreases; the tropics contain many more species than temperate or arctic areas. What causes these patterns of *species diversity* is not yet understood.

A *species* is a group of individuals that are capable of successfully reproducing with one another. New species are formed in at least three different ways:

1. When two populations of a species become so isolated from one another, usually by a *geographical barrier*, that movement between them no longer occurs, each population undergoes its own independent evolutionary sequence. In time, the isolated populations may become so different that even if the barrier is removed, they can no longer mate and reproduce successfully.

2. Through *allopolyploidy* in plants, sudden changes in chromosome numbers can prevent offspring from mating with individuals in the parent population.

3. Due to gradual changes in a population, biologists may classify populations as different species at different times. This classification system is called *linear speciation*.

Organisms are grouped into classifications according to their similarities, primarily in terms of their physical characteristics. From aggregates of most similar to aggregates of least similar organisms, these groups are the *species*, the *genus*, the *family*, the *order*, the *class*, the *phylum*, and the *kingdom*.

The History and Diversity of Life

10

Historical studies rarely reveal enough information to reconstruct past events completely. The history of life is a particularly difficult area of study because no written records—no logs or diaries—can be consulted. The only direct evidence of previous life forms is provided by fossils—skeletons or imprints of organisms that are embedded in rock. Relying on scientific techniques to determine the age of fossils and on Darwin's theory of evolution to interpret the relationships between different fossils in the progression of life forms, physicists, chemists, geologists, geographers, and biologists have reconstructed the major sequence of events in the history of life. The time intervals and the important features of each period in the development of life forms are provided in Table 10-1, which also includes the major geological changes in the earth's structure and the climate that occurred during each time interval.

The Long Era of Unicellular Organisms

A long period of chemical evolution from 4.6 to 3.5 billion years ago preceded the appearance of the first recognizable cells. Life began as a unique combination of these chemicals in the sea at some unknown place and time. The first life forms—simple unicellular organisms with only basic reproductive and food-processing structures—appeared in the fossil record about 3.5 billion years ago. Unicellular organisms dominated the oceans for a relatively uneventful period that lasted about 2.5 billion years.

Scientists have not yet completely reconstructed this period of unicellular domination. It is known that land masses were considerably smaller then and that at least five separate continents drifted across the surface of the earth. These continents apparently supported no life forms; their surfaces were bare rock, devoid of green plants or soil. In contrast, the oceans were teeming with tiny organisms.

Table 10-1
Major Features of the History of Life

Approximate starting date (millions of years before the present)	Period	Epoch	Major biological events	Major geological events
0.1		Recent (Holocene)	Human beings and higher animals	
3	Quaternary	Pleistocene	Early humans appear; many large mammals become extinct	Glacial advances and retreats
7		Pliocene	Large carnivores appear; many modern mammals appear	Cascades and Andes uplifted
26		Miocene	Grazers appear	Cooler climates; Himalayas and Alps formed
38	Tertiary	Oligocene	Many modern animal families appear	Warm climates; Pyrenees uplifted
53		Eocene	Modern mammals appear	Climates fluctuate
65		Paleocene	Modern birds appear; hoofed animals diversify	
136	Cretaceous		Angiosperms abundant; giant reptiles become extinct; insects abundant	Rocky and Sierra Nevada mountain ranges begin to form; Pangaea begins to break up
195	Jurassic		Primitive birds appear; reptiles abundant; conifers dominant	Warmer climate
225	Triassic		Primitive dinosaurs; first mammals appear	Warm, semi-arid climate

Characteristics of the First Unicellular Organisms

The first cells were simple organisms, probably similar to present-day bacteria, that fed on molecules from the surrounding waters. Almost 1 billion years after the first cells appeared, the next major type of cell evolved. This cell was able to photosynthesize—to make its own food molecules from carbon dioxide, water, and energy from the sun. Modern cells similar to this early type of cell are the blue-greens.

An important by-product of photosynthesis was the release of oxygen (O_2) into the oceans. The presence of oxygen greatly changed the environment and profoundly affected the subsequent evolution of life. When primitive cells acquired the ability to use oxygen to break down food molecules, they gained about 20 times more usable energy per food molecule.

Table 10-1 *(continued)*

280	Permian	Reptiles displace amphibians; insects abundant; gymnosperms displace ferns	Widespread glaciation
345	Carboniferous	Ferns, primitive reptiles, and primitive gymnosperms	Glaciation in some parts of the world
395	Devonian	First terrestrial insects and amphibians	Appalachian Mountains uplifted; Pangaea formed
440	Silurian	First terrestrial plants and arthropods	Appearance of shallow seas in North America
500	Ordovician	Fishes diversify; marine invertebrates dominant	
570	Cambrian	First vertebrates (primitive fish) appear; unicellular eukaryotes (protists) diversify; marine invertebrates abundant	
1,400	Periods not defined	First unicellular eukaryotes	
2,700		Simple photosynthesizing prokaryotic cells	
3,500		First cells (prokaryotes)	
4,600		Formation of the earth	

The eventual accumulation of oxygen in the atmosphere above the oceans also influenced evolution by providing a protective shield of ozone (O_3) against ultraviolet radiation from the sun. This type of radiation contains so much energy that it is disruptive to the organization of complex molecules and therefore to life itself. After the atmospheric oxygen accumulated, significantly reducing the amount of ultraviolet radiation that reached the earth's surface, life began to exist in the surface waters and on the continents.

Reproduction in Simple Cells

We have already learned that simple prokaryotic cells usually contain a single DNA molecule rather than chromosomes. These early cellular forms reproduced by replication of the DNA molecule followed by division of the cell. Each new cell received a single copy of the genetic code. This mode of reproduction, called *fission*, provides no mechanism for the recombination of genes. Genetic differences between individuals, which play such an important role in evolution, result entirely from mutations. Because mutations are rare, the offspring from these early cells were almost always precise copies of the parent.

Fission was the usual mode of reproduction in early cells, but was probably not the only method. Primitive cells may have occasionally achieved genetic recombination through crossovers (discussed in Chapter 7). Because each cell contained only a single DNA molecule, a second and different DNA molecule would have to be present for a crossover to take place. Something similar to sexual union does occur between different types of modern bacteria. Part of the DNA molecule leaves one bacterium and temporarily resides inside another bacterium, permitting crossovers between the two DNA molecules. The DNA molecules of primitive living systems may also have escaped into their surroundings occasionally and been incorporated into other cells.

The Origin of Viruses

The discovery that bacteria occasionally lose their DNA offers a possible explanation for the origin of viruses. Viruses are not cells and, for this reason, are considered to be a "gray area" of biology. A *virus* is not a complete organism, but it does have many of the basic properties of life (Figure 10-1).

A virus is simply a nucleic acid (DNA or RNA) molecule enclosed in a coat of protein, as shown in Figure 10-2(a). Because it is incapable of protein synthesis, a virus can only reproduce when it is inside a cell. Viruses may have originated as "escaped" nucleic acid molecules that gradually acquired their present form.

A virus can reproduce only by invading a host cell and using its host's molecules and structures. Every virus has a specific means of inserting its nucleic acid molecules into a host cell, as illustrated in Figure 10-2(b). The RNA and ribosomes of the host cell do not distinguish between their own nucleic acids and the RNA or DNA from the virus, as shown in Figure 10-2(c), and make proteins according to the instructions in the viral nucleic acid. In this way, new viruses form, eventually burst out of the host cell, as Figure 10-2(d) illustrates, and invade other cells.

Viruses enter all kinds of cells, and their activities are usually destructive to the host cell. Because viruses multiply and move from one organism to another, they cause many types of infectious diseases. The role that viruses play in causing human diseases will be considered further in Chapter 18.

The First Complex Cells—The Protists

At some point in the early evolution of life, simple cells became more complex. Fossils of the first eukaryotic cells appear in rocks that are about 1.4 billion years old. These fossil cells have a nucleus containing chromosomes and other special structures, such as chloroplasts and mitochondria. Exactly how this organizational complexity occurred is not yet understood. Two hypotheses explaining the

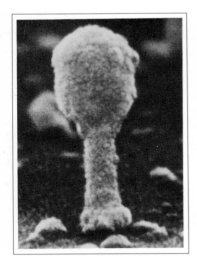

Figure 10-1
Highly Magnified Photograph of a Virus (×280,000)
Only the protein coat is visible; the DNA molecule is located within the coat.

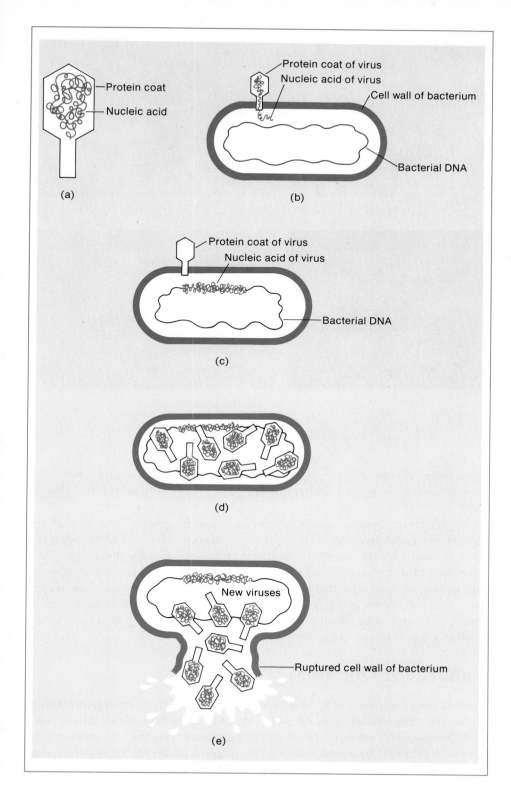

(a) Protein coat

Nucleic acid

(b) Protein coat of virus

Nucleic acid of virus

Cell wall of bacterium

Bacterial DNA

(c) Protein coat of virus

Nucleic acid of virus

Bacterial DNA

(d)

(e) New viruses

Ruptured cell wall of bacterium

Figure 10-2
Reproduction of a Virus

(a) The structure of a virus in its free form includes a nucleic acid molcule and a coat composed of one or more protein molcules. (b) The nucleic acid is injected into a cell (a bacterium here). (c) The nucleic acid of the virus is incorporated in the DNA molcule of the cell. (d) The cell makes more viruses according to the instructions within the viral nucleic acid. (e) The cell eventually ruptures and the newly formed viruses are released.

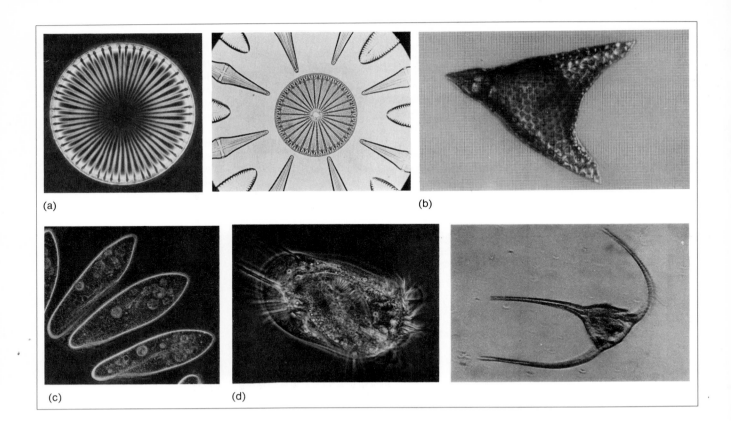

Figure 10-3
Various Types of Protists

(a) Diatom (×340) and enlarged diatom; (b) radiolarian (×382); (c) paramecium (×118); (d) marine ciliates (×273; ×428).

development of eukaryotes are discussed in Chapter 5. These complex, unicellular organisms—the *protists*—are still abundant today in a multitude of forms (Figure 10-3).

Many different types of protists evolved—and in far greater diversity than the earlier prokaryotic organisms. These protists contained several chromosomes and could generate genetic variability both by crossovers and by the independent assortment of chromosomes. But each protist was haploid and contained only one copy of the genetic code, so that these two forms of recombination could occur only when two individuals in a population briefly merged their nuclei. This form of sexual reproduction has been observed in modern protists and is similar to the union of a sperm and an egg in more advanced life forms.

Multicellular Organisms

The relative infrequency of genetic recombination in unicellular organisms may explain the comparatively slow rates of evolution during the first few billion years of the history of life on earth. Multicellular organisms were the first organisms to reproduce primarily by sexual means and to incorporate genetic recombination as a regular part of the life cycle. After the appearance of multicellular organisms, there was an explosive increase in the evolution of new forms of life.

The first multicellular organisms appeared less than 1 billion years ago. All multicellular organisms are composed of eukaryotic cells, which remain part of a single individual after cell division instead of separating as they do in unicellular organisms. A multicellular organism is therefore able to grow to a much larger size than can a unicellular organism, and this increase in the size of individual organisms has been one of the most pronounced trends in evolution. Because so many lines of evolution exhibit this trend, it can be deduced that there are advantages to large size.

Two major forms of multicellular organisms—animals and plants—evolved in the oceans. The trend toward an increase in animal size began shortly after the first multicellular animals appeared. As an individual increases in size, it can attack greater numbers of smaller animals; larger individuals therefore have a greater selection of food items. Large animals can also develop greater control over their movement, which is a requirement of a predatory way of life. To a large extent, unicellular organisms must travel where the currents carry them; larger animals are more capable of directing their movements toward food and away from danger.

A different adaptive mechanism produced an evolutionary increase in the size of plants. Plants compete for sunlight, and when they are present together, the taller plants acquire more of the sun's rays than the shorter ones. This selective force did not begin to operate on plants until they invaded the shallow coastal waters and eventually occupied land.

Early Multicellular Plants

The first photosynthetic cells to become multicellular organisms inhabited shallow waters on the edges of the oceans and were quite similar to modern seaweed. To permit aquatic, photosynthetic cells to take in mineral nutrients and exchange gases (oxygen and carbon dioxide) with the environment, water must flow over their surfaces. Early coastal plants could maximize the flow of water by remaining fairly motionless in the turbulent waters. Because unicellular organisms cannot control their movements, the first plants to occupy coastal waters successfully were multicellular. One of the cells attached itself to the bottom, and the remaining cells were connected to one another in a long strand. With the exception of the cell that formed the attachment, there was little specialization within the different cells of the plant. All of the cells continued to perform photosynthesis, convert energy, and construct molecules.

The exact type of life cycle undergone by the early multicellular plants is unknown, but it may have resembled the life cycle of today's marine coastal plants, in which plant forms alternate between generations. In one generation, all of the cells of a plant are haploid (each containing a single set of chromosomes) and the plant produces gametes by mitosis. Each gamete then unites with another gamete (fertilization) to form a zygote (containing two sets of chromosomes), which develops into a diploid plant after a number of mitotic cell divisions. Certain cells of the diploid plant undergo meiosis at maturity to produce haploid gametes. Each haploid gamete then leaves the parent diploid plant, multiplies by mitosis, and forms a multicellular haploid plant. This type of life history is diagrammed in Figure 10-4.

The early multicellular plants living in coastal waters were periodically exposed to dry conditions. Because they were attached to the bottom, these plants were

Figure 10-4
A Plant Life Cycle of
Alternating Generations
This type of life cycle probably
occurred in primitive multicellular
plants. Individuals alternate between
haploid and diploid types.

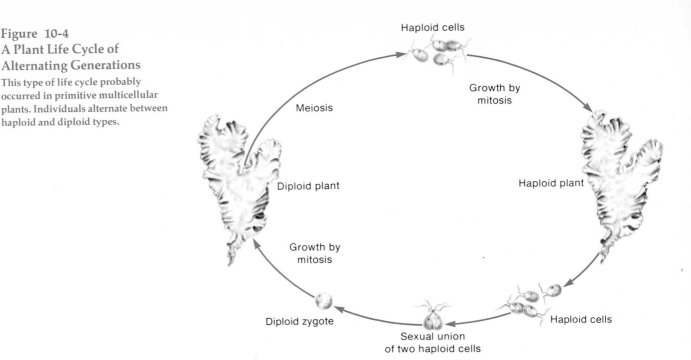

Haploid cells

Growth by
mitosis

Meiosis

Diploid plant

Haploid plant

Growth by
mitosis

Diploid zygote

Sexual union
of two haploid cells

Haploid cells

unable to move away from their environment when it became less hospitable. The progeny of the surviving plants eventually acquired adaptations to drier conditions and began to inhabit the barren land.

Early Multicelluar Animals

Very few fossils of early multicellular animals have been found, probably because they had soft bodies and lived within the muddy layers of the ocean floor. Few skeletons or imprints in rocks remain to provide evidence of their existence.

Sponges (Phylum Porifera)   The simplest multicellular animals alive today are the *sponges (phylum Porifera)*. The bodies of sponges resemble colonies of unicellular organisms that perform loosely coordinated cellular activities. Most sponges live in marine waters, and all adult sponges are *sessile* (fixed in place by attachment to the *substrate*—the ocean floor or some other solid object). A sponge obtains food by *filter-feeding*, or forcing a current of water through its body and filtering out small organisms and other food particles. Water enters the body through many small openings and leaves through one or more larger channels, as shown in Figure 10-5(a). The current of water is moved by tiny *flagella*, shown in Figure 10-5(b), which are whiplike projections on the cells lining the inner cavity of the body. The presence of flagella on some of the body cells suggests that sponges may have originated from colonies of flagellated protists—a group of unicellular organisms with flagella.

Most sponges regulate the amount of water flowing through their bodies by expanding and contracting their openings. Sponges also contract their openings when stimulated by contact or the presence of noxious materials. The ability of a sponge to alter the size of its openings in response to certain stimuli provides evidence of communication and integration among the cells and indicates that their organization is more complex than a colony of simple unicellular organisms.

Almost all sponges have internal skeletons. In some species, the skeleton is constructed from a substance called *collagen*—a protein material similar to the material found in human tendons and ligaments. Your household sponge is composed of this skeletal material.

Sexual reproduction in sponges is accomplished by the union of an egg and a sperm, usually from different individuals. The zygote develops into a *larva*, which is little more than a mass of cells. After floating free in the water for a while, the larva attaches itself to the ocean floor or some other solid object and develops into an adult sponge.

Jellyfish, Corals, and Sea Anemones (Phylum Coelenterata)   A somewhat more complex group of marine organisms than the sponges is the *phylum Coelenterata*, which includes the jellyfish, corals, and sea anemones. All coelenterates have a saclike digestive cavity with a single opening, which serves as both a mouth and an anus. This type of digestive tract is considered primitive and is certainly inefficient. Unless the animal takes in food when it is not in the process of digesting food, the undigested food will be mixed with the waste material, because both materials are transported through a single channel. Tentacles surrounding the opening of the digestive cavity are used to paralyze and hold prey, which range from unicellular organisms to fish. The coelenterates are composed of much more highly specialized and integrated assemblages of cells than any of the sponges.

The coelenterate body is *radially symmetrical* (circular, with no front or rear). This type of body is usually unable to direct its movement. The two basic types of coelenterate body—the *polyp* and the *medusa*—are related to two different ways of life. The polyp is sessile; the medusa is free-swimming. Individuals of a species may assume one of these forms exclusively, or both forms may occur at different times in the life of an individual. Sea anemones always occur in the polyp form, whereas jellyfish always occur in the medusa form (Figure 10-6). When both types of body structures are present at different times within individuals of a species, the medusa form produces the sperm and eggs. A fertilized egg develops into a larva, which settles to the bottom and becomes a polyp. Fragments of the polyp eventually break off (asexual reproduction), and each fragment develops into a medusa. It is not known whether the larva, the polyp, or the medusa appeared first in the course of evolution.

Flatworms (Phylum Platyhelminthes)   The *flatworms (phylum Platyhelminthes)* form a third group of relatively simple marine animals. Compared to the more recently evolved parasitic flatworms, such as tapeworms and flukes that have no digestive cavities, most free-living flatworms resemble the coelenterates in that they have a digestive cavity with a single opening. In the flatworms, this cavity is highly branched and spreads throughout most of the interior of the body.

The flatworms are more similar to advanced animals than either the coelenterates or the sponges. Flatworms are *bilaterally symmetrical* and, as shown in Figure 10-7(a), have a distinct head with an aggregation of nerve cells and primitive

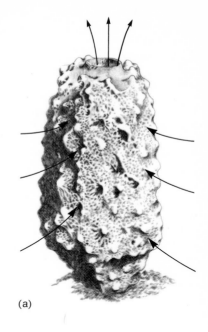

(a)

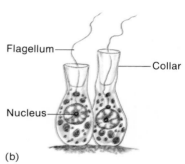

(b)

Figure 10-5
The Simplest Multicellular Animal—A Sponge

(a) The structure of a sponge. In this species, water enters through many small openings and leaves through a single large opening. (b) *Flagellated cells,* such as the one shown here, line the internal surface of the body and create a current of water with whiplike motions of the *flagella.* The collar collects small food particles from the moving water.

Figure 10-6
The Two Forms of Coelenterates

(a) The sea anemone—the sessile, polyp form. (b) The jellyfish—the free swimming, medusa form. The actual structures of these two coelenterates are quite similar: one is the inverted form of the other. In both forms, food enters and waste materials leave the body by the same pathway.

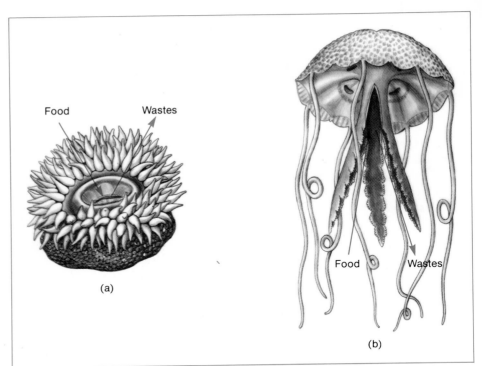

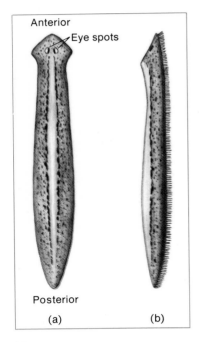

Figure 10-7
A Flatworm

(a) These animals are bilaterally symmetrical: they have a front (*anterior*) end and a rear (*posterior*) end, and the left side of the body is a mirror image of the right side. (b) Side view, showing sheets of cilia that permit the animal to move through water.

receptors for vision and smell. These features are considered to be adaptations for forward movement; the head advances first in response to stimuli detected by its sensory receptors. Free-living flatworms also have specialized cells for the excretion of waste materials and for reproduction. Locomotion is accomplished by broad sheets of *cilia*, shown in Figure 10-7(b)—projections from the cells that create a water current in a manner similar to the flagella. Many types of flatworms have lost these more complex adaptations due to their parasitic modes of life.

The sponges, coelenterates, and flatworms are believed to be similar to the first types of multicellular animals, because they all lack (1) a *body cavity*, or *coelom*, for the suspension of internal organs; (2) a *one-directional digestive cavity* with two openings for the specialization of digestive activities; and (3) a *circulatory system* for the transportation of materials from one part of the body to another. These characteristics of animals probably evolved later, after the sponges, coelenterates, and flatworms branched off from the main line of multicellular animals that led to the advanced forms.

Evolution of the Mollusca, Annelida, and Arthropoda

The Phylum Mollusca

There are two major groups of advanced animals: (1) the Mollusca, Annelida, and Arthropoda, and (2) the Echinodermata and Chordata (Figure 10-8). Both of these animal groups left abundant fossils, so we know considerably more about their history than we do about the earlier forms of multicellular animals.

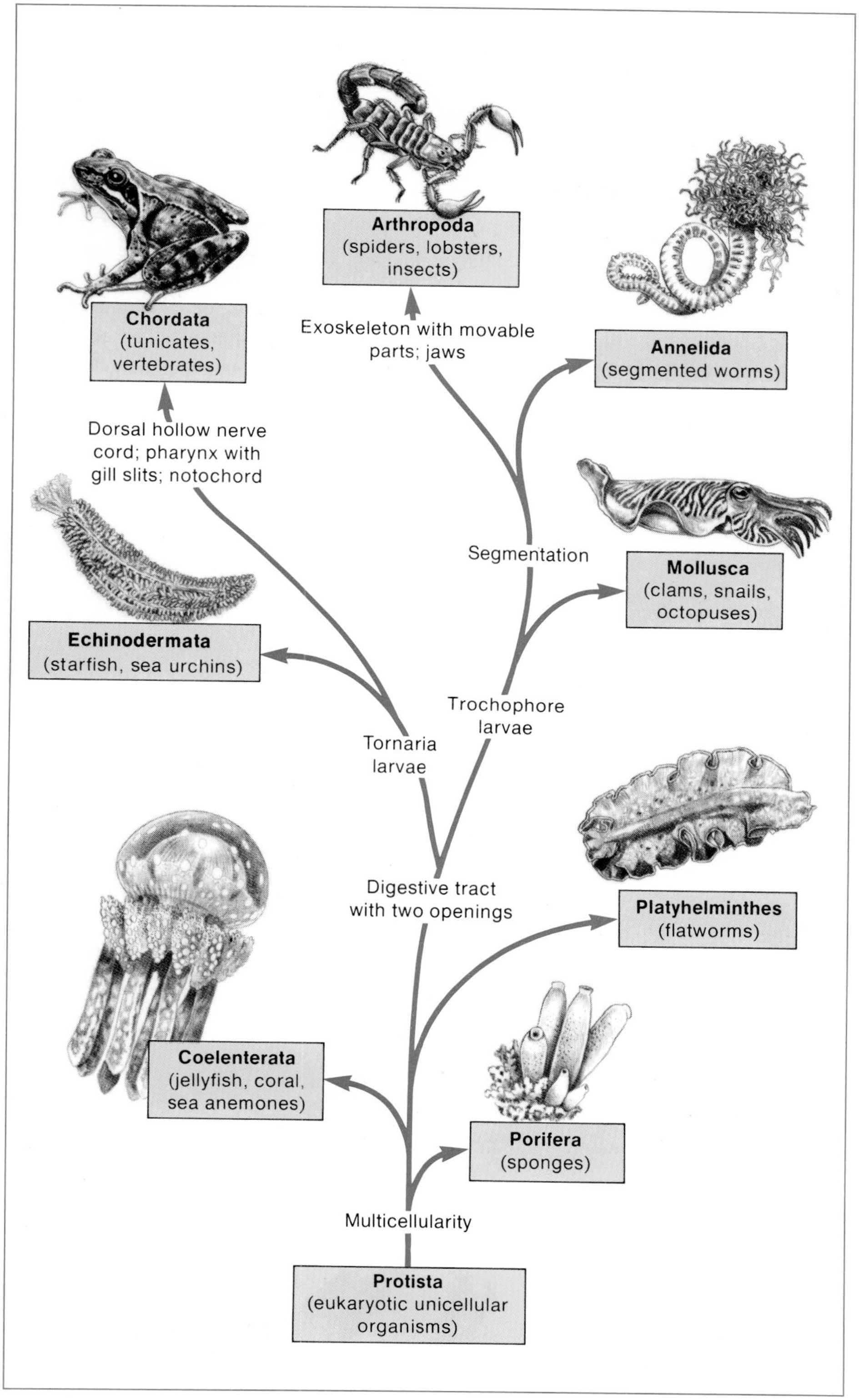

Figure 10-8
A Possible Evolutionary
Sequence of Multicellular
Animals

Arthropoda
(spiders, lobsters,
insects)

Exoskeleton with movable
parts; jaws

Chordata
(tunicates,
vertebrates)

Annelida
(segmented worms)

Dorsal hollow nerve
cord; pharynx with
gill slits; notochord

Segmentation

Mollusca
(clams, snails,
octopuses)

Echinodermata
(starfish, sea urchins)

Trochophore
larvae

Tornaria
larvae

Digestive tract
with two openings

Platyhelminthes
(flatworms)

Coelenterata
(jellyfish, coral,
sea anemones)

Porifera
(sponges)

Multicellularity

Protista
(eukaryotic unicellular
organisms)

The phylum Mollusca includes such animals as snails, clams, mussels, oysters, squid, and octopuses. Most *mollusks* are free-living (neither sessile nor parasitic) marine animals with soft bodies that are protected by a heavy shell formed from mineral deposits (calcium carbonate and calcium phosphate). Excellent mollusk fossils are abundant in Cambrian rocks, indicating that they were prevalent and in an advanced stage of evolution about 600 million years ago.

All mollusks have a one-directional digestive tract with specialized regions that form the mouth, stomach, intestine, and anus. Mollusks also have a circulatory system comprised of a heart and connecting blood vessels, as well as special organs for reproduction (ovaries and testes) and excretion. These organs are suspended within a body cavity, or coelom. A distinctive characteristic of the mollusk is a muscular region behind the mouth, called the *foot*, which permits the animal to move by crawling. Some mollusks, such as the octopus, have an elaborate head with a brain, eyes, and tentacles. Practically all of the basic systems found in the most advanced animals are present in the mollusk.

The Phylum Annelida

Like the Platyhelminthes, the phylum Annelida is a group of worms. Except for their general shape, however, these two types of worms are quite distinct. The annelid body is composed of many segments; each of these segments is a unit containing a set of nerve centers, muscles, blood vessels, coelom, and excretory organs. A familiar example of an annelid is the earthworm (Figure 10-9). The main nerve cords, digestive tract, and a few blood vessels extend throughout the entire

**Figure 10-9
The Structure
of an Earthworm**

The body is composed of many segments, each containing its own complete set of internal structures. Here, internal organs are shown at two different locations in the body. At the anterior end, the brain, nerve centers, and hearts are identified; in the middle, the excretory organs are indicated.

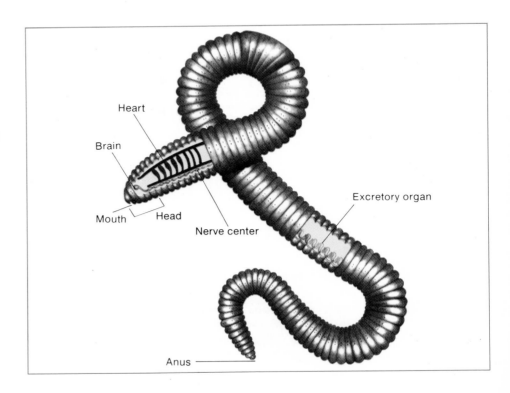

length of the body and coordinate all of the segments. The three most anterior segments of the body form the head, which has a mouth and sense organs; the last segment forms the anus. Appendages or "legs" attached to each segment of the marine annelids are used in locomotion. Segmentation occurs in almost every group of animals considered more advanced than the annelids.

Most of the annelids are marine worms, which are of particular interest in reconstructing the early evolution of animals in the oceans. Like the mollusks, these annelids exhibit several important evolutionary features: a circulatory system with at least one heart, bilateral symmetry, a coelom, and a digestive tract with two openings. The most important characteristic linking the marine Annelida and the Mollusca is that both groups undergo a similar multicellular larval stage, called a *trochophore* (Figure 10-10), which has a mouth and an anus. A band of cilia surrounding the midregion of the larva resembles a wheel ("trochophore" = wheel-bearing); the larva swims through the water by moving its cilia. The trochophore larva is a feeding stage of development that is not undergone by all annelids and mollusks. However, the fact that some animals in both phyla have a trochophore larva suggests that the marine Annelida and the Mollusca had a common ancestry.

The Phylum Arthropoda

The phylum Arthropoda includes the crustaceans (barnacles, lobsters, crabs), the arachnids (spiders, mites, ticks), and the insects. The ancestors of this large animal group were probably similar to marine worms. Arthropods do not undergo a trochophore larval stage, but the body segmentation of the arthropods definitely links them to the annelids. The entire group of mollusks, annelids, and arthropods is therefore indirectly linked: their larval form links the mollusks to the annelids; body segmentation links the annelids to the arthropods.

Some of the characteristics that distinguish arthropods from annelids and mollusks are:

1. A hard coating of polysaccharides on the outside of the body (mollusk coatings are composed only of mineral salts), which forms an *external skeleton*, or *exoskeleton*, with movable joints.
2. *Legs* divided into movable segments.
3. Distinct *groups of muscles* derived from many body segments that are connected to and move parts of the exoskeleton.
4. Fewer *body segments*.
5. Distinct *jaws*.

The first abundant arthropod fossils were the *trilobites* (Figure 10-11), which first appeared in the Cambrian period. These marine animals had fewer body segments than the annelids and some of the segments were fused together. Many of the anterior segments found in the annelid were condensed into an elaborate head in which parts of the nervous system were brought together to form a larger brain in the trilobite. The legs originally associated with the anterior segments of the body were converted into mouth parts. The trilobite body was shorter and thicker than the annelid body.

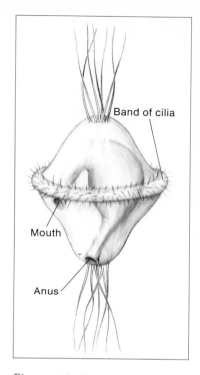

Figure 10-10
The Trochophore Larva
Some species of mollusks and annelids undergo this type of larval stage. This similarity suggests a common ancestry.

Early marine arthropods had the heavy outer skeleton seen in some modern arthropods, such as the lobster, in which mineral salts are impregnated into the underlying polysaccharide structure. Due to its thick body and its outer skeleton, the arthropod could not exchange oxygen and carbon dioxide between its cells and the surrounding water the way the annelid could. The arthropod therefore developed specialized organs which exposed large areas of soft tissue to the surrounding water. Such surfaces evolved in two different ways in the arthropods—as structures on the outer surface of the body (*gills*) and as structures on the inside of the body in the form of *lungs* or *trachea* (a network of rigid channels).

Because surfaces for gas exchange inside the body were largely protected from drying out, arthropods with lungs or trachea eventually evolved to the point that they could live on land. Arthropods with gills were restricted to living in the water, because gills dry out in air and lose their ability to exchange gases.

The Echinodermata and Chordata

The second group of more advanced animals is represented by two phyla with a larval form that differs distinctly from the trochophore larval form of the mollusks and annelids. The *tornaria larva* has long, winding bands of cilia rather than a single band around the midregion (Figure 10-12), but its digestive tract is similar to the tract in the trochophore larva. The tornaria larva is found predominantly in the phylum Echinodermata, which includes the starfish, brittle stars, and sea urchins.

Other differences between the two major groups of advanced animals occur in the early stages of development when the zygote first undergoes cell division to form a multicellular animal. After the zygote divides into two cells in the Echinodermata and the Chordata, two complete individuals develop if these two cells are separated. When the first two cells are separated in the mollusks, annelids, and arthropods, neither cell develops normally. The pattern formed during the first few cell divisions of the zygote also differs between the two groups of advanced animals (Figure 10-13). The early cells of the echinoderms and the chordates develop in a *radial pattern* in which the daughter cells lie directly above or adjacent to one another. In the mollusks, annelids, and arthropods, a *spiral pattern* develops in which the daughter cells lie at an oblique angle to one another, so that the cells are arranged in a spiraling pattern.

The Echinodermata

The echinoderms are relatively complex animals with a one-directional digestive tract, a coelom, and specialized organs for excretion and reproduction. Echinoderms also have nervous and circulatory systems, although neither system is as developed as it is in the other advanced animals.

Most adult echinoderms are radially symmetrical and protected by a hard, often rigid skeleton. Most older fossil forms were sessile as adults. The larval stage, however, is always mobile and bilaterally symmetrical.

A distinctive feature of echinoderms is a *water-vascular system*—a network of canals and appendages that permits movement and food capture (Figure 10-14). The basic unit of this system is the *tube foot*, a small structure that adheres to

Figure 10-11
A Fossil Trilobite from the Cambrian Period
The trilobites had fewer body segments than the annelids; many body segments were fused together.

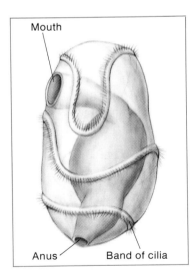

Mouth

Anus   Band of cilia

Figure 10-12
The Tornaria Larva
The tornaria larva that occurs in echinoderms and primitive chordates differs distinctly from the trochophore larva that occurs in mollusks and annelids.

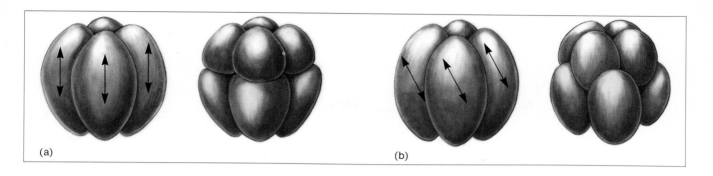

(a)   (b)

surfaces by suction and the secretion of a sticky substance. Hundreds of tube feet acting together in a starfish, for example, can exert an enormous force. They are used for walking and to open clam shells.

The Chordata

Most of the chordates are *vertebrates*—fishes, amphibians, reptiles, birds, and mammals. In addition to the Vertebrata, the phylum Chordata contains the Urochordata and the Cephalochordata. At some time in their lives, all chordates exhibit three characteristics:

1. A flexible *internal skeleton,* or *notochord*—a support structure in the form of a rod that extends through the length of the body. (In humans and other vertebrates, the notochord occurs only in the embryonic stage of life.)

2. A *dorsal hollow nerve cord,* or *spinal cord,* that lies above (dorsal to) the notochord.

3. A *pharynx* with gill slits—the first part of the *digestive tract,* or *throat.* (Gill slits are openings in the pharynx to the outside of the body. In most of the higher vertebrates, gill slits appear briefly and only develop partially in very young embryos.)

Figure 10-13
Two Basic Patterns of Cell Division in the Zygote

(a) The *radial pattern* occurs in the echinoderms and chordates. As each cell divides, one of the resulting cells lies directly above or to one side of the other cell. Arrows indicate the direction of the chromosomes during mitosis and therefore the two poles of the cell. Only the divisions from four to eight cells are shown. (b) The *spiral pattern* occurs in the mollusks, annelids, and arthropods. Here, the direction of chromosome movement during mitosis is oblique, so that each of the four new cells formed lies between two of the previously existing cells.

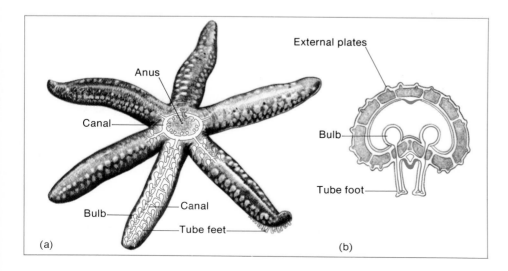

(a)   (b)

Figure 10-14
The Water-Vascular System of an Echinoderm—The Starfish

(a) The system of water canals, bulbs, and tube feet is shown in one of the rays. (b) Cross section of a ray, showing bulb and tube feet. When the bulb contracts, water is forced into the foot, causing it to become long and rigid. When the foot touches a surface, suction develops and the foot adheres. The bulb later expands as water is moved back into it, and the foot becomes short and limp. Hundreds of tube feet on a starfish act in synchronization to cause locomotion.

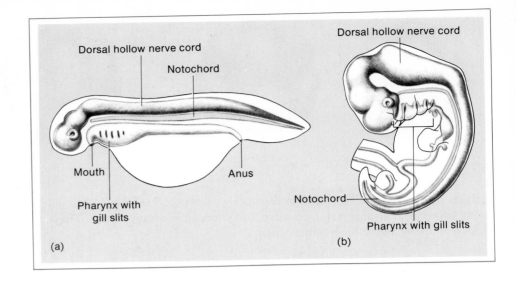

Figure 10-15
The Three Characteristics of All Chordates—A Notochord, a Dorsal Hollow Nerve Cord, and a Pharynx with Gill Slits

(a) Young form of primitive fish.
(b) Human embryo (about 4 weeks old).

The three characteristics of all chordates are diagrammed in Figure 10-15.

The *urochordates* are strictly marine animals. Although they resemble sponges as adults, their internal body organization is much more complex. Urochordates are sessile as adults but are mobile during the larval stage when the animal is similar in body structure to a small fish (Figure 10-16). Urochordate larvae are elongated and bilaterally symmetrical, with a notochord for support.

The *cephalochordates* (Figure 10-17) are small, fishlike marine animals that burrow in the sand and filter-feed by removing particles from the water that passes into the mouth and out through gill slits in the pharynx. Cephalochordates are capable of swimming, but spend most of their time buried in the sand, tail

Figure 10-16
Structure of a Urochordate

(a) The adult is sessile but has a fairly complex internal structure. (b) The larval form exhibits the chordate characteristics. It is free-swimming and bilaterally symmetrical.

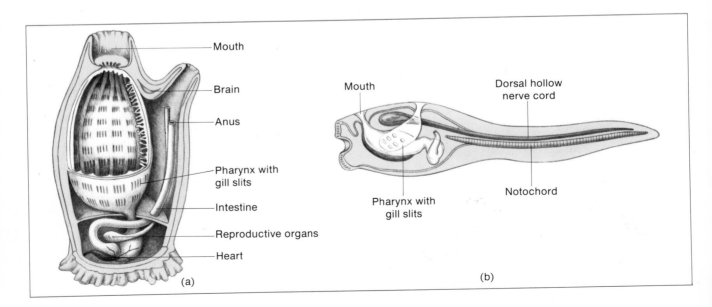

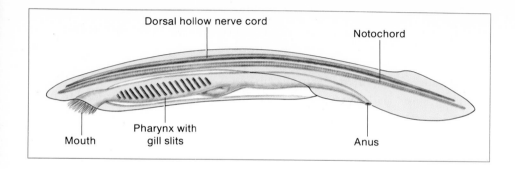

Dorsal hollow nerve cord

Notochord

Mouth

Pharynx with
gill slits

Anus

Figure 10-17
Structure of a
Cephalochordate
This fishlike animal has the simplest
body structure of the chordates.

downward, feeding in a sedentary manner. The evolution of the pharynx with its gill slits is believed to have been a selective response to this type of lifestyle.

In the *vertebrates,* which form the third group in the phylum Chordata, a series of segmented bones, the *vertebrae,* surrounds the nerve cord and replaces the notochord in adults. The vertebrates form a large and diverse group of animals, which we will examine more closely later in the chapter.

The chordates are placed in the same group of advanced animals as the echinoderms because their larval forms and their patterns of early development from the zygote are similar. The two phyla appear to have evolved from some common ancestral group of animals, now extinct and with unknown adult characteristics, which underwent developmental stages similar to those of the echinoderms.

The Origin of the Vertebrates

The most widely accepted interpretation of the origin of vertebrate animals is that the first fish developed from the larval form of a urochordate (see Figure 10-16.) In many ways, these larvae are similar in structure to primitive fishes. They are bilaterally symmetrical, elongated, and have a notochord that extends through the body above the digestive tract and serves to stabilize the body during swimming. A similar larval form may have gradually evolved the ability to reproduce and eliminated the adult stage from its life cycle. Thus, an immature form of invertebrate, having bilateral symmetry, a notochord, a dorsal hollow nerve cord, and a pharynx with gill slits, may have become the first fish. This explanation is highly plausible, because reproduction in the immature stage has occasionally been observed in other animals. The phenomenon is called *neoteny.*

The oldest vertebrate fossils are of fishes from the Cambrian period. The earliest vertebrates were covered with heavy armor plates and had no jaws. They must have been strong swimmers, because there is evidence of powerful muscles along the length of the body. Although many animals left fossils during the Cambrian period, it is difficult to find an obvious evolutionary link between the invertebrates and the vertebrates.

Many different forms of fishes gradually evolved. Most of the differences in body structure seem to have been related to the types of food the fishes ate. Like many fishes today, the early fishes fed on other multicellular animals—a diet that required the internalization of the process of holding and absorbing food. A predator must also be able to move in an oriented direction toward a specific prey. The major specializations seen in the more advanced forms of modern fishes are:

1. High rates of *energy utilization* in all cells.
2. *Sensory organs, jaws,* and a *neuromuscular system* to perceive and respond to prey.
3. A *digestive tract* to process large food items.
4. A *heart* and *blood vessels* to transport food, gases (oxygen and carbon dioxide), and wastes.
5. *Kidneys* to filter waste materials from the blood.
6. A surface *(lungs* or *gills)* to permit gas exchange.
7. Highly developed *nervous* and *hormonal systems* to integrate and regulate all of these specialized cells.

The evolutionary sequence from primitive to modern forms of fishes exhibits a gradual increase in the complexity and efficiency of all these systems. These specialized features and their evolution will be examined in detail in Part 3.

Invasion of the Land

The First Terrestrial Plants

The first organisms to come ashore were the descendants of the primitive multicellular plants that lived in coastal waters. As the coastlines rose and fell, these plants were periodically exposed to dry conditions. Generations of plants that survived these dry periods gradually accumulated features that allowed them to live entirely out of water.

The oldest fossils of land plants are found in rocks of the Silurian period (440 million years ago). Life on land was hazardous for these early plants. Their cells dried out, and the only abundant water was in the ground. As more and more plants successfully made the transition from water to land, some plants began to shade each other from the sun's rays.

A new type of plant with terrestrial adaptations eventually appeared. This plant had an outer covering that reduced evaporation and a network of tubes (the *xylem*) that brought water and minerals from the ground to the cells. The xylem also provided a rigid support that permitted the plant to grow tall enough to prevent it from being shaded by other plants. A second network of tubes (the *phloem*) transported food molecules and other materials among the different cells of the plant. Except for the xylem and the phloem, the cells of the early land plants remained unspecialized. These plants had no leaves or true roots. The entire plant was simply a stem (Figure 10-18), and most of the cells retained their primitive task of photosynthesis and energy conversion.

Early land plants reproduced in basically the same way their oceanic ancestors had reproduced—by alternating generations between diploid and haploid individuals (Figure 10-4). The diploid plant, called the *sporophyte,* was generally larger and lived for a longer time. Specialized cells on the stem of the sporophyte periodically underwent meiosis. The resulting haploid cells (called *spores*) were released

Figure 10-18
Model of an Early Land Plant
Early plants had no leaves or true roots and were little more than stems.

and dropped to the ground, where each spore divided several times and grew into a small plant, called the *gametophyte,* in which all of the cells were haploid. These small plants lived for a short time in the moist soil, eventually producing both eggs and sperm. The sperm swam through water in the soil to an egg, fertilization occurred, and the resulting diploid cell underwent many mitotic divisions to produce another sporophyte. This alternation of a sporophyte and a gametophyte stage remains the basic life cycle of plants today. A critical limitation to reproduction in these early plants was that the sperm could only reach an egg by swimming, just as it had in its ancestral waters, so that early plants could only live in very wet soils.

In the next geological period, the Devonian period, plants evolved more specialized structures, developing both roots and leaves. *Roots* are made of special cells that actively take in minerals and water from the soil; *leaves* perform photosynthesis. Devonian plants grew to the size of modern trees but were restricted to moist environments because they still required water to reproduce.

The First Terrestrial Animals—Arthropods

During the Silurian period, *arthropods* became the first animals to invade land. These early terrestrial forms were the ancestors of modern scorpions, spiders, ticks, mites, and daddy longlegs. All of these ancestral animals had four pairs of legs and were equipped with external skeletons and internal surfaces for gas exchange, so that they were partially adapted to terrestrial life. Arthropods were the first group of organisms to evolve reproductive methods that did not require water. A copulatory organ, or *penis,* in the male was inserted into the female where the eggs were retained; sperm and fluid were released, and the sperm then swam to meet the egg. After fertilization, the egg was surrounded with a waterproof material that protected the embryo from drying out. The difficulty of fertilization on dry land was overcome in this remarkable way.

More animal groups invaded land during the Devonian period. *Insects* evolved from the early oceanic arthropods and soon became the dominant class of arthropods on land. Insects have three pairs of legs and a thick, external skeleton that supports the body and prevents dehydration when the surrounding air is dry. Gas exchange is accomplished by a system of *tracheal tubes* that penetrate to every portion of the body. Insect mouth parts evolved into many specialized types for sucking, biting, chewing, and rasping, and insects were the first animals to develop wings for flight. No arthropods have adapted to living on dry land as successfully as the insects have. They remain an abundant and diverse group, comprising more than half of all the different types of living organisms found on earth today.

The Evolution of Amphibians

Another group of animals—the fishes—began to invade land successfully during the Devonian period. Many early fishes had lungs, but in most groups that inhabited deep waters, these inflatable structures were converted into swim bladders and gas exchange was restricted to the gills. The *swim bladder,* when filled with the appropriate amount of gas, allows a fish to maintain a particular depth in the water.

Figure 10-19
Skeletal Drawings of (a) a
Devonian Lungfish and
(b) an Early Amphibian

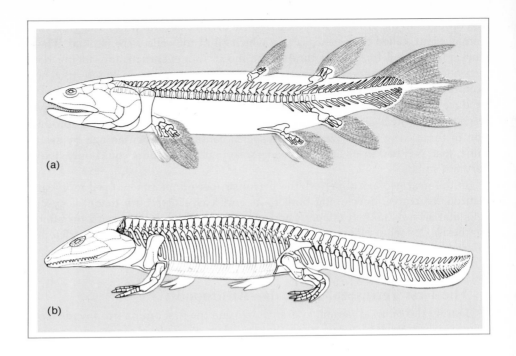

(a)

(b)

A few types of fishes lived in shallow ponds and retained their lungs. When oxygen concentrations in the pond water became very low, the fishes would occasionally rise above the water surface to fill their lungs with air. Although these fishes were primarily aquatic, they periodically walked from pond to pond on slightly modified fins. Some Devonian "lungfish" evolved true legs and became *amphibians* (Figure 10-19). The earliest amphibians had an elongated body with four legs and a tail and were similar to modern salamanders. The class Amphibia now includes frogs, toads, salamanders, newts, and a peculiar form of burrowing, wormlike vertebrate with scales embedded in the skin—the Apoda.

Early amphibians thrived on land and many new forms evolved during the Devonian period, which was exceptionally warm and humid. Amphibians fed on insects, which were already in plentiful supply. Their transition from water to land was facilitated by the presence of large plants and the moist areas provided by their shade.

These early amphibians, like their descendants today, depended on water in many ways. The amphibians retained the need for water to reproduce. Both eggs and sperm were released into the water, the sperm swam to the egg, and fertilization occurred. During the early stages of development from zygote to adult, the young amphibian forms lived in the water, just as frog tadpoles do today. The amphibians also required water to maintain a moist skin, which functioned as a second surface for gas exchange in addition to their simple lungs. Modern amphibians are similar to these early forms and are also restricted to life near water.

Fossils of early amphibians have been found in such unlikely and distantly separated places as Greenland and Antarctica. The interpretation of this distribution is that the continents were once connected and the environments on both land masses were warm and humid.

THE HISTORY AND DIVERSITY OF LIFE · 10

Further Evolution of Land Forms

The Ferns

During the Carboniferous period, which followed the Devonian period, a new form of plant—the *ferns*—emerged. These ferns grew to great heights and had much larger leaves than their Devonian predecessors. Most of the ferns we see today are quite short with underground stems. But ferns similar to those of the Carboniferous period still grow in the tropics. These ferns are as tall as trees and their stems are above ground.

Reproduction in ferns has remained essentially unchanged. Water is still required for the sperm to swim to the egg to accomplish fertilization. The fern you are familiar with is a sporophyte, which contains only diploid cells. Special cells in this plant undergo meiosis to produce haploid cells (spores), which can often be seen adhering in clusters to the undersides of leaves. Eventually, these spores drop to the ground, and each spore develops into a tiny haploid plant only a few millimeters long. These gametophytes produce eggs and sperm, which unite to form a zygote. After multiplication by mitosis, the zygote develops into a sporophyte (Figure 10-20).

Ferns were extremely abundant in moist areas during the Carboniferous period. When they died, many fell into the swampy waters and were not completely broken down by decomposers. After millions of years of warm temperatures and high pressures, these partially decomposed plants eventually became

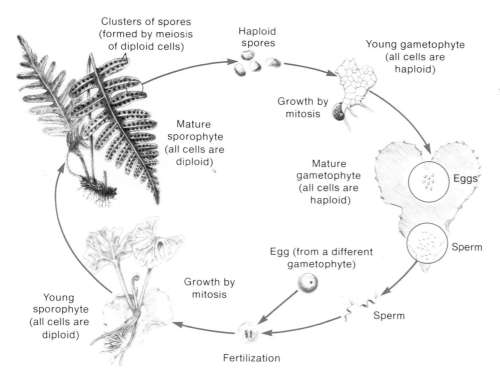

Clusters of spores (formed by meiosis of diploid cells)

Haploid spores

Young gametophyte (all cells are haploid)

Growth by mitosis

Mature sporophyte (all cells are diploid)

Mature gametophyte (all cells are haploid)

Eggs

Sperm

Egg (from a different gametophyte)

Growth by mitosis

Sperm

Young sporophyte (all cells are diploid)

Fertilization

Figure 10-20
The Life Cycle of a Fern

Special cells on the underside of the sporophyte leaf undergo meiosis to produce haploid cells, called *spores*. The spores drop to the ground, and each spore develops into a gametophyte. Eggs and sperm are produced by the gametophyte. The sperm swims to an egg, fertilization is accomplished, and the resulting diploid cell develops into another sporophyte.

Figure 10-21
Major Lines of Evolution in the Reptiles
Early reptiles were highly diversified, but only a few types of reptiles avoided extinction to become the ancestors of modern reptiles, birds, and mammals. (Boxes indicate time of origin.)

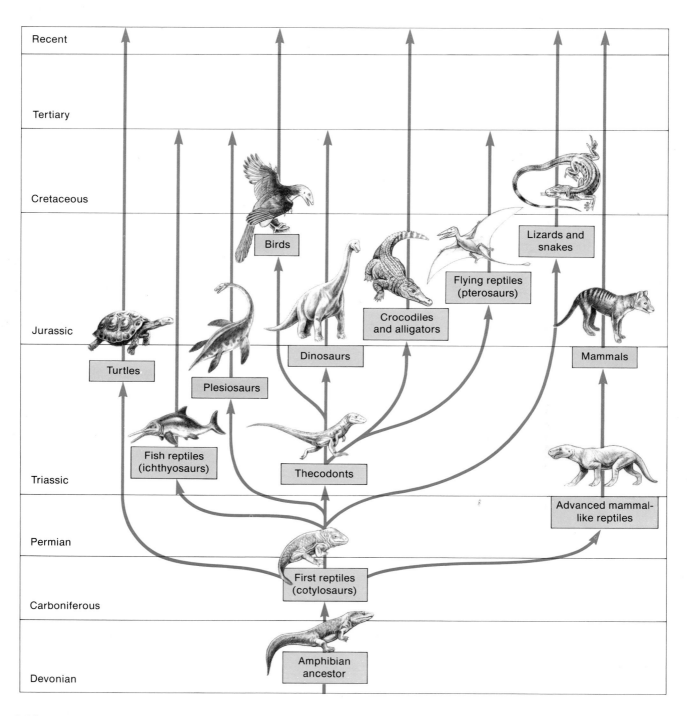

coal and oil—two of our major sources of energy today. Stored within the molecules of these ancient, compressed plants is chemical energy derived by the process of photosynthesis from solar energy that fell on the earth over 300 million years ago.

The Reptiles

A new group of animals, the *reptiles,* also appeared during the Carboniferous period. Reptiles were the first vertebrates to accomplish reproduction completely free of water. Like the female arthropod, the female reptile retained the eggs in her body until they were fertilized. As in the male arthropod, a penis evolved in the male, which released swimming sperm and fluid into the female. After fertilization, yolk within the egg provided enough food for the development of the embryo, and each egg was surrounded by a shell that held the stored food and prevented evaporation. The eggs were laid after the shells formed, and the young reptiles hatched when the embryos were completely developed. An external aquatic environment was therefore not required during fertilization or early development. Reptiles also had more efficient lungs than their predecessors, the amphibians, and no longer used their skin for gas exchange. The outer surfaces of reptiles were dry and became covered with scales that protected against water loss.

Having escaped the need to live near water, the early reptiles began to colonize the vast land areas. They became abundant and diverse, evolving into many forms—snakes, lizards, turtles, alligators, crocodiles, and dinosaurs (Figure 10-21). During most of the time that these reptiles were evolving, it is accepted that all of the earth's continents were connected to form the supercontinent Pangaea (Figure 10-22). Unhampered by water barriers, these animals moved across this single land mass, leaving fossil evidence of their presence on every subsequent continent.

The most fascinating of all reptiles—the dinosaurs—evolved into many diverse forms, some of which grew to enormous proportions (as much as 20 meters in length and 8 tons in weight). Most dinosaurs were land dwellers, but some of these reptiles could fly and others returned to the oceans. The dinosaurs dominated the earth from the end of the Triassic period to the end of the Cretaceous period and then mysteriously became extinct about 65 million years ago. These mass extinctions occurred just prior to the extensive evolution of mammals at the time the supercontinent Pangaea was breaking up into separate continents and the mountains were being formed. The most recently proposed explanation for the extinction of dinosaurs is that an *asteroid* (a small planet) hit the earth and created a thick dust cloud that darkened the earth's surface and suppressed photosynthesis for more than a decade. Under such conditions, the plants and the large plant-eating dinosaurs would have died. However, the plants would not have become extinct, because their seeds could have survived in a dormant state during this time. Any animal capable of eating seeds could also have survived this dark period.

During the Jurassic period, one type of dinosaur evolved into the first bird, developing wings from forelimbs and feathers from scales. At first, wings were adaptations that increased running speed; only later were they used for flight. Birds still retain a reproductive method similar to that of reptiles, involving internal fertilization and the development of the young within eggs.

Figure 10-22
Diagram of the Single Continent Pangaea

This continent existed 200–300 million years ago and later broke up into the several continents on the earth today. The continents were completely joined, but are shown here with their present coastlines.

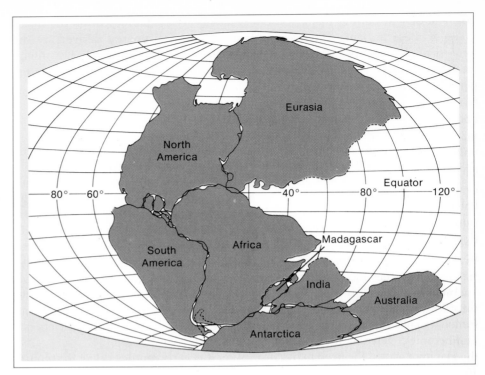

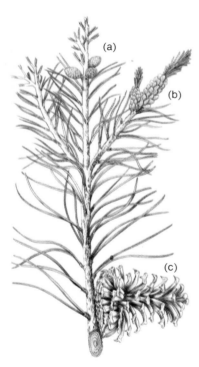

Figure 10-23
Reproductive Structures in a Pine Tree

(a) Female cone at the time of fertilization. (b) Male cone. (c) A female cone that was fertilized two years earlier and is now shedding seeds.

A New Form of Plant Reproduction—Seeds

While reptiles were evolving a method of reproduction that eliminated their dependence on water, plants were also evolving into forms that developed *seeds* and could therefore reproduce without water. The first abundant seed plants were the *gymnosperms*—a group that includes such common modern plants as the pine, cedar, fir, spruce, and juniper.

Unlike the reproductive adaptations in the reptiles and arthropods, the reproductive process in seed plants completely eliminated the swimming sperm. The reproductive pattern in a pine tree (Figure 10-23) is typical of the gymnosperms. The adult pine is a sporophyte. The gametophytes are small and live as parasites directly on the sporophyte, rather than independently as the gametophytes of the ferns do. The cones contain the gametophytes; both male and female cones grow on each pine tree. The male gametophytes are shed from the male cone as pollen grains and are carried by the wind; by a random process, some pollen grains land on the female cones of other trees, adhering to a sticky material produced by the female gametophyte. Each pollen grain contains a sperm nucleus and other parts that form a *pollen tube.* When contact is made between the pollen and the female cone, the pollen tube grows into the female cone to meet the ovule, which contains stored nutrients and the egg. The sperm nucleus is then discharged into the ovule, and fertilization of the egg takes place.

The fertilized egg develops within the female cone until it eventually forms a seed, which contains both the embryo and stored nutrients. These materials are enclosed within a protective *seed coat* that prevents evaporation. In time, the seed is released from the cone and drops to the ground. When the embryo has matured and environmental conditions are favorable, the seed coat bursts open and the young plant begins life as a small pine tree. The embryo may remain dormant for years before suitable environmental conditions trigger the change from seed to seedling, which is called *germination.*

Like the reptiles, the gymnosperms first appeared in the Carboniferous period and became abundant in the Permian period. The climate was very dry during the Permian period, and it is not surprising that both the reptiles and the gymnosperms largely replaced their ancestors—the amphibians and the ferns—during this time. Because they no longer required water to reproduce, the reptiles and gymnosperms were able to expand their ranges; because they still required water to reproduce, the amphibians and ferns were forced to contract their ranges under these dry conditions.

More Recent Forms of Life

In addition to the evolution of birds, two other major events in the history of life occurred between the Triassic and Cretaceous periods (Figure 10-24). A new form of plant—the angiosperm—evolved from a group of plants that were ancestral to the gymnosperms, and a new form of animal—the mammal—evolved from the reptiles. Mammals did not become diverse or abundant until the dinosaurs disappeared at the end of the Cretaceous period, but the new angiosperm was immediately successful.

The Flowering Plants

The *angiosperms,* or flowering plants, have several advantages over the gymnosperms. The majority of angiosperms depend on animals rather than the wind to carry pollen to the female gametophyte. This system is more efficient because most pollen carried by the wind is wasted and never reaches an egg. Such animals as bees, butterflies, moths, and hummingbirds transport pollen much more effectively.

Angiosperms produce colorful, scented *flowers* that contain *nectar*—a sugary solution that serves as food for many animals. *Pollinators* (animals that carry pollen) are primarily nectar-eaters and learn to associate the color or scent of a flower with the presence of food. While it gathers nectar within a flower, pollen adheres to parts of the animal's body and is then deposited on the sticky end of the female parts of another flower (Figure 10-25). As in the pine, the pollen grows in a tube that extends downward into the female gametophyte. The sperm nucleus within the pollen tube then enters the ovule and fertilizes the egg nucleus.

The pollinator must visit more than one flower to transfer the pollen from one plant to another. A plant provides enough nectar in each flower to attract but not enough to satisfy a pollinator, thereby ensuring that pollinators will visit several different flowers.

Each type of flowering plant exhibits characteristics that attract a different type of pollinator. Flowers that attract bees are usually brightly colored but not red, because bees do not see this color. Such flowers may have patterns that appear in

Figure 10-24
Evolutionary Sequences That Led to the Major Plant and Vertebrate Animal Groups

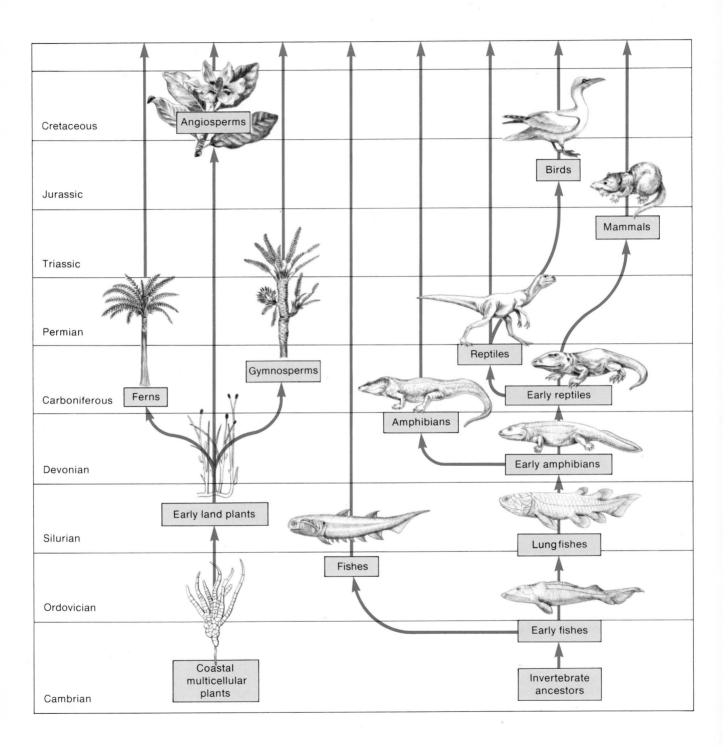

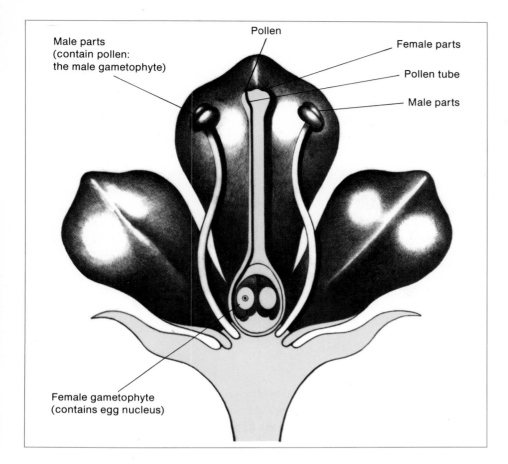

Male parts
(contain pollen:
the male gametophyte)

Pollen

Female parts

Pollen tube

Male parts

Female gametophyte
(contains egg nucleus)

Figure 10-25
The Basic Parts of a Flower
Pollinators transport *pollen* (the male gametophyte) from the male parts of one flower to the sticky ends of the female parts of another flower. A pollen tube then grows down into the female gametophyte, and the sperm nucleus from the pollen passes through this tube to the egg.

the ultraviolet range of light, which is visible to bees but not to humans. Bee-pollinated flowers also have a sweet fragrance and are open during the day. Some of their petals are often fused to provide a landing platform for the bees, as shown in Figure 10-26(a). Flowers that are moth-pollinated are only open in the evening when the moths are active. These flowers have a heavy odor and are white, because this color is most visible at night. Hummingbird-pollinated flowers are usually red or yellow and have no smell, because birds do not have an acute sense of smell. These flowers are open during the day and are often in the shape of long tubes that are just the appropriate length for a hummingbird's long beak, as Figure 10-26(b) illustrates. They do not have fused petals, because hovering birds do not require a landing platform. Flies also serve as pollinators. Fly-pollinated flowers often exude an odor like that of decaying material.

In addition to flowers, angiosperms also produce *fruits*—materials that form around seeds and aid in their dispersal to new sites. Depending on the type of plant and its mode of seed dispersal, the fruit may be fleshy or dry. Fleshy fruits like oranges or strawberries are brightly colored and sweet-tasting. The fruit is eaten by a variety of animals, but the seeds are not digested and drop to the ground in feces

Figure 10-26
Flowers That Attract
Different Pollinators

(a) Bee-pollinated flowers are
often blue with fused petals and
nectar near the surface.
(b) Hummingbird-pollinated flowers
are usually red or yellow with nectar
located deep within tubelike
blossoms.

(a)

(b)

later when the animal defecates. In this way, the seeds of fleshy fruits are dispersed
and develop into plants far from their parents. Dry fruits are structured so that
their seeds (for example, the seeds of the dandelion) float on the wind or adhere to
animals that carry them to new sites. An advantage of seed dispersal is that it
reduces the competition between parent and offspring for sunlight, water, and
nutrients.

A further advancement in plant evolution is that the angiosperms do not
provide an egg with food until it is fertilized, whereas the gymnosperms provide
each egg with food prior to fertilization. Since many eggs are never fertilized,
angiosperms conserve energy by developing nutrients only for the eggs that will
develop into offspring. Angiosperm reproduction will be discussed in greater detail
in Part 4.

The advent of flowering plants increased the diversity and abundance of insects
tremendously. Birds and mammals also relied on these new plant forms for food,
especially the fruits and seeds. In time, humans appeared and also found that the
flowering plants provided a nourishing food source. Eventually, the angiosperms
became the basis of modern agriculture.

The Mammals

Mammals first appeared in the Triassic period, but were not abundant or
diverse until the Tertiary period. Many different types of mammals may have
survived and evolved primarily because the dinosaurs became extinct; when these
large reptiles were abundant, they probably monopolized resources.

All mammals are members of the class Mammalia, which is divided into three subclasses—Prototheria, Metatheria, and Eutheria. Mammals share four main characteristics:

1. All mammals nurse their young.
2. All mammals regulate and maintain approximately constant internal body temperatures, despite fluctuations in environmental temperatures.
3. Most mammals are covered with hair, which provides insulation.
4. Mammals have much larger brains in proportion to their body sizes than other animal groups do.

The first true mammals were about the size of mice. They fed initially on insects and later on the leaves, seeds, fruits, and roots of angiosperms. These primitive mammals probably reproduced in the same way that their reptilian ancestors did. The first mammals existed for a long time without undergoing any major evolutionary changes.

Similar, although larger, mammals survive today on the isolated continent of Australia and its surrounding islands. These *monotremes* (subclass Prototheria) are the spiny anteater and the duck-billed platypus (Figure 10-27). Like the reptiles, these peculiar mammals lay eggs. Unlike the reptiles, however, after the young hatch, they suckle milk from *nipples,* which are modified sweat glands on the mother. In the anteater, the nipples and milk glands are located within a pouch, where the newly hatched young live until they are able to be independent.

Another type of mammal found primarily in Australia is similar to the mammals that evolved from the primitive ancestors of the spiny anteater and the duck-billed platypus. These *marsupials* (subclass Metatheria) first appeared in the fossil record during the early Cretaceous period. The young complete their early stages of growth within the mother but are highly underdeveloped when they are born. They attach themselves to nipples within a pouch on the mother, where they

Figure 10-27
The Monotremes—Australian Mammals That Lay Eggs
(a) The duck-billed platypus and (b) the spiny anteater have retained many of the primitive characteristics of the first mammals.

(a)   (b)

remain until they are mature enough to feed themselves. Except for the mono-tremes, all of the mammals that evolved in Australia—including the kangaroos, wallabies, wombats, and koala bears—are marsupials.

Elsewhere in the world today, almost all mammals are *placentals* (subclass Eutheria). In these mammals, a special structure, the *placenta,* develops within the uterus of the female and forms a connection between the maturing young and its mother's circulatory system. Placental young remain within the mother for a much longer time than marsupial young and suckle milk from the mother's nipples after they are born. Placental mammals first appeared during the Cretaceous period and became abundant and diverse in the Tertiary period.

During the time that the marsupials evolved, Pangaea was just beginning to break up into separate continents, so that the marsupials were distributed through-out the world. However, the placentals evolved later, appearing first in North America, and had just begun to expand their ranges to South America when Pangaea began to divide. The placentals did not reach Australia before it became isolated from the remaining continents, so that Australia contained only marsupial mammals. At the time of this separation, the mammals in South America were predominantly marsupials, but some placental mammals were present.

As the placentals expanded their ranges, they replaced the marsupials wher-ever the two forms occurred in the same area. Apparently, placentals are better competitors than marsupials, so that where competition was intense, the marsupi-als became extinct. Marsupials continued to exist in Australia but were gradually replaced by placentals in South America.

Placental mammals moved freely from North America into Eurasia over a land bridge that connected eastern North America and western Europe for millions of years. When this connection was finally broken early in the Tertiary period and the Atlantic ocean separated the two continents, a second land bridge across the Pacific between Siberia and Alaska joined the two continents. North America and Eurasia were therefore connected in some way throughout most of mammalian history.

Mammals also migrated freely between Eurasia and northern Africa until the middle of the Tertiary period, when the climate in North Africa became very dry. The Sahara Desert formed at that time, and few Eurasian mammals were able to pass through this large expanse of hot, dry land. Only after the middle of the Tertiary period did the early African migrants become isolated from their Eurasian ancestors and evolve into the unique mammalian forms found in Africa today.

Many mammals, including the camel, horse, and rhinoceros, first appeared in North America. At first, these mammals were quite small. Early camels were the size of rabbits, rhinoceroses were the size of dogs, and horses were the size of foxes.

Some of the early grazing mammals evolved into carnivorous mammals. All later species of carnivores evolved from two original groups—the cats and the dogs. These two types of carnivores can be distinguished by their hunting strategies. Cats are good sprinters; they wait for their prey to make a mistake and then pounce and kill swiftly. Dogs and their relatives are long-distance runners; they hunt in groups and chase their prey until it is exhausted. Many different kinds of mammals evolved from the primitive dog, including the weasel, skunk, badger, fox, otter, hyena, coyote, wolf, bear, and raccoon. Modern descendants of the early cats include the lion, tiger, cheetah, lynx, bobcat, ocelot, leopard, puma, and jaguar.

Most groups of mammals gradually increased in size and improved in running ability—an evolutionary trend that probably resulted from predator—prey interactions. The prey became larger and faster, because individuals with these characteristics tended to escape from their predators; the predators became larger and faster, because individuals with these characteristics tended to capture more food.

The Effects of the Ice Age

During the *Pleistocene epoch,* or the *Ice Age,* glaciers covered large parts of the continents (Figure 10-28). The Ice Age began about 3 million years ago and ended

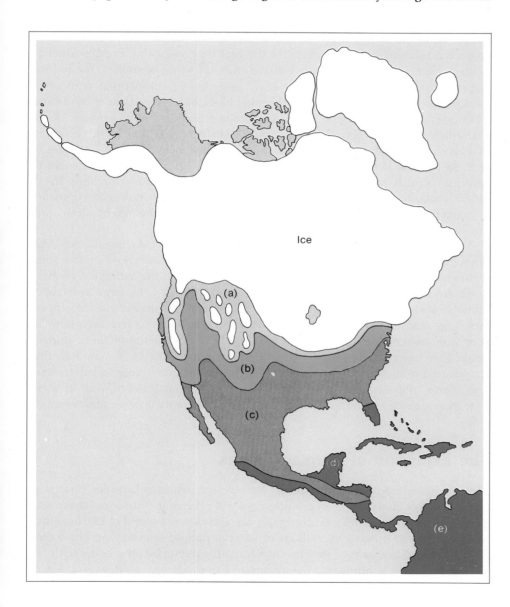

Figure 10-28
The Extent of the Glaciers in North America During the Pleistocene Epoch
Climatic zones: (a) arctic; (b) subarctic; (c) temperate; (d) subtropical; (e) tropical.

only 10,000 years ago. This cold period profoundly affected all life forms. One unusual group—the human species—first appeared during the Ice Age, thrived under these conditions, and became the dominant mammal on earth.

During the Pleistocene epoch, as many as eight major waves of glacial advances may have occurred. The two primary effects of these advances were that (1) such a great volume of water formed into glaciers that the level of the oceans was lowered, exposing previously submerged lands and providing migratory routes between continents, and (2) the glaciers occupied continents, reducing the amount of land available to plants and animals.

Two areas of exposed land greatly affected the animals of North America. The land bridge between Siberia and Alaska provided a migratory route for many mammals. Humans, elk, moose, and caribou entered North America over this bridge. Another land bridge permitted plants and animals to be exchanged between North and South America for the first time since the fragmentation of Pangaea. South America had been isolated from the other continents for so long that its animal populations had not been subjected to competition with other species. North American animals, on the other hand, had evolved under conditions of severe competition, because North America was connected with Eurasia. When a land bridge formed between North and South America, the North American animals competed successfully with the South American animals, causing most of them to become extinct. The only survivors were animals that had no North American counterpart, such as the monkey, anteater, and armadillo. North America acquired a few types of South American animals, including the opossum (a marsupial), armadillo, porcupine, hummingbird, vulture, giant sloth, and mockingbird.

Many large North American mammals became extinct during the Ice Age, although massive extinctions did not occur when glaciers advanced on the other continents. Two possible factors may have contributed to the extinction of these North American mammals: humans may have overhunted the large animals or habitable areas may have decreased in size to the point that they could not support such large mammals. Forced into smaller and smaller areas, where their populations were extremely reduced, the probability of the extinction of these animals would have been greatly increased by hunting. Some of the large mammals that disappeared from North America were the rhinoceros, mammoth, saber-toothed tiger, horse, and elephant. However, by this time, populations of most of these large mammals had migrated to other continents, where their descendants remain today.

The Evolution of Human Beings

All humans are members of the species *Homo sapiens,* which is Latin for "man who is wise." In this final section on evolution, we will present the history of these wise men and women—how they evolved their deep roots in the world of life, how they were formed and molded by millions of years of natural selection, and how they finally emerged to take their precarious place as the dominant species on earth.

The evolution of humans is the history of a new adaptive strategy—*culture,* or the total accumulation of achievements that are not directly determined by genes, including language; the tools and methods developed to feed, shelter, and organize groups of people; and achievements in philosophy, science, commerce, and the arts. This strategy represented a highly significant development in the evolution of life.

Culture is a product of biological evolution that did not emerge until human forms appeared about 2 million years ago with the following biological attributes: excellent vision, a large brain with a unique capacity for learning and reasoning, hands that could delicately grasp and manipulate objects, and a skeleton that supported an upright, *bipedal* (two-footed) posture. These biological characteristics appeared gradually over 70–80 million years of primate evolution.

Primates

Humans are members of the animal kingdom, chordate phylum, vertebrate subphylum, mammalian class, and primate order (Table 10-2). The primate order, in turn, is divided into two suborders, the Prosimii ("pro" = almost; "simians" = apes) and the Anthropoidea ("anthrop" = man; "oid" = like). The living prosimians include the lemurs, lorises, and tarsiers; the living anthropoids are the monkeys, apes, and humans. Because apes and humans are the most closely related anthropoids, both are placed in the *superfamily* Hominoidea ("homo" = man; "oid" = like). Past and present *hominoids* that have uniquely human features are placed in the family Hominidae ("homo" = man; "id" = member of the family). Although there have been at least two species of the genus *Homo,* the only living survivor is *Homo sapiens.* The major stages in primate evolution are diagrammed in Figure 10-29.

The primates evolved before most other groups of modern mammals and retained many of the characteristics common to the earliest mammals throughout their subsequent evolution. While other mammalian groups evolved specialized

Table 10-2
The Taxonomy of the Modern Human Compared to the Taxonomy of the Domestic Dog

Taxonomic category	Category that includes humans	Category that includes dogs
Kingdom	Animalia	Animalia
Phylum	Chordata	Chordata
Subphylum	Vertebrata	Vertebrata
Class	Mammalia	Mammalia
Order	Primates	Carnivora
Suborder	Anthropoidea	—
Superfamily	Hominoidea	Canoidea
Family	Hominidae	Canidae
Genus	*Homo*	*Canis*
Species	*sapiens*	*familiaris*

Figure 10-29
The Evolutionary Sequence That Led to the Living Primates

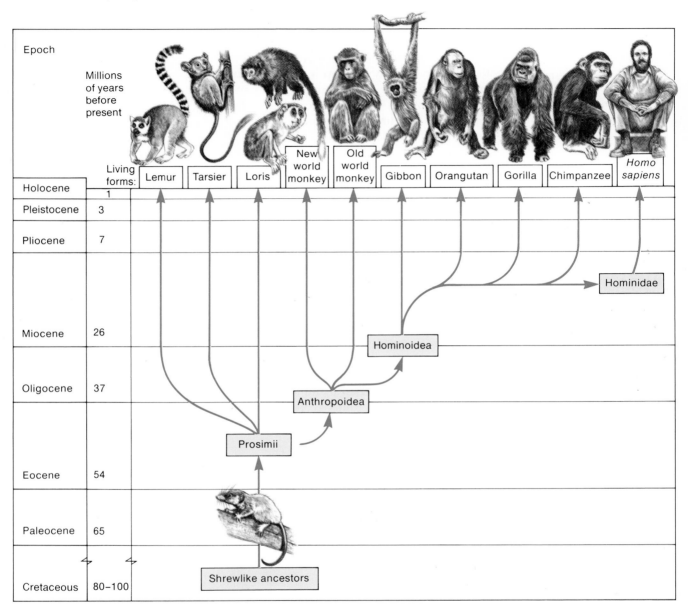

Epoch	Millions of years before present										
	Living forms:	Lemur	Tarsier	Loris	New world monkey	Old world monkey	Gibbon	Orangutan	Gorilla	Chimpanzee	*Homo sapiens*
Holocene	1										
Pleistocene	3										
Pliocene	7										
Miocene	26										
Oligocene	37										
Eocene	54										
Paleocene	65										
Cretaceous	80–100										

Hominidae

Hominoidea

Anthropoidea

Prosimii

Shrewlike ancestors

limbs adapted for flying, swimming, running, or killing, the primates retained a limb structure similar to that of the earliest vertebrates (Figure 10-30).

As a group, primates have so few distinctive specializations that many biologists prefer to define the order by the set of biological traits that gradually developed as primate evolution progressed. The major trends in primate evolution are considered to be:

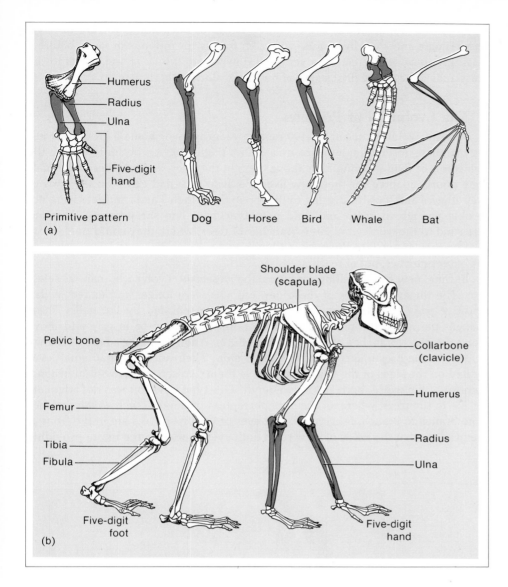

Figure 10-30
A Comparison of the Forelimbs of Primates and Other Vertebrates
(a) Vertebrate forelimbs, modified by evolution for flying, swimming, running, or killing, are all derived from the primitive pattern (shown on the left) found in the earliest vertebrates. (b) The primate forelimb has retained the unspecialized features of the limb structure of the early vertebrates. A monkey skeleton is shown here.

1. The development of *prehensile* (adapted for grasping and holding) hands and feet.
2. An increase in the flexibility of limb movement.
3. A tendency toward upright posture.
4. A decreased dependence on smell and a related increased dependence on vision as the primary means of obtaining information about the environment.
5. A progressively larger brain, particularly the outer layer (called the *cortex*).
6. An increased life span and longer periods of childhood dependence.
7. The development of a complex social life that involved an increasing proportion of learned behaviors.

Figure 10-31
A Modern Tree Shrew

Figure 10-32
Prosimians
(a) The tarsier from southeast Asia moves by jumping and holding onto tree branches. (b) The gray slow loris moves along by gripping and pulling itself along branches. Note that the eyes are in the front of the head in both forms.

All of the trends in primate evolution began as adaptations for life in trees. Our primate ancestors lived and evolved in forests for millions of years before a small group of semi-erect apes ventured down from the trees and out onto the grasslands to begin the final stages of human evolution.

The Evolution of Primates

Primate evolution began 80–100 million years ago when small, ground-dwelling mammals resembling modern tree shrews (Figure 10-31) began to exploit the rich stores of food growing high above them in the trees. At first, these mammals were poorly adapted for their new lives. Because they had claws on the ends of their fingers and toes for digging in the forest floor, their hands and feet were not flexible enough to grasp and hold onto branches. The shrews were therefore restricted to the trunks and large branches of trees, where they could hold on by digging their claws into the bark, although most of the fruit and leaves were on the outer branches, out of reach.

In time, new tree-dwelling creatures, the *prosimians,* evolved by natural selection. The fingers and toes of these new animals were longer and ended in flat, sensitive pads; the claws of the shrews had been replaced by flat nails. These changes permitted the prosimians to jump and hold onto or to grip and move along the outer branches of trees (Figure 10-32). Another major adaptive change evident in the prosimians was improved vision. Their ancestors had small eyes located on the sides of the head, which gave them two separate, two-dimensional pictures of the world, both lacking in depth. This type of vision was not adequate for life in the trees, where an accurate perception of depths, such as the distance from branch to branch, is essential. The eyes of the prosimians are in the front of the head, so that the fields of view from both eyes overlap in the brain, providing

(a)

(b)

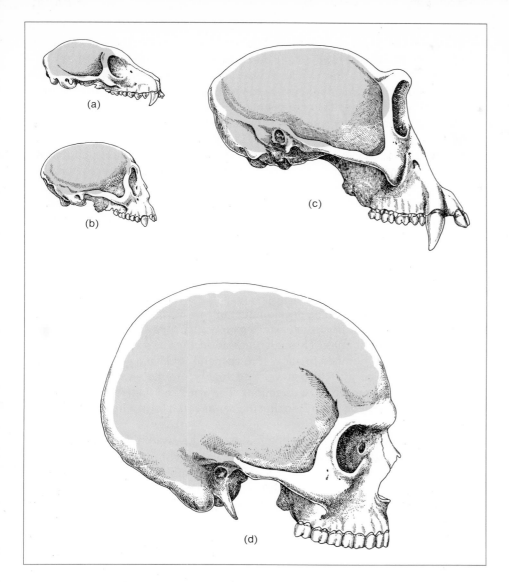

Figure 10-33
Brain Expansion During Primate Evolution
The increase in brain size that occurred during primate evolution can be seen in this comparison of the brains (brown areas inside skulls) of (a) a prosimian, (b) a monkey, (c) an ape, and (d) a modern human. The skulls shown here are approximately one-third their normal size.

three-dimensional depth perception. Mechanisms also developed in the eyes that permitted the prosimians to see details more precisely and allowed some prosimians to perceive color.

The first clearly recognizable anthropoids to appear in the fossil record—the *monkeys*—were advanced over the prosimians in all major areas of primate evolution. The monkeys were the product of millions of years of selection for improved adaptations to life in the trees, such as running, leaping, and swinging from branch to branch. The most significant adaptive change occurred in the monkey's nervous system, especially in the brain. Compared to the prosimian brain, the monkey brain is greatly expanded and has a much greater capacity for learning and for precise movement control (Figure 10-33). For example, the prosimians move all of

(a) Tree shrew

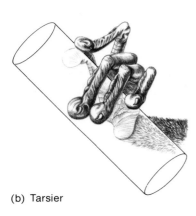

(b) Tarsier

(c) Macaque (monkey)

Figure 10-34
Evolution of the Primate Grip from Ancestral Shrew to Monkey

their fingers together in one gross movement, whereas the monkeys can control the movement of each finger, permitting the thumb to move against the other fingers in a delicate grip that can be used to hold and manipulate objects (Figure 10-34).

Like the prosimians, the monkeys walk and run on all four feet. However, unlike the prosimians, the monkeys spend much of their time sitting in an upright posture, particularly while they are resting and grooming. Perhaps in response to this mode of behavior, the trend in primate evolution has been for the head to rotate downward in relation to the spine. In the human, the head is rotated 90° downward from the position of the head of the ancestral tree shrew, so that the human jaw is perpendicular to the spine. If this rotation had not occurred, today we would be walking upright with our faces pointing toward the sky.

A new form of primate—the *apes*—emerged during the late Oligocene and early Miocene periods. The apes differed from previous tree-dwelling primates in many respects, exhibiting a new pattern and structure of teeth, a major increase in brain size, and a much larger body size, modified to allow a radically new form of swinging movement.

Monkeys move along the tops of branches on all four limbs, whereas apes move along the undersides of branches by grasping them with their forelimbs and swinging from grip to grip. The development of this movement, called *brachiation* (Figure 10-35), freed the apes from balancing on branch tops and permitted them to evolve into very large animals. Through brachiation, the ape's arms, shoulders, and upper body became modified to allow it to suspend itself in a semi-erect posture for long periods. The apes also developed highly mobile shoulder joints that permitted the arms to reach in all directions. The arms and fingers became longer; the wrists did not bend backward, so that the hands could form a hooklike grasp on tree branches.

The three families of modern hominoids are the Hominidae (the family of man), the Hylobatidae (gibbons), and the Pongidae (chimpanzees, gorillas, and orangutans). Each family has evolved separately for millions of years. No modern ape is ancestral to the human; each family evolved from a different ancestral group of apes or "near-apes" that existed during the Miocene period. Our closest relatives among the living primates—the chimpanzees and gorillas (Figure 10-36)—are strikingly similar to humans in anatomy, behavior, body chemistry, and even in susceptibility to infectious diseases. Some anthropologists contend that humans are so similar to gorillas and chimpanzees that the divergence between humans and apes must have occurred as recently as 4–5 million years ago. Most anthropologists, however, believe that the fossil record indicates a much earlier divergence as long ago as 10–15 million years.

The Physical Differences Between Apes and Humans

Although chimpanzees and gorillas are descended from brachiating apes and have retained many adaptations for this type of movement, they now spend much of their lives traveling on all four limbs along the ground. They walk with their feet flat on the ground and bend their hands forward, supporting some of their weight on the knuckles. Modern humans, in contrast, stand and move in an entirely upright posture.

Figure 10-35
The Supreme Primate Acrobat—A Brachiating Gibbon
Ralph Morse from *Photographing Nature*, TIME-LIFE BOOKS, © Time Inc.

Figure 10-36
The African Great Apes: Chimpanzees and Gorillas

Chimpanzee

Gorilla

Figure 10-37
A Comparison of the Most Important Structural Features of Humans and Apes
These side and front views of the skeletons of a human and a gorilla show the major adaptive changes that have occurred during the evolution of human *bipedal posture* (walking on two feet rather than four).

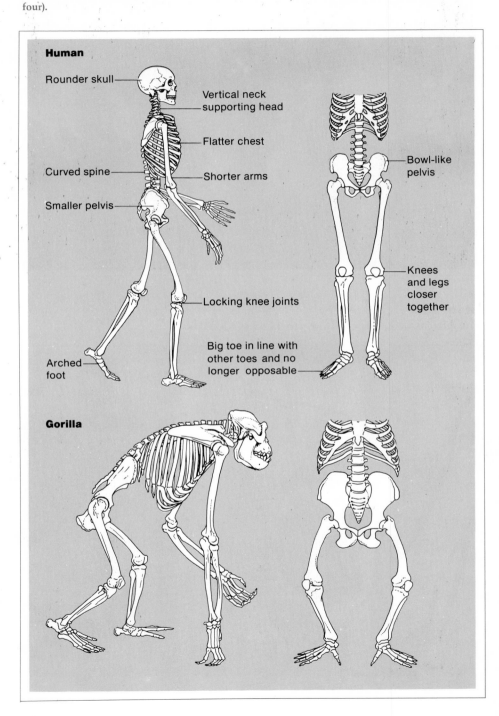

Human

Rounder skull

Vertical neck supporting head

Flatter chest

Curved spine

Shorter arms

Smaller pelvis

Bowl-like pelvis

Knees and legs closer together

Locking knee joints

Big toe in line with other toes and no longer opposable

Arched foot

Gorilla

The skeleton of a gorilla is compared to the skeleton of a modern human in Figure 10-37. The human skeleton is a balanced, vertical frame that transmits the weight of the body to flexible, platform-like feet. The skull is round and balanced on the top of a vertical column of neck vertebrae. The chest is flattened, and the bones of the spine are locked together to form a flexible rod that curves backward at the base of the spine, where the spinal column is balanced by a broad, bowl-like pelvis. The gorilla pelvis is long and thin, so that the weight is thrown forward onto the knuckles. Gorilla legs are bent and far apart; human legs are straight and close together, having knee joints that lock in place in the upright position. The gorilla foot is flat with long toes; the big toe can be moved in opposition to the other toes to permit grasping and holding. The human foot is arched and flexible with short toes, and the big toe is not opposable.

An increase in brain size is one of the more important differences between apes and humans (see Figure 10-33). Modern humans have the largest brains of all primates, even though they are not the largest primates in terms of body size. In general, the largest animal in related species has the largest brain: a larger animal has larger and more numerous muscles, a greater surface area, and more sensory receptors, so that its brain must be larger to coordinate and direct body movements. In primate evolution, however, brain size has increased beyond the size proportional to body size. This proportionally "extra brain" in the primates, which produces an unusually high level of intelligence, occurs to a greater degree in humans than in apes. Also, the internal organization of the brain is not the same in apes and humans. For example, structures in the human brain that permit spoken and written language are not found in the ape brain.

An unusual adaptation in both apes and humans is growth of the brain after birth, which results in a larger adult brain without greatly increasing the size of the birth canals and pelvic openings in females. In the human, the brain is 25% of adult size at birth and does not reach full adult size until the child is 12–14 years old. In chimpanzees, the brain is 65% of adult size at birth and reaches maturity by 8–9 years.

Another important difference between apes and humans is the relationship of the thumb to the other fingers of the hand. In humans, the thumb is longer, straighter, and more flexible and can be moved in opposition to the other fingers, as shown in Figure 10-38(a). These features allow the human hand to have both a power and a precision grip, as shown in Figure 10-38(b). Apes also have both of these hand grips, but their precision grip is much less delicate. These developments in the human hand were essential adaptations for making and using sophisticated tools.

Both apes and humans have 32 teeth (16 on the upper jaw and 16 on the lower jaw), as shown in Figure 10-39(a). Four chisel-shaped *incisors* in the front of each jaw are used for biting and cutting food; two pointed, cone-shaped *canines* just behind the incisors are used for holding and tearing food. Behind the canines are 10 broad and flat *cheek teeth*—4 premolars and 6 molars (including the "wisdom" teeth)—that are used for crushing and grinding food.

Although gorillas and chimpanzees have the same number and pattern of teeth as humans, there are significant structural differences. The larger and heavier ape jaw is rectangular, whereas the human jaw is arched. Apes also have much larger canines and a gap between the canines and the other teeth, called the *diastema,*

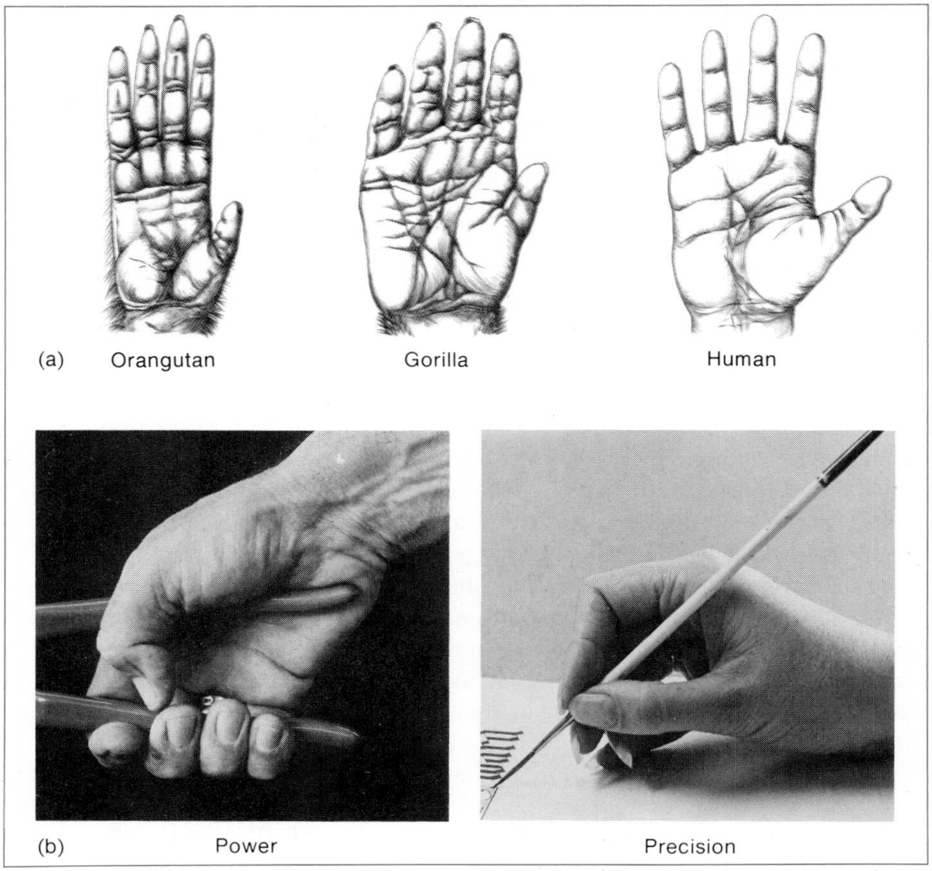

**Figure 10-38
Ape and Human Hands**
(a) Compared to the ape thumb, the human thumb is longer, straighter, more flexible, and opposable to the other fingers. (b) The human hand has both a power and a precision grip.

shown in Figure 10-39, that allows the canines to slide together when the jaw is closed. The human jaw has no diastema.

These differences in teeth and jaw structures reflect differences in diet between apes and humans. The gorilla is a *vegetarian* and eats large quantities of hard and tough plants. The chimpanzee, an *omnivore,* eats the same plants as the gorilla, but its diet also includes a variety of such animal materials as insects, lizards, eggs, and small mammals. Both the gorilla and the chimpanzee need large canines for ripping and tearing and big cheek teeth for grinding their food. The early humans were also omnivores, but because they could process their food with cutting tools and fire, their teeth could be smaller and less specialized. The hominids probably began to lose their large canines when they started to eat grass seeds. Considerable sideways grinding of the cheek teeth is necessary to consume this type of food, and large canines would have interfered with this movement.

From these anatomical comparisons, it is clear that chimpanzees, gorillas, and modern humans are merely different end products of the same general evolutionary trends. No fundamentally new structures have emerged since the divergence of

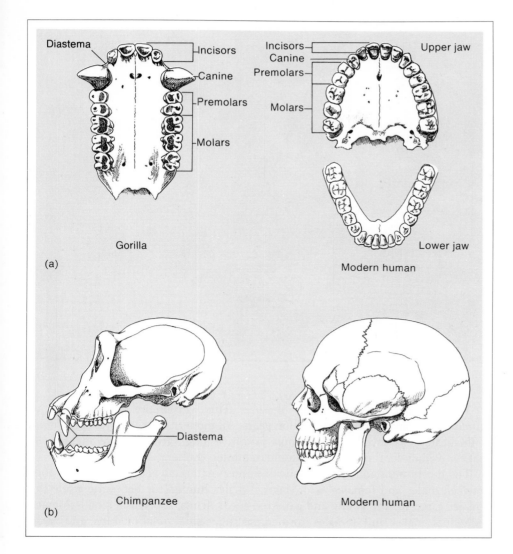

Figure 10-39
**The Jaws and Teeth
of Apes and Humans**

(a) The upper jaw of a gorilla is shown next to the upper and lower jaws of a human. Although both forms have the same number and type of teeth, the human jaw is more arched with smaller canines and no diastema between the canines and the other teeth. (b) When the chimpanzee closes its jaw, the canines slide into the diastemas of each jaw. In contrast, when human jaws close, the teeth all come together.

these three primate forms. Instead, existing structures have been gradually modified by natural selection to better adapt each primate form to its specific environment.

Behavioral Differences Between Apes and Early Humans

The *Cro-Magnon people* were the first *Homo sapiens* to appear in the fossil record that were anatomically identical to modern humans. Cro-Magnon men and women existed between 30,000 and 35,000 years ago in Europe, Africa, and Asia. They were so similar to modern men and women that if they were alive today and were given the proper clothing and training, they could walk among us unnoticed (Figure 10-40). All of the significant changes in human populations since the evolution of the Cro-Magnons have been cultural rather than biological.

Figure 10-40
Cro-Magnon People
Cro-Magnon hunters returning to
their cave shelter, where the animals
they have killed will be cooked over
the continuously burning hearth fire.

The Cro-Magnons were *hunter–gatherers.* This way of life has been part of
human evolution for the past 2 million years. All modern humans—from the most
sophisticated city dwellers to the most primitive jungle tribes—have adaptations
that evolved under the conditions found in hunter–gatherer communities.

The most important characteristic of hunter–gatherers is the division of labor
between males and females. Cro-Magnon males hunted for big game while the
females cared for the young and gathered seeds, fruits, roots, and small animals.
Individual males and females formed lasting sexual relationships and lived
together with their children and other relatives as a family. Several families joined
together to form *nomadic hunting bands* that moved with the game herds during their
seasonal migrations. Each band established a base camp, creating shelters in caves
or constructing tents of animal skins. The women stayed in the camp, where stores
of food were stockpiled and the wounded and sick members of the band were
nursed back to health.

Gorillas and chimpanzees also live in social bands, but these bands are not
highly organized and exhibit little family structure. Gorilla bands of 8–30 individuals include one dominant male, several other adult males, and the females and
young. Mating males and females form temporary bonds that last two or three
days. Gorilla bands do not establish camp sites; they move continually through
large territories, nesting each night in a new place.

Chimpanzees form bands of 20–60 individuals that are also composed of a
dominant male, other adult males, and the females and young. Males and females
form no pair bonds; instead, receptive females mate with a series of males in a
frenzy of sexual activity. Although chimpanzees are not hunter–gatherers, they do

exhibit some of the basic characteristics of this lifestyle. Several adults join together in cooperative hunting bands to trap and kill animal prey, and the group shares the meat after a successful hunt. Unlike the Cro-Magnons, however, who skillfully planned and carried out hunting trips over great distances, chimpanzees are opportunistic and hunt only when some small or weakened animal wanders into range and can be easily killed. The Cro-Magnons hunted game of all sizes, including the now extinct giant woolly mammoth, which weighed up to 8 tons. The chimpanzees never hunt animals larger than 20 pounds. Like gorillas, chimpanzees wander through a large area and do not establish base camps.

The Cro-Magnons must have developed a complex language to coordinate their elaborate hunting expeditions. Human language is called an *open system*, because it places no limit to the thoughts that can be communicated. In contrast, the apes have a *closed system* of language that consists of sounds and facial expressions that have fixed meanings, such as anger, terror, or begging for food. There is evidence that chimpanzees communicate over long distances to inform their social group of danger or of the presence of food. Chimpanzees have been seen using sticks to beat on trees that contain ripe fruit; this behavior appears to have no effect other than to bring the remaining members of the band to the trees. Captive chimpanzees and gorillas have been trained to use a sign language comparable to the hand signs used by hunters in primitive cultures today (Figure 10-41).

Anthropologists have unearthed more than 100 different types of tools in Cro-Magnon campsites. These tools are made of stone, wood, bone, and antlers and include spears, axes, many types of knives and choppers for cutting and processing food, and scrapers and needles for making clothing from animal skins. These highly sophisticated tools required planning and several stages of preparation.

Gorillas have never been observed using tools, but chimpanzees make and use several forms of tools. Chimpanzees fashion cups from leaves, which they use to drink water and to scoop out the brains of the animals they kill. Chimpanzees also use a specially formed twig to extract termites from their nests in dirt mounds. When a chimpanzee finds a termite mound with open exit holes, it searches for a twig of the appropriate length, strips the leaves and branches from the twig, inserts the twig into the mound, and licks off the termites it pulls out. Compared to Cro-Magnon tools, chimpanzee tools also require planning and preparation but do not reflect as high a degree of intelligence. Tools were critical to the survival of the Cro-Magnons, whereas tools play only an occasional role in chimpanzee life.

The Family of Man: The Evolution of Hominids

We have just reviewed the more significant anatomical and behavioral changes that occurred during the evolution from ape to human. The numerous fossils of "ape-men" and "near-men" that have been discovered indicate that all of these changes did not occur simultaneously. Instead, adaptations accumulated gradually, and each anatomical or behavioral change led to subsequent changes. Some form of *bipedal posture* (walking on two feet) probably developed first, which then freed the hands to carry helpless infants, transport food, and make and use tools, and also elevated the head to improve vision. The primary selection pressure in

Figure 10-41
Sign Languages Being Used by a Human and a Chimpanzee
(a) Hand signs used by African Bushmen in the Kalahari Desert during hunting expeditions. (b) A chimpanzee using symbols from the American Sign Language.

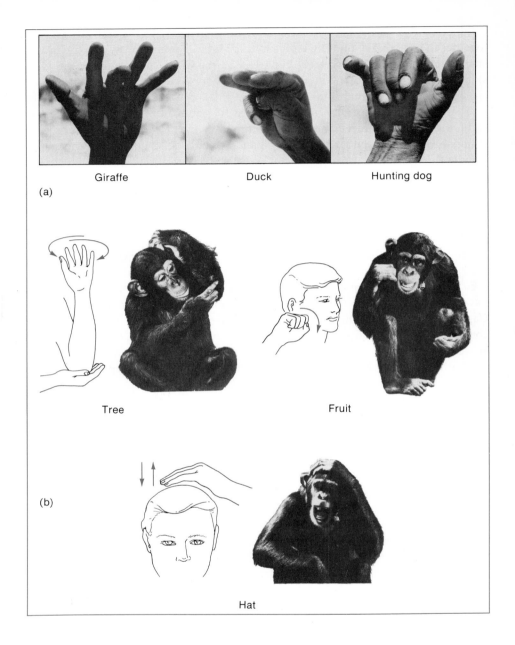

Giraffe   Duck   Hunting dog

(a)

Tree   Fruit

Hat

(b)

hominid evolution from the ape to the Cro-Magnon—the dietary shift toward larger and larger animals—caused adaptive changes in hunting methods and the use of tools that eventually led to the hunter–gatherer lifestyle.

Our knowledge of existing hominid fossils is summarized in Figure 10-42, where the solid lines represent accurately dated fossils that most authorities have assigned to the indicated hominid group. The dashed lines represent fossils that probably fall into the indicated hominid group but that have not been identified with certainty. No attempt is made here to show ancestral relationships. These

Figure 10-42
Summary of the Fossil Evidence to Support Hominid Evolution

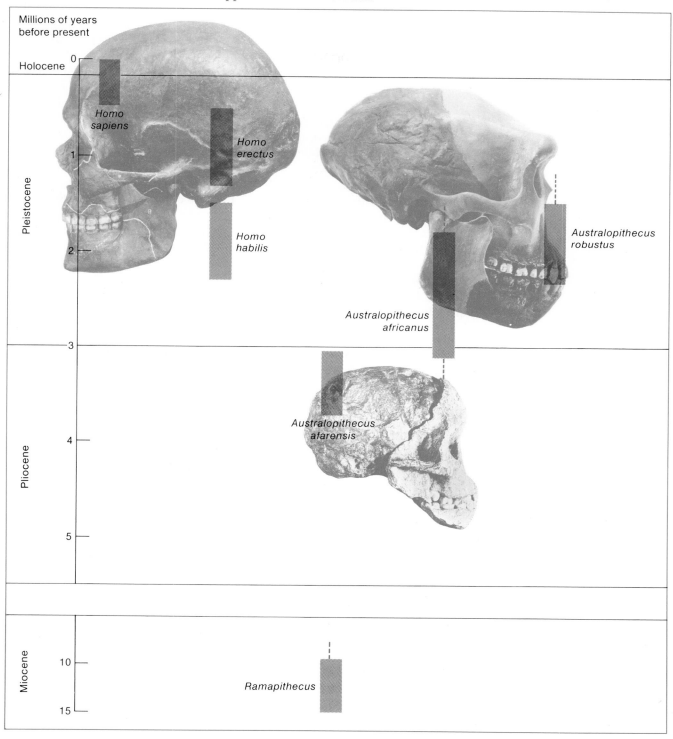

THE EVOLUTION OF HUMAN BEINGS   235

relationships have not been firmly established and remain an area of controversy and mystery which contains almost as many hypotheses as authorities.

Scientists generally concur that humans evolved from apelike creatures that lived during the Miocene period, when climatic conditions allowed tropical forests to develop over much of Europe and Asia. These forests were inhabited by a population of apes that walked in a semi-erect posture on the ground and were more bipedal than the other apes. However, these changes in posture and diet were not of evolutionary significance until the climate become more arid in the Pliocene period and the tropical forests began to decrease, giving way to grasslands. This new form of ape was then prepared or *pre-adapted* to take full advantage of these new conditions. When these apes moved out of the forests to live in mixed areas of trees and open grasslands, hominid evolution probably began.

A type of fossil ape found throughout Asia and Europe may represent the remains of these first hominids. This fossil was first discovered in India and given the genus name *Ramapithecus* ("Rama" = the hero of Indian legend; "pithecus" = ape). This creature definitely lived between 10–15 million years ago, and evidence suggests that it may have existed as recently as 8 million years ago.

The next form of known hominid, *Australopithecus afarensis*, appeared about 3.7 million years ago in Africa and provides the first evidence of an entirely bipedal form of primate. Although it was only about $3\frac{1}{2}$ feet tall with an ape-sized brain and not yet human in form, this hominid possessed some distinctly human traits, including small molar teeth, human-like hands with short fingers, and pelvic and leg bones almost identical to those of modern humans. The size differential between males and females was greater than it was in any of the later hominid forms. Despite the fact that it walked erect and had human-like hands, no evidence suggests that this hominid used tools.

Australopithecus afarensis was probably the ancestor of several later forms of "ape-men," also placed in the genus *Australopithecus*, as well as of the genus *Homo*, which eventually produced the modern human. *Australopithecus africanus* first appeared in the fossil record about 3 million years ago. It weighed 40–60 pounds, walked in an entirely erect posture, and had human-like teeth, but it had more of an ape brain than a human brain. The lifestyle of this small, wiry omnivore was similar to that of a chimpanzee. *Australopithecus africanus* was found in various parts of Africa, where it apparently existed until about 1.5 million years ago. A similar form of hominid, *Australopithecus robustus*, which appeared somewhat later in the fossil record, was much larger, weighing 150–200 pounds. Because of its heavier jaws and more massive molars, this hominid probably adhered to a more vegetarian diet than *Australopithecus africanus*. It also walked erect and had a brain the size of an ape brain.

Another type of creature seems to have lived during the same period and often in the same areas of Africa as the australopithecines. This hominid form was also bipedal and medium in size (weighing about 95 pounds) with hominid teeth, but it had a much larger brain than any australopithecine. Due to its larger brain, authorities believe that this hominid was a human form and have placed it in the genus *Homo*. Its full name is *Homo habilis* ("habilis" = mentally skillful). The first evidence of human culture—primitive, chipped-stone tools found in campsites about 2 million years old—was included with some of the more recent remains of these hominids. The animal bones found at these sites also indicate that these first tool users hunted and ate much larger animals than other hominids of the same period. Brain expansion and the use of tools are apparently associated in the evolution of humans.

Still another type of hominid lived between 1.5 million and 200,000 years ago. *Homo erectus* possessed all of the human characteristics: a large brain (although it was still smaller than the Cro-Magnon brain), bipedal posture, human teeth and hands, and a fully developed hunter–gatherer lifestyle. The most significant differences between this hominid and the Cro-Magnons, in addition to certain skeletal features, were that *Homo erectus* had a somewhat smaller brain and a less complex and sophisticated culture. Unlike *Homo habilis,* which was restricted to the tropical climates of Africa, *Homo erectus* roamed the temperate regions of Europe and Asia. However, *Homo erectus* had not developed adequate cultural tools to move into the coldest regions of the earth; this movement did not occur until the emergence of the Cro-Magnons.

Fossils with the brain size of the modern human first appeared in the record of hominid evolution about 250,000 years ago. Although they differed from the fully modern hominid form, having heavier jaws, thicker skulls, and an unsophisticated culture compared to the Cro-Magnons, these hominids were sufficiently similar to the modern human to be classified as *Homo sapiens.* Many different forms of *Homo sapiens* have been discovered throughout Europe and Asia; the most famous are the *Neanderthals (Homo sapiens neanderthalensis),* who lived between 100,000 and 45,000 years ago.

A gap exists in the fossil record until about 35,000 years ago, when the Cro-Magnon people finally emerged. When they first appeared, the Cro-Magnons were hunter–gatherers with superb hunting skills. Within 30,000 years of their emergence, they had become the farmers and metal workers of Egypt, the philosophers of Greece, the Indians of the Americas, the Eskimos of the Arctic, and all of the other peoples of recorded history. All humans who exist today are Cro-Magnons (*Homo sapiens*) — the only surviving hominid species.

Summary

Each major evolutionary step in the history of life has apparently followed a particular environmental change or a major reorganization of some type of organism. The evolution of *photosynthesis* — a method of utilizing the sun's energy to manufacture food within a cell — set off a dramatic chain of events in existing organisms. Photosynthesis increased the speed of food production and altered the environment by releasing oxygen gas (O_2) into the oceans and, eventually, the atmosphere. The presence of abundant oxygen facilitated the more complete release of energy from food molecules (*aerobic respiration*), leading to the evolution of *multicellular organisms* and, in turn, to *sexual reproduction.* Much greater genetic variability within populations resulted from sexual reproduction, increasing the rates of evolution. In time, the accumulation of atmospheric oxygen and the appearance of multicellular plants permitted life forms to invade the land.

Organisms were frequently unable to adapt completely to their new land environments, because their life cycles were irrevocably tied to their previous ocean environment. The reproductive phase of the life cycle most often resisted alteration. Early terrestrial plants and animals were restricted to aquatic environments, because their sperm had to swim through water to fertilize their eggs. For millions of years, the swimming sperm prevented both plants and animals from totally adapting to terrestrial life. Sperm movement in the absence of free water was eventually achieved by the evolution of a *pollen tube* in the seed plants—the *gymnosperms* and *angiosperms*—and by internal fertilization in the arthropods, reptiles, birds, and mammals.

Another trend common to the history of life is the recurrent interrelationship between the evolution of plants and animals: a major reorganization in plants is usually followed by a rapid evolutionary change in animals. Competition among early land plants for sunlight resulted in the evolution of structural changes that provided plants with the support they needed to grow higher above the ground. Taller land plants in turn provided a shaded, moist environment for the first terrestrial vertebrates, the *amphibians*. A major condition necessary for the later evolution of insects, birds, and mammals appears to have been the evolution of seed plants, which provided an abundant and reliable source of food.

The complex coevolution of major groups of plants and animals occurred in a setting of great geological change. Dramatic upheavals of the earth's crust, advances and retreats of inland oceans, widespread glaciations, and movements of land masses from one place to another all profoundly influenced the evolution of the earth's creatures.

Humans, or *Homo sapiens*, are classified *taxonomically* within the *primate order*. We share with other members of this order, including the lemurs, monkeys, and apes, a number of biological adaptations for life in trees. Primate evolution began about 100 million years ago when small, ground-dwelling mammals first exploited resources in trees. In time, these animals developed *prehensile* hands and feet for grasping tree limbs, greater flexibility of arm and leg movement, a more upright posture, improved vision, enlarged brains, longer life spans, and a complex social life. These adaptations are the major trends in primate evolution.

The next major stage in the evolutionary pathway that led to the development of the human was a return to the ground and the evolution of characteristics that were appropriate to this new way of life. At this point, the primates were substantially different from their earlier ground-dwelling ancestors that had existed prior to life in trees. One of the groups that left the trees for a more terrestrial way of life developed a completely upright posture, thereby freeing the hands for making and using tools. This line of evolution led to humans.

In the gradual evolution from ancestral apes to humans (the *Cro-Magnons*), the most significant changes were: (1) the evolution of a skeleton and musculature for *bipedal posture* (walking on two feet), (2) brain expansion, (3) the development of hands with long, *opposable* thumbs, (4) a decrease in the size of the teeth and jaws, and (5) the development of *culture* and a *hunter—gatherer lifestyle*.

The main selection pressure in the evolution from ape to Cro-Magnon men and women was the ability to capture larger and larger animal prey. Adaptive changes that improved hunting methods led to the use of tools, a sophisticated means of communication, and the hunter–gatherer lifestyle. This selection for hunting efficiency produced the biological adaptations that resulted in modern humans. Today, we are basically the same as our Cro-Magnon ancestors, who lived about 35,000 years ago.

The Individual Life:
Adaptations for
Survival

3

Organizations for Survival

11

Until this point, we have regarded the individual organism as little more than a mechanism for transporting genes across time from generation to generation. Our goal has been to provide a basic understanding of how life originated and subsequently evolved on earth, and we have been concerned with processes that involve large numbers of organisms and vast reaches of space and time.

Another central focus of the science of biology is to gain an understanding of how each kind of organism functions. In the remainder of this book, we will turn our attention to the briefest moment in evolutionary time—the life span of the individual organism.

The individual organism inherits a set of genes that develop and organize all aspects of its life to achieve reproductive success. The selection of genes throughout the hereditary line gives each individual a reasonable chance to survive and reproduce as long as the conditions in its environment are similar to those experienced by its ancestors. The inherited set of genes includes instructions that determine the components of body structures, the way that these structures will function together, and the sequence of developmental changes that will form the course and nature of the individual's life.

In Part 3, we will examine the adaptations that allow the organism to make use of specific environmental resources and to overcome environmental dangers. In Part 4, we will view the entire sequence of the individual life from the first moments of embryonic existence through each stage of development, growth, and reproductive maturity to the final, inevitable stages of aging and death. We will therefore be examining the same phenomena of individual life from two different perspectives, although adaptations for survival and development are actually integrated and synchronized with environmental conditions throughout the life of each organism.

Life in a Hostile Environment: The Problem of Survival

All organisms are fragile containers filled with high concentrations of uniquely organized molecules. These molecules are continually synthesized from materials collected in the environment, and other living organisms and nonliving forces act just as persistently to return these molecules to the physical environment.

The molecules that have accumulated within an organism are valuable sources of energy and nutrients for other organisms. All animals and fungi and many microorganisms depend on other living organisms for food. A primary problem that the living organism faces is to avoid being eaten. Even humans are vulnerable to other organisms—ranging from the attacks of bloodsuckers, such as mosquitoes, which land directly on the skin, to the more subtle invasions by microorganisms, which thrive and multiply on the body's accumulated molecules if not combated.

Nonliving physical forces in the environment are also continuously acting to reclaim the organism's stored molecules. These forces, which organize and disorganize all physical matter in the universe, include the electrical forces that attract and repel atoms and molecules, the continual diffusion of molecules away from areas of high concentration, and the tendency of energy to move toward the maximum state of disorganization, or *entropy*. The internal chemistry of the organism differs greatly from its surrounding environment. Nonliving forces act at all times to reduce these differences—to disrupt the organization of the individual and to disperse its contents throughout the environment (Figure 11-1).

However, a living organism cannot survive in a closed, sealed container. We have seen that an organism must regularly exchange materials with its environment to maintain the chemical reactions necessary to provide energy and structural materials for all cellular activities. Vital substances must be taken in; an organism must be continually supplied with energy, water, oxygen, and a variety of other materials. Waste products such as heat, carbon dioxide, and nitrogenous materials from chemical reactions, which would kill the organism if they were allowed to accumulate, must also be continually released back into the environment.

Although this regular exchange of materials with the environment is essential to maintain an organism's vital activities, it is also a constant threat to the organism's survival. An organism that has sufficient access to the environment to receive substances from and release substances into its surroundings is vulnerable to the same degree to forces acting to incorporate the organism back into the environment.

Life has persisted in many forms for billions of years because all organisms inherit solutions to their environmental problems—specific *adaptations* that permit each organism to deal successfully with the particular physical and ecological conditions it is likely to encounter. These adaptations are anatomical, physiological, and behavioral. *Anatomical adaptations* are the parts of an organism and the ways in which they are connected. *Physiological adaptations* are the ways in which these parts function; physiology includes essentially all of the chemical activities that take place within an organism. *Behavioral adaptations* are the organism's responses to the environment that can be externally observed.

Figure 11-1
The Loss of Organization in a Dead Organism
Only a dark silhouette and a few pieces of wood remain where a dead tree once lay. The structure of the tree has been returned to the environment as a result of physical forces and the activities of other organisms. The stump and log in the photograph are in early stages of decay.

Why Some Organisms Live Longer Than Others

Although all organisms inherit adaptations to environmental problems, not all individuals live out their full life spans. The four major factors contributing to an organism's premature death are competition for resources, the vulnerability of very young and old organisms, environmental changes, and accidental damage.

When the materials that individuals must acquire from the environment are not sufficient to support all members of a population, *competition for resources* takes place. The individuals that win this competition acquire sufficient materials to survive. The losers, weakened by a lack of resources, die prematurely from starvation, dehydration, exposure, attack by other organisms, and many other causes. Any type of adaptive advantage, in the form of variations in speed, strength, specialized feeding structures, intelligence, and many other traits, is an important factor in determining who wins and who loses in the competition for resources. Competition for resources is one of the main ways in which new genotypes are tested by natural selection.

Another important factor in determining survival is the *age of the organism.* Adaptations that enhance survival often are not fully developed in young individuals. The eggs and larvae of aquatic animals, for example, are unable to recognize or evade predators. Young birds that cannot fly and young mammals that are awkward and uncoordinated are easy prey for other animals. The young are also inexperienced in competing for resources. Similarly, old individuals are more vulnerable to predators and less successful in the competition for resources, because adaptive mechanisms are slowly deteriorating due to the aging process.

Each inherited adaptation functions within a restricted range of environmental conditions. If an *environmental change* occurs that is outside the range of previous ancestral experience, the organism may not be equipped to survive. Consider an example of what can happen when a new type of predator enters an organism's environment—a predator with hunting techniques and weapons that have not been encountered in the organism's recent genetic past. When a canal was dug around Niagara Falls in 1931, the new waterway provided the first opportunity for the sea lamprey to move into the Great Lakes of North America (Figure 11-2). The resident

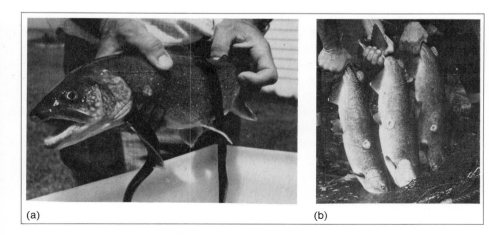

(a)   (b)

Figure 11-2
The Effects of a New Predator
(a) Sea lampreys attacking lake trout.
(b) Fishermen displaying lake trout with scars where sea lamprey had been attached.

populations of lake trout had inherited few defenses against this new predator, which feeds by attaching itself to a fish and sucking out its body fluids. Consequently, the lamprey almost eliminated trout from the Great Lakes. An anti-lamprey campaign on the part of concerned humans has reversed this process to some degree, but these lake trout populations are still vulnerably small.

Any alteration in the physical dimensions of an organism's habitat that disrupts its exchange of vital materials with the environment can also be life-threatening. An unusually heavy snow storm that buries the winter food of grazing elk and antelope can starve or weaken these animals to the point that they are highly vulnerable to predators and disease. Drought conditions can deprive plants and animals of water until they are weakened or killed. An unusually heavy rain storm on a small pond can dilute the chemical concentrations in the water to the degree that many resident organisms are unable to maintain essential body fluids.

Another form of disruption in the delicate balance between organisms and their environments is *accidental damage.* Injury is so common that most organisms inherit mechanisms for repairing at least some of the damage that can occur. In humans, a cut on the skin quickly heals and a broken leg mends. Humans are further protected by cultural procedures for healing and protecting wounded individuals. However, the prospects for survival for most organisms, even those that inherit repair mechanisms, are so precarious that severe accidental damage will probably be fatal. A running deer that steps into a hole and breaks its leg has little chance of survival. Although the deer has the same mechanisms as the human for mending the break, it does not have time to rest the leg. In its wounded state, it is easy prey for a predator and will be unable to find sufficient food and water to survive the repair period. Similarly, a plant stepped on by a grazing animal has mechanisms to overcome the damage but may be so weakened that it cannot successfully compete for the resources it needs to grow.

Organizations for Survival in Unicellular Organisms

In unicellular organisms, all of the capabilities that are required to survive environmental dangers and to acquire essential resources are contained within a single cell. These unicellular organisms are members of the kingdoms Monera and Protista in the five-kingdom system of classification. The kingdom Monera includes all of the bacteria, blue-greens, and methanogens, which are prokaryotic unicellular organisms (Figure 11-3). The kingdom Protista includes the unicellular algae and protozoa (Figure 11-4), which are eukaryotic unicellular organisms.

Most unicellular organisms are invisible to the naked eye but are highly significant components of the world of life. Unicellular organisms occur everywhere and are essential to the continued existence of every ecological community. They are an important food source for larger organisms, particularly in lakes and oceans. Their activities also help to maintain the supply of usable nitrogen, and (together with the fungi) they are the principal decomposers of dead organisms. In the absence of unicellular life forms, biological materials would not be recycled, no nutrients would be available for plant growth, and the earth would be thickly carpeted with dead organisms. Unicellular forms of life are also the primary cause of disease in all other living organisms.

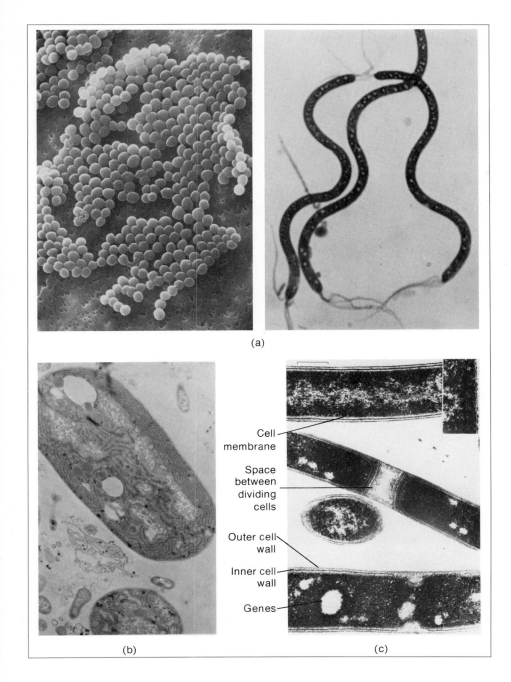

(a)

(b)   (c)

Cell membrane

Space between dividing cells

Outer cell wall

Inner cell wall

Genes

Figure 11-3
Organisms in the Kingdom Monera

Each organism within this kingdom is a single, prokaryotic cell. (a) Two types of bacteria: individuals of the spherical or *coccus* variety on the left ($\times$3,818) and of *spiral* bacteria on the right ($\times$5,500). (b) The internal structure of a blue-green ($\times$28,000). (c) The internal structure of several methanogens ($\times$14,000).

Unicellular organisms provide remarkable examples of successful organizations for survival. After facing billions of years of environmental challenges, unicellular life forms remain abundant. An understanding of the requirements of and the adaptations for unicellular survival is critical to an understanding of multicellular organizations for survival. If the multicellular individual is to survive, it must have adaptations that ensure the survival of its component cells.

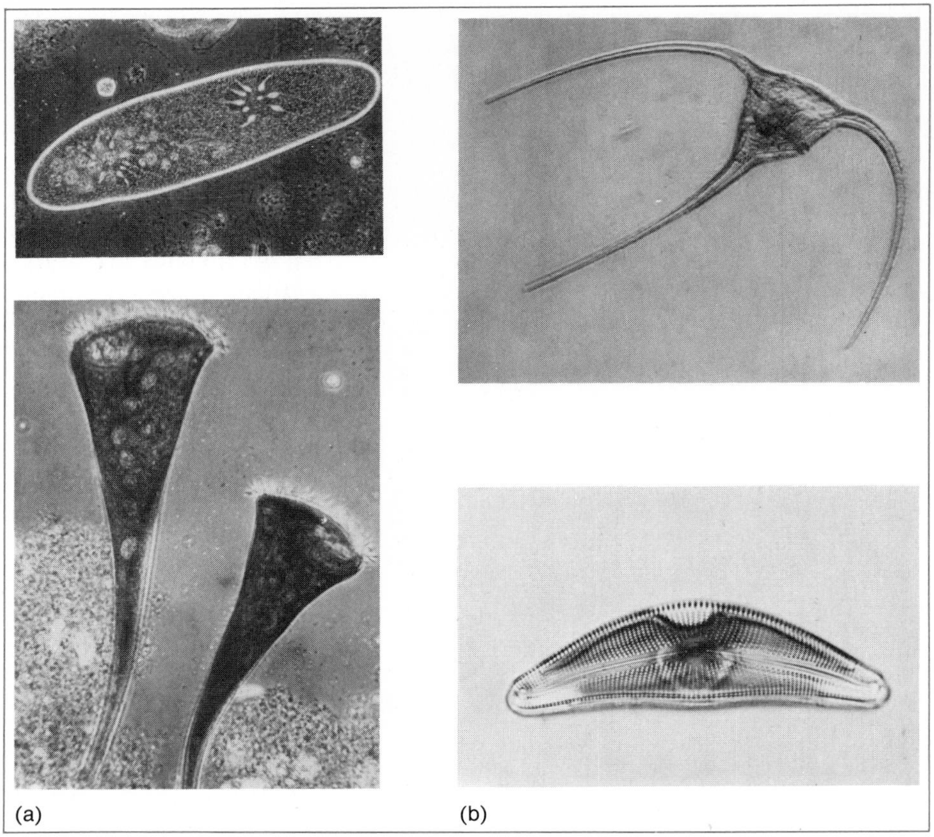

Figure 11-4
Organisms in the Kingdom Protista

Each organism within this kingdom is a single, eukaryotic cell. (a) Two protozoa: a *paramecium* on the top, and a *stentor* on the bottom. (b) Two kinds of unicellular algae: a *dinoflagellate* on the top, and a *diatom* on the bottom.

Packaging Life in a Single Container

In the unicellular organism, all of the structures and chemical reactions required to maintain the activities of life are packaged within a single, membrane-enclosed container. This form of packaging places severe limitations on the adaptations that can evolve in these individuals.

One consequence of being unicellular is that the organism is small. In all cells, the essential exchange of materials with the environment is controlled by a selectively permeable outer membrane. As a cell increases in size, its total chemical activity increases proportionately, which, in turn, increases the need for materials to be transported into and out of the cell. It is a geometric law that as a three-dimensional structure (in this case, a cell) increases in volume, its outer surface area increases at a slower rate. The ratio of surface area to volume therefore decreases as a cell becomes larger. If a cell continued to increase in size, at some point it would no longer have a sufficient surface area to exchange materials with the environment at a fast enough rate to meet its internal chemical demands. An upper limit is therefore placed on cell size; in a unicellular organism, this means that an upper limit is placed on the size of the entire individual.

The size of a unicellular organism is also limited by the methods that are available to transport materials from place to place within the cell. Materials can be transported within a cell by diffusion, movement through the endoplasmic reticulum (when these internal membranes exist), or the continuous flow of fluids within the ground substance of the cytoplasm (called *cytoplasmic streaming*). All three of these methods of transport are slow, inefficient, and adequate only for short distances. They can therefore only meet the transport needs of unicellular life forms when the cell remains small.

Another restriction of packaging life in a single cell is that the individual's small size does not provide much room for complex structures. Certain structures must isolate and concentrate chemicals and other structures must control the movement of materials if elaborate chemical reactions are to occur simultaneously within a cell. The only such structures available to unicellular organisms are internal membranes, which form compartments where chemicals can be concentrated to prevent them from diffusing throughout the cell. The minimal internal structures of the Monera, shown in Figure 11-5(a), permit only relatively simple chemical reactions and slight specialization of functions to occur within the different parts of the cell. However, the Monera can utilize and synthesize fairly complex molecules by making maximum use of the membranes they do have, particularly by concentrating and moving materials along the inner surface of their enclosing cell membrane. Members of the kingdom Protista, on the other hand, have the most highly complex cell structures ever to evolve. Although the protists are unicellular organisms, they contain specialized membrane-bound *organelles,* shown in Figure 11-5(b), that perform such unique functions as synthesizing RNA and DNA, releasing molecular energy, taking in food, storing and excreting wastes, locomotion,

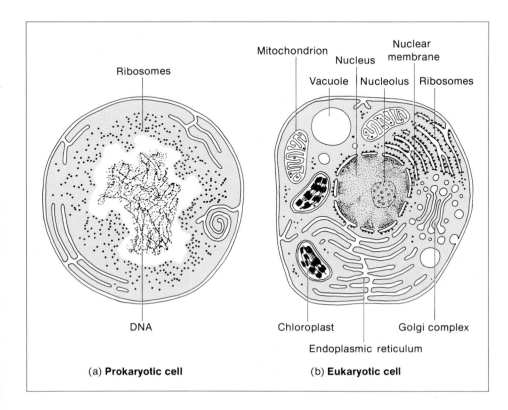

(a) **Prokaryotic cell**   (b) **Eukaryotic cell**

Figure 11-5
The Contrast Between Eukaryotic and Prokaryotic Cell Structures

The *prokaryotic cell* (a) is much less complex than the *eukaryotic cell* (b), which contains a nuclear membrane and many other membrane-bound organelles. The eukaryotic cell shown here contains a chloroplast, within which photosynthesis occurs, although not all eukaryotic cells have this structure. Both cells are the same size here, although prokaryotic cells are usually smaller than eukaryotic cells.

and sensing environmental changes. Although this degree of structural complexity permits chemical complexity, it is much simpler than the structural complexity possible in a multicellular life form in which each type of cell may have its own specialized structure and function.

Acquiring Resources

Although each unicellular individual is somewhat restricted in terms of the activities it can perform, unicellular organisms as a group exhibit a great variety of cellular activities. Almost every type of cellular function performed in multicellular organisms is also performed in unicellular organisms, and many chemical reactions are unique to unicellular organisms.

Within the kingdom Monera, *bacteria* are particularly diverse in function. Some bacteria are *autotrophs* (self-feeders) that obtain energy from sunlight or from inorganic molecules. However, most bacteria are *heterotrophs* (other-feeders) that obtain energy and nutrients from the molecules within other organisms. The largest group of bacteria—the *decomposers*—recycle biological materials back into the food chain by feeding on dead organisms. Another large group of bacteria—the *parasites*—live within other organisms and feed on their accumulated materials.

Most *blue-greens* are *photosynthetic autotrophs* that use sunlight as a source of energy and release oxygen as a by-product of the reaction. These autotrophs have the simplest nutritional requirements of any living organism: the sun's energy, carbon dioxide, nitrogen gas, a few minerals, and water. Blue-greens live in soil, fresh water, and oceans in all parts of the world. Some blue-greens live within protists in symbiotic relationships (Figure 5-13, page 96) in which the blue-greens provide photosynthesized sugars for the protists and the protists provide protection and nutrients for the blue-greens.

Little is known about the third group of monerans—the *methanogens*. These unicellular organisms were once considered to be a form of bacteria, but the methanogens perform such unique basic chemical activities that they are now categorized as a separate evolutionary line. The methanogens are strictly *anaerobic*, living only in habitats that lack oxygen gas, and acquire energy by converting hydrogen gas (H_2) and carbon dioxide (CO_2) into methane (CH_4). These unusual forms of prokaryotes live in sediments at the bottom of lakes, ponds, and marshes; they also occur in the digestive tracts of herbivorous mammals.

The kingdom Protista includes the unicellular algae and protozoa. Almost all *algae* are photosynthesizers containing chloroplasts and are exceedingly abundant in fresh or marine waters. The *protozoa* obtain their resources in a variety of ways. Some protozoa, like the *amoeba* and the *paramecium*, are predators with adaptations for attacking, capturing, and ingesting other unicellular and even small multicellular organisms. Other protozoa feed exclusively on small particles of biological material dissolved in liquid solutions. A large and diverse group of protozoa is parasitic on or within other organisms. Most protozoa have a well-developed means of locomotion and occur everywhere—in the air, soil, and water as well as inside other forms of life.

To understand the capabilities of a unicellular organism, we will examine one form of protist—the paramecium—in greater detail. The paramecium is found in fresh water everywhere. This microscopic organism moves through water by rhythmically beating its hairlike organelles, called *cilia*, which extend outward

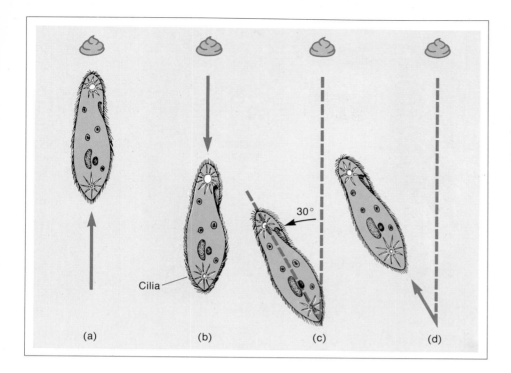

**Figure 11-6
Avoidance Movements
in a Paramecium**

(a) The paramecium detects a toxic substance ahead. (b) It backs up by reversing the movements of its cilia and (c) makes a directional change of 30°. (d) The paramecium then readjusts its ciliary movement and proceeds in a forward direction again.

30°

Cilia

(a)   (b)   (c)   (d)

from the cell membrane. The cilia not only move the paramecium but are also sensitive to such conditions as the temperature and chemistry of the surrounding water and the presence of a solid object ahead.

The paramecium is adapted to live within a very narrow range of environmental conditions, which the cilia are genetically programmed to recognize. If the cilia detect a suitable environment ahead, they will beat in a way that propels the organism forward; if the cilia detect unsuitable conditions ahead, they will automatically reverse their movements to propel the organism backward. The paramecium then makes a 30° turn and moves in a forward direction again (Figure 11-6). This forward—backward, 30°-turn pattern is repeated until the cilia detect and propel the organism toward a suitable environment, where the organism remains in place.

The permanent, cilia-lined infold of the paramecium's outer membrane, shown in Figure 11-7(a), is comparable to the mouth and digestive tract in multicellular animals. If the paramecium encounters food, the cilia within this infold drive a stream of water containing the food down into the organism's digestive tract. Part of the tract then indents to form a pocket around the food. This pocket eventually pinches off from the digestive tract, so that the food is contained within a membrane-enclosed sac, or *food vacuole.* As this vacuole moves through the cytoplasm, digestive enzymes diffuse into it and dissolved food particles diffuse outward into the surrounding cytoplasm. By the time the vacuole reaches the *anal pore*—a permanent opening comparable to the anal opening in higher animals—it contains only unusable materials, which are expelled into the external environment.

Because it lives in fresh water, a paramecium must continually resist the inward

Figure 11-7
**Digestion and Water Control
in a Paramecium**

(a) Movement of *cilia* surrounding the
mouth causes food items to enter the
digestive tract. Part of the tract then
encloses the food within a membrane
sac, or *food vacuole*, which pinches off
from the rest of the digestive tract. As
the food vacuole moves through the
interior of the organism, enzymes
enter it and the food is digested.
Usable molecules leave the vacuole;
waste molecules remain in the vacuole
and are transported to the *anal pore*
and out of the paramecium. (b) The
contractile vacuole functions in the
paramecium to remove excess water
from the interior of the cell. On the
left, the contractile vacuole is being
filled with water from channels that
extend throughout the cell. On the
right, the vacuole is contracting,
forcing water out through the
opening. At the same time, the walls
of the channels contract to force more
water into the vacuole to be pumped
out during the next contraction.

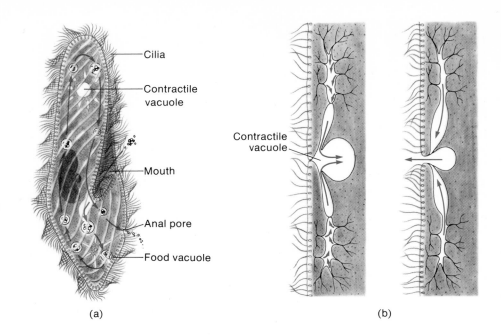

(a)

(b)

diffusion of water, which it accomplishes by means of specialized *contractile vacuoles*.
These vacuoles expand as they take in excess fluids from the cytoplasm and then
contract when they are full to pump the excess water back into the external
environment. This process is diagrammed in Figure 11-7(b).

The paramecium is only one example of the many protists that have specialized
structures to promote division of labor despite the severe limitations imposed by
unicellular organization.

Defenses Against the Environment

Every cell—whether it is a unicellular organism or a component of a multicellular organism—can maintain its vital activities only within a narrow range of
conditions. The unicellular organism is able to continue its internal chemical
activities, grow, and reproduce only when the environment matches its narrow
range of genetically determined optimal conditions.

One solution to this restriction is to live in an environment that is both optimal
and relatively unchanging, such as an ocean or some other large body of water.
Therefore, it is not surprising that vast numbers of unicellular organisms inhabit
large bodies of water. Different regions of the oceans are highly stable with regard
to temperature, salt concentrations, amount of light, and so on. Populations of
unicellular organisms have evolved in each of these regions and consequently
function optimally under these specific conditions.

Another type of relatively stable environment is found within multicellular
organisms. An enormous variety of unicellular organisms live on or in multicellular forms and have been widely distributed over land and through water by
their multicellular hosts.

Unicellular organisms are also able to inhabit changing environments by reducing their rate of chemical activity and entering a resting or *dormant state* when conditions are unfavorable. Many unicellular forms secrete a thick outer coating around the cell membrane to prevent the exchange of materials with the environment, remaining in this condition until the environment can support their chemical activities again. Some bacteria are able to remain alive within protective coatings for many years and can even survive such extreme environments as boiling water.

Organizations for Survival in Multicellular Organisms

Many biologists believe that if multicellular organisms had not evolved, life might still be confined to the world's oceans. The invasion and colonization of dry land required a complex of adaptations that could not be acquired under the severe restrictions imposed by unicellular organization. These elaborate adaptations were made possible by a new form of living organization—the multicellular individual. This new fundamental unit of life, which first appeared in the fossil record approximately 1 billion years ago, was truly revolutionary in terms of its enormous potential for adaptation.

The first multicellular organisms were shallow-water plants that evolved the ability to live simultaneously in the moist, dark, mineral-rich soil and the dry, brilliantly illuminated air. The complex of adaptations that permitted these pioneering individuals to survive in two such vastly different environments included mechanisms for extracting nutrients and water from the soil, taking in gases from the air, releasing waste products into a dry environment without excessive water loss, supporting the plant against the forces of wind and gravity, and transporting materials through the enlarged body mass.

The adaptive potential of these new organisms did not result from the development of new and superior types of cells. The cells of the most complex multicellular organisms have fundamentally the same organization and biochemical capabilities found in eukaryotic unicellular forms; no single cell in a multicellular organism is as complex and self-sufficient as a paramecium. Instead, the revolutionary potential of multicellular organization was the result of a new biological phenomenon—the *division of labor* among the many specialized cells within a single individual.

In unicellular organization, a single cell contains all of the structures and chemical activities of the organism; in multicellular organization, these tasks are divided among groups of cells. This potential for division of labor freed the evolutionary process from the binding limitations of unicellular organization. Single cells must remain quite small, but progressively larger organisms can evolve as aggregations of small cells in which each cell performs a single task, such as digestion, reproduction, or exchanging gases with the environment.

The Dual Organization

An individual cell within a multicellular organism is faced with many of the same survival problems of a unicellular organism. Both cells are aquatic "organisms" that must maintain a regular exchange of materials with a liquid environment. Like unicellular organisms, individual cells within a multicellular organism can only function within a narrow range of environmental conditions and are also subject to attack by microorganisms and larger predators.

Due to their specialized nature, the cells in a multicellular organism are less self-sufficient than the cells in unicellular organisms. *Cellular specialization* is accomplished by reducing the diversity of the chemical activities that take place within a cell to those required for basic maintenance and the performance of the specialized function. Vertebrate liver cells, for example, produce and secrete bile; nerve cells are restricted in function to the transmission of electrochemical messages. This reduction of chemical activities deprives specialized cells of many unicellular capabilities. They can no longer obtain food, move away from danger, or seek an optimal environment; they cannot even enter a dormant state, because their activities are essential to the survival of the entire organism.

These essentially helpless cells can only survive if all of their needs are met by the multicellular organism within which they live. Each multicellular organism therefore has a *dual organization.* On one hand, the organism is a coordinated unit that responds to dangers and opportunities in the external environment; all of the cells work together to solve the survival problems of the entire organism. On the other hand, each multicellular organism provides a relatively constant and optimal liquid internal environment for its cellular components. This dual organization—to act as an integrated unit and to sustain individual cells—is basic to all multicellular organisms.

The Extracellular Fluid

The fluid that surrounds and nourishes every living cell within a multicellular organism is called the *extracellular fluid* ("extra" = outside of). In the first multi-cellular organisms, this fluid was probably seawater from the marine environment, which was moved through the spaces between the cells by the ebb and flow of the ocean. The evolution of increasingly complex plants and animals was accompanied by the evolution of new, specialized mechanisms that allowed individual organisms to produce and regulate their own extracellular fluids. These fluids now exist

Figure 11-8
The Extracellular Fluids of a Human

Two kinds of fluids surround and nourish the cells within animal tissue—*blood plasma* and *lymph.* Plasma forms the liquid part of the blood and is located within the arteries, veins, and capillaries. The cells of the tissue are immersed in lymph, or *interstitial fluid.* Both fluids regularly exchange materials through the capillary walls.

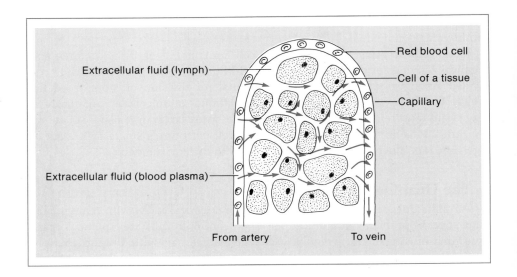

Extracellular fluid (lymph)

Extracellular fluid (blood plasma)

Red blood cell

Cell of a tissue

Capillary

From artery

To vein

Table 11-1
Similarities Between Extracellular Fluids and Seawater

Concentrations of the various ions are relative to sodium (Na^+). For example, for every 100 sodium ions in seawater, there are 3.6 potassium ions (K^+). The king crab is considered to be almost identical to its Triassic ancestors and is therefore one of the more ancient organisms; its extracellular fluids are most similar to seawater. Although the extracellular fluids of the codfish and the human have apparently undergone some evolutionary adjustments in terms of ionic concentration, they remain remarkably similar to seawater.

	Na^+	K^+	Ca^{++}	Mg^{++}	$SO_4^=$	Cl^-
Seawater	100	3.6	3.9	12.1	20.9	181
King crab	100	5.6	4.1	11.2	13.4	187
Codfish	100	9.5	3.9	1.4		150
Human	100	6.8	3.1	0.7		129

Source: J. W. Kimball, *Biology*, Third Edition (Reading, MA: Addison-Wesley, © 1974), Figure 6-3. Reprinted with permission.

in many diverse forms, including the sap of plants, the blood of insects, and the blood and lymph of mammals (Figure 11-8).

Although a variety of extracellular fluids can be found in modern animals, all of these fluids have many properties in common with seawater, particularly in terms of their concentrations of such minerals as sodium, potassium, and calcium (Table 11-1). Clearly, when multicellular animals invaded the land, they brought with them an internal sea to which their cells were accustomed. A human is actually composed of billions of cells immersed in a watery salt solution living within a thin-walled container in the dry air. The extracellular fluids of terrestrial plants are less similar to seawater; nevertheless, each plant is also an aquatic community of cells.

Homeostasis

The chemical process of life takes place within a narrow range of conditions. All organisms have some capacity to maintain their internal conditions so that the chemical reactions inside their cells can continue without interruption no matter what changes are occurring in the external world. With the exception of a small percentage of specialized reproductive cells, all of the cells in a multicellular organism contribute to the maintenance of optimal conditions in their common fluid environment. The maintenance of a state of internal constancy is referred to as *homeostasis* ("homeo" = steady; "stasis" = standing).

Forces such as the climate, the stress of physical activity, and the disruptions of eating a heavy meal act continuously to destabilize the organism's internal environment. Each disturbance in an organism's internal environment triggers a response that tends to reverse the disturbance and return the internal fluids to their optimal state.

All organisms maintain some degree of homeostasis, but multicellular organisms have a much greater degree of control over their internal environment than unicellular organisms. Homeostasis is most highly developed in birds and mam-

**Figure 11-9
Temperature Regulation
in a Plant**

The skunk cabbage is able to emerge early in the spring, because it has the capacity to melt the surrounding snow with the heat of its tissues. The structures shown in the photograph enclose the flowers of the cabbage, which maintain a nearly constant internal temperature of 22.4°C. The plant generates this high temperature, which is much warmer than the surrounding snow and air, by maintaining a rapid rate of cellular respiration, which produces heat as a by-product.

mals. These animals can regulate various aspects of their extracellular fluids, including temperature, ratio of acids to bases, and concentrations of water, oxygen, minerals, sugars, and cellular waste products.

Plants have also evolved adaptations to maintain a constant internal environment and even to control temperature to some extent. Each leaf of a vascular plant contains a great number of valvelike openings that control both the temperature and the water content of the plant's internal fluids by regulating how much water evaporates from the leaf surface (evaporation removes heat). The skunk cabbage (Figure 11-9) produces flowers that are warm enough to melt through snow and ice. The plant maintains the temperature of its flowers at 22.4°C (72°F) by sustaining a rapid rate of cellular respiration, which produces heat as a by-product.

Anatomical, physiological, and behavioral adaptations all play a role in homeostasis. For example, anatomical adaptations for the regulation of temperature include variations in the outer surfaces of the organism from thick to thin, depending on the amount of heat that has to be released. Physiological adaptations include the amount of heat produced by varying the rates of cellular chemical activities that give off heat. Behavioral contributions to temperature regulation include seeking shade or sun and adopting postures that facilitate heat intake or loss (Figure 11-10).

The degree to which an organism can control its internal environment is also the degree to which the organism can live independently of changing environmental conditions. Evidence to support this fact can be found by contrasting life in a northern field or pond in the summer and in the winter. In the summer, all of the organisms are fully active. In the winter, mammals and birds, such as foxes, rabbits, and sparrows remain active, but frogs, insects, and plants must become dormant to survive the cold.

Basic Features of Plant and Animal Organizations

There are three kingdoms of multicellular organisms in the five-kingdom system of classification—Plantae, Animalia, and Fungi. In this section, we will consider the survival problems and adaptive solutions of the plants and animals. We will conclude this chapter with an examination of the multicellular fungi, which have characteristics of both plants and animals.

Over a billion years of multicellular evolution, natural selection has preserved many similar solutions to the problems of survival faced by a multitude of diverse forms of plants and animals. This selection process enabled both plants and animals to invade land and to withstand all of the new environmental challenges that accompanied their transition from relatively stable aquatic conditions to the harsh, dry, and variable terrestrial world. An increase in size provided both lines with selective advantages. Larger plants had greater access to the available light and could prevent light from reaching their competitors. Larger animals could capture a greater array of prey and were less accessible to predators. Many adaptations also evolved in both the plant and the animal kingdoms to promote a relatively stable, liquid, internal environment for the increasingly specialized and helpless cells buried deep within these larger organisms.

By examining the physical characteristics of a multicellular organism, an expert

Figure 11-10
Temperature Regulation in Animals

Every animal must maintain an optimal temperature to maintain its cellular activities. Animals other than birds and mammals attain this internal temperature by seeking areas in the environment where the optimal temperature prevails. These animals are positioned at their optimal temperatures on the graph. Natural selection has modified their body chemistry so that chemical reactions occur best at the temperatures found in their external environments. Note that the optimal temperature differs greatly for the various animals shown. The enzymes in the Antarctic icefish, for example, function most rapidly and efficiently at much lower temperatures than the enzymes in the desert pupfish.

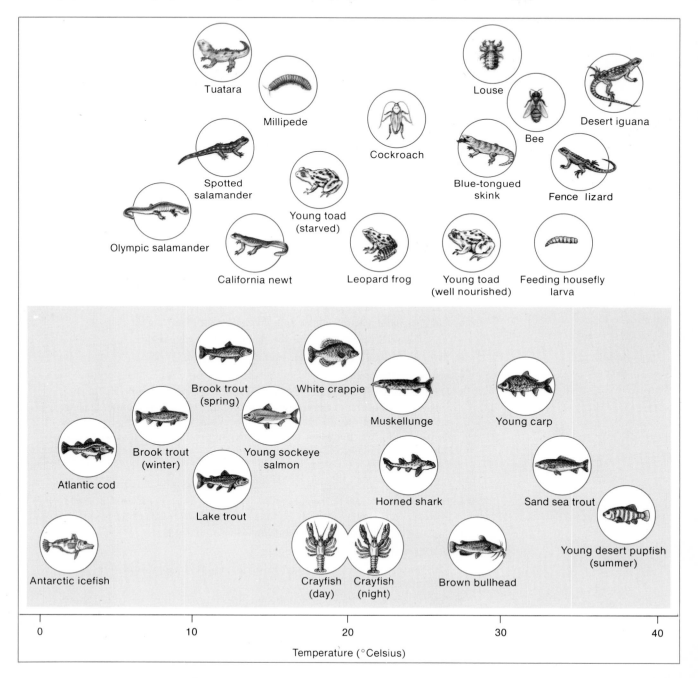

can learn about its way of life, its line of evolutionary descent, the range of environmental conditions experienced by its ancestors, its method of exchanging materials with the environment, its enemies, and its sources of energy and nutrients. The three primary characteristics that determine these factors are the types of cells, the structures for acquiring and processing energy and nutrients, and the mobility of the organism (Table 11-2).

Table 11-2
Summary of the Characteristics Used to Distinguish Among Plants, Animals, and Fungi

Kingdom Plantae

- *Type of organism:* Multicellular with division of labor among the cells.
- *Cell structure:* Eukaryotic with plastids and rigid cell walls.
- *Nutrition:* Mostly autotrophic through photosynthesis.
- *Movement:* Fixed in place as adults; some capacity to move in response to environmental changes by growth and by altering internal liquid (turgor) pressure.
- *Trends in evolution:* Specializations of body parts, such as leaves for light absorption and gas exchange, channels for transporting materials, roots for acquiring water and nutrients, and outer surfaces for protection. Control and integration of vital activities entirely by chemical messengers.

Kingdom Animalia

- *Type of organism:* Multicellular with division of labor among the cells.
- *Cell structure:* Eukaryotic not enclosed in cell walls; no plastids.
- *Nutrition:* All animals are heterotrophic.
- *Movement:* Controlled mobility; some are sessile in certain phases of life.
- *Trends in evolution:* Great variety of highly complex, specialized cells for such functions as movement, sensing and responding to the environment, digestion, gas exchange, transport of materials, homeostasis, structural support, and the formation of protective outer surfaces. Integration of bodily activities by both chemical messengers and a nervous system.

Kingdom Fungi

- *Type of organism:* Most are multicellular with little division of labor, except during the reproductive phase of life.
- *Cell structure:* Eukaryotic with rigid cell walls; no plastids.
- *Nutrition:* All fungi are heterotrophic.
- *Movement:* Each cell is fixed in place, but the entire organism can enter new environments by growth.
- *Trends in evolution:* Rapid growth by moving materials from one cell to another; efficient digestion; production of toxic materials as a form of defense; dormancy during food shortages; tolerance for high concentrations of waste materials; the formation of mutually beneficial relationships with other kinds of organisms.

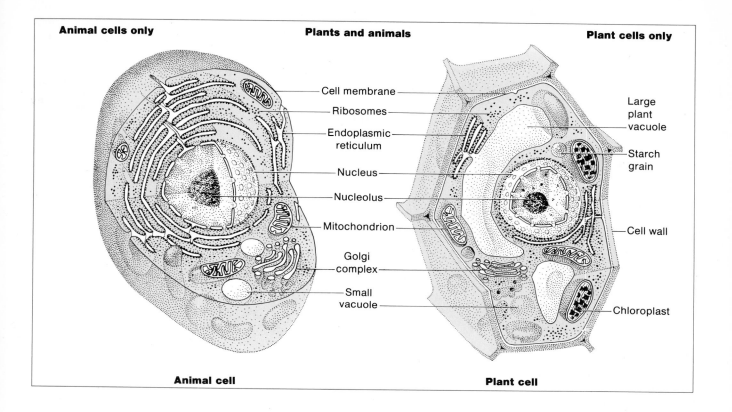

Animal cells only   **Plants and animals**   **Plant cells only**

Cell membrane

Ribosomes

Endoplasmic reticulum

Nucleus

Nucleolus

Mitochondrion

Golgi complex

Small vacuole

Large plant vacuole

Starch grain

Cell wall

Chloroplast

Animal cell   **Plant cell**

Two fundamental differences between plant and animal cells reflect very different evolutionary trends in these two kingdoms (Figure 11-11). First, virtually all plants have at least some cells that contain light-absorbing pigments within plastid organelles, whereas these structures are not present in animal cells. Second, a rigid outer wall surrounds the cell membrane of all plant cells; animals cells lack these structures.

These differences in cellular structure reflect the fact that plants and animals acquire energy and nutrients in quite different ways. All plants are autotrophs and can synthesize their own molecules from inorganic nutrients and sunlight. The plant leads a slow-paced life in a fixed position. Its survival problems have been solved by the evolution of thick-walled cells organized into leaves to provide surfaces for light absorption and gas exchange, roots to anchor the plant and extract water and minerals from the soil, rigid columns to transport extracellular fluids and provide mechanical support, areas to use for food storage, and outer surfaces to protect against desiccation and mechanical damage. All plants use chemical messengers, or *hormones,* to coordinate cellular activities and to synchronize the plant's activities with the changing seasons.

All animals are heterotrophs and can obtain energy and nutrients only by eating other organisms. Because they make use of the biological molecules synthesized by other organisms, animals must have some mechanism to convert these molecules into their own unique forms. Most animals ingest large food

Figure 11-11
The Contrast Between Animal and Plant Cells
Structures found only in plant cells are listed on the right; structures found in both plant and animal cells are listed in the center. Only a few structures that exist during cell division (not shown here) are unique to animal cells.

particles and store them in internal cavities, where the food particles are sub-sequently broken down into simpler molecules, absorbed into the internal environment, and eventually converted by the cells into other molecular forms.

A unique characteristic of animals is that they have *controlled mobility* and are able to move their entire bodies from place to place. Some animals do not have this capacity and, like plants, are sedentary, or *sessile.* Most animals are mobile throughout their lives, although a variety of animals have sessile and mobile phases of life. Clams, coral, and barnacles are mobile during their immature phases but are fixed in place as adults.

Many selection pressures have influenced the evolution of controlled mobility in animals. The greatest pressure has been the nutritional specialization of animals for eating other animals. To survive, the predator must be able to overtake and kill the prey; to survive, the prey must avoid capture. This ongoing predator–prey struggle has produced adaptations that have increased rapid movement in both

Figure 11-12
Cells That Are Unique to Animals

Nerve cells coordinate and control movement by receiving information at the *dendrites* and transmitting this information to other cells via the *nerve axons. Muscle cells* have the unusual ability to contract (become shorter) and to relax (become longer). Each muscle cell contains several nuclei. *Sensory cells* receive information directly from the environment and transmit it to nerve cells. These cells are highly specialized in structure and function. The rod cell of the eye shown here detects information about visual aspects of the environment.

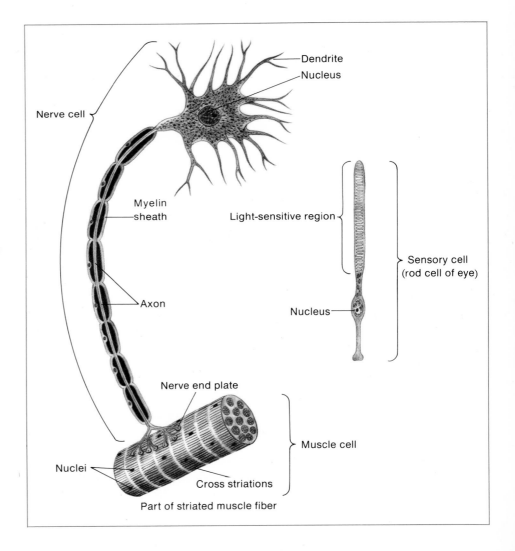

predators and prey. These adaptive trends include the capacity to scan and process environmental information, the precise and coordinated movement of body parts, and the development of mechanisms to monitor and regulate internal fluid composition to meet the demands of increased cellular activity during rapid movement. Three types of specialized cells—the muscle, nerve, and sensory cells (Figure 11-12)—which appeared early in animal evolution and are found today in all except the simplest animals, played an integral part in these predator–prey adaptations.

Muscle cells are specialized to contract or shorten and to relax or lengthen, thereby altering the position of the rigid body parts to which they are connected and permitting the animal or some of its parts to move. *Nerve* and *sensory cells* are specialized in the property of irritability, or the capacity to respond to environmental stimuli. Sensory cells, most of which are modifications of nerve cells, occur in many forms. Each type of sensory cell is sensitive to a specific type of information, such as light, sound, or pressure. Nerve cells coordinate and control body activities. Movement and almost all other aspects of animal physiology and behavior involve the muscle, nerve, and sensory cells.

Plants have no adaptations for rapid whole-body movement through the environment and lack the specialized muscle, nerve, and sensory cells found in animals. Because plants can manufacture their food from materials in the soil and energy from the sun, they can remain in a fixed position, so that there has been no selection for mobility in plants. However, plants do have the capacity to coordinate the movement of their body parts in response to environmental changes in light, gravity, temperature, touch, and humidity. These movements are caused by cell growth and by changes in pressures within the cells.

Plants and animals therefore acquire energy and nutrients in fundamentally different ways. These differences have led to a divergence of adaptations during the evolution of these two forms of multicellular organization.

The Basic Features of Organization in the Fungi

The fungi (see Table 11-2) are a peculiar group of organisms. They resemble plants in many ways, but they are heterotrophs. They are similar to animals, but they have cell walls and do not have specialized muscle, nerve, or sensory cells to provide mobility. Most individual fungi are aggregates of eukaryotic cells, but there is little division of labor among the cells except during a brief reproductive phase. The kingdom Fungi includes the mushrooms, yeasts, and molds (Figure 11-13).

The development of a multicellular fungus begins when a single cell divides by mitosis. As the cells form, they become elongated and join at the ends to produce a long thread of cells, called a *hypha.* At various points, the hypha branches to form a meshlike structure. Cell division occurs only at the ends of the hypha, causing one-directional horizontal growth outward from the mesh of cells. The older portions of the body do not undergo cell division. The meshlike body structure permits each cell to come into direct contact with the external environment.

An unusual feature of this type of multicellular organization is that connections between the adjoining cells allow fluid and organelles to flow freely from one cell to the next. From a microscopic viewpoint, the interior of the fungus is teeming with

Figure 11-13
Types of Fungi

(a) Mushrooms occur in a variety of beautiful colors and forms. The reproductive body of a tropical species that grows on dead wood appears on the top; the reproductive body of a poisonous mushroom that grows on dead leaves on the forest floor appears on the bottom. (b) Yeasts are unicellular forms of fungi that are used in brewing and baking. (c) Molds require damp conditions for growth. These very small slime molds are reproducing on the surface of dead vegetation.

activity. Materials flow from one part of the body to another in response to chemical and physical forces. New materials are synthesized in the older cells and then stream to the growing tip, resulting in very rapid growth. Each cell at the growing tip usually contains several nuclei that have been formed by mitosis and are ready to be incorporated into the new cells.

Each fungal cell obtains energy and nutrients by releasing digestive enzymes onto other organisms. These enzymes convert large molecules into smaller molecules, such as simple sugars and amino acids. Each cell is able to synthesize a variety of digestive enzymes, so that the fungus can make use of many different kinds of foods. After external digestion is complete, the smaller molecules diffuse through the cell wall and move across the cell membrane by active or passive transport. Water is essential for diffusion, and most fungi can survive only under conditions of high humidity. Once the smaller molecules are absorbed into the cell, intracellular enzymes either break them down to release the energy they contain or incorporate them into new molecular forms. The waste products of these reactions then leave the cell by diffusion.

Fungi feed on a variety of other life forms. Like the bacteria, many fungi feed on dead organisms, especially the wood and leaves of plants, helping to recycle nutrients within every ecosystem. Other kinds of fungi are parasites that feed on the accumulated materials within living plants and animals. These forms of fungi destroy crops and cause human disease. Some fungi are active predators that capture small worms (Figure 11-14) and protists by trapping prey with a sticky secretion or by entangling them within a network of threadlike cells. The fungus then penetrates the body of the prey and digests and absorbs its contents.

Many fungi establish mutually beneficial relationships with other organisms, including algae, insects, and plants. A *lichen*, for example, is a mutually interdependent association between a fungus and algae (Figure 11-15); neither the fungus nor the algae can survive alone. The fungus obtains nutrients from the photosynthesizing algae; the algae gain the ability to inhabit a greater variety of environments from the fungus. Lichens can survive extreme environmental conditions that cannot be tolerated by the majority of multicellular organisms.

Like all other cells, the cells of each type of fungus can operate only under a specific set of physical conditions. Fungi are able to seek these conditions by growth and by bending their multicellular bodies.

The primary threat to the survival of an individual fungus is *starvation*. The outward expansion of the body when food is abundant facilitates the discovery of new food sources. But if the fungus cannot find food to sustain further growth, the life of the individual is threatened. One solution to the problem of inadequate food sources is dormancy, and most fungi are able to enter such a resting state. When

Figure 11-14
Predation by a Fungus

This small roundworm (×132), or *nematode*, has been captured by the sticky secretions on the threadlike cells of a fungus. The fungal cells will eventually grow into the worm, secrete digestive enzymes, and absorb the dissolved molecules.

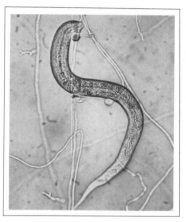

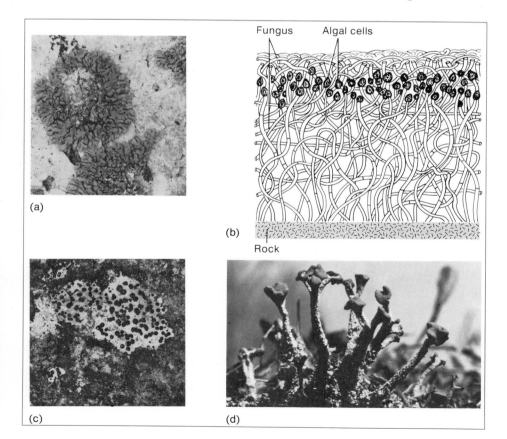

(a)

(b)

Fungus   Algal cells

Rock

(c)   (d)

Figure 11-15
Lichens—Associations Between Fungi and Algae

(a) Lichen growing on a rock. (b) Diagram of the internal structure of a rock lichen, showing the arrangement of the unicellular algae and the elongated cells of the fungus. (c) Two kinds of lichens growing on the bark of a tree. (d) Brilliant lichen (3 millimeters in height)—called "the British soldier" because of its bright red color.

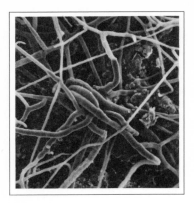

Figure 11-16
A Network of Fungal Hyphae Feeding on the Tissues of a Bean Plant (×265)
Each cell releases digestive enzymes into the tissue and absorbs the liquid, digested material into its interior.

food becomes available again, dormant fungal cells in the immediate vicinity of the food burst into activity (Figure 11-16), converting food molecules into the materials required to enlarge the network of hyphae. When the food source is exhausted, the fungus becomes dormant again. Dormancy also occurs when fungal cells are surrounded by actively feeding bacteria, indicating that there is competition for food between these two forms of decomposers.

Competition for food is a crucial problem for the fungi. Each individual is usually surrounded by other fungi of the same and different species. All fungi ingest the same small molecules of sugar and amino acids, although the evolution of dietary specializations between species with regard to the larger food molecules has decreased the intensity of the competition for food.

Many of the characteristics of fungi are adaptations to help the organism compete for food. One characteristic is its rapid growth rate, which facilitates the movement of materials from the older, nondividing cells to the younger, dividing cells at the tips of the hyphae (exemplified by the rapid outward growth of mold on bread). A rapid growth rate permits the fungus to discover new food sources and convert them into body cells before the food is located by competing individuals. A second form of fungal adaptation to competition for food is a high degree of efficiency in the production of appropriate digestive enzymes. As each new type of food is encountered, the specific enzymes required to break down its molecules are synthesized by the fungal cells. The speed and efficiency of these enzymes has been favored by natural selection, because individuals that can convert food efficiently are able to grow and survive when competition is intense. A third characteristic of the fungus that arises from the competition for food is its ability to produce substances that are toxic to other organisms. Some of these substances—*antibiotics*—are effective against bacteria, the chief competitor of some fungi. Additional substances released by fungal cells act specifically against other species of fungi.

Fungi also face the problem of the *pollution* generated by their own waste materials. Because fungi are unable to move their entire bodies from one place to another, wastes from their own chemical activities may become so highly concentrated within their immediate environment that survival is no longer possible. To some extent, growth away from these areas allows the actively dividing cells to exist in relatively waste-free areas. Another adaptive characteristic is a tolerance for these wastes. Many fungi are able to survive in the presence of high concentrations of their own waste materials as well as the waste products released by other fungi in the environment.

A final threat to the survival of the fungus is attack by *parasitic* or *predatory organisms.* In particular, bacteria often invade the cells of a fungus, rupturing the cell membranes. To defend against this form of attack, the fungi have rigid cell walls and the ability to produce antibiotics.

Summary

Each organism inherits characteristics that promote its survival and reproductive success in a basically hostile environment. The environment is hostile to the individual organism in two ways. First, many organisms feed on other organisms to obtain energy and nutrients. Second, nonliving forces—including electrical forces, diffusion, and tendency toward entropy—act continuously to disrupt the organization of living material and to disperse its contents throughout the environment.

An organism cannot be a completely closed container; it must exchange materials with the external environment, which makes it vulnerable to the loss of desired materials and to the gain of undesirable materials.

Adaptations are inherited solutions to the problems of survival. The three basic forms of adaptations are *anatomical, physiological,* and *behavioral adaptations.*

Although all organisms inherit adaptations to environmental problems, many organisms die before achieving their maximum life spans. Most premature deaths can be attributed to (1) failure in the competition for resources, (2) vulnerability of very young and old organisms, (3) environmental changes, and (4) accidental damage. By removing certain genotypes from the population, these deaths contribute to the process of natural selection and to the further evolution of adaptive characteristics.

The organisms in the kingdoms Monera and Protista are packaged within a single cell, which severely restricts the evolution of complex adaptations. The major restriction of the unicellular organism is its *small size.* A single cell cannot become very large, because its volume increases at a faster rate than its surface area. If a single cell were permitted to grow indefinitely, the area of the cell membrane would become too small relative to the cell's volume at some point and it would no longer be able to exchange materials with the external environment at an adequate rate. In addition, as a cell becomes larger, it is no longer able to move materials from place to place within its interior rapidly enough to remain fully active.

A second disadvantage of unicellular organization is that it *limits the diversity of the chemical reactions* that can occur within the organism at the same time. Although unicellular organisms contain cell membranes and the Protista also contain *organelles* to compartmentalize many chemical activities, this unicellular organization does not permit the diversity of chemical reactions that occur in a multicellular organism.

A third restriction on the unicellular organization is that it must live in an essentially *stable liquid environment.* The organism must be able to exchange materials across its delicate outer membrane to maintain its internal environment within a narrow, optimal range of physical conditions. Although unicellular organisms have adapted to dry and highly variable environments, their life cycle under such conditions usually involves a *dormant state,* during which the organism is surrounded by a protective coating.

Despite these limitations, the unicellular organism has survived remarkably well throughout the history of life. The waters, soils, and multicellular organisms of the earth are teeming with unicellular organisms, and many forms have persisted essentially unchanged for billions of years. Clearly, within the environmental conditions to which it has adapted, the unicellular organism is an almost perfect organization for survival.

Multicellular organisms have a greater ability to adapt to their environments than unicellular organisms do. Multicellular organization, involving a *division of labor* between cells, permits the *specialization of structure and function* among various cells within the organism, so that it can survive and reproduce in a very different environment from its ancestral ocean. Each multicellular organism has a *dual organization:* it must provide for the survival of its individual cells, and it must also respond as a unit to the external environment. To ensure the survival of each cell, many specialized cells maintain the chemistry and temperature of an appropriate *extracellular fluid* at optimal conditions for cellular activities. *Homeostasis* is the term used to describe this state of internal constancy, which has developed to the greatest degree in birds and mammals.

All plants and animals are composed of eukaryotic cells that contain many common structures and biochemical pathways. However, two fundamental differences reflect the separate evolutionary heritages of plants and animals. First, plant cells contain light-absorbing pigments arranged inside organelles called *plastids*; no animal cells contain these structures. Second, plant cells have rigid outer walls, which are not found in animal cells. Plants and animals also differ strikingly in their capacity for movement. Most animals have *controlled mobility*, whereas most plants are *sessile*. Mobility in animals is made possible by specialized *muscle, nerve*, and *sensory cells* that are not found in plants. The major differences in the evolution of the kingdoms Plantae and Animalia reflect basic differences in the ways in which these two types of multicellular organisms acquire energy and nutrients. Plants are *autotrophs*, synthesizing their own molecules from inorganic nutrients and sunlight. Animals are *heterotrophs*, acquiring energy and nutrients from the molecules of other organisms.

The *fungi* are placed in a separate kingdom in the *five kingdom system of classification*. Fungi have cell walls, but they are not plants because they cannot synthesize their own food. Fungi are also heterotrophs, but they cannot be classified as animals because they have no specialized muscle, nerve, and sensory cells to provide mobility. The fungi are usually multicellular organisms, but they exhibit a division of cellular labor only during the reproductive phase of their lives.

The fungi face four serious problems of survival—starvation, competition for food, waste pollution, and predation. Adaptations for acquiring food include *rapid growth*, which is facilitated by the movement of organelles and materials from older to younger parts of the body (the growing tips), and the ability to enter resting *(dormant)* stages. Some fungi acquire food by forming mutually beneficial relationships with other organisms. Competition for food is alleviated by rapid growth, dietary specializations to process food efficiently, and the production of materials that are toxic to competitors. Fungi reduce the effects of their own waste pollution by growing away from previously occupied areas or having a tolerance for high levels of waste materials. Finally, rigid cell walls and an ability to produce *antibiotics* are adaptations to survive predators—especially bacteria.

Chemical Control Systems

12

An organism must coordinate its internal activities with the events that occur in its external environment if it is to survive. In unicellular organisms, most responses to environmental changes are controlled by the DNA molecules. This direct genetic control is possible because the distances within a cell are small and the vital activities that must be coordinated are relatively simple.

In multicellular organisms, the activities of so many diverse cells in the body must be coordinated that each specialized cell cannot be controlled directly by its genes. Multicellular organisms therefore have special control systems that monitor changes in the external and the internal environments and coordinate the appropriate responses to these changes in all body cells. In Chapters 12 and 13, we will examine the control systems that perform this specialized function in animals and plants.

The Problem of Control: A Human Example

Human control systems have evolved gradually over hundreds of millions of years. When we consider only the modern human, we are impressed with a single dominant aspect of control—intelligence. A closer study, however, reveals many equally impressive but less conspicuous aspects of control—the complex systems of intercellular communication that coordinate internal body events with changing conditions in the external environment.

The differences between intelligence and the more basic control systems are evident if we consider what happens during an ordinary moment of life—a game of tennis on a hot afternoon (Figure 12-1). While the player's intelligence is planning strategy and coordinating body movements with the flight of the ball, most of the human control systems are occupied with the more important problem of maintaining an optimal internal environment for the survival of the organism.

Figure 12-1
Bjorn Borg at Wimbledon
Most of the cells in Borg's body are
working to promote his survival
rather than his success at the game.

The player's cells must remain immersed in an optimal fluid environment, but the increased demands placed on body cells during strenuous activity on a hot day can cause extreme, even lethal, changes in the body's internal fluids. During exercise the rate of cellular respiration increases, requiring greater amounts of sugar and oxygen and producing more carbon dioxide and heat than when the body is at rest. These cellular demands create excesses and deficiencies in the internal fluids that must be corrected immediately if the player's cells are to survive. These are only a few of the numerous internal changes that are monitored and controlled without the player's conscious effort.

Elements of a Control System

Because control systems are among the most complex phenomena in biology, it will be helpful to examine the elements and functions of a nonliving control system first—the control of the water temperature in a fish tank. Although the details of this system differ markedly from a living control system, all control systems are based on the same underlying principles.

The five components of the water-temperature control system in a fish tank are a sensor, an input message wire, a control center, an output message wire, and a water heater (Figure 12-2).

The Sensor and Input Messages

The *sensor* transmits information about temperature changes through the *input message wire* to a *control center*. In this system, the sensor is a thermometer with an internal column of mercury that expands when heated and contracts when cooled. The thermometer is connected to a circuit that sends electrical currents through the input message wire to the control center. As the water temperature in the tank

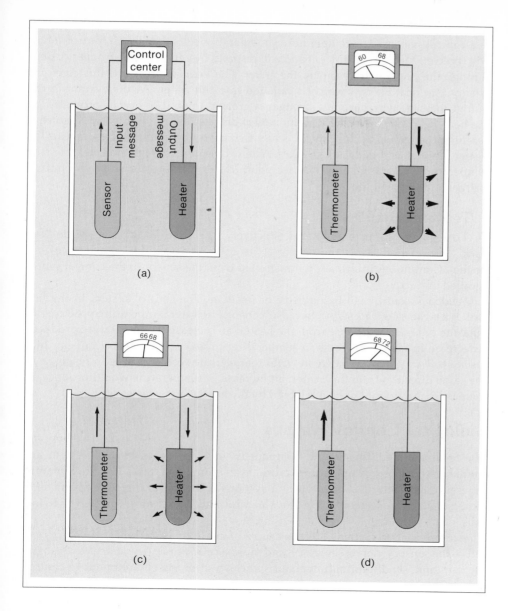

Figure 12-2
The Components of a Temperature-Control System in a Fish Tank

(a) The control system contains a sensor, an input message wire, a control center, an output message wire, and a heater. (b) The *set point* in this system is 68°F. When the temperature drops below the set point, the mercury in the thermometer (the *sensor*) contracts, decreasing the amount (thin arrow) of electrical current in the *input message wire*. The *control center* (the *thermostat*) responds by increasing the amount of current (wide arrow) in the *output message wire* to the heater, causing it to release more heat. (c) As the temperature approaches the set point, more current is transmitted through the input message wire and the control center responds by sending less current to the *heater*, causing it to release less heat. (d) When the temperature rises above 68°F, the control center stops sending current through the output message wire, which turns off the heater.

increases, the thermometer column expands, increasing the amount of electrical current transmitted through the input message wire. The temperature of the water is transformed into an electrical current language; greater input current reflects higher water temperature.

Output Messages

The control center evaluates the electrical input messages (the strength of the current) that it receives from the sensor and sends output messages to the heater, controlling the amount of heat that it generates.

The control center in Figure 12-2 is designed to maintain the water temperature in the fish tank as close to 68°F as possible. Such a fixed value is called the *set point* in a control system. At a temperature considerably below 68°F, the control center will receive a lower input current and will respond by generating a stronger current through the *output message wire* to the *heater*. The heater transforms this electrical energy into heat energy, which is radiated into the water. As the current in the output message wire becomes stronger, a greater amount of heat is produced by the heater. When the sensor (thermometer) detects that the water temperature is reaching the set point, the control center decreases the amount of current to the heater, so that it produces less heat. When the input current indicates that the temperature has risen above the set point, the control center stops the output current and shuts off the heater.

Feedback Systems

To regulate water temperature effectively, the control center must be "fed back" information about the effects of its commands on the regulation of the system. Continuous information transmitted from the sensor to the control center is called *feedback.*

Feedback control can be negative or positive. The control system in the fish tank is a *negative-feedback system,* because a change in water temperature produces an opposite or *negative* response from the heater: an increase in temperature produces a decrease in heat; a decrease in temperature produces an increase in heat. In a *positive-feedback system,* a change in water temperature would produce a response in the same direction from the heater: an increase in temperature would produce an increase in the amount of heat released by the heater, and vice versa.

Biological Control Systems

The feedback mechanisms and components of a biological control system are similar to those of a temperature-control system in a fish tank. The fundamental difference between these two types of systems is that all of the components of a biological control system utilize the electrical and chemical properties of living cells.

Animal control systems are constructed from four basic types of specialized cells—the sensory, nerve, secretory, and muscle cells. *Sensory cells,* like the sensor in the fish tank, send information about some aspect of the environment to control centers. In the fish tank, information about water temperature is tranformed into electrical currents. In animal control systems, environmental information is translated into neural or hormonal language. The *neural language* is composed of coded patterns of electrochemical impulses that are passed along chains of nerve cells. The *hormonal language* is communicated by chemical messengers—specific molecules called *hormones* that are synthesized by certain secretory cells called *endocrine cells.* Hormones are carried by the moving fluids of the body to *target cells,* where they direct changes in cellular activities.

In an animal control system (Figure 12-3), sensory cells detect some environmental stimulus, such as a touch on the animal's skin or the increased pressure of blood on the animal's vessels, and transform this information into a neural or hormonal message. In biology, the transformation of information is called *transduction.* The transduced environmental information is then carried by nerve cells

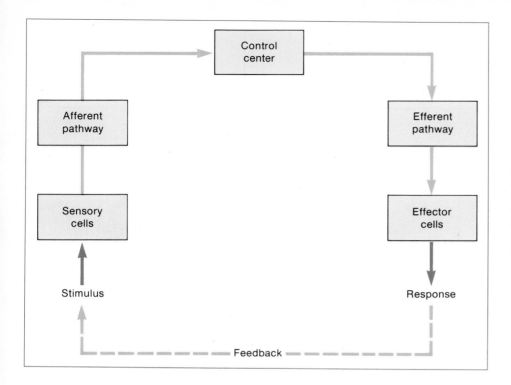

Figure 12-3
The Components of a
Biological Control System

An environmental change stimulates the sensory cells, which transform it into a *neural* or a *hormonal message*. The message moves through an *afferent pathway* to the control center, where it is evaluated. The control center then issues a command that is carried along the *efferent pathway* via a neural or a hormonal message to the *effector cells*. These muscle or secretory cells generate the appropriate response to the stimulus, and information about the response is fed back to the control center.

(neural message) or body fluids (hormonal message) to a control center. The path of the input message from the sensory cells to the control center is called the *afferent pathway.*

The biological control center evaluates the afferent message and issues a command that causes the animal to respond to the environmental stimulus. An example is the temperature-control centers in the human. Regardless of temperature changes in the external environment, the internal temperature of a human is maintained within about 2°F of 98.6°F by specialized groups of nerve cells in a region of the brain called the *hypothalamus* (Figure 12-4). These cells are sensitive to the temperature of the blood flowing through the hypothalamus and also receive afferent neural messages from sensory cells in the skin that are sensitive to temperature.

The hypothalamic control cells continuously compare afferent messages about the state of the body's temperature with a fixed set point of 98.6°F. If body temperature deviates from 98.6°F, the control cells send appropriate messages to areas of the body that can change the body temperature. For example, if the control cells are informed that body temperature has decreased below the set point, they stimulate such activities as shivering, an increase in cellular respiration, a decrease in sweating, and a reduction in blood flow to the surface of the skin.

The responses commanded by control cells are carried out by specialized *effector cells,* and the paths from control centers to the effector cells are called *efferent pathways* (see Figure 12-3). An efferent pathway is analogous to the output message wire from the control center to the heater in the fish tank; effector cells are analogous to the heater in that they can effect changes in the environment.

Figure 12-4
The Location of Several Important Control Centers in the Human Brain

(a) The left side of a human brain viewed from the outside. (b) A section through the center of the brain.

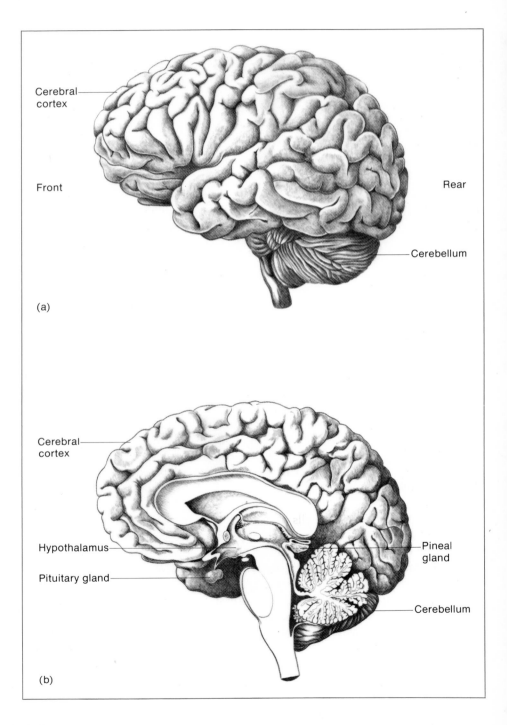

CerebraI cortex

Front

Rear

Cerebellum

(a)

Cerebral cortex

Hypothalamus

Pituitary gland

Pineal gland

Cerebellum

(b)

Animals have two types of effector cells—muscle and secretory cells. *Muscle cells* are specialized for contraction and produce movements of part or all of the organism. In the temperature-control example, body temperature is raised (1) when the muscles of the blood vessels in the skin contract to reduce the amount of blood flow to the surface of the body and (2) when the *skeletal muscles* (attached to the bones) rapidly contract to produce shivering movements, releasing heat (a by-product of cellular respiration in the muscle cells). Because blood carries the heat produced by body cells, less heat is released from the body surface if less blood reaches the skin.

Secretory cells synthesize and release, or *secrete,* special molecules. In addition to the endocrine cells in the afferent pathway that secrete hormonal messengers to control centers, endocrine cells in control centers and at various points along the efferent pathways secrete hormones that travel to the effector cells, where they alter cellular activities. Body temperature is increased when hormonal messengers from endocrine cells in the thyroid gland command all body cells to increase their rate of cellular respiration. Another type of secretory cell, the *exocrine cells,* are effectors in control systems. The exocrine cells do not secrete hormonal messengers; instead, they release materials that are part of the response to the stimulus. For example, exocrine cells in the sweat glands are commanded to reduce sweat secretions when the body temperature falls below the set point.

Control centers usually transmit commands to effector cells in the form of both neural and hormonal messages. Thus, the hypothalamic control cells command muscle and exocrine activities via neural messages and increase cellular respiration via hormonal messengers.

Most biological control systems make use of negative feedback. When the hypothalamic centers receive information that the body temperature has reached 98.6°F, they direct the reduction of heat-producing and heat-retaining activities. Control centers in turn are controlled by the activities they regulate via feedback.

The sensory, nerve, secretory, and muscle cells in animal control systems are organized into tissues and organs. *Tissues* are groups of the same kinds of cells; *organs* are groups of various types of cells and tissues that work together as a unit to perform a specific function. The human eye is a sensory organ that is composed of many sensory, neural, muscle, secretory, vascular, and other types of cells that are organized to receive visual information and transduce it into neural messages. Nerve cells are grouped together to form a variety of neural tissues and organs, the most important of which is the *brain*—the dominant control center in most animals. Secretory cells are organized into tissues and into organs called *glands.* The nature and function of the two principal types of secretory cells—the endocrine and the exocrine cells—will be discussed in the next section. The characteristics of muscle tissues and organs will be examined in Chapter 19.

Tissues and organs that are linked together anatomically or functionally to perform a particular task are called a *system.* The two principal control systems in animals are the chemical system and the nervous system. In vertebrates, the *chemical control system* is composed entirely of endocrine cells and is referred to as the *endocrine system.* This system is composed of endocrine tissues and glands that are distributed throughout the body and coordinated by hormonal messengers carried by the bloodstream. The *nervous system* consists of interlocking pathways and control centers constructed from neural tissues and organs.

The neural control system in animals evolved as part of the adaptive solution to the problems of movement. This system responds rapidly and precisely to environmental stimuli. By contrast, endocrine control systems are much slower and less specific, because hormones travel at the speed of moving body fluids and usually control a broad range of cellular activities. The endocrine system prepares the animal to respond by mobilizing body resources, maintaining states of activity, and regulating the rhythms and cycles of animal life.

Neural control systems are unique to animals, but hormones and other chemical control agents are found in all organisms. Even nerve cells utilize aspects of chemical control; messages are passed very short distances from one nerve cell to the next by chemical messengers called *neurotransmitters*. Chemical messengers are the fundamental and universal mode of communication within and between the cells of all organisms.

Chemical Control in the Vertebrates

The chemical components of vertebrate control systems include exocrine and endocrine tissues and glands (Figure 12-5). *Exocrine* (external secretory) *glands*

Figure 12-5
Two Types of Glands in a Vertebrate Control System
(a) The cells of an *endocrine gland* secrete hormones directly into the bloodstream. (b) The cells of an *exocrine gland* secrete materials through a duct into some chamber of the body or onto the body surface. A human salivary gland is shown here.

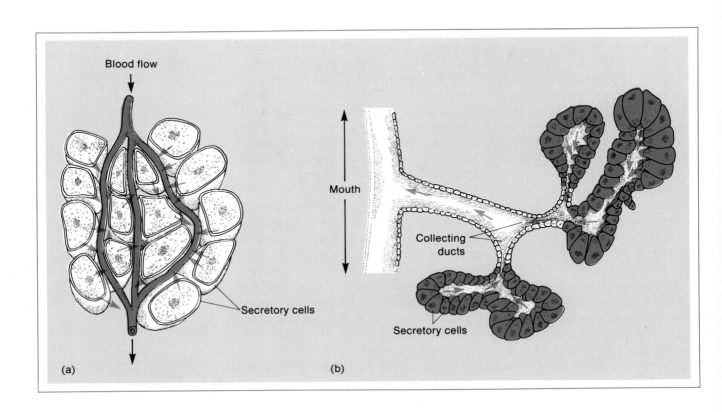

Table 12-1
Some Exocrine Glands in Humans

Gland	Secretion	Function
Lachrymal	Tears; secreted onto the inner surface of the eyelids	Lubricate and cleanse the eyes
Sweat	Sweat; secreted onto the body surface	Removes heat from the body
Pancreas and liver	Digestive enzymes; secreted into the small intestine	Convert food from large to small molecules
Salivary	Saliva; secreted into the mouth	Converts carbohydrates into simple sugars
Mammary	Milk; secreted from nipple on breast	Provides natural food for infants

release their chemicals into ducts or tubes that lead directly into chambers of the body or out onto the body surface. Table 12-1 provides some examples of exocrine glands in humans.

In this chapter, we will concentrate on *endocrine* (internal secretory) *glands,* which release their secretions—the hormones—directly into the bloodstream.

What Is a Hormone?

All cells are chemical factories in which a great variety of reactions occur in different sequences at different rates. In animals, the activities of individual cells are coordinated so that all of the cells in the body can function in an interrelated way. Much of this coordination is accomplished by hormonal messengers.

The existence of chemical messengers, or hormones, was first demonstrated conclusively in the 1840s by the German physician A.A. Berthold. Prior to Berthold's work, it had long been known that if the testes of male farm animals were removed, anatomical and behavioral changes would occur. The castrated animals became fatter and noticeably less aggressive, and their meat became more tender and had a better flavor. Berthold investigated the mechanisms that produced these changes by castrating roosters at different ages and then reimplanting testes in some of them. Berthold found that without reimplantation, adult roosters lost their male characteristics (aggressive behavior, brightly colored comb, etc.) and preadult roosters never developed these characteristics. However, if the testes were reimplanted—seemingly anywhere in the body—the adult rooster retained its male characteristics and the young bird developed them.

After each reimplantation, Berthold killed the bird and examined the testes in their new location within the body. In each case, he found that a unique system of blood vessels had developed that connected the transplanted testes with the bird's bloodstream. Berthold concluded that in reimplanted (and normal) birds, the testes produce some sort of chemical that is carried by the circulating blood to all parts of the body, where it stimulates different activities in specific tissues.

However, a chemical messenger was not actually isolated and identified until 1902, when the English physiologists William M. Bayliss and Ernest H. Starling discovered that a substance secreted by endocrine tissue in the lining of the small intestine plays an essential role in the digestion of food. This substance, *secretin*, is released into the bloodstream when food is present in the intestine and travels throughout the body, but only affects target cells in the *pancreas*—a gland composed of both exocrine and endocrine cells. The primary function of the pancreas is to manufacture and secrete digestive enzymes into the small intestine when stimulated by secretin (Figure 12-6). Starling suggested that secretin—and other chemical messengers yet to be discovered—be given the name "hormone," which is Greek for "to stimulate" or "to excite."

Since the discovery of secretin, a variety of hormones have been isolated that regulate an enormous range of activities in vertebrates, invertebrates, and plants. Paradoxically, as we have learned more about hormones, it has become harder to define a hormone precisely. The term itself is misleading. Although many hormones stimulate cellular activities, other hormones inhibit cellular activities. In addition, not all hormones are produced by endocrine glands. Hormones are secreted by a variety of cells in plants and the simpler invertebrates that have no specialized endocrine glands. Even in vertebrates, nonendocrine tissues secrete hormone-like messengers. Another complicating factor is that hormones are so structurally different that they cannot be defined according to their chemical composition.

Figure 12-6
Control of Digestion by the Hormone Secretin

The presence of food in the digestive tract stimulates endocrine cells in the lining of the small intestine to release *secretin*. This hormone travels via the bloodstream to the pancreas, where it stimulates the release of digestive enzymes.

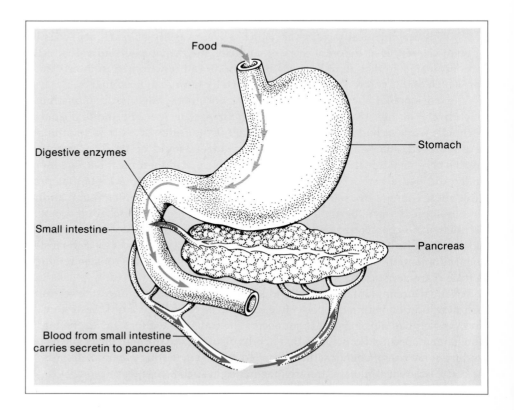

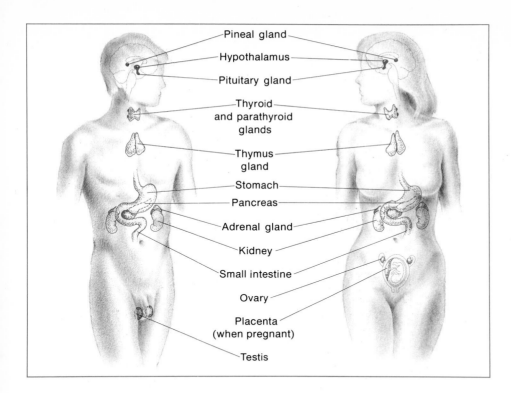

Figure 12-7
The Human
Endocrine Glands

Pineal gland
Hypothalamus
Pituitary gland
Thyroid
and parathyroid
glands
Thymus
gland
Stomach
Pancreas
Adrenal gland
Kidney
Small intestine
Ovary
Placenta
(when pregnant)
Testis

The modern definition of a hormone is necessarily quite general. In essence, a *hormone* is a biological molecule that is produced by one set of cells and then carried by the organism's body fluid to a set of target cells, where it influences some form of chemical activity.

The endocrine systems of the human male and female are diagrammed in Figure 12-7. Some of the more important human hormones and the activities they influence are provided in Table 12-2. Vertebrate hormones can be categorized as trophic or nontrophic.

Trophic Hormones

Trophic hormones regulate activities within the endocrine system. They are produced by one endocrine structure and affect the activities of other endocrine structures. The *anterior pituitary gland* (Figure 12-8), for example, secretes four trophic hormones: thyroid-stimulating hormone (TSH), which controls the secretion of hormones by the thyroid gland; adrenocorticotrophic hormone (ACTH), which stimulates hormone secretion by the cortex of the adrenal gland; and two gonadotrophic hormones, follicle-stimulating hormone (FSH) and luteinizing hormone (LH), which influence the secretion of sex hormones by the reproductive glands, or *gonads*.

Because the trophic hormones released by the anterior pituitary gland regulate so many other glands, it has been called the "master gland" of the human body. However, the body's real master "gland" is the hypothalamus of the brain, which

Table 12-2
Principal Components of the Human Endocrine System

Gland or tissue	Hormone	Major effect
Hypothalamus	Releasing hormones, release inhibiting hormones	Regulate secretion by the anterior pituitary gland
Anterior pituitary	Growth hormone	Stimulates growth
	Thyroid-stimulating hormone	Controls the thyroid gland
	Adrenocorticotrophic hormone	Controls the adrenal cortex
	Follicle-stimulating hormone	Increases the growth and activity of the ovaries and testes
	Luteinizing hormone	Increases secretion of sex hormones by the ovaries and testes; controls menstrual cycle
	Prolactin	Produces milk
Posterior pituitary	Oxytocin	Contracts uterine muscles; releases milk
	Antidiuretic hormone	Retains water in kidneys and blood; contracts smooth muscles
Pineal	Melatonin	Regulates secretion by the hypothalamus
Thyroid	Thyroxine	Increases cellular respiration
Parathyroid	Parahormone	Controls calcium and phosphorus levels in the body fluids
Adrenal cortex	Many hormones, including: Cortisol (hydrocortisone)	Controls synthesis of glycogen; maintains blood sugar levels
	Aldosterone	Controls ionic concentrations in the body fluids
Adrenal medulla	Epinephrine (adrenalin)	Augments the fight-or-flight response
	Norepinephrine (noradrenalin)	Augments the fight-or-flight response
Pancreas	Insulin	Makes glucose in the blood available for cell use and for the synthesis of glycogen
	Glucagon	Stimulates the breakdown of glycogen into glucose
Small intestine	Secretin	Stimulates the release of digestive enzymes from the pancreas
Ovary	Estrogens	Express female secondary sex characteristics; regulate menstrual cycle
	Progesterone	Expresses female secondary sex characteristics; regulates menstrual cycle
Placenta	Estrogens	Maintain pregnancy
	Progesterone	Maintains pregnancy
Testes	Testosterone	Expresses male secondary sex characteristics
Thymus	Thymosin	Aids in the development of the immune system

lies above the pituitary gland (see Figure 12-8) and controls the secretions of the anterior pituitary by means of chemical messengers, called *releasing hormones* and *release inhibiting hormones* (also called *releasing factors* and *release inhibiting factors*), that are transported by the bloodstream. Although the hypothalamus secretes hormones, it is composed entirely of nerve cells and forms a direct link between the nervous system and the endocrine system. Human body temperature is controlled in part by releasing hormones produced by nerve cells in the hypothalamus, which stimulate the anterior pituitary gland to release TSH, which in turn stimulates the

Figure 12-8
The Pituitary Gland, Its Hormones, and Their Target Organs

The *anterior pituitary gland* produces the growth hormone (GH), prolactin, and four trophic hormones: thyroid-stimulating hormone (TSH), adrenocorticotrophic hormone (ACTH), follicle-stimulating hormone (FSH), and luteinizing hormone (LH). The *posterior pituitary gland* produces antidiuretic hormone (ADH) and oxytocin. The release of hormones from the pituitary gland is controlled by the hypothalamus of the brain.

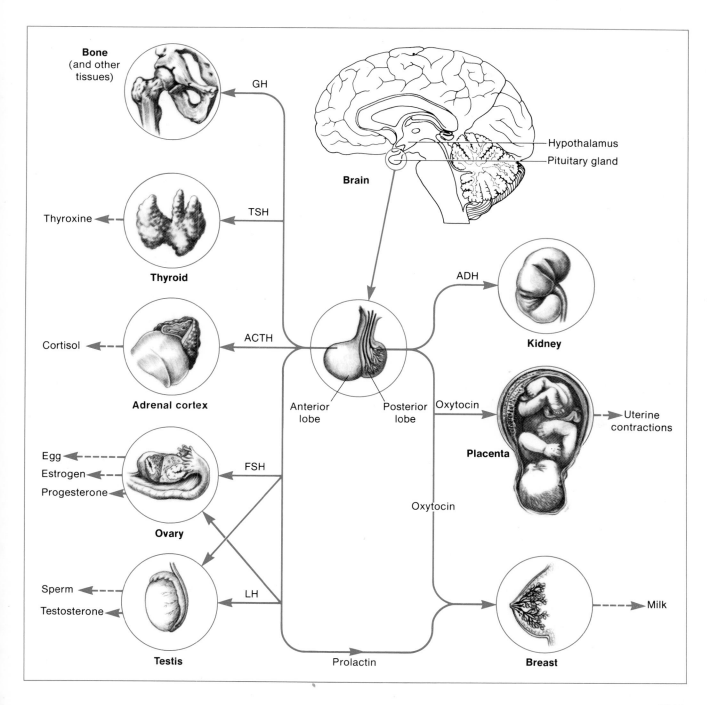

(a) When the concentration of water in the blood drops below the set point, the hypothalamus commands the posterior pituitary gland to release more ADH, which causes more water to return from the urine to the blood. (b) When the concentration of water in the blood is at or above the set point, the hypothalamus commands the posterior pituitary gland to stop releasing ADH.

thyroid to release thyroxine thereby increasing cellular respiration in all body cells.

Trophic hormones control the secretions of other endocrine structures; *nontrophic hormones* act directly on specific target cells outside the endocrine system. The effects produced directly by nontrophic hormones and indirectly by trophic hormones can be categorized as regulatory or developmental.

Regulatory Hormones

Regulatory hormones adjust and control cellular activities to maintain normal homeostatic conditions in the internal body fluids, temporarily shift these conditions to abnormal emergency states in response to stress, and regulate the daily and seasonal rhythms and cycles of vertebrate life.

An example of the hormonal regulation of homeostasis is the effect of *antidiuretic hormone* (ADH) on the concentration of water in the blood. ADH is the efferent messenger for the negative-feedback control system diagrammed in Figure 12-9. A control center in the hypothalamus is sensitive to changes in the concentration of water in the blood. Whenever the cells in this center sense a

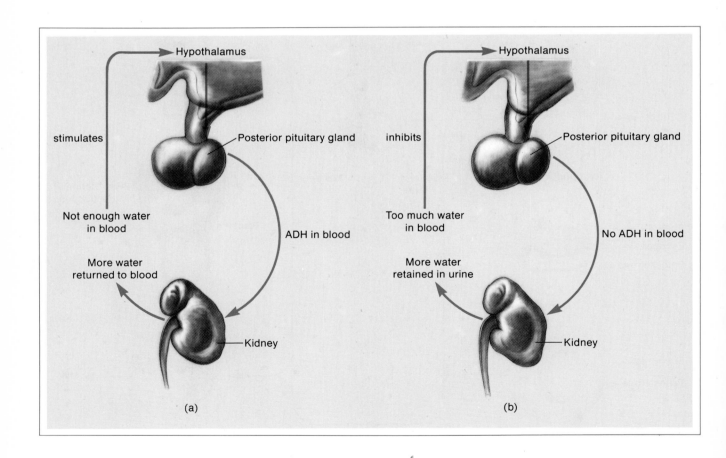

(a)

(b)

concentration below the set point, they transmit a neural message to the *posterior pituitary gland* (see Figure 12-8), which releases ADH into the bloodstream. The ADH travels throughout the body but only affects target cells in the kidney, instructing them to take back water that is being released in the urine until an optimal concentration of water in the blood is restored. When the hypothalamic control center senses that the blood is too dilute, it signals the posterior pituitary gland to stop secreting ADH, so that more water is excreted in the urine.

The posterior pituitary gland is a physical extension of the hypothalamus. ADH and another hormone, *oxytocin*, which influences contractions of the uterus during childbirth, are actually synthesized in the hypothalamus and then travel down the nerve cells to storage areas in the posterior pituitary, where they are secreted into the bloodstream in response to neural commands from the hypothalamic control centers. In contrast, the anterior pituitary gland, which is not physically connected to the hypothalamus, manufactures and secretes its own hormones on the command of hormonal messengers traveling from the hypothalamus via the bloodstream.

The hypothalamic-pituitary control of water concentration in the blood is one example of how synchronized activities of neural and endocrine control systems maintain homeostasis. Hormonal messenger systems are involved in regulating essentially all aspects of vertebrate physiology and behavior. Through the continual adjustment of cellular activities, a stable internal environment is maintained despite external changes.

Most vertebrates exhibit daily physiological and behavioral patterns of activity called *circadian rhythms* ("circa" = about; "dian" = a day), which are also regulated by coordinated neural and endocrine activities. Human body temperature, for example, is not maintained at a perfectly constant 98.6°F, but varies ±2° in a consistent and predictable cycle each day. Body temperature is usually lowest in the early morning and peaks sometime between midday and late afternoon. Each individual has a unique daily rhythm of body temperature as well as of heart rate, blood pressure, blood sugar concentration, and many other bodily functions.

The mechanisms that regulate vertebrate circadian rhythms are not fully understood. These rhythms are correlated in part with the external rhythms of light and dark; however, they continue after an animal has been placed in continuous light or dark, indicating that some internal clock must maintain these rhythms in the absence of external cues. In vertebrates, the mechanism may be located in the *pineal gland,* shown in Figure 12-4(b). This small endocrine gland receives information about external light conditions through the eyes and, in some species, directly through the skull. The pineal gland secretes a hormone, *melatonin,* in daily cycles that are synchronized with the external light cycles. Melatonin, in turn, stimulates the release of hormones by the hypothalamus and other neural and endocrine control centers.

Many other cycles in vertebrate life are also regulated by hormones. The rhythms of vertebrates living in or near coastal waters are often synchronized with the ocean tides. Hormones in the human female regulate the 28-day menstrual cycle. In nonhuman vertebrates, all aspects of reproduction—from cycles of sexual receptivity in females to seasonal patterns of migration—are controlled and synchronized by hormonal messengers.

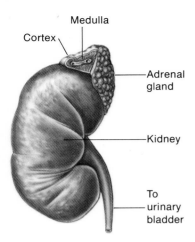

Figure 12-10
Location of
the Adrenal Glands

Birds and mammals have two adrenal
glands, one located on top of each
kidney. Each adrenal gland has an
outer structure, the *cortex*, and an
inner structure, the *medulla*.

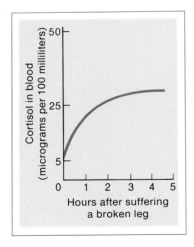

Figure 12-11
The Effect of Stress on the
Concentration of Cortisol
in the Blood of a Human

During the first few hours after a leg
has been broken, the body increases
its production of cortisol and releases
it rapidly into the blood.

The Regulation of Emergency Responses

Vertebrate regulatory hormones play a critically important role in mobilizing the body to respond to sudden emergencies. Many neural and endocrine systems participate in these responses, especially the hormones secreted by the adrenal glands.

All vertebrate animals have two adrenal glands. In birds and mammals, one is located on top of each kidney (Figure 12-10). Each adrenal gland has two components—an *adrenal cortex*, which forms the outer layer of the gland, and an *adrenal medulla*, which forms the inner layer. The cortex and the medulla are actually two separate glands; they develop in the embryo from different tissues, are controlled by different systems, and manufacture and secrete different types of hormones.

The adrenal cortex secretes a variety of hormones that are essential to the organism's survival. If the adrenal cortex is surgically removed, leaving the medulla intact, the animal will die. Many systems would malfunction simultaneously, resulting in lowered blood pressure, kidney failure, severe changes in blood chemistry, and a loss of appetite and subsequent starvation.

In humans, the hormone *cortisol*, or *hydrocortisone*, is secreted into the blood by the adrenal cortex whenever the individual is stressed due to illness or injury (Figure 12-11), extreme physical effort, pain, radical changes in body temperature, or intense emotion. Cortisol stimulates the breakdown of body proteins into sugars to provide energy, liberates vital amino acids for use in tissue repair, combats inflammatory reactions that accompany injuries, increases the capacity of muscles to remain contracted, and is involved in a variety of other life-protecting functions.

The hypothalamus controls the secretion of cortisol. When the hypothalamus receives information indicating the body is undergoing some sort of stress, it secretes *corticotrophin-releasing factor* (CRF), which is carried by the blood to the anterior pituitary gland, where it stimulates the secretion of stored *adrenocorticotrophic hormone* (ACTH). The ACTH is then carried by the bloodstream to the adrenal cortex, where it stimulates the release of cortisol. This control system is regulated by negative feedback, because when cortisol concentrations in the blood reach a certain level, the secretions of CRF and ACTH are inhibited (Figure 12-12).

Unlike the adrenal cortex, the adrenal medulla is not essential to the organism's survival. This gland is a component of the *sympathetic nervous system* and serves only to supplement and reinforce the activities of this system. If the adrenal medulla is surgically removed, the animal appears to recover completely.

The sympathetic nervous system produces a general emergency response called the *fight-or-flight response*. As a result of neural messages, more sugar is released into the blood, the blood pressure, breathing rate, and heart rate all increase, blood is shunted from the digestive tract to the skeletal muscles, blood-clotting time is reduced, and a variety of other changes occur to prepare the animal for extreme effort.

The adrenal medulla is stimulated by the sympathetic nervous system during the fight-or-flight response to secrete two hormones—*epinephrine* (also called *adrenalin*) and *norepinephrine* (also called *noradrenalin*)—into the bloodstream. These hormones sustain or increase emergency activities in all organs that are stimulated by the sympathetic nervous system. The secretion of epinephrine and norepinephrine from the adrenal medulla continues as long as the sympathetic nervous system commands the fight-or-flight response.

Developmental Hormones

In addition to the regulation of the internal environment, the other major function of hormones is the implementation of the genetic plan for the growth and development of the organism. The anterior pituitary gland secretes a *growth hormone* that ensures normal growth by stimulating the accumulation of protein in almost all body cells. The secretion of this growth hormone is controlled by hormones released by the hypothalamus. When too little growth hormone is secreted during childhood, a human develops into a *midget,* or a *pituitary dwarf.* When too much growth hormone is secreted during childhood, a human develops into a *pituitary giant.* Both conditions are illustrated in Figure 12-13(a). If an excess amount of growth hormone is secreted during adulthood, only certain body tissues will grow—a condition known as *acromegaly,* which is shown in Figure 12-13(b).

Other developmental hormones in humans are the two *gonadotrophic hormones* secreted by the anterior pituitary gland—*follicle-stimulating hormone* (FSH) and *luteinizing hormone* (LH)—which control the growth and function of the reproductive glands (the ovaries in females and the testes in males). Prior to the onset of puberty (roughly before 11 years of age in girls and 13 years of age in boys), cells in the hypothalamus prevent the release of FSH and LH and the ovaries or testes remain immature. At puberty, the function of the hypothalamic cells changes and the hypothalamus releases hormones directing the anterior pituitary to secrete FSH and LH.

In males, FSH stimulates the testes to grow and begin producing sperm and LH stimulates certain cells in the testes to begin producing the male sex hormones, including *testosterone,* which stimulate the development of such secondary sex characteristics as deepening of the voice and growth of hair on certain parts of the face and body. In females, FSH and LH stimulate maturation of the ovaries and of the eggs they contain and production of the female sex hormones (*estrogens* and *progesterone*), which stimulate such secondary sex characteristics as the development of the breasts and fat deposits in certain areas of the body.

How Vertebrate Hormones Work

Most hormones command their target cells to perform more or less of their specialized activities. These commands can be very specific, as is true of the effect of secretin on the pancreas or of ADH on the kidneys, or they can produce more widespread effects, as is true of the effects of cortisol or epinephrine. These hormones alter two basic cellular activities—the rate at which enzymes and other proteins are synthesized and the permeability of the cell membrane.

Hormones regulate the activities of the enzymes that enhance certain chemical reactions in their target cells. For example, epinephrine controls the activity of the enzymes in liver cells that convert stored glycogen molecules into glucose—the form of sugar that can be used by cells. The control of the production and activation of enzymes permits hormones to regulate the speed and direction of

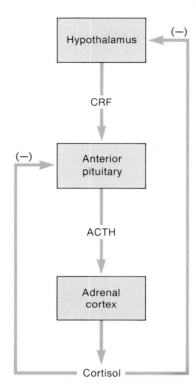

Figure 12-12
The Release of Cortisol Is Regulated by Negative Feedback
As the concentration of cortisol in the blood increases, it causes a decrease (indicated by the minus sign) in the amount of cortisol-releasing factor (CRF) secreted by the hypothalamus and in the amount of adrenocorticotrophic hormone (ACTH) secreted by the anterior pituitary gland.

Figure 12-13
The Effects of Abnormal Amounts of Growth Hormone

(a) Too little growth hormone during childhood produces a *pituitary dwarf*; too much growth hormone during childhood produces a *pituitary giant*.
(b) *Acromegaly* develops when abnormally large amounts of growth hormone are secreted after the individual has reached adulthood. In this sequence of photographs from (1) to (4), the bones of the face become progressively thicker as the individual ages from 24 to 42 years.

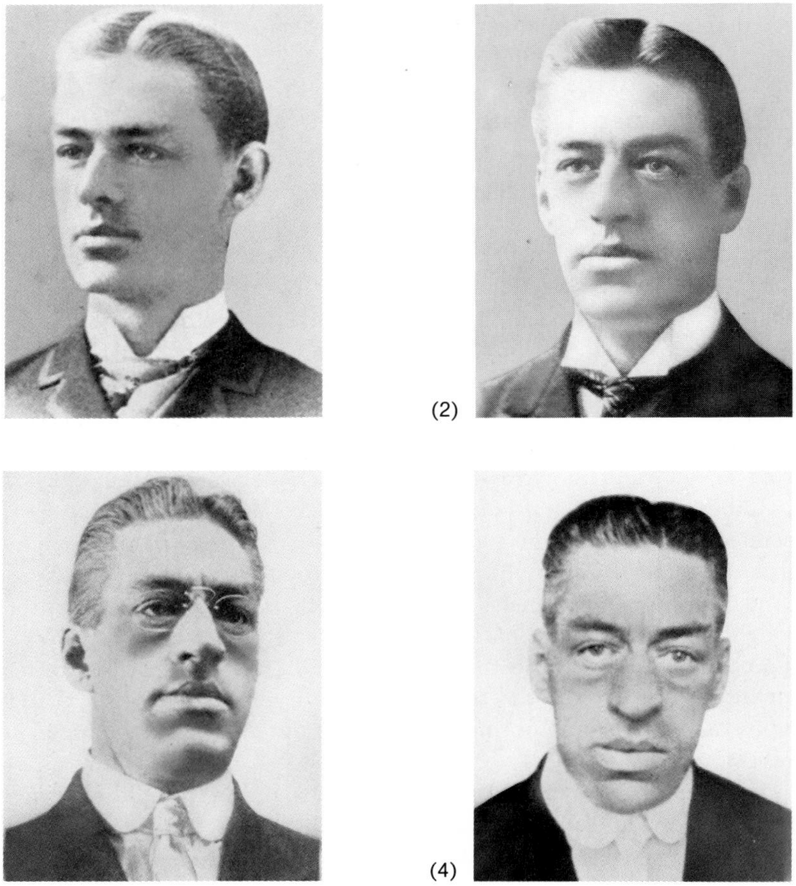

(1)

(2)

(3)

(4)

(a)

(b)

chemical reactions in the cells. Almost all vertebrate hormones also have some effect on the rate at which water and dissolved substances (ions, amino acids, sugars) move through the cell membrane. For example, ADH increases the permeability of kidney cells, so that water can pass into these cells at a faster rate. Some hormones actually change the chemical activities within cells by controlling gene expression. These hormones stop the activity of some genes or stimulate inactive genes into action, as occurs in humans at puberty when FSH and LH from the anterior pituitary gland stimulate the growth and development of the ovaries and testes.

Hormones control cellular activities either by entering the cell directly or by transferring commands to specific chemical messengers within the cell. Whether a hormone enters a cell or initiates changes from outside the cell depends on its chemical nature—whether or not it is a *steroid*.

Steroid hormones are produced by only two types of endocrine glands—the adrenal cortex and the reproductive glands. All adrenocortical hormones and all sex hormones are steroids. Although their functions vary enormously, all steroid hormones have similar molecular structures. Steroids can enter cells directly because they are small lipid molecules, synthesized from cholesterol, that are able to pass freely through the lipid structure of the cell membrane.

Steroid hormones pass into all cells, but only their target cells contain specific *receptor molecules* (large proteins) in the cytoplasm that allow the hormones to regulate cellular activities. Each type of hormone interacts with a different type of receptor molecule, and each target cell contains only receptors that interact with the hormones to which the cell is sensitive. The hormone and the receptor form a complex (Figure 12-14), which moves from the cytoplasm into the nucleus and attaches itself to a specific site on a DNA molecule of a chromosome. The hormone–receptor complex exerts an influence on gene activity at the site by triggering a transcription of the genetic code into messenger RNA, which then migrates to a ribosome where its specified protein is synthesized. After the protein is produced, the hormone–receptor complex returns to the cytoplasm, where the hormone is rendered inactive and the receptor molecule is made available for reuse.

Nonsteroid hormones, which are derived from proteins, exert control over their

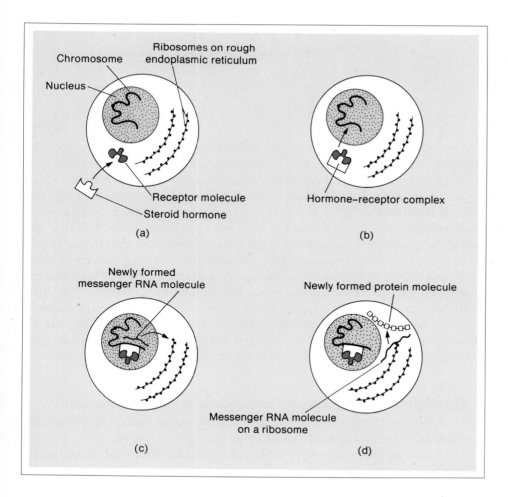

**Figure 12-14
Steroid Hormones Enter Cells and Directly Influence Gene Activity**

(a) *Steroid hormones* are small lipid molecules that can pass freely through the lipid structure of a cell membrane. Each target cell contains *receptor molecules* that combine with the specific hormones that regulate the cell's activities. (b) Inside the cell, the hormone and the receptor molecule form a complex that attaches itself to a particular site on the DNA molecule of a chromosome. (c) The genetic code at the site of the complex is transcribed into a messenger RNA molecule. (d) The newly formed messenger RNA molecule carries the genetic code to a ribosome, where it is translated into a protein molecule.

(a) A *nonsteroid hormone* attaches itself to a receptor molecule on the target cell membrane. (b) The attachment activates the enzyme adenylate cyclase, which promotes the formation of *cyclic AMP* from ATP. The cyclic AMP then acts as a *second messenger* in influencing cellular activities.

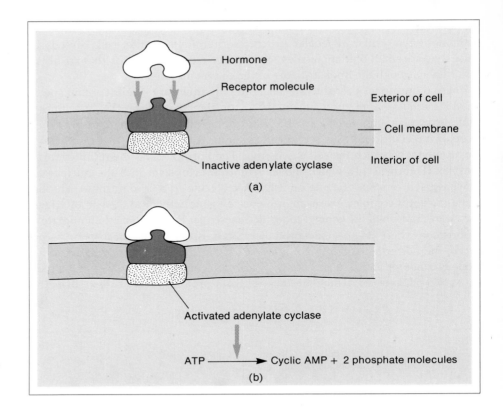

target cells by attaching themselves to receptor sites on the cell membrane. The binding of a hormone to its receptor activates the enzyme *adenylate cyclase*, which is present within the membrane. Once activated, this enzyme promotes the synthesis of *cyclic adenosine monophosphate* (cyclic AMP) from *adenosine triphosphate* (ATP), which is illustrated in Figure 12-15. The word monophosphate in the term cyclic AMP indicates that this molecule has only one phosphate group; the other two phosphate groups that are present in ATP have been removed (Figure 12-16). Because the remaining phosphate group is arranged in a ring, this molecule is called *cyclic AMP*. The hormone is released very rapidly after it is bound to the receptor on the cell membrane and is quickly destroyed by enzymes. The cyclic AMP that is synthesized in response to the hormone initiates a series of chemical reactions within the cell that influence the activity of the genes in the nucleus. Because this command information is relayed, the nonsteroid hormone is called the *first messenger* and the cyclic AMP is called the *second messenger*.

Hormone-like Substances: Prostaglandins and Histamine

In vertebrate biology, the term "hormone" is usually reserved to describe chemical messengers that are secreted by components of the endocrine system. However, research over the past 15 years has shown that most types of body cells will synthesize and secrete hormone-like messengers if appropriately stimulated.

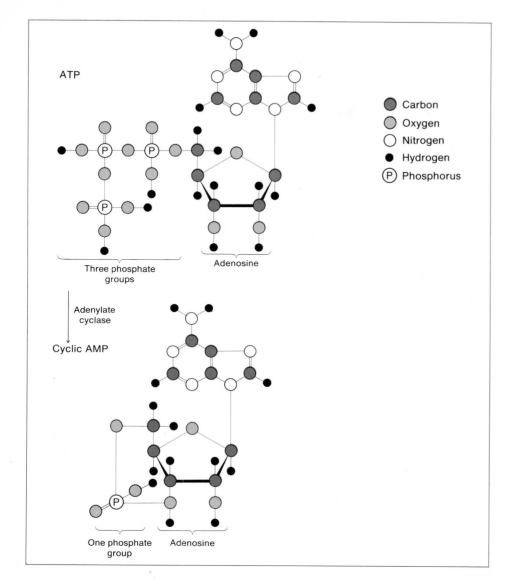

ATP

Three phosphate groups

Adenosine

- Carbon
- Oxygen
- Nitrogen
- Hydrogen
- (P) Phosphorus

Adenylate cyclase

Cyclic AMP

One phosphate group

Adenosine

Figure 12-16
The Synthesis of Cyclic AMP
In the presence of the enzyme adenylate cyclase, the two outer phosphate groups are removed from an ATP molecule. The remaining phosphate group is reorganized so that it is attached by two bonds to the adenosine group, thereby forming a "cycle."

Like hormones, these messengers are transported by the bloodstream and affect cellular activities. Unlike hormones, these messengers are released by cells that are not specialized secretory cells. Two hormone-like substances of current interest are the prostaglandins and histamine.

Prostaglandins were first discovered in human *semen*—the fluid containing sperm that is released from the penis at ejaculation—and were so named because they were believed to be secreted only by the prostate gland. It is now known that although semen does contain the highest concentrations of these substances, virtually all mammalian cells have the capacity to secrete prostaglandins. There are at least 16 different molecular forms of prostaglandins, and each one performs a unique function.

Prostaglandins are the "miracle drugs" of the 1980s. Certain prostaglandins that produce contractions in the smooth muscles of the uterus can be administered to initiate labor during childbirth, to induce abortion, and to speed the movement of sperm toward the fertilized egg to increase the chance of fertilization. Drugs that counteract the effects of these prostaglandins, *antiprostaglandins,* relieve the pain of menstrual cramps by decreasing the contractions of the uterine muscles. Other prostaglandins relieve asthma attacks by relaxing the smooth muscles of the respiratory passages, and others allow ulcers to heal by reducing the secretions of gastric juices into the stomach. One type of prostaglandin raises blood pressure; another type lowers blood pressure. One type inhibits the formation of blood clots; another type increases the probability of blood-clot formation.

Not all prostaglandins produce beneficial effects. One type of prostaglandin contributes to the development of such conditions as fevers, headaches, and inflamed joints. Certain hormones secreted by the adrenal cortex, such as cortisol, inhibit the synthesis of these prostaglandins. The drug aspirin also counteracts their effects.

Prostaglandins are lipid molecules that are synthesized within cells from fatty acids stored in the cell membrane. These substances are the only regulatory chemicals known to be derived from fatty acids. Almost any type of perturbation of the cell membrane initiates the synthesis of prostaglandins, including the reception of a nonsteroid hormone, ultraviolet light, a bee sting, or even mechanical stimulation of the membrane. The exact mechanisms by which prostaglandins achieve their effects are not known, but some evidence suggests that they may control the production of cyclic AMP.

Another important hormone-like messenger, *histamine,* is a small molecule synthesized from common amino acids that is produced and secreted by any type of cell when it is injured. Histamine aids in the repair of damaged cells by relaxing muscles in the walls of blood vessels to permit more blood to flow to the injured area. Histamine also increases the permeability of the blood vessels to antibodies and white blood cells—specific materials that combat illness and injury.

Histamines probably produce the symptoms of the common cold, asthma, and allergic reactions. For example, the presence of plant pollen in the nose or sinuses of a hay fever sufferer stimulates the secretion of histamine. These molecules facilitate the influx of body fluids and defense materials into the localized area to produce the allergic reaction (congestion, inflammation, runny nose and eyes, and so on). Prostaglandins probably act to enhance the effects of histamine during this reaction. Antihistamines and the hormones secreted by the adrenal cortex counteract the effects of histamine.

Chemical Control in the Invertebrates

Among invertebrate animals, the chemical control systems of insects have been studied the most extensively and will be used here as a basis for comparing the invertebrate and vertebrate control systems. The chemical and neural control systems in insects are not completely separated into endocrine and neural structures. In addition to separate endocrine glands that secrete hormones, the brain and other aggregations of nerve cells throughout the insect's central nervous system produce a variety of chemical messengers, all apparently proteins, that function in a similar manner to hormones.

The release of all chemical messengers by endocrine or neural tissues is controlled largely by the brain, which receives and responds to stimuli from both the external and the internal environments. As in vertebrates, these chemical messengers regulate the activities of their target cells by entering a cell directly and interacting with the genes to alter the types of proteins that are synthesized or by attaching themselves to receptor molecules on the cell membrane and initiating the formation of cyclic AMP. Whether a chemical messenger acts directly or indirectly is not known in most cases. Messengers that govern the profound changes during the insect's life cycle probably influence gene activity directly, whereas messengers that are involved in the daily adjustments of homeostasis probably act indirectly through cyclic AMP. Chemical messengers control three basic types of activities in insects: homeostasis, development, and social communication.

Chemical Control of Homeostasis

The chemical and neural control systems in insects provide a feedback system that maintains the constancy of internal body fluids by conveying information about excesses and deficiencies of materials to parts of the body capable of correcting these imbalances. The roles of chemical messengers in homeostasis are not well known because most of these chemicals have not yet been isolated and identified. Until a chemical messenger can be obtained and applied in pure form, its effects on cellular activity cannot be revealed.

Insects do not appear to secrete a single hormone, comparable to thyroxine in vertebrates, that regulates the overall rate of cellular respiration. Instead, this role is shared by many different chemical messengers released by the insect's glands and nervous system, which control such activities as the excretion of water and salts, the storage of fat, the synthesis and breakdown of proteins, fats, and carbohydrates, the rate of cellular respiration, the heart rate, and the daily and seasonal rhythm of the insect's life, including dormancy.

Chemical Control of Development

Animals that have an external skeleton, or *exoskeleton*, face a special growth problem, because their rigid outer surfaces cannot expand as their body size increases. These animals grow by periodically shedding and replacing their exoskeletons with larger ones—a process called *molting*, which occurs in all arthropods (insects, crustaceans, and arachnids) as well as in some other forms of invertebrates. Growth is coordinated with molting in all arthropods by two hormones—ecdysone and juvenile hormone.

In the insect's life cycle, the young form emerges from an egg and passes through a series of immature stages before becoming an adult. Each stage begins and ends with a molt, during which the developing insect is released from the confines of its exoskeleton so that it can continue to grow. Each immature stage, called an *instar*, is a period of feeding during which the insect accumulates nutritional reserves for the next growth spurt and molt. Once the adult stage is reached, the insect's growth stops.

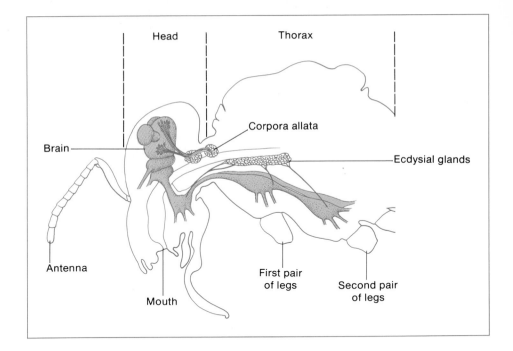

Ecdysone stimulates the cellular activities required for growth and molting. In the presence of this hormone, the cells beneath the exoskeleton separate from the rigid exterior, divide by mitosis, and form a new exoskeleton beneath the old one. When the old exoskeleton is shed, the insect rapidly increases in size before the new, larger exoskeleton becomes hard and rigid.

The periodic synthesis of ecdysone begins in the *ecdysial* (or *prothoracic*) *glands,* which are located in the head or the thorax, depending on the type of insect (Figure 12-17). A releasing hormone secreted by cells in the brain travels via the bloodstream to the ecdysial glands, stimulating the release of a precursor molecule into the blood that is converted into ecdysone elsewhere in the body.

The second hormone affecting growth and molting in the insect, *juvenile hormone,* is synthesized and released into the blood by a pair of glands in the head called the *corpora allata* (Figure 12-17). What stimulates the corpora allata to release this hormone is not known. Juvenile hormone controls the nature of the molt— either to another immature stage or to the adult stage (Figure 12-18). In the presence of high concentrations of juvenile hormone, the insect will molt to another instar; in the absence of juvenile hormone, the insect will molt to the adult stage. Juvenile hormone prevents the expression of genes that produce the insect's adult characteristics.

Ecdysone plays no part in the chemical control system of the adult, and the ecdysial glands degenerate after maturity. However, the corpora allata release juvenile hormone again during reproduction in the adult. Juvenile hormone controls the maturation of eggs in the female, the secretion of the reproductive glands in the male, and many other aspects of reproductive physiology and behavior.

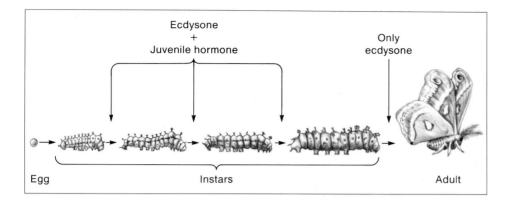

Figure 12-18
Hormonal Control of Insect Development
An insect hatches from an egg in the form of a small *instar*. In the presence of both ecdysone and juvenile hormone, each *molt* produces a larger instar. In the presence of only ecdysone, the molt produces an adult.

The Role of Chemical Messengers in Social Communication

Insects influence the behavior of other members of their species by releasing *pheromones*—chemicals that the insect secretes into the external environment to elicit specific responses from other individuals. These "social hormones" initiate a wide range of insect behaviors.

Sex pheromones are produced by most female butterflies, moths, and beetles. Males of the same species perceive these pheromones from a distance and move upwind toward the source. Once near a female, the male is then stimulated to mate by the pheromone. In some insects, the male also produces a sex pheromone that sexually excites the female, making her more receptive to him and increasing the rate at which her eggs mature.

In many beetles, feeding triggers the release of an *aggregation pheromone*, which attracts other beetles to the site to feed and mate. Sometimes an ingredient in the food is the chemical precursor in the synthesis of the pheromone, as is true in the case of the pine bark beetle. *Alarm pheromones* are warning signals secreted primarily by social insects—termites, ants, bees, and wasps—to alert and disperse other members of the species.

Termites and many types of ants release *trail-marking pheromones* that serve as navigational aids, directing other members of the colony to a food source or to a place in the nest in need of repair. Ants that forage along trails are usually following a pheromone. After finding food, this type of ant lays a chemical trail as it returns to the nest; the trail is then followed by other foragers.

Other pheromones produced by social insects control how the young develop. A pheromone released by the queen bee suppresses the sexual development of young females in the colony, so that they become workers rather than reproducing queens. When the queen dies, the lack of her pheromone allows new queens to develop. In some insect species, the colony consists of a caste system of workers that have different body forms adapted to perform different functions. Each caste produces a unique pheromone; when its concentration rises to a certain level, no more members of that caste are produced, thereby allowing the colony to maintain appropriate numbers of each caste.

Chemical Control in Plants

A variety of structurally distinct tissues and organs, each performing a unique function, contribute to the survival and reproduction of a plant. The activities of all of the different parts of a plant are coordinated with each other and with the external environment by chemicals. Plants do not have specialized nerve cells to transmit information. A plant responds to its environment primarily through its growth, which is controlled entirely by hormones.

In this section we will examine how new tissues are added to an adult plant and how this growth is controlled by hormones. The growth and germination of seeds will be discussed in Part 4, which deals specifically with reproduction and early development.

The Growth of a Plant

A plant increases its size by cell division and cell enlargement in the *meristem tissues.* Elsewhere in the plant, growth is achieved only through the enlargement of already established cells. We will limit this discussion to the formation and development of new cells in the meristem tissues of flowering plants. The three areas of meristem tissue in an adult plant are the shoot apical meristem, the root apical meristem, and the cambium.

The *shoot apical meristem,* which is located at the tips of plant stems, forms three types of new plant structures—stems, leaves, and lateral buds (Figure 12-19). At first, the new cells produced in the meristem by cell division are very small and unspecialized. The cells in the outer part of the meristem elongate first, developing into immature leaves that extend above the central meristem and protect it from injury and desiccation, as shown in Figure 12-20(a). The central area of the meristem then grows rapidly as the newly formed cells take in water and elongate. As the growth process continues, small pieces of meristem tissue are left behind between the immature leaves and the new stem, as illustrated in Figure 12-20(b). These *lateral buds* may develop into branches from the main stem. Each point on the stem where leaves and lateral buds occur is called a *node,* and the area of stem between two nodes is called the *internode* (Figure 12-19).

The shoot apical meristem is dominant over the lateral buds because the lateral buds will not develop into branches when the meristem at the tip of the stem is active. Branches can be forced to develop in houseplants and decorative shrubs by pinching or cutting off the tips of the stems. Removing the shoot apical meristem allows the lateral buds to develop and a bushier plant to form.

The *root apical meristem* regularly produces new cells that elongate and form the specialized cells of the root (Figure 12-21). *Lateral roots* form from pieces of meristem tissue that remain deep within the main or *tap root* as it grows downward. The delicate meristem tissue at the tip of each root is covered by a layer of protective cells called the *root cap.*

Most plants also grow by increasing the diameters of their stems and roots. Two layers of meristem within the stem and the root—the vascular cambium and the cork cambium (Figure 12-22)—regulate this growth process. The *vascular cambium* is an area of actively dividing cells that lies between and increases the widths of the xylem and the phloem (the channels through which fluids are transported).

The *cork cambium*, a layer of meristem lying near the outside of the stem and the root, produces new cork cells that increase the width of these structures. The outer layer of cells formed by the cork cambium prevents water loss and protects the stem and the root against mechanical damage.

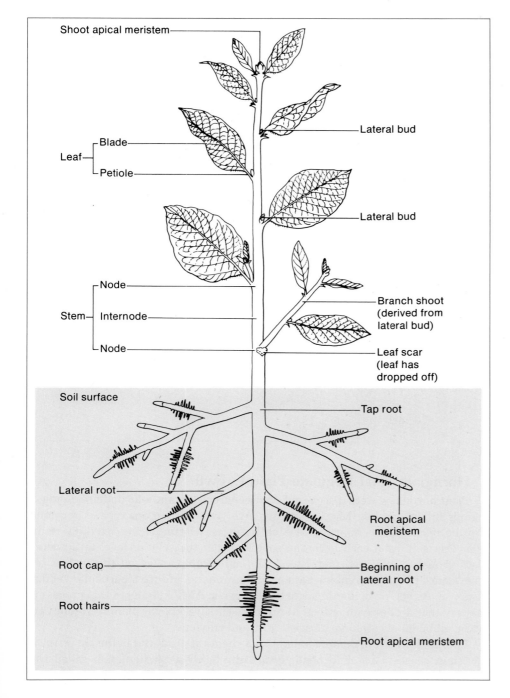

Figure 12-19
External Anatomy of a Plant
Two areas of cell division in a plant are the *shoot apical meristem* and the *root apical meristem. Lateral buds* form from small pieces of shoot apical meristem tissue on the outside of the stem. *Lateral roots* form from pieces of meristem tissue deep inside the *tap root.*

**Figure 12-20
Growth in the
Shoot Apical Meristem**

(a) The growth sequence is from top to bottom. The top figure shows the cells of the central meristem surrounded by primitive leaves. The middle figures show the central meristem growing above the primitive leaves and eventually forming two more outgrowths on the sides of the central meristem. In the bottom figure, these two additional outgrowths form primitive leaves, and the process continues. (b) Photograph of a section through a shoot tip, showing the apical meristem (AM), the most recently formed primitive leaves (P_1), and successively older primitive leaves (P_2, P_3, and P_4). The lateral buds (L) are developing on each side between the stem and the primitive leaves.

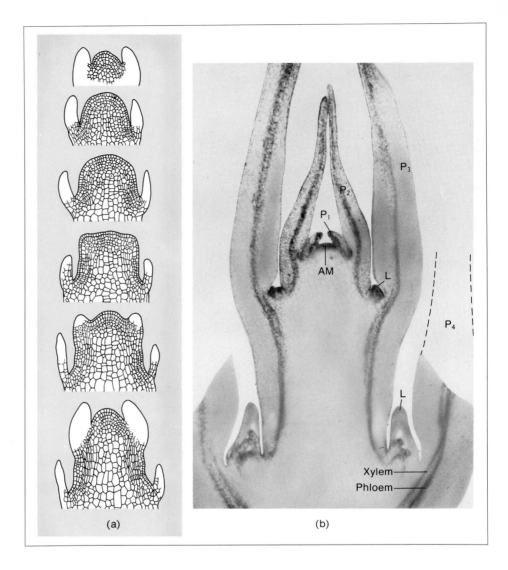

(a)

(b)

Hormones That Regulate Plant Growth

Plant hormones were discovered relatively recently and our understanding of them is still incomplete. Many responses involve more than one type of hormone, and the relative concentrations of each hormone determine the activities of the cells. Precisely what each hormone contributes at the cellular level to influence plant growth is not well understood. Special receptor sites on the plant's cell membranes probably combine with each hormone to induce a particular cellular response. As yet, no evidence suggests that cyclic AMP influences this process in the way that it does with animal hormones. The five known plant hormones are auxin, gibberellin, cytokinin, abscisic acid, and ethylene.

In 1928, Fritz Went demonstrated that a chemical produced in the shoot apical meristem causes a plant to bend toward the light—a phenomenon called *phototropism* ("photo" = light; "tropism" = orientation in response to a stimulus). In a

Figure 12-21
Growth in the Root Apical Meristem

Cell division occurs in the root tip, just beneath the *root cap*. As the meristem moves downward, the newly formed cells elongate and become specialized in structure and function.

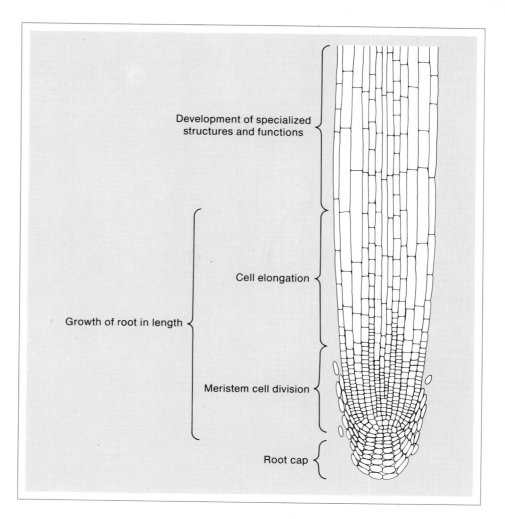

Development of specialized structures and functions

Growth of root in length

Cell elongation

Meristem cell division

Root cap

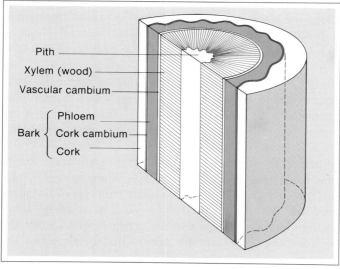

Pith

Xylem (wood)

Vascular cambium

Phloem

Bark { Cork cambium

Cork

Figure 12-22
Growth in the Cambium

Two areas of growth increase the diameter of a stem or a root. *Vascular cambium* adds new xylem and phloem cells. As new phloem is formed, the old phloem channels are sloughed off; however, old xylem is retained, so that an addition of new xylem causes the stem, branch, or root to enlarge. *Cork cambium* adds new cells to the outer protective layer of the stem and the root. This layer is called *cork* in woody plants.

Figure 12-23
Phototropism Is Caused by Auxin

(a) The steps in Fritz Went's experimental demonstration of a plant hormone are shown from left to right. Went cut the tip off the *coleoptile* (a protective sheath around the new leaves) of a 3-day-old grass seedling, placed it on a block of gelatin, and pulled up the leaf beneath the coleoptile. After Went rested the tip of the coleoptile on the gelatin block for an hour and then replaced the block on one side of the leaf, the coleoptile bent away from the side where the gelatin block had been placed. Went concluded that a chemical released onto the gelatin from the tip of the coleoptile had caused one side of the remaining part of the coleoptile to elongate. (b) The distribution of auxin at various times (left to right) after the exposure of a plant to light on only one side. The dots represent auxin molecules.

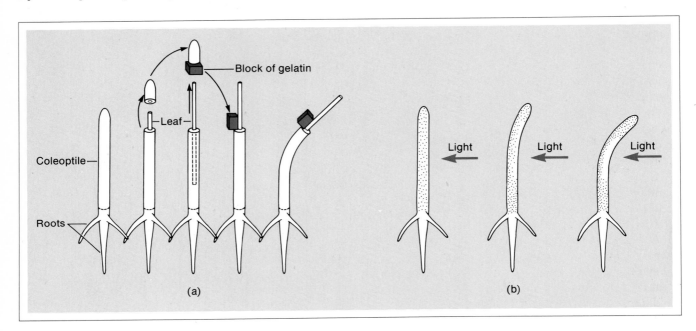

cleverly designed experiment, illustrated in Figure 12-23(a), Went showed that when light hits one side of a plant, a chemical from the tip of the stem is transported to the shaded side of the plant, where it stimulates the cells of the stem to elongate. As the shaded side of the stem grows, the plant bends toward the light, as shown in Figure 12-23(b). Went considered this chemical to be a plant hormone and named it *auxin.*

Auxin causes the cells of the stem to produce enzymes that soften the cell walls so they can expand. At the same time, auxin alters the cell membranes to permit the cells to take in both sugars and ions, especially potassium. Thus, water tends to move into the cells by osmosis, and the cell walls expand to allow growth, resulting in cell elongation.

Auxin is probably responsible for the unequal growth of the stem or the root in other *plant tropisms*—growth patterns that are oriented to some environmental stimulus. For example, *geotropism*—growth toward or away from gravity—causes the stem to grow upward and the root to grow downward. In addition, auxin activates growth in the cambium, causes *apical dominance* (which occurs when the tips of the stem inhibit the development of the lateral buds), stimulates the development of fruit, and promotes the formation of roots from a stem that has been cut from an adult plant.

During the 1930s in Japan, a second plant hormone, *gibberellin,* was found to be a product of a fungus that causes foolish-seedling disease in rice. When it is associated with a rice plant, the fungus releases gibberellin, which causes the stem of the plant to grow so tall that the entire plant falls over and dies. It is now known that plants produce gibberellin naturally.

An important role of gibberellin under natural conditions is to alter the internodal lengths between the nodes on stems to create different plant forms. Many plants undergo a stage of life during which they produce leaves without much stem growth. These leaves become stacked on top of each other, forming a rosette. A common example of slight stem growth is the nonreproductive form of a dandelion, in which many leaves grow close to one another and near the ground. The internodal lengths are short when the plant produces very little gibberellin. Just before the flowers mature, the stem elongates rapidly, so that the flowers and the seeds will be sufficiently high above the ground to facilitate pollination and seed dispersal. Rapid growth of the stem, which increases the internodal lengths, occurs when large amounts of gibberellin are present.

Gibberellin is produced in the roots, young leaves, and developing seeds of the plant. It enhances both stem and leaf growth, promotes flowering in some plants, and plays a role in breaking winter dormancy. The natural effects of gibberellin may result from interactions with auxin.

A third type of plant hormone, *cytokinin,* was discovered in the 1950s. Its primary function is to promote cell division in the meristem tissues by stimulating the synthesis of proteins and nucleic acids. Cytokinin also promotes cell enlargement in developing leaves, controls dormancy during unfavorable environmental conditions, and negates apical dominance so that the lateral buds can develop.

The fourth type of plant hormone, *abscisic acid,* inhibits plant growth. This hormone is involved in the death of plant organs and in many aspects of the preparation of plants for dormancy. Abscisic acid prevents cell division and elongation, causes leaves and fruits to detach from the plant, and promotes the development of winter buds. These effects appear to result from an inhibition of protein synthesis.

The regulatory activity of the most recently discovered plant hormone, *ethylene,* was first described in the 1960s. Apparently, ethylene is produced in the meristem tissues and diffuses out of the plant as a gas. The best known effect of ethylene is its promotion of the ripening of fruit. One rotten apple will cause all of the apples in a barrel to spoil, and green fruits will ripen more rapidly when placed in the same container with ripe fruits. Both of these effects are due to the release of ethylene gas.

The Reception of Environmental Information

Clearly, plants have mechanisms that receive information from the environment and translate it into stimuli that generate the appropriate responses to environmental conditions. However, plants do not appear to have sensory cells. Instead, specialized molecules interpret a plant's environmental conditions. The

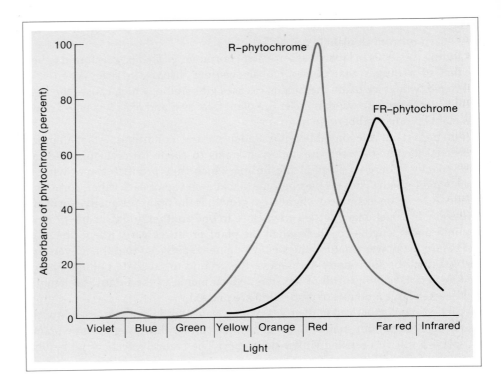

only known receptor molecule in plants is *phytochrome*—a special pigment that responds to changes in light and then triggers adjustments within the plant.

Phytochrome is a protein molecule, located within the membranes of leaf cells, that is extremely sensitive to certain forms of light (Figure 12–24). This pigment occurs in two molecular forms—R-phytochrome and FR-phytochrome. R-Phytochrome absorbs the red light present in sunlight, which causes the molecule to change to the FR-phytochrome form (Figure 12–25). FR-Phytochrome absorbs the far-red light that exists in shade, which causes the molecule to change back to the R-phytochrome form. In complete darkness, FR-phytochrome also reverts to R-phytochrome. The relative amounts of these two molecular forms are determined by light conditions and govern many cellular activities within the plant.

FR-Phytochrome appears to regulate the permeability of cell membranes to

Figure 12-25
**The Molecular Conversion
of Phytochrome**

The conversion of R-phytochrome to
FR-phytochrome, and vice versa, acts
as an on–off switch governing a
plant's cellular activities in response
to environmental changes.

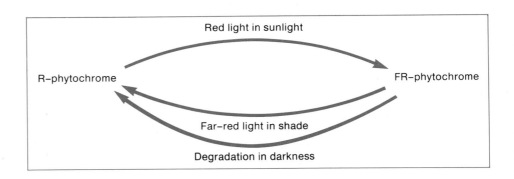

certain materials, including potassium and other ions—an important factor in osmosis. FR-Phytochrome also prevents the synthesis of gibberellins and controls the production of other hormones and enzymes.

A plant growing in darkness or shade is exposed to far-red light and its phytochrome changes from the FR form to the R form, thereby promoting the synthesis of gibberellin. Large amounts of gibberellin cause the development of a tall, spindly plant with only a few leaves. In contrast, a plant grown in bright sunlight receives more red than far-red light and much of its phytochrome is converted to the FR form, which inhibits the synthesis of gibberellin. In the absence of this growth hormone, the plant becomes short and bushy.

The phytochrome system plays an important role in leaf and stem growth, seed germination, the onset of flowering, the daily rhythms of leaf movement, the formation of plastids within the cells, and several other processes.

Evidence suggests that plants may contain a second receptor pigment, since certain growth phenomena and phototropisms occur in response to blue light. However, a pigment that absorbs blue light has not yet been discovered.

Plants also respond to gravity, electricity, the presence of solid objects, and water, but little is known of the receptors that stimulate these responses.

Seasonal Cycles of Growth and Dormancy

Plants undergo profound preparatory changes well in advance of unfavorable seasons. Trees and shrubs in northern climates survive the winter cold and drought by entering into a resting or *dormant state. Dormancy* is not simply a quiescent period that occurs in direct response to cold temperatures or water shortages; it is a special physiological property—a genetically programmed change that enables the plant to adapt to predictable periods of adverse environmental conditions.

Dormancy occurs in seeds, trees, and shrubs, and in the roots and underground stems of some plants. Here, we will consider the entry into and the release from dormancy in deciduous trees in northern climates.

The most conspicuous sign of dormancy in deciduous trees, such as the maple, aspen, and apple, is the loss of leaves in autumn. Leaves are delicate structures that contain much water. If they were not shed before winter, they would freeze and die and the materials they contained would be lost from the tree. Instead, valuable materials are removed from the leaves and are stored in the interior of the tree before the leaves are shed.

The process of leaf death actually begins in early August (Figure 12–26), when conditions seem optimal for cellular activities. Well in advance of cold temperatures, a decline occurs in the production of the enzymes used to synthesize glucose from carbon dioxide and water, which is followed several weeks later by a decline in the production of chlorophyll—the green pigment that captures light energy. Once a leaf stops producing the enzymes and chlorophyll required for photosynthesis, it loses its ability to retain water and becomes wilted. Biological molecules and valuable nutrients, including nitrogen, potassium, phosphorus, magnesium, sulfur, manganese, amino acids, and carbohydrates, are then transported from the leaves to the woody parts of the trees, where they are stored until dormancy ends in the spring.

In the absence of chlorophyll production, the green color gradually disappears from the leaves and the red and yellow pigments that are masked at other times by

Figure 12-26
**Leaf Changes in Preparation
for Winter Dormancy**

(a) In early August, the enzymes
involved in photosynthesis and the
concentrations of protein and
nitrogen in the leaves begin to
decline. (b) By early October,
chlorophyll production declines, and
the leaves lose their ability to retain
water and become wilted.

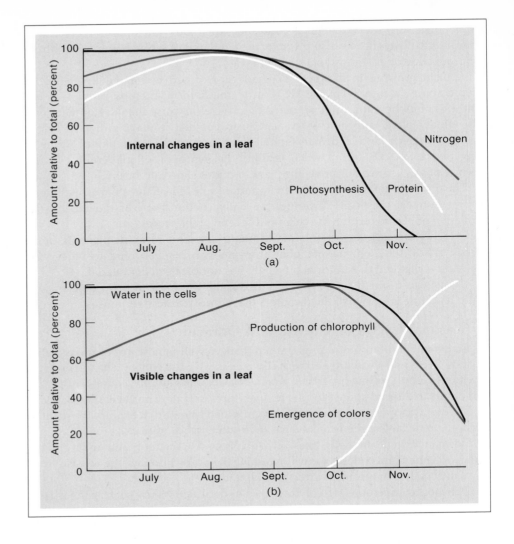

the green chlorophyll become visible. Some new pigments are also produced in
response to the cool temperatures of autumn. Eventually, the autumn leaves are
shed and a *leaf scar* forms over the area where each leaf was attached, sealing the
wound and protecting the tree from cold, desiccation, and the invasion of parasitic
fungi and insects.

The environmental cue that initiates dormancy is the *photoperiod,* or the relative
length of day compared with the relative length of night. The photoperiod alters
the production of hormones that stimulate seasonal changes in cellular activities.

Leaf death is controlled by the amount of abscisic acid compared with the
amounts of auxin, cytokinin, and gibberellin in the leaf. A basic change in hormone
production occurs at the time of the summer solstice (June 21 in northern hemispheres). After that date, the days become increasingly shorter and the nights
become increasingly longer until the winter solstice (December 21), when the

process is reversed. Beginning with the summer solstice, the tree produces increasingly smaller amounts of the growth hormones (auxin, cytokinin, and gibberellin) and an increasingly greater amount of abscisic acid. By late July or early August, these changes in hormone concentrations stimulate the shedding of the leaves.

When nutrients are transported from the leaves to the woody portions of the tree in early autumn, abscisic acid is transported as well. This hormone promotes physiological changes that permit the tree to withstand winter conditions—strengthening the cell walls, producing different kinds of enzymes, and increasing the concentrations of solutes in the internal fluids of the tree. *Solutes* are ions and molecules that interfere with the formation of hydrogen bonds between water molecules. Because water cannot freeze unless these bonds form, an increase in the concentration of solutes lowers the temperature at which freezing occurs, thereby preventing ice from damaging the cell membranes.

Abscisic acid also inhibits meristem activity, so that the tree stops growing. In early autumn, the shoot apical meristem and the lateral buds produce *winter buds.* Small, scaly leaves called *bud scales,* shown in Figure 12–27(a), develop instead of normal leaves and stems. These bud scales enclose the meristem and protect it from cold and desiccation. A cotton-like material beneath the bud scales acts as insulation. The meristem cells produce miniature leaves or flowers and then enter a period of low maintenance activity. The bud scales also release substances that further inhibit meristem activity.

A plant emerges from dormancy in the spring in response to a variety of environmental signals, including moisture and temperature levels and the photoperiod. Plants then produce increasing amounts of auxin, gibberellin, and cytokinin, and very little abscisic acid as long as the days increase in length, or until about June 21. This particular balance of hormones causes plant growth.

The buds respond to the gradually rising temperatures by swelling up and pushing the bud scales apart. As the inhibitory effect of the bud scales on growth is removed, the miniature leaves or flowers begin to grow, as shown in Figure 12–27(b).

The stores of minerals, amino acids, and carbohydrates that were removed from the leaves the previous autumn contribute to the growth and development of new leaves in the spring. New structures and molecules have to be manufactured in the absence of leaves, which are the main organs of synthesis. Similarly, trees such as cherry and dogwood that produce flowers before their leaves are formed in the spring must use materials extracted from the autumn leaves before they were shed.

Seasonal Cycles of Reproduction

Plants reproduce on roughly the same date each year. Certain kinds of flowers appear at specific times—violets and daffodils in the early spring; asters and goldenrod in the autumn. The environmental signal that initiates reproduction in most plants is the photoperiod. The response of each type of plant to a unique photoperiod is programmed into its genes.

The aspect of the photoperiod that initiates reproduction is the length of the night. By keeping plants indoors, where the length of light and dark periods can be manipulated, researchers found that a plant that normally flowers when there are 8

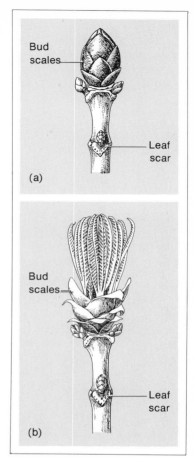

Figure 12-27
Winter Buds

(a) Protective bud scales enclose the shoot apical meristems during winter. A *leaf scar* covers the area from which a leaf has been detached. (b) In spring, the bud swells up and pushes the scales apart, so that the new leaves can grow.

hours of light and 16 hours of dark will not flower if it is briefly exposed to bright light in the middle of the dark period. The disruption of the long dark period is interpreted by the plant's phytochrome system as the end of one short night and the beginning of another.

Phytochrome in the leaves stimulates the production of hormones that carry the message from the leaves to the shoot apical meristem and the lateral buds, where flowers are produced. Two hormones are probably involved in flowering. The first, gibberellin, causes the rapid elongation of the stems of some plants. The second hormone, tentatively called *florigen*, has not yet been isolated. It is probable that this hormone causes meristem tissue to develop into flowers rather than leaves by shifting the emphasis of the maturation process from growth to reproduction.

Summary

Some of the cells within each multicellular organism are organized into *control systems* that monitor changes in the external and internal environments and coordinate the appropriate responses to these changes in the body cells. Animals have both chemical and neural control systems; plants have only a chemical control system. Control systems are composed of *tissues* (groups of similar cells) and *organs* (groups of various types of cells and tissues that function as a unit) that are linked together to perform a specific task.

Animal control systems are composed of sensory, nerve, secretory, and muscle cells. *Sensory cells* translate environmental information into chemical or neural messages that are then sent to a *control center*, usually the *brain*. The path from the sensory cells to the control center is called the *afferent pathway*. From the control center, the chemical or neural message is sent to the cells that can effect an appropriate response to the environmental change. Responses are carried out by *effector cells*, which are usually muscle or secretory cells. The path from the control center to the effectors is called the *efferent pathway*.

The brain and neural pathways are formed of *nerve cells. Muscle cells* move part or all of the body. *Secretory cells* synthesize and secrete special materials that participate directly in a physiological response or alter the activities of other types of cells. Once a response to an environmental change has occurred, information about the altered state of the body is fed back to the control center. Most biological control systems are *negative-feedback systems;* as the effect of the response increases, the control center acts to decrease the response. In a *positive-feedback system,* an increase in the effect of a response causes the control system to increase the response further.

The chemical control systems of vertebrate animals are composed of secretory cells that are organized into tissues and into organs called *glands.* These structures are located throughout the body, and their activities are coordinated by chemical messengers, called *hormones,* that are transported by the bloodstream. There are two categories of glands—*exocrine* (external secretory) *glands,* which release chemicals into body chambers or onto the body surface via ducts, and *endocrine* (internal secretory) *glands,* which release hormones directly into the bloodstream.

A *hormone* is a biological molecule that is produced by a set of secretory cells and carried by the organism's body fluids to a set of *target cells,* where it influences a variety of cellular activities. In vertebrates, most hormones are produced by the endocrine glands and transported by the bloodstream. *Trophic hormones* are produced by one endocrine structure and influence the activities of other endocrine structures. The *anterior pituitary gland*, controlled by *releasing hormones* and *release inhibiting hormones* from the *hypothalamus* of the brain, secretes four trophic hormones: thyroid-stimulating hormone, adrenocorticotrophic hormone, follicle-stimulating hormone, and luteinizing hormone. *Nontrophic hormones* act directly on target cells outside the endocrine system. The effects of hormones are either regulatory or developmental. *Regulatory hormones* adjust and maintain the internal states of the body by controlling homeostasis under normal conditions, elevating cellular activities during times of stress and generating the daily *circadian* and seasonal rhythms and cycles of vertebrate life. An example of a regulatory hormone is *antidiuretic hormone*, which is secreted by the *posterior pituitary gland* and regulates the concentration of water in the blood. Some regulatory hormones govern emergency responses. *Cortisol*, secreted by the *adrenal cortex*, participates in the regulation of physiological adjustments to stress. *Epinephrine* and *norepinephrine*, both secreted by the *adrenal medulla*, enhance the *fight-or-flight response*. *Developmental hormones* implement the genetic plan for the growth and development of the organism. The *growth hormone* regulates growth during childhood, and the two *gonadotrophic hormones* (the follicle-stimulating and luteinizing hormones) control sexual development at puberty.

Most hormones command their target cells to increase or decrease their normal chemical reactions by altering the rates at which enzymes are synthesized and by influencing the permeability of the cell membrane. Some hormones actually change cellular activities by activating certain genes and deactivating other genes in the cell, thereby causing the synthesis of different enzymes and other proteins.

Some vertebrate hormones are *steroid hormones*, which gain control of cellular activities by entering the cell and forming a complex with specific large proteins, or *receptor molecules*, in the cytoplasm. The complex then attaches itself to a particular site on a DNA molecule, where it influences gene activity. Other vertebrate hormones do not enter their target cells; instead, these *nonsteroid hormones* attach themselves to receptors on the cell membrane, where they promote the synthesis of *cyclic adenosine monophosphate* (cyclic AMP), which initiates changes in cellular activities.

Two types of hormone-like substances of current interest in vertebrate biology are the prostaglandins and histamine. Both substances are released by cells that are not specialized for secretion and are transported to other cells via the bloodstream. *Prostaglandins* are fatty acid molecules of similar structure that perform a unique function. Their effects on the smooth muscles in the linings of the uterus and the respiratory and digestive tracts are most notable. Some prostaglandins contribute to fevers, headaches, and inflammations of the joints. *Histamine* is a small molecule synthesized from common amino acids that is produced and secreted by any type of cell that has been injured. Histamine increases fluids within the area and probably produces the symptoms of colds, asthma, and allergic reactions.

Like vertebrate animals, invertebrate animals have both neural and chemical control systems. The chemical control system in insects is comprised of secretory cells in the central nervous system and in the endocrine glands that produce and release chemical messengers that control homeostasis, development, and social communication. Many chemical messengers regulate homeostasis in insects, but most of them have not yet been identified and their specific effects are unknown. Two hormones regulate growth in insects—ecdysone and juvenile hormone. An insect has a rigid *exoskeleton* and can only grow during periods when this outer covering is shed—a process called *molting*. *Ecdysone* stimulates and coordinates growth and molting in the insect. *Juvenile hormone* determines whether the insect will molt to a larger immature form (an *instar*) or to the adult form. A *pheromone* is a chemical messenger that an insect releases to communicate information to and elicit a specific behavioral response from one or more insects of the same species. Pheromones govern such insect behaviors as mating, aggregation, alarm, and navigation, as well as the development of castes in the colonies of social insects.

The responses of a plant to internal and external environmental changes are coordinated by hormones and are primarily achieved by growth in the form of cell elongation and cell division. Cell elongation can occur in any plant cell, but cell division occurs only in the *shoot apical meristem*, the *root apical meristem*, and the meristem of the *cambium*. Five hormones are known to regulate plant growth. *Auxin* causes *plant tropisms*—unequal growth of the stem that enables the plant to bend toward or away from an environmental stimulus. *Gibberellin* generates rapid growth of the stem by stimulating the shoot apical meristem. The main function of *cytokinin* is to promote cell division in the meristem tissues. *Abscisic acid* is a growth inhibitor, and *ethylene* promotes the ripening of fruits. In addition to their primary effects, each hormone influences many other plant activities.

Specialized molecules in plants receive information about environmental conditions. Only one type of receptor molecule, *phytochrome*, has been identified in plants. Phytochrome has two molecular forms—R-phytochrome and FR-phytochrome—and changes from one form to the other in the presence of light or darkness. The relative abundances of the two forms of phytochrome reflect light conditions and alter the activities of the plant accordingly.

Many plants enter periods of *dormancy*, or reduced activity. A deciduous tree in a northern climate begins to prepare for winter dormancy in summer. From the time of the summer solstice, the tree starts producing more abscisic acid and less growth-promoting hormones (auxin, gibberellin, and cytokinin). By August, the hormonal balance of the tree has shifted in favor of dormancy rather than growth. Abscisic acid induces the removal and storage of nutrients from the leaves and physiological changes in the remainder of the tree to enable it to withstand the winter conditions of freezing and drought. Abscisic acid also causes the meristem to form *winter buds* instead of leaves and stems. As spring approaches, the tree produces growth-promoting hormones in abundance, and its abscisic acid content decreases. The tree emerges from dormancy and enters a phase of new growth, making use of the nutrients that it stored during the preceding autumn.

Each type of plant reproduces at a unique time of the year, as dictated by the *photoperiod* (the relative length of day compared to night). The appropriate photoperiod creates a specific balance between the two forms of phytochrome, which in turn produces hormonal changes in the plant. Gibberellin causes stem elongation, and a hypothesized hormone called *florigen*, which has not yet been chemically isolated, seems to cause the development of the buds to shift to the production of flowers instead of leaves and stems.

Neural Control Systems

13

Chemical control systems are found in all plants and animals. Although these systems communicate relatively slowly (delays of seconds or even minutes can occur between an environmental stimulus and the response), they have provided adequate solutions to the control problems that have developed during plant evolution. Throughout animal evolution, however, chemical control systems have been supplemented by much faster and more precise *neural control systems.*

The fundamental element in a neural control system is the *nerve cell* or *neuron,* which performs the specialized functions of irritability and conductivity. *Irritability*—a property found to some degree in all cells—is the capacity to respond to an environmental stimulus. In neurons, the responses are electrochemical impulses, which are conducted from neuron to neuron at great speed (up to 200 miles per hour in some pathways in vertebrate nervous systems).

Information about events in the external and internal environments is translated by specialized *sensory cells* into a neural language consisting of patterns of electrochemical impulses. This information is conducted through *afferent neural pathways* to *neural control centers,* where it is processed and evaluated and responses to the information are initiated. Commands from the neural control centers are transmitted as patterns of electrochemical impulses along *efferent neural pathways* or as *hormones* in moving fluids to *effector cells* in the muscles and glands. The entire sequence from stimulus to response can occur within *milliseconds* (thousandths of a second).

Although nerve cells and nervous systems initially evolved in animals as adaptations for rapid and precise movements in response to environmental changes, they also provided animals with a remarkable potential—the evolution of *intelligence.* This complex and mysterious property of the nervous system has produced learning, thinking, remembering, creating, and language.

In this chapter, we will examine the nerve cell and the electrochemical impulse it transmits, the coding of environmental information into electrochemical impulses by sensory cells, and the form and evolution of intelligent nervous systems.

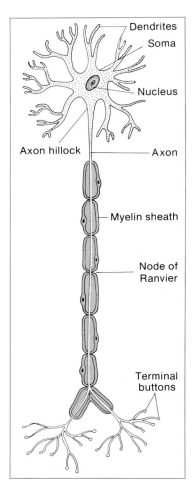

Figure 13-1
The Structure of a Typical
Neuron

The Nerve Cell

Nerve cells, or neurons, differ considerably in size and shape but usually have three distinct regions—the soma (or cell body), dendrites, and an axon (Figure 13-1). All of these regions are enclosed within a single, continuous, semipermeable membrane, and the cell is completely immersed in *interstitial fluid* (the extracellular fluid that surrounds cells in tissues). Adjacent neurons are never in direct physical contact; there is always a fluid-filled space between the cells.

The *soma*, which contains the nucleus and other organelles characteristic of eukaryotic cells, performs basic cellular activities such as protein synthesis and respiration as well as certain unique functions associated with receiving and conducting electrochemical impulses. The membrane of the soma is specialized for receiving impulses from other neurons, and molecules called *neurotransmitters*, which convey electrochemical impulses across the fluid-filled space to the next neuron, are synthesized in the soma.

The *dendrites* are branchlike extensions ("dendron" = tree) of the neuron that also receive electrochemical impulses.

Most neurons have a single, long *axon* that extends from the soma on the opposite side from the dendrites and transmits impulses along its length to narrow branches that end in knobs called *terminal buttons.* The axons of some neurons are covered with special insulating materials. In vertebrates, this material—*myelin*—is composed of about 80% lipids and 20% proteins. The *myelin sheath* is discontinuous and is regularly interrupted by exposed areas of the axon called the *nodes of Ranvier* (Figure 13-1).

The junction between the terminal button and the next neuron or effector cell in the pathway is called a *synapse,* and the fluid-filled space between the two cells is called a *synaptic cleft.* The cleft is extremely narrow—approximately 200–500 angstroms wide (1 angstrom = 100 millionths of a centimeter).

The impulse conducted along the cell membrane and transmitted across the synaptic cleft is usually an electrochemical impulse in that it involves both electrical and chemical changes. The electrical changes are brought about by the selective movement of ions across the cell membrane. When these changes reach the terminal buttons, neurotransmitters flow across the synaptic cleft and generate impulses in adjacent neurons or such specialized activities as muscle contractions in effector cells. Most neurons receive and conduct electrochemical impulses in a sequence of four events: the resting membrane potential, the action potential, the propagation of the impulse along the axon, and the transmission of the impulse across a synapse.

Resting Membrane Potential

As is true of all other cells, the fluid inside the neuron is quite different from the interstitial fluid surrounding the cell. This difference is maintained by active and passive transport mechanisms in the cell membrane that allow certain atoms and molecules to enter and other atoms and molecules to leave the cell. Because many of these particles are ions, the selective permeability of the membrane produces a forced separation of charges between the inside and the outside of the membrane. A *membrane potential* is a measure of the charge inside the cell relative to the charge outside the cell. A *resting membrane potential* is measured when the neuron is not actively receiving, generating, or conducting impulses. For any segment of the membrane (dendrite, soma, or axon), this resting potential is usually −70 millivolts (mV; 1mV = 1/1000 volt), which means that the interior of the cell is 70 mV more negative than the exterior of the cell.

Three types of ions play significant roles in determining the resting potential of the axon: sodium (Na^+), potassium (K^+), and large protein molecules with negative charges. The inside of the resting axon contains approximately 30 times more potassium ions and about 10 times fewer sodium ions than the surrounding interstitial fluid. All of the negatively charged protein molecules are located inside the axon. The different concentrations of these three types of ions are maintained by an interplay of several factors: diffusion, electrical attractions and repulsions, active transport across the cell membrane, and the selective permeability of the axon membrane to these three ions.

Channels in the axon membrane permit the movement of materials (Figure 13-2). The negatively charged protein molecules are too large to fit through these channels and always remain inside the axon. The sodium ions move through a channel in the membrane with a gate that is sensitive to the electrical state of the membrane. At the resting membrane potential of −70mV, the voltage-sensitive gate is essentially closed, although some sodium ions can leak inward toward the internal negative charge and the area of lower sodium ion concentration.

The potassium ions move through two types of channels—one that is permanently open and another, comparable to the sodium channel, with a voltage-sensitive gate. In the resting state, the voltage-sensitive gate is closed, but the potassium ions can still move freely out of the cell along the permanently open channel. Potassium ions tend to diffuse out of the axon slowly and continuously, but their movement is opposed by their attraction to the negatively charged protein

Figure 13-2
The Axon Membrane During the Resting State

The resting membrane potential is due to a greater concentration of sodium ions (squares) outside the membrane than inside, a greater concentration of potassium ions (circles) inside the membrane than outside, and negatively charged protein molecules inside the membrane. The gates in the voltage-sensitive channels for sodium and potassium ions are essentially closed, but potassium ions are free to move through the permanently open potassium channels. The sodium–potassium pump returns ions that have moved through the channels.

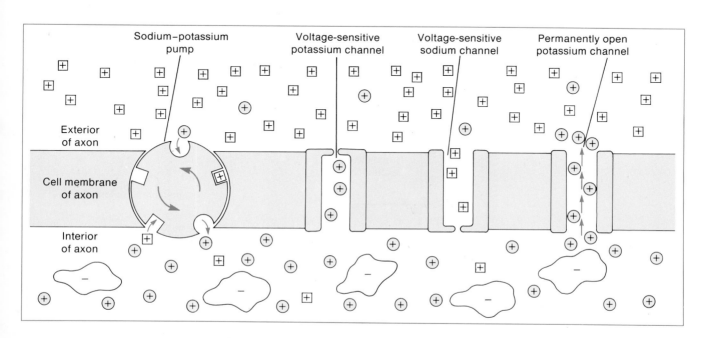

molecules and their repulsion by the positively charged sodium ions outside the membrane.

If these channels were the only active mechanisms during the resting phase, the interior of the axon would become more and more negatively charged as potassium diffused out much more rapidly than sodium diffused into the axon. However, an active transport mechanism, called the *sodium–potassium pump* (Figure 13-2), is continuously moving sodium ions out of the cell and returning potassium to the cell interior. This pump—a protein molecule within the membrane—operates from energy stored in ATP molecules.

The Action Potential

The *action potential* is a very rapid change in the axon membrane potential from negative (−70mV) to positive (+40mV) and back to negative (−70mV). This change always occurs in exactly the same pattern (Figure 13-3). Initially, the internal negativity gradually decreases until a *threshold of excitation* is reached at approximately −60mV. A rapid change in the membrane potential from negative to

Figure 13-3
An Action Potential

An action potential begins with a slow decrease in the negativity of the membrane potential until it reaches a threshold of excitation at −60mV. The voltage-sensitive sodium gates then open and the sodium ions rush into the axon, causing a rapid decrease in negativity. When the membrane potential reaches +40mV, the voltage-sensitive sodium gates close and the voltage-sensitive potassium gates open. As potassium ions move out of the axon, the potential becomes more negative than it is in the resting state. The voltage-sensitive potassium gates then close, and the sodium–potassium pump returns the membrane to its resting state (−70mV).

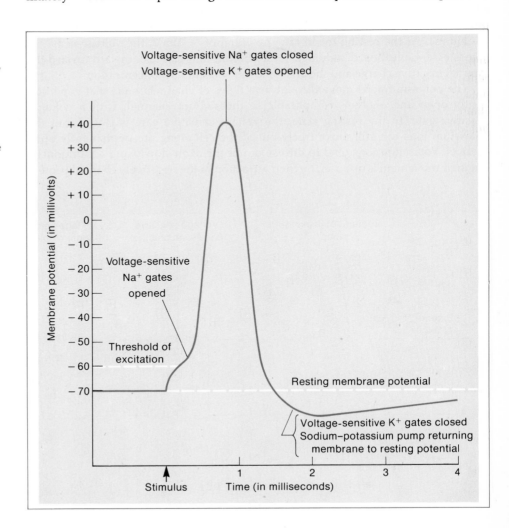

positive to negative, called the *spike potential* (also called the *absolute refractory period*), is then completed within 1 millisecond. The spike potential is followed by a *relative refractory period* lasting several milliseconds during which the potential, which has become more negative than the resting potential, gradually returns to the resting state of $-70mV$.

During the action potential, the sodium–potassium pump removes sodium from and returns potassium to the cell interior. During the relative refractory period, this mechanism gradually returns the membrane to the resting potential within about 10 milliseconds of the end of the spike potential.

Propagation of the Impulse

The action potential that rises and falls in a limited area of the axon membrane is only one stage in a neuron's reception and transmission of electrochemical impulses. The next stage in this process—the movement of a sequence of action potentials down the axon from the *axon hillock* (Figure 13-1) to the terminal buttons—is called the *propagation of the impulse.*

An action potential can be produced artificially at any unmyelinated site on the axon membrane by applying an electrical stimulus that makes the membrane potential 10mV more positive. In the organism, this decrease in internal negativity is usually stimulated by the receipt of neurotransmitters from other neurons by the dendrites or soma. These neurotransmitters produce electrical changes in the cell membrane that flow toward the axon. An action potential results if these changes produce a decrease in the membrane potential to $-60mV$ when they arrive at the axon hillock, where the axon emerges from the soma.

If the membrane at the axon hillock is stimulated to produce an action potential, the resulting change in potential will affect adjacent sections of the membrane. At the peak of the spike potential, the interior of the axon is positively charged ($+40mV$) in relation to the cell exterior. Inside the membrane, positively charged ions (Na^+ and K^+) are driven away from the site of the action potential by repulsion of like charges. Sodium ions have moved into the cell simultaneously, making the region just outside the membrane less positive in charge so that positive ions move along the external surface of the membrane toward the site of the action potential (Figure 13-4). These internal and external electrical currents affect adjacent sites of the membrane; the interior becomes more positive, and the exterior becomes less positive. If this change reaches the threshold of excitation ($-60mV$), an action potential identical to the first action potential develops in the adjacent site on the membrane. This new potential, in turn, produces an identical action potential in an adjacent site farther down the axon membrane.

Although the ionic currents produced by an action potential flow toward and away from the soma, the sequence of action potentials always moves toward the terminal buttons at the end of the axon. The sequence does not move back into the soma because its membrane does not have a sufficient density of voltage-sensitive ion channels to propagate action potentials. Farther down the axon, an action potential only affects the membrane on the side nearer the terminal buttons, because a membrane site that has just generated an action potential cannot immediately produce another. During the spike potential (the absolute refractory period), no stimulus of any strength can generate an action potential at that site.

Figure 13-4
Ionic Movements at the Site of an Action Potential

The high density of positively charged sodium and potassium ions inside the membrane at the site of an action potential causes them to move away from the site. Outside the membrane, the relatively low density of positive ions at the site of an action potential causes the ions to move toward the site. Ionic movements on adjacent sites decrease the negativity of the membrane potential.

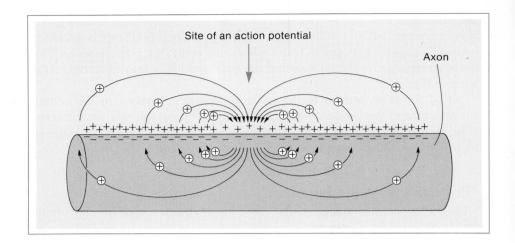

During the relative refractory period—the 10-millisecond period after the spike potential—when the membrane is more negative than it is in the resting state, only an unusually strong stimulus can generate an action potential. Due to these refractory periods, the series of action potentials moves only in the direction in which no recent action potentials have occurred—away from the soma.

This propagation of the impulse obeys the *"all-or-none" law* of neural conduction; all of the sequence or none of the sequence occurs. If the axon hillock membrane is stimulated to produce an action potential, then a sequence of identical action potentials will travel down the axon to the terminal buttons.

The sequence of action potentials moves down the axon at a constant velocity dependent on the temperature and diameter of the axon and whether or not the axon has a myelin sheath. In axons that have no myelin sheaths, the speed of propagation depends entirely on the temperature and diameter of the axon. Speed of propagation increases as the temperature rises to a critical point beyond which the proteins in the membrane become inactive. Propagation also speeds up as the diameter of the axon increases because the internal current of positive ions moves faster in larger axons. As the ions move away from the site of the action potential, some leakage through the membrane slows the current flow inside the axon. Less leakage occurs in larger axons, so that adjacent areas of the membrane rapidly accumulate sufficient positive ions to reach the threshold of excitation. The axons in invertebrate animals have no myelin sheaths, and some of them are very large in diameter. These large axons carry electrochemical impulses to the muscles involved in fast responses, such as the burrowing movements of a worm or the escape movements of a squid.

In vertebrate animals, rapid impulse propagation is achieved by myelination. Axons in all but the most primitive vertebrates are covered with myelin sheaths, which permit very small axons to conduct impulses rapidly, thereby conserving space and energy. For example, an unmyelinated squid axon and a myelinated frog axon propagate impulses at the same speed, but the squid axon has a diameter 40 times greater and expends 5,000 times more energy to propagate each impulse than the frog axon. If vertebrate evolution had selected for larger axons rather than myelination, the human spinal cord would have had to be several meters in diameter to achieve the same speed of response.

The myelin sheaths around axons are produced by two types of non-neural cells—the oligodendroglia and the Schwann cells—both of which are types of *glial cells.* There are roughly ten times as many glial cells as there are neurons in the human brain. *Oligodendroglial cells* are located near neurons in the brain and the spinal cord of vertebrates. Their principal function is to produce myelin sheaths; oligodendroglia also hold the neurons in place and maintain the composition of surrounding fluids. During the development of the brain, each oligodendroglial cell sends out many paddlelike extensions that wrap tightly around the axon (Figure 13-5), squeezing the cytoplasm out of the paddle so that the sheath is composed entirely of cell membrane. Each paddle covers an area of about 1 millimeter along the axon; the short, exposed areas of the axon between the paddles are the *nodes of*

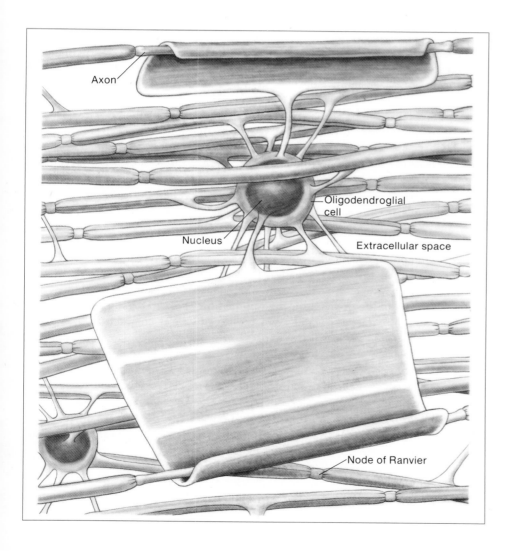

Axon

Oligodendroglial cell

Nucleus

Extracellular space

Node of Ranvier

Figure 13-5
Myelin Sheath Formation by an Oligodendroglial Cell
Each oligodendroglial cell sends out paddle-like extensions that wrap tightly around an axon, squeezing the cytoplasm out of the paddle so that each sheath is composed only of cell membrane. The areas of exposed axon between myelin segments are the *nodes of Ranvier.*

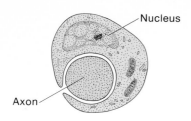

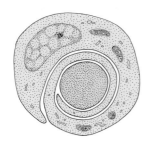

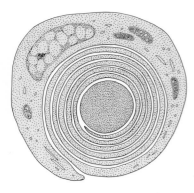

Figure 13-6
Myelin Sheath Formation by a Schwann Cell

From top to bottom, a Schwann cell wraps itself around an axon many times. Each Schwann cell forms one segment of the myelin sheath.

Ranvier. Schwann cells produce myelin sheaths around the axons of neurons that are not located in the brain or the spinal cord, such as the nerves in the arms and legs. These myelin segments are also about 1 millimeter long with nodes of Ranvier between the segments. Each segment is produced by one Schwann cell, which wraps itself, including the soma, tightly around the axon (Figure 13-6).

Myelin sheaths are excellent insulators. Ions flow more rapidly away from the site of an action potential because they cannot diffuse through the axon membrane where it is surrounded by myelin. Action potentials can form only at the nodes of Ranvier. The action potential jumps from node to node along the axon, traveling at a much faster rate than it does in unmyelinated axons because action potentials occur at fewer sites. Less energy is also required to produce the action potential, because only the sodium—potassium pumps at the nodes of Ranvier must work to restore membrane conditions.

Transmission Across a Synapse

Propagated action potentials are the fundamental units of information that pass through a nervous system. Neural messages consisting of coded patterns of action potentials are utilized in all aspects of neural control from monitoring and interpreting environmental events to the initiation of rapid, coordinated responses.

Action potential messages are transmitted from cell to cell across a *synapse*—a junction between two cells formed by two membranes, one from each cell, and by a fluid-filled space (the synaptic cleft) between the membranes. The *transmitting membrane* is from a terminal button of an axon; the *receiving membrane* is a specialized receptor surface on a neuron (the dendrites or soma) or an effector cell. Because action potentials travel only from the soma to the terminal buttons, the transmitting membrane is called the *presynaptic membrane* and the receiving membrane is called the *postsynaptic membrane.*

There are two basic types of synapses—electrical and chemical. In an *electrical synapse,* action potentials move from a presynaptic to a postsynaptic membrane in a way that is similar to the propagation of the impulse down an axon. In such a junction, the synaptic cleft is so narrow that electrical currents generated by the action potentials in the presynaptic membrane flow across the gap to the postsynaptic membrane, where they affect voltage-sensitive ion channels and produce action potentials. These electrical synapses, also called *tight junctions,* are common in invertebrates but are not so common in vertebrates. In mammals, electrical synapses do not occur between neurons but do occur between the muscle cells of some organs, such as the heart and the walls of the digestive tract.

All synapses in the human nervous system are *chemical synapses.* The synaptic cleft is too wide for electrical currents to pass between the membranes. Instead, the action potential message is transmitted to the postsynaptic membrane by specific neurotransmitter molecules that diffuse across the synaptic cleft.

Two types of structures are prominent in the terminal buttons of a chemical synapse—mitochondria and synaptic vesicles (Figure 13-7). As in all cells, the *mitochondria* provide usable energy for cellular activities. The small, rounded, membrane-enclosed *synaptic vesicles* contain thousands of neurotransmitter mole-

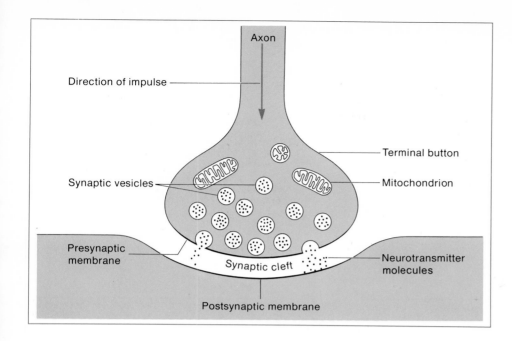

Figure 13-7
A Chemical Synapse
Information is conveyed by *neurotransmitter molecules* from the *presynaptic membrane* of one neuron to the *postsynaptic membrane* of the next neuron or effector cell. The *synaptic vesicles* are membrane-enclosed packages of neurotransmitter molecules.

cules. These molecules are synthesized in the soma, where they are also packaged in the vesicles. The vesicles are then transported down the axon to the terminal buttons, where they are stored. By some mechanism not yet fully understood, action potentials arriving at the terminal buttons trigger the release of these stored neurotransmitters into the synaptic cleft. Several vesicles move to the presynaptic membrane, fuse with it, and then rupture, spilling their contents into the synaptic fluid (Figure 13-8). The released neurotransmitter molecules then diffuse across the cleft to the postsynaptic membrane.

The transmitter molecules become attached to receptor molecules embedded in the postsynaptic membrane. These attachments trigger changes in the permeability of the postsynaptic membrane that, in turn, produce changes in the *postsynaptic membrane potential.* These changes, called *postsynaptic potentials,* can be excitatory or inhibitory, depending on the type of receptor molecules to which the neurotransmitters become attached. An *excitatory postsynaptic potential* (EPSP) is a decrease in the internal negativity of the postsynaptic membrane from its resting state of -70 mV that increases the probability that the receiving neuron will be stimulated into generating and propagating an action potential. An *inhibitory postsynaptic potential* (IPSP) is an increase in this internal negativity that decreases the probability that an action potential will occur.

The postsynaptic membranes of neurons are specialized receptor surfaces on the dendrites and somas. These membranes do not have sufficient voltage-sensitive sodium and potassium channels to produce action potentials; instead, they contain ion channels with gates that are sensitive to the presence of specific neurotransmitters. The joining of a neurotransmitter molecule to an excitatory receptor molecule produces an EPSP by initiating a process that briefly opens an

Figure 13-8
Movement of Neurotransmitter Molecules Through the Synapse
(a) Synapses between a motor neuron and muscle cells. (b) Close-up of a cross section through the synapse boxed in (a). (c) Photograph of a nerve–muscle synapse (×230,000) showing two synaptic vesicles discharging neurotransmitter molecules into the synaptic cleft.

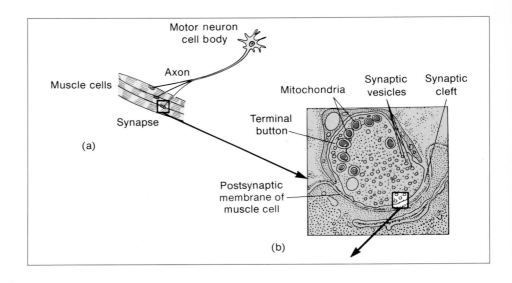

(a)

(b)

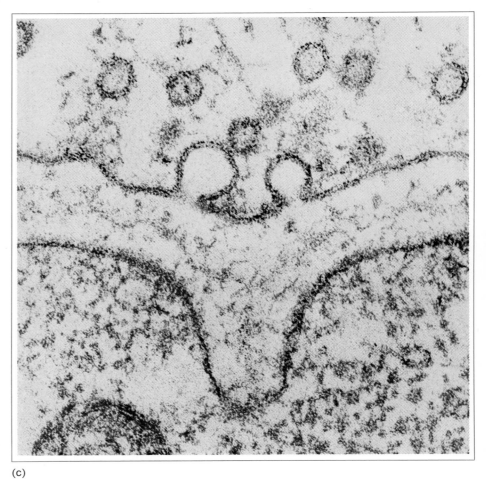

(c)

ion channel to allow the diffusion of sodium and potassium ions. Because the inflow of sodium ions is greater than the outflow of potassium ions, the post-synaptic membrane potential becomes more positive. The joining of a neurotransmitter molecule to an inhibitory receptor molecule to produce an IPSP is not completely understood but probably opens an ion channel that allows an increased outflow of potassium.

The receptor surfaces of neurons are covered with synapses (Figure 13-9). In the human brain, each neuron typically has 1,000–10,000 synapses; in some brain areas responsible for the coordination of movement, a single neuron may have as many as 70,000 synapses. Not all of these synapses are releasing neurotransmitters at any given moment, but the synapses that do are producing a great many simultaneous EPSPs and IPSPs in a neuron's receptor membranes. All of these changes in potential generate ionic currents that flow along the membranes of the dendrites and somas toward the axon hillock. If, at any given moment, the axon hillock membrane is in an excitable state and the sum of all of the currents reaching it from the receptor surfaces is sufficient to reduce its membrane potential to the threshold of excitation (-60mV), then the receiving neuron will generate and propagate an action potential.

Immediately after the change in membrane potential has occurred, the neurotransmitter molecules are removed from the receptor molecules on the postsynaptic membrane. If this did not occur, no new information could bridge the synapse; a continuous "busy signal" would result. The released neurotransmitter molecules are either broken down into nonfunctional components or taken back into the terminal buttons, where they may be reused.

Most animal nervous systems utilize a large variety of synapses, which provide precision and flexibility of control in at least three ways. Different synapses can regulate different control functions, synapses can vary from slow-acting to fast-acting in terms of response time, and synapses can perform specialized activities that produce a variety of long-term changes in cellular function.

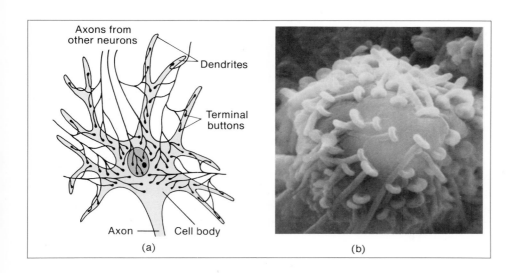

Figure 13-9
A Neuron is Covered with Synapses

(a) Diagram of axons connected by synapses to a single neuron.
(b) Terminal buttons of synapses ($\times 1,500$) on the surface of a neuron of a sea hare (*Aplysia californica*).

Chemically unique synapses are found in functionally different parts of a vertebrate nervous system. Acetylcholine synapses occur between neurons and muscle cells. Synapses in which the neurotransmitter molecules are derived from amino acids (the biogenic amines) are found in areas of the human brain that deal with motivation and emotion. Examples of such neurotransmitters are dopamine, serotonin, and norepinephrine (also called noradrenaline and identical to the hormone of the same name). Synapses with serotonin are important in the regulation of sleep. The brain also produces morphinelike substances called *endorphins* that regulate pain and the emotional responses to pain. The *enkephalins*, one type of endorphin, are believed to function as neurotransmitters in neural systems that carry pain information.

Unique neurotransmitters and receptor molecules in the synapses of different control systems prevent these systems from interfering with each other and allow neurotransmitters and specific regulating hormones to have precise effects on a cell. The specific chemistry of synapses has permitted humans to use drugs to intervene in the activities of specific parts of the nervous system.

The speed and duration of synaptic activities depends on the particular transmitter and receptor. Postsynaptic potentials persist for different lengths of time, from a few milliseconds in fast-acting synapses to as long as one second in some slow-acting synapses in the brain. The duration of a postsynaptic potential is a dimension of the neural language that influences whether or not action potentials result from new synaptic stimulation and therefore whether or not information is passed on through the nervous system.

In some fast-acting synapses, such as the nerve–muscle junctions commanded by acetylcholine, the coupling of transmitter and receptor molecules has an immediate effect on membrane permeability, somehow directly opening the appropriate ion channels that produce postsynaptic potentials. However, the receipt of most types of transmitters at a postsynaptic membrane initiates a sequence of chemical reactions involving cyclic AMP that can produce short-term changes in membrane permeability as well as long-term changes in cellular function. In these types of synapses, the neurotransmitter acts as the *first messenger,* much like the nonsteroid hormones discussed in Chapter 12. The neurotransmitter joins with its specific *receptor molecule,* activating the enzyme *adenylate cyclase* in the membrane and causing cyclic AMP (adenosine monophosphate) to form from ATP (adenosine triphosphate). The cyclic AMP acts as a *second messenger* inside the neuron, initiating short-term changes in membrane permeability that produce postsynaptic potentials and, it is now believed, a variety of long-term changes in cellular activities. These long-term cellular changes, which may last for years—even for the lifetime of a human—could play an important role in the formation and retention of memories.

We already know that hormone-activated cyclic AMP influences the rate and type of proteins that are synthesized in the cell. In the neuron, many proteins are utilized in the reception, generation, and propagation of action potentials. By producing long-lasting changes in the synthesis of these proteins, neurotransmitters and their second messengers can produce long-term changes in the function of the neuron. These functional changes may be a basic mechanism for storing information in the nervous systems.

The Sensory Cell

We have learned that the five components of a *biological control system* are sensory cells, afferent pathways, a control center, efferent pathways, and effector cells (see Figure 12-3, page 271). The sensory cells of a neural control system detect an environmental change (the stimulus) and transmit information about this change along an afferent neural pathway to the control center. From the control center, which is an aggregation of interconnected neurons, commands are transmitted along an efferent neural pathway or by hormones to the effector cells, which initiate a response to the environmental stimulus. Now we will examine the first two components of a *neural control system*—the sensory cells and the afferent neurons—in more detail.

An animal's response to a change in its external or internal environment is coordinated by its neural control system. Environmental information must therefore be converted into neural messages before a response can occur. An environmental change always involves some form of energy. The conversion of environmental energy into the electrochemical energy of neurons is called *transduction* and is performed by sensory cells.

Transduction and the conveyance of transduced information along afferent neural pathways may be performed by a single sensory cell or by a group of sensory cells organized into a *sensory organ.* When a single sensory cell performs both of these functions, it is a *sensory neuron.* As Figure 13-10(a) illustrates, the cell membrane enclosing a dendrite responds to a particular form of environmental energy by altering its permeability to ions, so that the negativity of the membrane potential is reduced to the point that it becomes a *generator potential.* If the generator

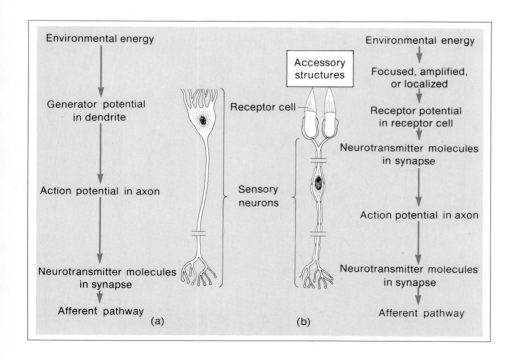

Figure 13-10
Structures That Transduce and Convey Environmental Information

(a) A single *sensory neuron* may transduce environmental energy and convey it to an afferent neural pathway. (b) A *sensory organ* contains a variety of sensory cells; each type of cell performs a unique function in the transduction and conveyance of environmental information.

potential is large enough, it initiates an action potential in the axon that travels directly to the control center or is passed from neuron to neuron along an afferent pathway in the form of electrochemical impulses. The generator potential may range from much smaller to much larger than the action potential, which has a fixed value.

The specialized functions performed by the component cells of a sensory organ are shown in Figure 13-10(b). Some of these cells form accessory structures that focus, amplify, or localize the environmental energy. (For example, the lens of a vertebrate eye bends the light so that it converges onto the retina.) The *receptor cells* of a sensory organ actually transduce information. The cell membrane of a receptor cell responds to a particular form of environmental energy by changing its permeability to ions, which generates a *receptor potential.* Like the generator potential, the receptor potential may be much smaller or much larger than the action potential. If it is large enough, the receptor potential alters the membrane potential in the sensory neuron to which it is connected by a synapse. This change may initiate an action potential in the axon of the sensory neuron, which is then conveyed along the afferent pathway to the control center.

Everything an animal detects about the environment that is external to its nervous system is accomplished by its sensory cells. Each type of sensory cell is sensitive to a certain form of energy, such as heat, light, or sound. An animal can receive information only if it has sensory cells that are capable of transducing the form of energy accompanying the environmental event. An animal's sensory cells result from natural selection and vary for different species. Humans lack suitable sensory cells for many forms of energy, including X-rays, radar, and radio or television waves; we detect such phenomena only after they have been converted by mechanical devices into forms of energy that our sensory cells can recognize.

Figure 13-11
Properties of a Wavelength of Radiant Energy
The wavelength shown here is the distance traveled by radiant energy in $\frac{1}{2}$ second. The frequency of this wave is two cycles (or wave peaks) per second.

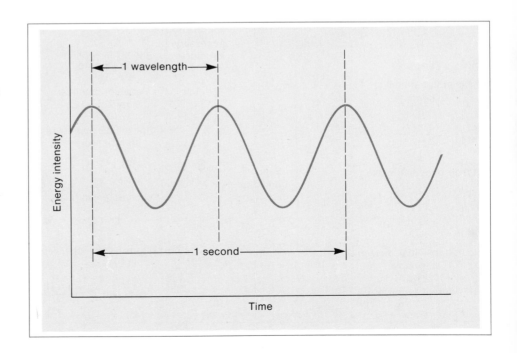

Many animals can detect environmental phenomena that we cannot. Bees see ultraviolet radiation, bats and porpoises hear very high-pitched sounds, snakes locate prey by sensing the heat of their bodies, and some fishes detect prey and mates by means of electrical currents. Sensory cells in most animals receive some forms of radiant energy (*photoreceptors* and *thermal receptors*), mechanical energy (*mechanoreceptors*), and chemical energy (*chemoreceptors*).

Photoreceptors

Radiant energy is described in terms of wavelengths (Figure 13-11), or regular oscillations in the intensity of the energy. A *wavelength* is the distance between two successive peaks of energy intensity and can vary from thousands of meters (radio waves) to fractions of a millimeter (gamma rays), as shown in Figure 13-12. *Light* is

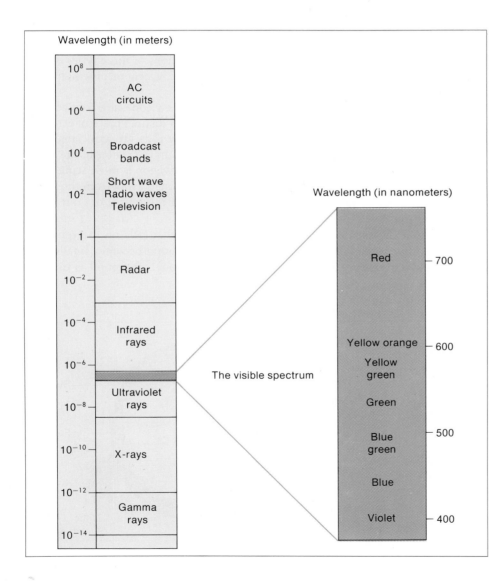

Figure 13-12
Wavelengths of the Different Forms of Radiant Energy

the range of wavelengths visible to humans, which extends from about 400 to 750 nanometers (1 nanometer = 1 billionth of a meter). Most organisms can detect the presence or absence of light, but not all organisms can actually see detailed images. Sensory organs that produce visual images of the external world are found in the vertebrates, arthropods, and certain mollusks (the squid and the octopus).

The transduction of light energy by *photoreceptor cells* is brought about by pigment molecules in the cell membranes. When exposed to light, the pigments change from one molecular form to another, and this change alters the permeability of the membrane so that the interior of the cell becomes less negative. If the receptor potential that forms is large enough, it initiates an action potential in an associated sensory neuron.

The photoreceptors in the human eye are located in the *retina,* a thin layer of neural tissue lining the back of the eyeball (Figure 13-13). The eye contains two basic types of photoreceptor cells—rods and cones (Figure 13-14). The *rods* are so sensitive to light that they can detect a single photon but cannot distinguish fine details or colors except shades of gray. Rods play a particularly important role in vision in dim light. Sensory information from combinations of three types of *cones* produces color vision; each cone is sensitive only to red, green, or blue light. Few mammals other than primates are able to see color.

Cones distinguish fine detail because only one or a few cones are connected to one sensory neuron; each cone provides the brain with a distinct piece of information. Rods do not distinguish fine detail because many rods are associated with each sensory neuron. Most cones are located in an area in the center of the retina called the *fovea* (see Figure 13-13), so that detailed color vision is most acute when the eye looks directly at an object. Most rods are positioned away from the fovea, so that vision in dim light is better on the periphery of the retina, or "out of the corner of the eye."

The size of the *pupil* controls the amount of light that falls on the retina. In bright light, the muscles of the *iris* contract, so that the pupil becomes smaller and less light enters the eye; the light that does enter through the pupil is directed

Figure 13-13
The Structures of the
Human Eye

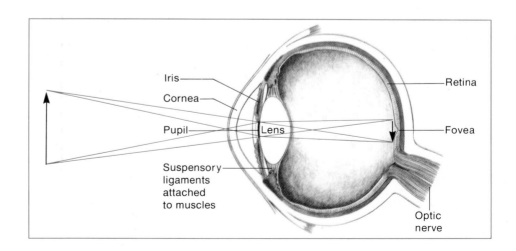

Iris

Cornea

Pupil

Lens

Suspensory
ligaments
attached
to muscles

Retina

Fovea

Optic
nerve

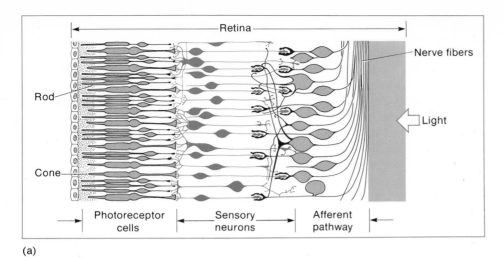

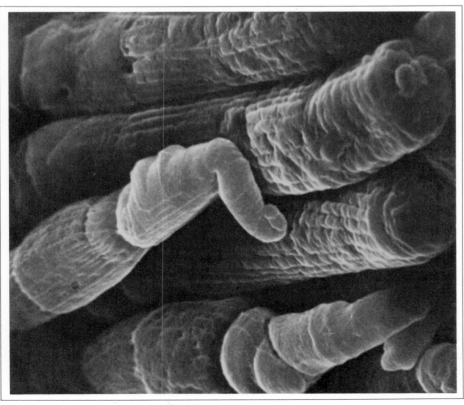

**Figure 13-14
Photoreceptors in the
Vertebrate Eye**

(a) A portion of the human retina,
showing the positions of the rods and
cones and their connections to the
afferent pathway. (b) Photograph of
rods (cylinders) and cones (pointed
cones) in a salamander retina (×875).

toward the fovea. In dim light, the muscles of the iris relax, the pupil becomes larger, and incoming light is spread over the entire retina. The *cornea* and the *lens* of the eye bend the incoming light so that it is focused on the retina. The cornea covers the outer surface of the eyeball. The lens is located just behind the iris; muscles attached to the lens can change its shape, so that light is focused in accordance with the distance of the object being viewed. The eyeball itself is attached to muscles that move it continually, so that the image is shifted from one set of receptor cells to another. If this movement did not occur, the image of a stationary object would gradually fade from view, because photoreceptor cells respond only to changes in light.

In contrast to the vertebrate eye, the arthropod eye is a *compound eye* composed of many subunits, the *ommatidia*, all of which are identical in structure and function (Figure 13-15). Each ommatidium consists of a lens, a transparent crystalline cone, photoreceptor cells, and pigment cells. The lens and crystalline cone are accessory structures that direct light to the photoreceptor cells, where transduction takes place. The pigment cells, which absorb light, line the sides of each ommatidium and prevent light from traveling to adjacent ommatidia. By absorbing light that enters an ommatidium from an angle, the pigment cells also ensure that only light entering parallel to the sides of an ommatidium reaches the receptor cells. Each ommatidium receives light from a single point in space, so that the environment is perceived as a mosaic of separate, nonoverlapping images, making the compound eye particularly sensitive to moving objects.

Many insects have color vision. A bee cannot distinguish red but can see all of the other colors that a human sees and can also distinguish ultraviolet. Bee-

Figure 13-15
Compound Eyes

(a) The compound eye of a fly. (b) A compound eye is composed of thousands of *ommatidia*, each containing all of the structures necessary for transducing the image within a limited area of the environment. (c) Cross section of an ommatidium, showing the locations of the pigment and photoreceptor cells.

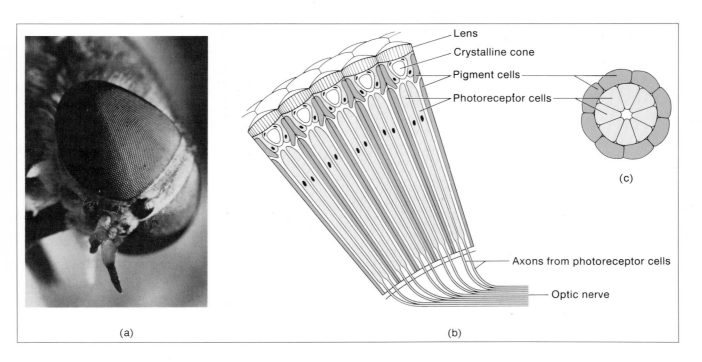

Lens
Crystalline cone
Pigment cells
Photoreceptor cells

(c)

Axons from photoreceptor cells

Optic nerve

(a)   (b)

(a)

(b)

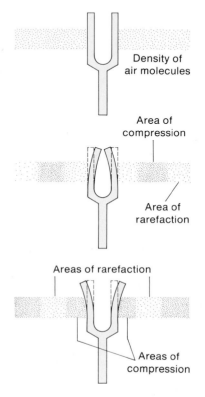

Figure 13-16
Ultraviolet Patterns on Marsh
Marigolds
(a) Photograph taken in ultraviolet
rays, showing the pattern visible to
bees. (b) Photograph of the same
flower taken in visible light.

Density of
air molecules

Area of
compression

Area of
rarefaction

Areas of rarefaction

Areas of
compression

Figure 13-17
The Properties of Sound
When the prongs of a tuning fork
vibrate, they change the random
distribution of air molecules to areas
of compression and rarefaction that
travel away from the tuning fork.

pollinated flowers have distinctive ultraviolet patterns that attract and guide these
insects to the nectar within (Figure 13-16).

Thermal Receptors

Radiant energy with wavelengths somewhat longer than visible light is *infrared,*
or *heat energy* (see Figure 13-12). Most animals are able to detect changes in temperature. *Thermal receptors* are especially well developed in birds and mammals
because their internal body temperatures must be maintained at specific set points,
which requires continual feedback between their internal control centers and body
temperatures.

The human skin contains two types of thermal receptors—one to detect temperatures warmer than and the other to detect temperatures colder than the set
point. Each receptor sends information to the brain along an afferent neural pathway. In humans, thermal receptor cells in the hypothalamus also respond to slight
changes in the temperature of the blood flowing through that region of the brain. It
is not yet known how a change in temperature is transduced into a neural message.

Mechanoreceptors

A *mechanoreceptor* responds to a mechanical stimulus that stretches or presses on
its cell membrane. The distortion of the membrane alters its permeability, producing a change in membrane potential. Mechanoreceptors transduce sound,
touch, pressure, and movement of the body into neural messages.

Sound is the rearrangement of a random distribution of air molecules into *areas
of compression,* where the molecules are densely packed together, and *areas of rarefaction,* where the molecules are far apart (Figure 13-17). These two areas move
away from the source of the disturbance in alternating waves. Although these air
molecules do not travel far, they generate new waves by disturbing nearby molecules. As the frequency of the waves increases, the pitch of the sound increases; as
the difference in the density of air molecules between the areas of compression and
rarefaction becomes greater, the sound becomes louder. The sound dies out when
so much energy has been dissipated that a wave of compressed molecules can no

Figure 13-18
The Structures of the
Human Ear

(a) The outer, middle, and inner ear.
(b) The relationship between the
middle ear and the cochlea, drawn as
it would look if it were uncoiled. The
cut through the cochlea indicates the
location of the cross section shown in
(c). (c) Cross section through part of
the cochlea, showing the position of
the tectorial and basilar membranes.
(d) Enlarged view of the tectorial and
basilar membranes, showing the
position of the hair cells.

longer disturb the surrounding air molecules. Sound traveling through water exhibits the same characteristics, except that water molecules are displaced rather than air molecules.

The human ear is composed of many accessory structures (Figure 13-18) that carry sound in the form of vibrations to receptor cells. The *outer ear* consists of a funnel, which channels sound waves into the ear canal and onto the *tympanic membrane*, or *ear drum*, causing it to vibrate as the compressed air bends it inward and the rarefied air bends it outward. The vibrations are amplified as they are transmitted along the three small, linked bones of the *middle ear*. The *malleus* presses against the tympanic membrane and passes the vibrations on to the *incus*, which passes them on to the *stapes*. The stapes presses against another membrane, the *oval window*, which forms the entrance to the *cochlea*—a fluid-filled, coiled tube in the *inner ear*. Vibrations of the oval window are transferred to the fluid in the cochlea.

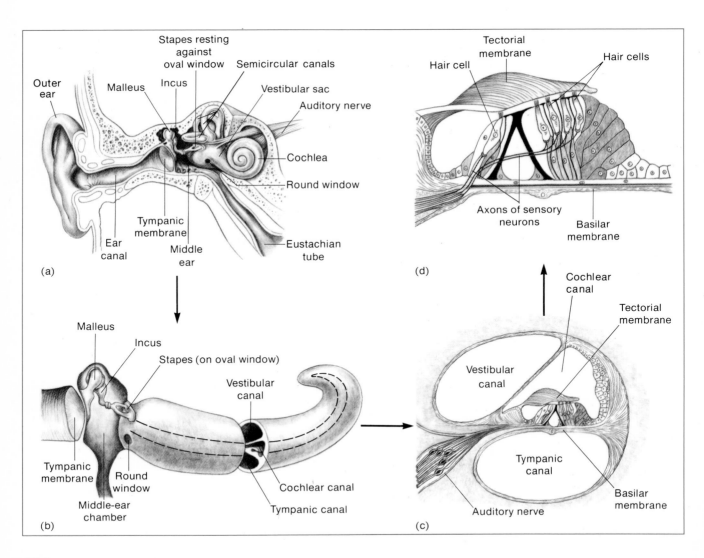

The sound receptors are *hair cells* (so called because they have long projections, or *cilia,* resembling hairs) attached to the *basilar membrane* on one end and in contact with the *tectorial membrane* on the other end. Most of the mechanical energy in the vibrations of the oval window is transferred via the fluid to the basilar membrane, which vibrates back and forth. When the hair cells on the basilar membrane move, they rub against the virtually stationary tectorial membrane. The membranes of the hair cells stretch and bend, creating receptor potentials that may initiate action potentials in associated sensory neurons. High-frequency vibrations displace the region of the basilar membrane nearest the oval window, whereas vibrations of lower frequencies displace portions of the membrane farther from the oval window. Thus, a different set of hair cells is affected by each type of sound.

The inner ear also contains receptors that respond to gravity and to changes in the rotation of the head. Movement of the head in relation to gravity is detected by hair cells in the *vestibular sacs*—two fluid-filled, interconnected cavities lying just above the cochlea, as shown in Figure 13-18(a). The cilia that project from the hair cells lining these sacs are embedded in a gelatinous material containing small crystals of calcium carbonate, heavy enough to be moved by gravity when the head is tilted, thereby causing the gelatin to shift position. As the gelatin shifts, it pulls on the cilia, distorting their cell membranes and creating receptor potentials.

Rotation of the head is detected by receptor cells in the three *semicircular canals* at the top of the inner ear, also shown in Figure 13-18(a). Each canal is oriented along one of the three planes of space—forward–backward, up–down, and left–right. The canals are filled with fluid, and one end of each canal consists of a small chamber lined with hair cells. When the canal rotates, the fluid tends to remain stationary. The movement of the canal against the stationary fluid distorts the membranes of the hair cells and causes receptor potentials.

Distinct mechanoreceptors in the skin generate potentials when their cell membranes are bent or distorted by *touch* or *pressure.* The dendrites of some sensory neurons are in contact with body hairs on the skin; when these hairs move, they press on the dendrites and initiate generator potentials, which may cause action potentials in the axon. Dendrites of sensory neurons that detect the associated sensation of pain are located in the skin, internal organs, cornea of the eye, teeth, and in tissues surrounding the muscles and bones. The exact cause of pain remains unknown and appears to be complex. Some forms of pain may be caused by the response of sensory neurons to substances released from damaged cells.

Mechanoreceptors inside the body are sensory neurons that respond to stretch or pressure and inform the brain of such changes as gas in the intestines, expansion of the lungs, and fullness of the urinary bladder. *Kinesthesia,* or *movement sensation,* is produced by sensory neurons in the skeletal muscles, tendons, and joints. The dendrites of these neurons respond to changes in muscle length and the force exerted on tendons and ligaments. They produce generator potentials which may become action potentials in the axon and provide information about the position of the limbs.

Chemoreceptors

The capacity to respond to chemical changes is a fundamental property of cells. It is not known how the presence of particular molecules is transduced into action

Figure 13-19
Taste Receptors

(a) A map of the taste receptors on the human tongue. (b) Position of a taste receptor inside a papilla.

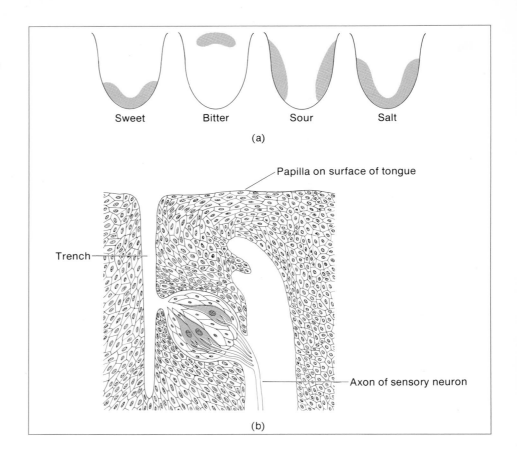

Sweet   Bitter   Sour   Salt

(a)

Papilla on surface of tongue

Trench

Axon of sensory neuron

(b)

potentials. *Chemoreceptors* in humans provide the senses of taste and smell and are important in maintaining homeostasis.

Taste receptor cells are located in the mouth, primarily on the upper surface and edges of the tongue, as shown in Figure 13-19(a). Taste receptors provide four basic tastes—sweet, bitter, sour, and salty—as well as various combinations of these tastes. More complex taste sensations depend on odor. The tongue is covered with small protuberances called *papillae.* Each papilla is surrounded by a *trench,* shown in Figure 13-19(b), that traps saliva containing dissolved food molecules. The chemically sensitive cilia of the taste receptor cells located in the walls of these trenches extend into the saliva.

Smell is detected by *olfactory receptor cells* located in a small area of the membrane on top of the nasal cavity. Olfactory receptor cells undergo changes in membrane permeability when certain molecules come into contact with them. These receptors respond to molecules carried in the air that become dissolved in the fluid that coats the nasal membranes. Sniffing forces more air onto the area where the receptors are located. The olfactory receptors in humans can detect thousands of different odors.

Many chemoreceptors inside the body monitor internal chemical events and provide feedback to the control centers of the nervous system. Receptors sensitive to carbon dioxide are located in a region of the brain and in some arteries. Chemoreceptors in the hypothalamus respond to changes in concentrations of water, sugar, salt, and other materials in the blood.

Sensory Coding

A *code* is a set of rules that permits information to be translated from one form into another. In *sensory coding*, information in the form of environmental energy is translated into information in the form of patterns of action potentials. Several different kinds of information about an environmental event, including the quality, intensity, location, and temporal order of the event, are translated into the action potential language.

The *quality of the event* refers primarily to the form of energy being received— sound, light, pressure, temperature, and so on. There are also qualitative differences within each form of energy, such as the pitch of a sound or the color of light. Qualitative characteristics are encoded by the specific sensitivities of individual sensory cells. Each sensory cell responds to only one form of energy and is most sensitive to a limited aspect of that energy. Each cone cell of the retina is most sensitive to only one of three colors; each sound receptor cell of the ear is attached to a different region of the basilar membrane and responds only to vibrations of a certain frequency. Each type of sensory cell translates the distinct type of environmental information into action potentials in specific afferent neurons. The brain distinguishes between the different afferent pathways and recognizes the quality of the event.

The *intensity of the event,* such as the brightness of light or the warmth of heat, is encoded in the rate at which the action potentials occur in the afferent pathway, as illustrated in Figure 13-20(a). The generator and receptor potentials are graded responses. As the intensity of the event increases, a greater change in membrane

Figure 13-20
Sensory Coding
Coded action potential messages recorded from axons of single sensory neurons over comparable time periods (horizontal lines). The vertical lines are the spike portions of the action potentials recorded at the same place on the axon during the time period. (The time scale is much shorter than the one shown in Figure 13-3.) (a) The frequency of identical action potentials from a sensory axon in the eye of a horseshoe crab reflects three different light intensities being presented to the eye. (b) The three lines of action potentials represent three types of temporal coding of information (patterns of action potentials) that have been recorded from sensory axons in various animals.

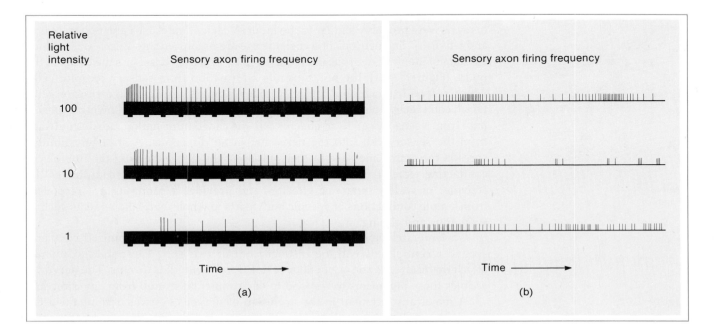

potential occurs in the sensory cell and higher frequencies of action potentials are elicited in the afferent neural pathway. Moreover, each sensory neuron may be connected by synapses to many receptor cells in a sensory organ, and each receptor cell responds to a given event with a somewhat different magnitude of receptor potential. More receptor cells respond to a more intense event, generating more action potentials in the sensory neuron. The number of afferent neurons responding to the sensory event is proportional to the intensity of that event. A stronger event affects more sensory cells; when these cells are associated with different afferent neurons, more neurons are stimulated.

The *location of the event* is the spatial pattern over which it occurs, such as regions of stimulation on the retinas of the eyes or of temperature on the body surface. This pattern is encoded in the distribution of activities throughout a population of afferent neurons. The particular afferent neurons that have action potentials and the frequency of the potentials in each neuron in the population provide the brain with a spatial pattern of the environmental event.

The *temporal order of the event* is the change in the intensity of the event over time. Patterns of speech or of moving visual images stimulate the same sensory cells in an irregular temporal pattern. This information is encoded in the pulses of action potentials in each afferent pathway, as illustrated in Figure 13-20(b).

The Evolution of Nervous Systems

Because neural tissues are rarely preserved in fossils, our knowledge about the evolution of nervous systems has been derived primarily from different forms of living animals. Although no living animal form is ancestral to another form, the uniquely specialized nervous system of each existing animal species is representative of the major stages in an evolutionary sequence.

Neurons appeared early in multicellular evolution. The first animals to have neurons were probably similar to the modern coelenterates (sea anemones, hydra, and jellyfish). The neurons of a coelenterate are organized into a *nerve net* in which essentially identical neurons interconnected by synapses crisscross the body. In the hydra (Figure 13-21), the net transmits information from sensory receptor cells, which respond to chemical and mechanical stimuli, to musclelike effector cells, which contract when stimulated by neural messages. Instead of following a specific path from a sensory cell to an effector cell, the coded information radiates outward from the sensory cell into the nerve net in all directions. A stronger stimulus spreads the neural message farther out along the net, increasing the number of contracting effector cells. Although this system is primitive, it coordinates the activities of widely separated effectors and provides a continuum of responses from simple contractions of specific body parts to whole-body movements such as swimming and gliding.

Coelenterates have radially symmetrical bodies that are organized circularly around a central axis with no distinct front or rear end. These marine animals attach themselves to the ocean floor or float freely with the currents. The nerve net enables these organisms to respond to environmental stimuli from any direction.

A major advancement in the evolution of nervous systems accompanied the appearance of animals with bilaterally symmetrical bodies organized on a longi-

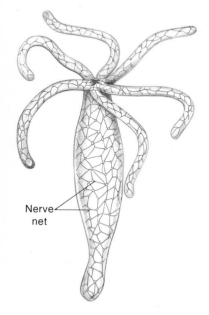

Nerve net

Figure 13-21
The Nerve Net of a Hydra

tudinal axis whose right and left sides are mirror images of each other. This type of body has a distinct *anterior* (front) *end, posterior* (rear) *end, dorsal* (top) *surface,* and *ventral* (bottom) *surface.* These dimensions are illustrated in Figure 13-22(a) for the first and simplest animal to exhibit this type of organization—the flatworm.

A modern form of flatworm is the *planarian.* The nervous system of a planarian is structurally different from the nerve net in coelenterates. Instead of being spread diffusely throughout the body, the neurons are gathered together to form pathways and control centers, as shown in Figure 13-22(b). Bundles containing the somas and axons of neurons form two parallel *nerve cords,* which extend along the full length of the body. In the anterior end, or *head,* the nerve cords fuse to form a *ganglion,* or aggregation of neurons (plural = *ganglia*). Sensory cells on the body surface and the head transmit information to the nerve cords and the ganglion, which act as control centers, coordinating the activities of the effector cells.

In nervous systems of more advanced groups of animals, neurons are also organized into *interior nerve cords* and an *anterior ganglion.* Biologists believe this pattern originated in a common ancestor of all existing forms of bilaterally symmetrical animals. In the planarians, we see the beginnings of the two most important trends in the evolution of nervous systems—centralization and cephalization.

Centralization

The first distinct separation between a *central nervous system* and a *peripheral nervous system* is exhibited in the planarians. The nerve cords and ganglion form the central nervous system; all other neural tissues form the peripheral nervous system. The planarians and all of the more complex animals have three types of neurons—sensory, motor, and interneurons. The *sensory neurons* receive information first, either directly from the environment or from receptor cells. The somas of sensory neurons lie outside the central nervous system and their axons lead into the system. The axons are usually bound together into sensory nerves. (A nerve is a bundle of axons, like a telephone cable.) *Motor neurons* make direct synaptic contact with effector cells. The somas of most motor neurons lie within the central nervous system; their axons extend outward to the effector cells and are usually bound together to form motor nerves. Mixed nerves, such as vertebrate spinal nerves, contain both sensory and motor axons. The neurons that connect the sensory and motor neurons are called *interneurons,* or *association neurons.* Their somas and axons are usually within the central nervous system.

The central nervous system consists of ganglia and nerve cords formed by the somas of motor neurons, parts of sensory and motor axons, and the interneurons. The peripheral nervous system consists of sensory cells, the somas of sensory neurons, parts of sensory and motor axons, and a variety of peripheral ganglia in more complex animals.

Cephalization

Unlike the radially symmetrical animal that roots itself to the ocean floor or floats freely with the currents, the bilaterally symmetrical animal moves in one direction with the head leading the rest of the body. Because the head comes into contact with elements in the animal's environment first, selection has favored the

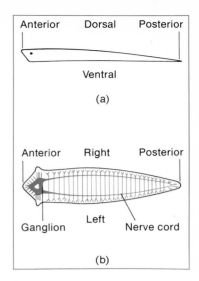

Figure 13-22
A Planarian

(a) Viewed from one side, the body has an *anterior end* with eye spots and a *posterior end.* It also has a *dorsal surface* and a *ventral surface.* The ventral surface makes contact with the ground. (b) The inside of a planarian viewed from above, showing that the left and right sides are mirror images of each other. The nervous system (the only internal structure shown here) has a central system comprised of a *ganglion* at the anterior end and two *nerve cords,* one on each side of the body.

aggregation of neurons and sensory cells in this part of the body—an evolutionary trend called *cephalization*. A related trend, called *cephalic dominance*, has been the increasing control acquired by the ganglion in the head over all aspects of physiology and behavior. In more complex animals, this dominant control center of the central nervous system is called the *brain.*

The Nervous Systems of Advanced Invertebrates

In invertebrates, brain and behavior reach the peak of complexity in the *hymenopterous insects* (ants, bees, and wasps) and in the *cephalopod mollusks* (the squid and the octopus). The nervous systems of hymenopterous insects are composed of paired longitudinal cords near the ventral surface of the body with ganglia in each segment (Figure 13-23), which control such segmental activities as movement of the legs and wings. The dominant ganglion in the head (the brain) contains hundreds of thousands of interneurons (more than 800,000 in the worker honeybee), which coordinate the complex social behavior and navigational learning characteristic of these insects.

Figure 13-23
The Nervous System of a Bee
Viewed from above, the central nervous system is comprised of a *brain*, a *ganglion* in each of the seven body segments, and two parallel *nerve cords.* The ganglia and the nerve cords lie near the ventral surface of the body.

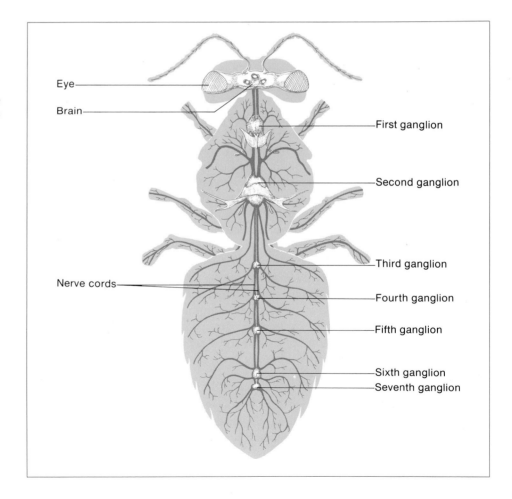

The Nervous Systems of Vertebrates

The evolution of vertebrate nervous systems also tends toward centralization, cephalization, and cephalic dominance. All vertebrate animals have a central and a peripheral nervous system, illustrated in Figure 13-24 for the human. In many advanced invertebrates, the central nervous system consists of paired, solid, ventral nerve cords with ganglia in each segment and a dominant ganglion in the head.

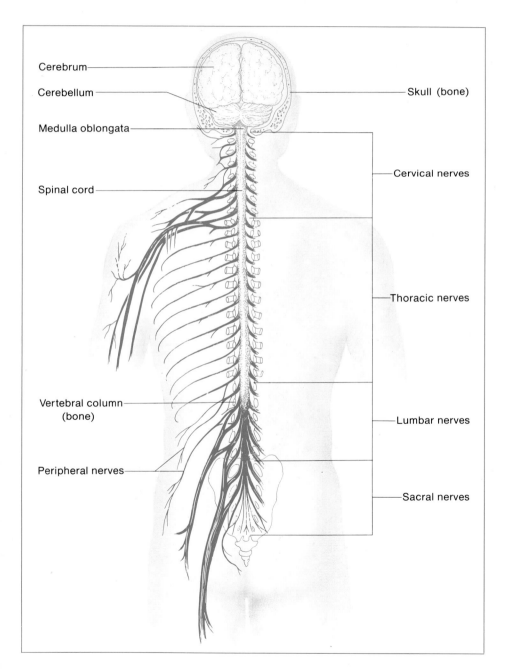

Cerebrum

Cerebellum

Medulla oblongata

Spinal cord

Vertebral column
(bone)

Peripheral nerves

Skull (bone)

Cervical nerves

Thoracic nerves

Lumbar nerves

Sacral nerves

**Figure 13-24
The Nervous System
of a Human**

The *central nervous system*, comprised of a brain and a spinal cord, and the *peripheral nervous system* of a human (shown only for the left side and viewed from the back). There are 31 pairs of spinal nerves (the cervical, thoracic, lumbar, and sacral nerves) that carry electrochemical impulses from sensory cells to the spinal cord and from motor neurons in the spinal cord to various effector cells in the body.

In vertebrates, the central nervous system has a different arrangement, consisting of a single, hollow, *dorsal nerve cord* (the *spinal cord*), with no conspicuous segmental ganglia, and a large, dominant brain in the head. Both the brain and the spinal cord are enclosed in fluid-filled bony structures; the *vertebral column* encloses the spinal cord and the *skull* encloses the brain. Sensory and motor axons enter and leave the spinal cord at regular intervals along its length (see Figure 13-24). The organization of these afferent and efferent pathways will be discussed in the next section.

The *central nervous system* of a vertebrate animal consists of all neurons and parts of neurons located in the skull and the vertebral column, including the interneurons, the somas of motor neurons, and parts of the sensory and motor axons. The *peripheral nervous system* of a vertebrate animal consists of all neurons and parts of neurons lying outside the skull and the vertebral column, including the somas of

Figure 13-25
The Brains of a Primitive Vertebrate and a Human

(a) A schematic representation of a primitive vertebrate brain, showing the positions and relative sizes of the forebrain, midbrain, and hindbrain. (top) The brain viewed from the left side; (bottom) the brain viewed from the left with the left side removed to show the interior. (b) A human brain viewed from the left with the left side removed. The midbrain is dark brown.

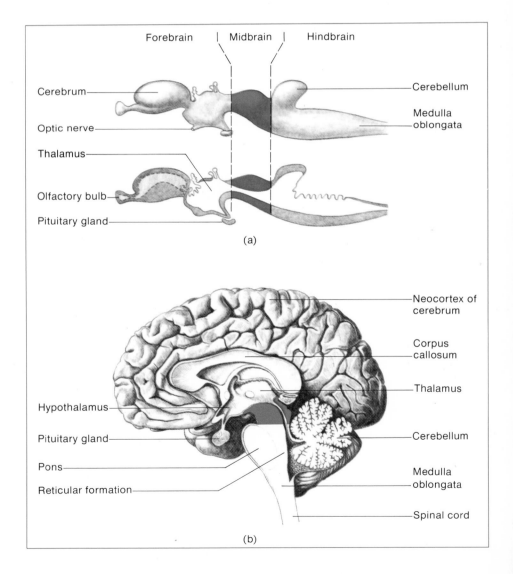

sensory neurons, parts of the sensory and motor axons, and a variety of ganglia that form part of the autonomic nervous system.

An important trend in vertebrate evolution has been an increase in the size and complexity of the brain. In such early vertebrates as the jawless fishes, the brain consisted of only three distinct swellings at the anterior end of the nerve cord—the hindbrain, midbrain, and forebrain, shown in Figure 13-25(a). These three enlargements appear during the early stages of human embryonic development but are much more elaborate and greatly altered in the adult human brain, shown in Figure 13-25(b).

Evolutionary changes in the brain from early vertebrates (represented by a modern fish) to advanced mammals are illustrated in Figure 13-26. The two noticeable structures in the *hindbrain* of a fish are the medulla oblongata and the cerebellum. In all vertebrates, the *medulla oblongata* controls feedback systems that regulate breathing and heart rates and help to control blood pressure levels. Afferent pathways from the spinal cord pass through the medulla oblongata, transmitting sensory information to the brain. Efferent pathways from other regions of the brain pass through the medulla oblongata on their way to motor neurons in the spinal cord. The *cerebellum* receives and integrates feedback information from the sensory cells that monitor muscular coordination, rhythms of movement, balance, and the position of the body in space. The great increase in the relative size and complexity of the cerebellum from fishes to mammals parallels an increase in the speed and precision of coordinated movements.

The hindbrain also contains the *reticular formation,* shown in Figure 13-25(b)—a diffuse structure present in fishes that becomes more elaborate and complex in the advanced vertebrates. In mammals, the reticular formation extends from the spinal cord through the hindbrain to the midbrain and plays a critical role in maintaining brain activities. The reticular formation receives input from all sensory systems as they pass to the forebrain and from all efferent systems as they pass to the spinal cord. This structure is also called the *reticular activating system* because it monitors the inflow and outflow of neural messages, determining which messages the nervous system should react to and which messages it can ignore. The reticular formation transmits emergency messages even when an individual is asleep. A mother will instantly awaken to a baby's cry, while the father's reticular formation will often interpret the cry as an irrelevant stimulus and allow him to continue sleeping.

In addition to the medulla oblongata, cerebellum, and reticular formation, the hindbrain of a mammal has a *pons,* also shown in Figure 13-25(b). The pons is involved in the control of many activities, including muscular coordination, feeding, facial expressions, breathing, and sleeping.

The *midbrain* of fishes and amphibians consists of the primary centers for the integration of sensory information and the control of behavior. The midbrain receives information about the internal environment from the spinal cord, information about odors from the forebrain, and information about vision directly from the eyes to its *optic lobes* (see Figure 13-26). Later evolutionary changes resulted in a steady decrease in the relative size and importance of the midbrain (see Figures 13-25 and 13-26) as the expanding forebrain acquired an increasing number of integration and control functions.

The most significant changes in the evolution of vertebrate nervous systems

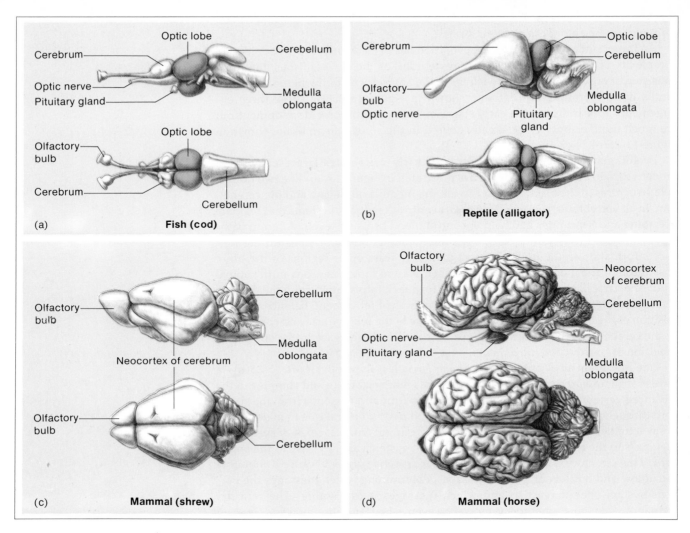

Figure 13-26   Evolutionary Changes in the Brain

The brains of vertebrate animals reflect major stages in the evolutionary sequence. Each type of brain is viewed from the left side in the upper drawing and from above in the lower drawing. The *midbrain*, shaded brown in the fish and reptile brains shown in (a) and (b), has progressively decreased in relative size during vertebrate evolution; in mammal brains, shown in (c) and (d), the midbrain has become completely covered by the greatly expanded cerebrum. Although the brains shown here appear to be roughly the same size, they actually differ greatly in size. (The horse brain is much larger than the others.)

occurred in the *forebrain* and eventually produced the structures responsible for human intelligence. There are four structures in the forebrain of fishes and amphibians—the olfactory bulbs, cerebrum, hypothalamus, and thalamus.

The *olfactory bulbs* receive information about odors from olfactory receptors in the nose, organize this information, and pass it on to the cerebrum. The *cerebrum* of a fish is a pair of smooth-surfaced lobes, called the *cerebral hemispheres,* that further integrate and organize olfactory information and transmit it to control centers in the midbrain. This form of cerebrum is referred to as a *smell brain.* The cerebrum of an amphibian is also a smell brain, but it has an additional structure. In the evolution from the fishes to the amphibians, the somas of neurons migrated from the interior of the cerebrum to form a thin outer covering or *cortex* ("cortex" = bark). This aggregation of somas is called *gray matter.* The interior of the cerebrum consists principally of axon pathways and is called *white matter,* because the myelin sheaths are white in color.

The cerebrum of the early reptiles remained essentially unchanged from the amphibian cerebrum, but this part of the brain in the mammal-like reptiles changed significantly. The forward, upper part of the cortex received and processed other kinds of sensory information in addition to olfactory information. During the evolution of mammals, this new cortex, or *neocortex,* changed profoundly, greatly increasing in size until it eventually covered the ancestral smell brain completely. In more advanced mammals, the surface of the neocortex is deeply folded and convoluted, as shown in Figures 13-25(b) and 13-26—an evolutionary change that permitted the surface area of the neocortex to expand without increasing the size of the skull.

The gray matter of the human neocortex is about $\frac{1}{4}$ inch thick and contains 70% of the roughly 15 billion neurons in the human brain. Many of the neural mechanisms that produce human intelligence are located in the neocortex. Like the mammals, the birds also evolved from the reptiles, but they never developed a neocortex and its capacities for intelligence.

In modern humans, many structures in the forebrain lie below the neocortex, including the limbic system, hypothalamus, thalamus, and corpus callosum. The *limbic system* is primarily an elaboration of the ancestral smell brain that has become covered over by the neocortex. This complex system of structures in the interior of each cerebral hemisphere lies above and essentially surrounds the hypothalamus. The limbic system receives olfactory information and is also involved in all aspects of emotional behavior and motivations. This system is also the main pathway from the rest of the cerebrum to the hypothalamus.

In Chapter 12, we discussed the role the *hypothalamus* plays in the control of the pituitary gland and, consequently, of the endocrine system. The hypothalamus also controls the *autonomic nervous system,* which regulates most internal organs. Control centers in the hypothalamus monitor a variety of activities and sensations, including body temperature, water balance, blood pressure, reproductive behaviors, thirst, hunger, aggression, pleasure, and pain.

Throughout vertebrate evolution, the *thalamus* has formed part of the forebrain. In the modern human, most sensory information passes through the thalamus on its way to the cortex. The thalamus also participates in the coordination of information being transmitted to motor neurons in the spinal cord along efferent pathways.

The *corpus callosum* coordinates activities in the two cerebral hemispheres. In the human brain, this structure consists of about 200 million axons that pass information from one hemisphere to the other.

The Human Nervous System

The human nervous system is the most complex control system ever observed. We know how sensory cells transduce environmental information and how this information is passed through the central nervous system to the neocortex. Similarly, we know how command messages are passed from the neocortex and other brain centers to the motor neurons of the spinal cord and how these motor neurons stimulate activities in effector cells. However, we know comparatively little about the neural mechanisms lying between the afferent and efferent pathways—the mechanisms that underlie the human consciousness and mind. For example, we do not know how the neural messages from the eyes are transformed into the sensations of visual experience or how the conscious mind initiates messages in the efferent pathways that produce body movements.

One reason so little is known about the human brain is that very little research is conducted on the human brain. Much of what scientists do know about brain function has been derived from research with other mammals. Although electrical activity in the human brain can be recorded from the scalp by taking an *electroencephalogram*, or EEG, doctors enter the living human brain only to perform surgical procedures, such as the removal of a tumor or a blood clot. Very limited but significant information has been obtained from these operations.

Another reason why comparatively little is known about the neural mechanisms of human mental functions is the immense complexity of the problem. Over 98% of the roughly 100 billion neurons in the human central nervous system are *interneurons*—neurons that do not directly receive sensory information or directly issue commands to effector cells. The majority of these neurons are involved in the processing and evaluation of information and the coordination of responses. Moreover, many parts of the brain participate in the most complex mental functions. Much of the neocortex, the cerebellum, the reticular formation, and other brain structures help to coordinate sensory information with body movements. Although many structures in the brain perform specialized functions, the brain is actually a vast network of synaptically interconnected neurons. It is estimated that the human brain contains more than 100 trillion (100,000 billion) synapses.

The Spinal and Cranial Nerves

The human central nervous system has two channels of communication with the organs of the body—the bloodstream and the spinal and cranial nerves. In Chapter 12, we examined some aspects of the bloodstream as a channel of communication. Here, we will examine the spinal and cranial nerves.

There are 31 pairs of *spinal nerves* connected to the spinal cord at regular intervals along the vertebral column (see Figure 13-24). One member of each pair extends to the left side of the body; the other, to the right side. Over much of their length, the spinal nerves are mixed nerves that carry both incoming afferent messages and outgoing efferent messages. Individual sensory neurons transmit afferent messages from receptor cells in the skin, muscles, tendons and joints, and internal organs to the spinal cord (Figure 13-27). The somas of the sensory neurons in each spinal nerve are located in one of the *dorsal root ganglia*, which are near the back of the spinal cord but outside the cord itself. The axons of sensory neurons enter the spinal cord through the *dorsal root*. The first synapse of each sensory neuron is located in the gray matter of the spinal cord. The somas of motor neurons are located in the gray matter of the spinal cord, and their axons extend through the *ventral root* to the spinal nerves and then to the appropriate effector cells.

Under the direction of the brain, the 31 pairs of spinal nerves control segments of the human body (see Figure 13-24). Each pair receives information from and

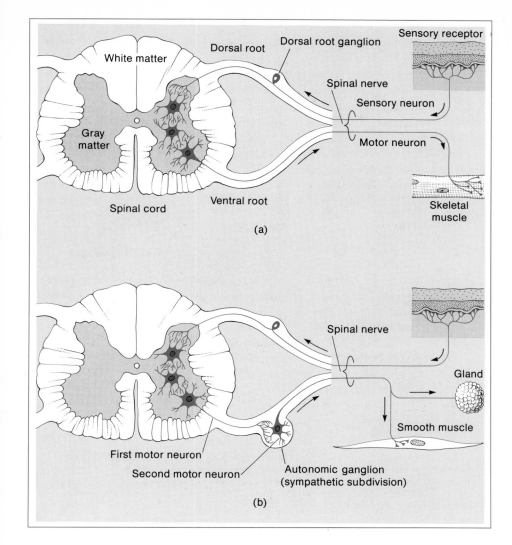

Figure 13-27
Connections Between the Spinal Cord and Components of the Spinal Nerves
(a) The arrangement in the somatic division. (b) The arrangement in the autonomic division (sympathetic subdivision).

sends efferent commands to a single region of the body. The *lumbar* and *sacral nerves* control the back, the *thoracic nerves* control the chest, and the *cervical nerves* control the neck.

The head is controlled directly by the brain through 11 of the 12 pairs of *cranial nerves* that are connected to the brain along its lower surface. Table 13-1 lists the function of each cranial nerve and indicates where it originates or synapses in the brain. Some cranial nerves are purely sensory, some are purely motor, and some are mixed. One cranial nerve—the *vagus* (number X)—does not serve the head but instead provides a direct channel from the brain to the internal organs. When brain communication with the spinal cord is cut off, as it can be when the neck is broken, the vagus nerve will still control many internal body functions, even though the individual is paralyzed from the neck down and insensitive to skin, muscle, and joint sensations.

Table 13-1
The Cranial Nerves

Number	Name	Functions (s, sensory; m, motor)	Origin or end in the brain
I	Olfactory	(s) Smell	Cerebral hemispheres (ventral part, or ancestral smell brain)
II	Optic	(s) Vision	Thalamus
III	Oculomotor	(m) Eye movement	Midbrain
IV	Trochlear	(m) Eye movement	Midbrain
V	Trigeminal	(m) Masticatory movements	Midbrain and pons
		(s) Sensitivity of face and tongue	Medulla oblongata
VI	Abducens	(m) Eye movement	Medulla oblongata
VII	Facial	(m) Facial movement	Medulla oblongata
VIII	Auditory vestibular	(s) Hearing (s) Balance	Medulla oblongata
IX	Glossopharyngeal	(s, m) Tongue and pharynx	Medulla oblongata
X	Vagus	(s, m) Heart, blood vessels, viscera	Medulla oblongata
XI	Spinal accessory	(m) Neck muscles and viscera	Medulla oblongata
XII	Hypoglossal	(m) Tongue muscles	Medulla oblongata

Source: Richard F. Thompson, *Introduction to Physiological Psychology* (New York: Harper & Row, 1975), p. 99, Table 3.4. Copyright © 1975 by Richard F. Thompson. Reprinted by permission of Harper & Row, Publishers, Inc.

The Peripheral Nervous System

The *peripheral nervous system* of the human, like other vertebrates, consists of all the neurons and parts of neurons that lie outside the skull and vertebral column. This system is separated into the somatic and autonomic divisions.

The *somatic division* consists of all sensory neurons from the skin, skeletal muscles, and joints and all motor neurons extending from the spinal cord to the skeletal muscles. This division, which is directed by the brain and reflex centers in the spinal cord, controls *voluntary activities*—the movements and behaviors produced by the conscious mind.

The *autonomic division* (Figure 13-28) controls *involuntary activities*—the regulation of internal body functions and the physical responses associated with strong emotions, such as sweating and increased heart rate. This division consists of all sensory neurons from the internal organs and all motor neurons to the smooth muscles of the internal organs, the cardiac muscles of the heart, the exocrine glands, and some endocrine glands. The efferent pathways of the autonomic nervous system are in turn divided into the sympathetic and parasympathetic subdivisions.

Figure 13–28
The Autonomic Nervous System

Efferent pathways in the sympathetic subdivision include synapses between motor neurons within a chain of sympathetic ganglia. There are two of these chains, one on each side of the spinal cord, but only one is shown here. Efferent pathways in the parasympathetic subdivision also include synapses between motor neurons within ganglia, but the ganglia are located near or in the individual organs and are not in direct communication with one another. The sympathetic and parasympathetic subdivisions typically produce opposite effects in the same organs.

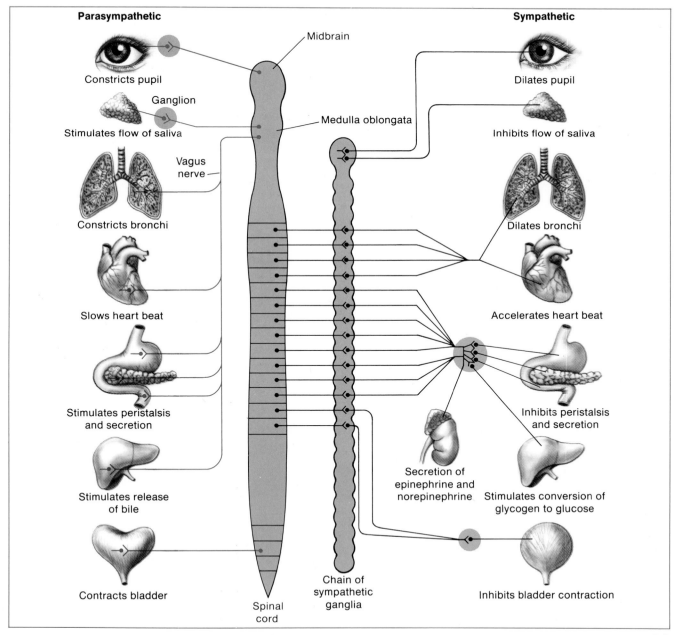

The *sympathetic subdivision* is active primarily during times of stress or strong emotions, producing the fight-or-flight response discussed in Chapter 12. The *parasympathetic subdivision* is most active during periods of rest or recovery from stress, returning the body to a state of homeostasis.

Motor neurons in the sympathetic subdivision leave the spinal cord in the thoracic and lumbar spinal nerves. However, instead of extending directly to the organs they control—as somatic neurons, shown in Figure 13-27(a), do—they form synapses with a second set of motor neurons that extends to the body organs from two parallel chains of *sympathetic ganglia* lying on each side of the spinal cord, shown in Figures 13-27(b) and 13-28. The sympathetic ganglia are in direct communication with one another and provide integrated, whole-body responses to stress.

Motor neurons in the parasympathetic subdivision leave the hindbrain and the midbrain in the cranial nerves and leave the spinal cord in the sacral nerves. There are no integrated chains of ganglia for parasympathetic motor neurons, but they also synapse with a second set of motor neurons in small ganglia near the organs they control.

Most internal organs receive motor neurons from both the sympathetic and the parasympathetic subdivisions of the autonomic nervous system, and these neurons usually produce opposite, or *antagonistic* effects (see Figure 13-28). The internal organs are continually stimulated and controlled by an interplay of these two subdivisions.

In general, the somatic division regulates voluntary activities and the autonomic division regulates involuntary activities. However, some somatic activities are involuntary, and the control of certain autonomic responses, such as urination and defecation, can be learned. It has recently been demonstrated that many unconscious autonomic functions can be consciously controlled after training.

Reflexes

When there are only a few synapses between a sensory and a motor neuron, the pathway is called a *reflex*. A reflex with only one synapse is rare in humans; one example, diagrammed in Figure 13-29, is the *knee-jerk reflex* used in medical examinations to test the state of the central nervous system. The knee-jerk reflex plays an active role in the fine control of posture and balance.

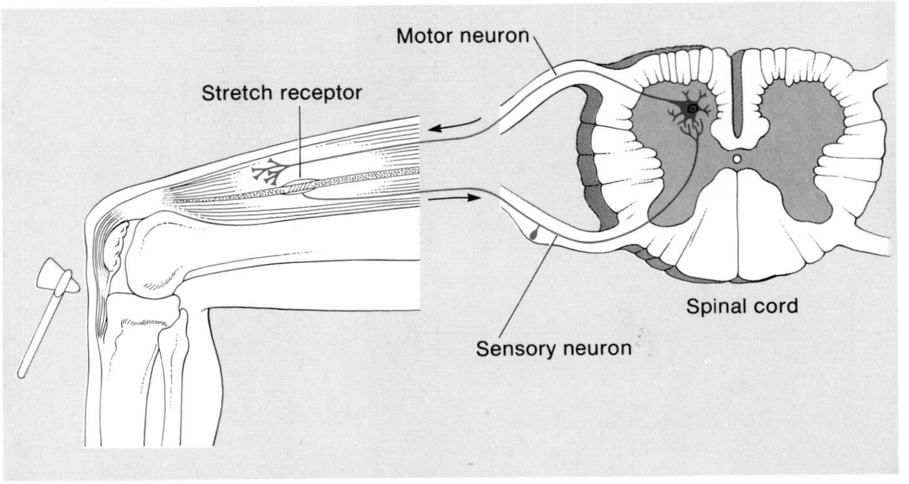

Figure 13-29
The Knee-Jerk Reflex

When the tendon in the knee is tapped, stretch receptors in the thigh muscle are stimulated and electrochemical impulses are conveyed along a sensory neuron to the spinal cord. There, the sensory neuron forms a direct synapse with a motor neuron. The impulse is transmitted to the motor neuron, and an involuntary contraction of the thigh muscle causes the lower leg to jerk forward. (Here, the spinal cord is in the reverse orientation relative to Figure 13-27.)

Many somatic and autonomic reflexes are controlled and integrated at each level of the spinal cord and in the hindbrain and the midbrain. Most of these reflexes have one or two interneurons between the sensory and motor neurons. Other reflexes involve several different spinal nerves and are connected by pathways within the white matter of the spinal cord. Although there are involuntary reflexes in humans, activities in the spinal cord are also relayed through the myelinated axons of the white matter to the brain.

Afferent-to-Efferent Connections in the Brain

In the autonomic division of the peripheral nervous system, information from both the bloodstream and the peripheral nerves is transmitted to control centers in the hindbrain, the midbrain, and the hypothalamus. The hypothalamus is the coordinating center for all autonomic responses and is influenced by the rest of the brain through the limbic system.

Regarding the somatic division of the peripheral nervous system, little is known about the pathways in the brain that connect the sensory information with the motor commands. Between the input and the output lie all of the brain mechanisms that produce the human mind and consciousness.

Clearly, the mechanisms that produce human mental states and capacities are not organized like a spinal reflex. Sensory pathways do not end in the brain where motor pathways begin. Instead, before sensory information reaches the motor pathways, it is processed simultaneously through many specialized structures and through vast areas of interneurons with unknown synaptic connections. The neocortex (more often referred to as the cerebral cortex in mammals) is unquestionably involved in complex human abilities, but the rest of the brain below the cortex also participates in these capacities.

Although much about the human brain remains an unsolved mystery, certain aspects of brain function have been clearly established. Here, we will briefly review some of the more significant discoveries and prevailing hypotheses about the human brain.

The Cerebral Cortex

The cortex of the left cerebral hemisphere is diagrammed in Figure 13-30. Much of what is known about the external structure and the function of the cortex is summarized in this figure. Structurally, the left side of the cortex appears to be an identical, mirror image of the right side of the cortex. However, in the majority of humans, the cortex of the left hemisphere contains unique areas (not visible to the eye) involved in the learning and use of language that are not present in the cortex of the right hemisphere.

Two principal landmarks distinguish the different regions on each side of the cortex—the central and temporal fissures. The *frontal lobe* lies anterior to the *central fissure* and is subdivided into a *motor cortex*, a *premotor cortex*, and a *prefrontal area*. Posterior to the central fissure lies the *somatosensory cortex*, the *parietal lobe*, and the *occipital lobe*. The *temporal lobe* lies below the *temporal fissure*.

Motor Pathways

In order to produce voluntary movements of the skeletal muscles, all contributing brain systems must influence the motor neurons of the spinal cord and the

Figure 13-30
The Left Side of the
Human Brain

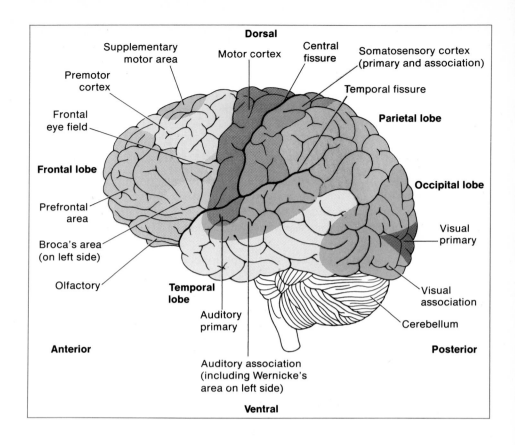

hindbrain and the midbrain. Various regions of the cortex and other brain centers have pathways leading to these motor neurons.

The *motor cortex* controls and organizes complex patterns of movement. Neurons with somas in the motor cortex have axons (some more than two feet in length) that synapse directly with all the motor neurons in the hindbrain, the midbrain, and the spinal cord. When regions of the motor cortex are artificially stimulated by applying an electrical current, movements occur in specific muscles and groups of muscles. The regions of the motor cortex that control the activity of specific muscles can be "mapped" using this technique. A map of the human motor cortex is shown in Figure 13-31(a), with control of the feet near the dorsal side and control of the face and tongue near the ventral side. Note that certain areas of the body have much more extensive control regions than others, particularly the face, tongue, and hands. This emphasis reflects the evolution of the human brain during the hunter–gatherer period of cultural development when there was selection for individuals who were superior in the use of tools and communication with group members.

The axons from the motor cortex to the motor neurons cross over to the opposite side of the body, so that the motor cortex of the left hemisphere controls muscles on the right side of the body and the motor cortex of the right hemisphere controls muscles on the left side of the body.

The motor cortex is influenced by many areas of the brain, including other

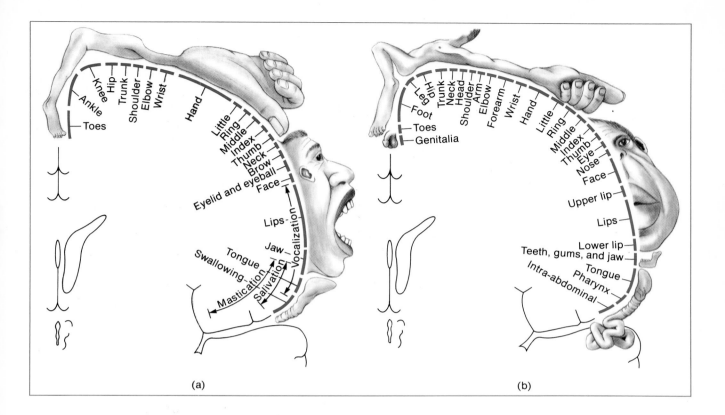

(a)   (b)

areas of the cortex, the cerebellum, and the reticular formation. For example, the *supplementary motor areas* and the *premotor cortex* (see Figure 13-30) are involved in the coordination of complex sequences of activity, such as playing a musical instrument. Various brain structures also have pathways to the motor neurons that principally serve to modify the commands from the motor cortex. Exactly how all of these structures and pathways function together to produce voluntary movements is not known.

Sensory Pathways

All sensory systems have fairly direct pathways with few synapses to the cerebral cortex. Touch, pressure, and stretch responses from the skin and internal structures are transmitted to the *somatosensory cortex* (Figure 13-30). Visual information travels from the eyes to the *visual primary cortex* in the *occipital lobes,* olfactory information travels from the nose through the limbic system to the regions of the *prefrontal areas* that are cortical extensions of the limbic system, and sound information travels from the ears to the *auditory primary cortex* in the *temporal lobes.* Temperature information reaches the hypothalamus before it reaches the somatosensory cortex; pain information travels to the reticular formation and then to the somatosensory cortex; and taste information probably travels to the somatosensory cortex.

The cortical areas that receive information from the skin, muscles, joints, eyes, and ears can also be mapped, just as the areas of the motor cortex were in Figure

Figure 13-31
Maps of the Motor Cortex and the Somatosensory Cortex of the Human Brain

(a) A cross section through the cerebrum just anterior to the central fissure. The surface of this area is the *motor cortex.* Each region of the motor cortex controls the muscular activities of a particular part of the body; the regions for some body parts are much larger than the regions for others.
(b) A cross section through the cerebrum just posterior to the central fissure. The surface of this area is the *somatosensory cortex.* Each region receives sensory information from specific muscles, joints, and areas of the skin. Again, the regions for some body parts are much larger than the regions for others. The motor cortex and the somatosensory cortex are present in both hemispheres of the brain, although only one hemisphere is shown here for each cortex.

13-31(a). When sensory cells in the skin, muscles, or joints are stimulated and the electrical activity in the somatosensory cortex is measured, each area of the cortex responds to the stimulation of a particular part of the body. Responses to stimulation of the feet occur in the dorsal part of the somatosensory cortex, and the areas that respond to stimulation of the legs, hands, face, and tongue are progressively more ventral, as shown in Figure 13-31(b). As in the motor cortex, the hands, face, and tongue have disproportionately large representations on the somatosensory cortex, indicating a greater sensitivity in these parts of the body. The afferent pathways also cross over, so that information from the left side of the body travels to the cortex of the right hemisphere and information from the right side of the body travels to the cortex of the left hemisphere.

The retinal surfaces of the eyes are represented on the surfaces of the *visual primary cortex* (Figure 13-32). Information from the left half of the retina in the right and left eyes is transmitted to the visual primary cortex of the left hemisphere, and information from the right half of the retina in the left and right eyes is transmitted to the visual primary cortex of the right hemisphere. Note that due to the cross-overs of information between the visual field (the environmental area seen by the eyes) and the retinas, the visual primary cortex of the left hemisphere receives in-

Figure 13-32
Pathways of Visual Information from the Visual Field to the Visual Primary Cortex

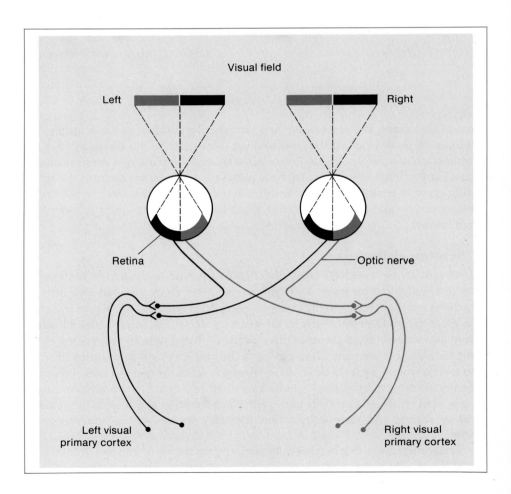

formation about the right side of the visual field and the visual primary cortex of the right hemisphere receives information about the left side of the visual field.

The surfaces of the basilar membranes in the ears are represented on the *auditory primary cortex*. High-frequency sounds from hair cells on the membrane nearest the oval window produce electrical activity in the anterior part of the auditory primary cortex, and low-frequency sounds from hair cells on the membrane farthest from the oval window produce electrical activity in the posterior part of the auditory primary cortex. Most of the sound information in the cortex of one hemisphere is transmitted from the ear on the opposite side.

Individual neurons within each of the sensory areas of the cortex respond to different aspects of sensory information. For example, neurons in each area of the somatosensory cortex, which represents a particular area of the body, respond to only one of the somatic sensations (touch, pressure, and so on). Specific neurons in the retinal maps in the visual primary cortex respond to shape, movement, or color. Similarly, specific neurons in each region of the auditory primary cortex respond to different qualities of sound, such as intensity and changes in temporal pattern.

All of the sensory pathways are connected to many subcortical regions before they reach the cortex. It is not yet known how all of these structures and pathways act together to produce conscious sensations.

Learning, Memory, and Language

The areas of the human cerebral cortex that receive sensory information and the areas that send commands to the motor neurons of the hindbrain, the midbrain, and the spinal cord represent only a portion of the entire cortex. The rest of the cortex is the *association cortex* (see Figure 13-30).

In the evolution of the human brain from shrewlike ancestors to modern humans, the relative size of the association cortex has increased consistently. Almost all of the cortex in a simpler form of mammal is sensory and motor cortex; most of the cortex in a human is association cortex. This trend has been accompanied by marked increases in mental capabilities—learning, thinking, remembering, and communicating. In humans, the neural mechanisms responsible for the most unique and complex mental abilities are found in the association cortex.

Most neurobiologists agree that information is processed (linked together in associations), remembered (stored in neural pathways), and understood (integrated with previous information) within the posterior association cortex of the somatosensory areas, the parietal lobes, the occipital lobes, and the temporal lobes. Activities within the association cortex of the frontal lobes permit the processed and stored information to influence an individual's behavior. Such elements of human personality as the planning and initiation of behavior and the perception of the emotional context of information appear to be governed by the frontal lobes. The evidence to support this far-ranging summary of the functions of the association cortex is too extensive to review here, but we will consider a few important aspects.

Learning is a change in behavior that results from experience the organism acquires during its lifetime. The entire brain may be involved in learning because it includes so many different human capabilities (emotions, drives, motor commands to muscles, memory). A critical aspect of learning is *memory*—the storage of information in the brain. The neural mechanisms of memory seem to be located in the posterior association areas—the somatosensory cortex, the parietal lobes, the occipital lobes, and the temporal lobes—and in some subcortical structures of the temporal lobes. Both forms of memory—short term and long term—appear to take place in these regions.

Short-term memory, which lasts from seconds to hours, is believed to occur in the cortical association regions surrounding the primary sensory areas. An example of short-term memory is remembering a telephone number only long enough to dial it. Short-term memory probably occurs as electrical impulses traveling from neuron to neuron through a pathway. If the memory involves a variety of sensations, these neural pathways must extend through different association areas.

Long-term memory, which lasts from several hours to a lifetime, probably occurs in the same association area pathways but involves permanent changes in the structure and function of the synapses. What these changes are and how they occur is not yet known. The most widely accepted hypothesis is that they are changes in protein synthesis. When neurotransmitters join with a receptor molecule on the cell membrane, they cause the synthesis of cyclic AMP, which travels to the chromosomes and causes a change in protein synthesis. Such changes could influence all aspects of synaptic activity.

Some of the evidence to support the conclusion that information is processed, stored, and understood in the posterior association cortex is also derived from knowledge about the neural mechanisms that produce the speaking, reading, and writing of language. In the late nineteenth century, two European doctors discovered, from examinations of the brains of deceased patients who had had language disorders, that diseased tissues in two regions of the association cortex in the left hemisphere produced language disorders. These regions are now named after their discoverers—*Broca's area,* located in the frontal lobe just anterior to the face region of the motor cortex, and *Wernicke's area,* located in the temporal lobe and extending from the auditory primary area to the borders of the association cortex in the parietal and the occipital lobes. Damage to Broca's area produces a specific type of speech problem; people who are afflicted with this problem understand what they want to say but have great difficulty saying it. Broca's area controls and coordinates the motor cortex cells that govern the muscles in the throat, tongue, and lips—the muscles that produce speech. When Wernicke's area is damaged, the language disturbance is more profound because the understanding of language is affected. The afflicted person may speak comprehensively but may not be able to understand written material, or may speak distinctly and clearly but may speak only nonsense. Considerable understanding of language pathways in the cortex has been derived from such discoveries. When you speak a word that you have just read, the information travels along the cortex of the left hemisphere from the visual primary area through the visual association cortex to Wernicke's area (where it is understood), through the cortex to Broca's area (where speech commands are formulated), and then to the motor cortex (where speech commands are issued to the appropriate motor neurons).

Speech mechanisms are present within the left hemisphere of the brain in 99% of all right-handed people and in 66% of all left-handed people. (Roughly 9% of the human population is left-handed.) A recent series of studies with split-brain patients has further confirmed that language function exists in the left hemisphere of most humans. In a split-brain operation, a cut is made through the corpus callosum—the main communication channel between the left and right hemispheres, shown in Figure 13-25(b). This operation is performed on patients with severe *epilepsy*—a disease of the cortex affecting a small area of tissue in one hemisphere that stimulates abnormal electrical activity that spreads over wide areas of the cortex and can produce muscle convulsions and unconsciousness. This electrical activity may extend across the corpus callosum to produce abnormal activity in both hemispheres. The split-brain operation is performed when epileptic seizures have persisted for years and occur many times each day. The split-brain operation does not cure epilepsy, but it decreases the symptoms greatly.

At first, aside from a decrease in the symptoms of epilepsy, split-brain patients were believed to retain the same personalities and intelligences that they had prior to the operation. However, when these patients were given sophisticated tests by the American psychologist Roger Sperry and his colleagues, a remarkable discovery was made—the operation had created two separate but not identical human minds within each patient. The testing procedure made use of the fact that information from one side of the body crosses over within the nervous system to the cortex of the opposite hemisphere. In Figure 13-33(a), the word "hatband" is presented in such a way that the right half of each retina sees "hat" and the left half of each retina sees "band." When asked to say what they saw, the split-brain patients said "band"; the cortex of the left hemisphere, which contains the mechanisms for speech, sees only "band." In Figure 13-33(b), only the cortex of the right hemisphere, which controls the left hand, perceives the word "nut," permitting the patients to pick out a nut from a group of objects behind a screen. When the split-brain patients were asked to describe what their left hands had selected, their left hemispheres did not know.

The results of these and other tests of split-brain patients indicate that the two hemispheres of the brain perform very different functions in the majority of humans. The left hemisphere usually contains the mechanisms for verbal abilities

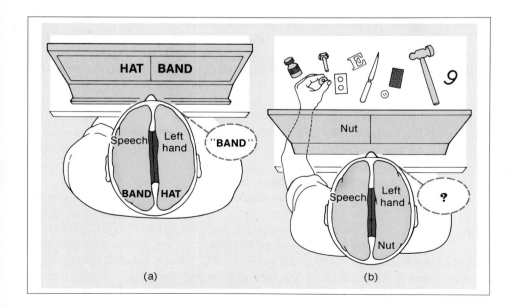

(a) (b)

Figure 13-33
Testing Split-Brain Patients
After the corpus callosum is cut, the patient is given tests during which information is transmitted to only one hemisphere of the brain. (a) The word "hatband" is presented so that the left visual primary area sees "band" and the right visual primary area sees "hat." The left cortex, which contains the mechanisms for speech, "says" that the word "band" is presented. (b) Sensory information in the form of the written word "nut" and tactile information from the left hand reach only the right cortex, which commands the selection of a nut from a group of objects behind a screen. The left cortex is ignorant of what has happened.

(reading, writing, and speaking) and for abstract and logical thought. The right hemisphere usually contains the mechanisms for understanding simple language but not for speech and seems to specialize in such nonverbal spatial capabilities as understanding three-dimensional space, recognizing faces, and perceiving certain aspects of music.

In normal humans, both hemispheres function together, each perceiving one-half of the surrounding environment. Continuous communication between the cerebral hemispheres through the corpus callosum enables the brain to achieve an integrated, conscious perception of the world. At birth, language can develop in either the left or right hemisphere; if language areas in the left hemisphere are damaged in a very young child, normal language will develop in comparable areas in the right hemisphere. Language appears to develop within areas of the brain, left or right, that produce basic primate abilities to perceive and act in three-dimensional space.

States of Consciousness

What is consciousness, what neural mechanisms produce consciousness, and what animals other than humans possess consciousness? Interaction between areas of the reticular formation and the pons regulates states of brain activity from sleep to full alertness, but other areas of the human brain also contribute to the phenomenon of consciousness. These areas are found in the brains of other animals as well, but without direct experience, we cannot determine whether they too possess consciousness.

One area of research that has produced insights into consciousness is the use of the *electroencephalogram* (the EEG) to study patterns of brain activity. Sensitive electrodes are glued to the scalp of a human or other animal, and the electrical activity occurring during different states of consciousness (in humans) or different behaviors (in other animals) is recorded. What the EEG records is not known with certainty but is probably the total electrical activity of action potentials and postsynaptic membrane potentials occurring in millions of neurons within the cortex.

The EEG has revealed that there are three fundamental states of consciousness in humans—awake, sleep, and REM (rapid eye movement) sleep. Figure 13-34 provides typical EEG patterns for these three states of consciousness over 10-second intervals. In Figure 13-34(a), the person is awake, alert, and performing some sort of mental activity, such as counting. The EEG indicates irregular patterns of activity of very low voltage. In Figure 13-34(b), the person is awake and relaxed, usually with eyes closed and mind drifting. The EEG pattern reveals waves of activity with peaks and troughs that recur 8–12 times per second. Waves that occur in this pattern are called *alpha waves.* The neuronal activity is synchronized; that is, electrical potentials increase and decrease in synchrony.

Four stages of sleep are shown in parts (c) to (f) of Figure 13-34. Neuronal activity becomes increasingly synchronized as the brain waves become progressively slower. In REM sleep the person is deeply asleep, but the EEG pattern in Figure 13-34(g) is almost identical to the EEG pattern of the fully awake and alert person shown in Figure 13-34(a). However, in REM sleep, the individual's vital signs and body temperature vary greatly and are highly irregular, whereas during stages 1–4 of sleep, the individual's vital signs (heart rate, breathing rate, and so

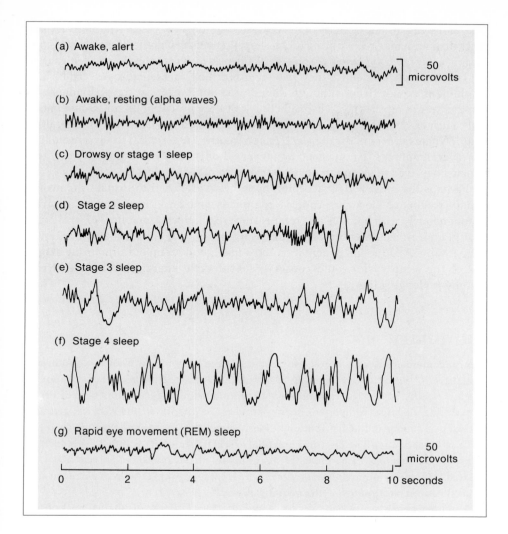

(a) Awake, alert

50 microvolts

(b) Awake, resting (alpha waves)

(c) Drowsy or stage 1 sleep

(d) Stage 2 sleep

(e) Stage 3 sleep

(f) Stage 4 sleep

(g) Rapid eye movement (REM) sleep

50 microvolts

0   2   4   6   8   10 seconds

Figure 13-34
EEG Records of the Three States of Consciousness
Electrodes attached to the scalp have recorded brain electrical activity over 10-second intervals during the three fundamental states of consciousness—awake, sleep, and REM sleep. The vertical scale on the right measures the intensity of the activity in microvolts. (a) An EEG of an awake and alert subject. (b) An EEG of an awake and relaxed subject, with eyes closed and mind drifting. Alpha waves are prominent in the record. (c) to (f) The EEG patterns of sleep stages 1–4, showing increased synchrony, which produces higher voltages, and slower waves. (g) The EEG of rapid eye movement (REM) sleep.

forth) are slow and regular and body temperature gradually decreases. In REM sleep, the muscles are relaxed but the eyes move rapidly under closed lids and the limbs twitch and move. A person will almost always report dreaming if awakened from REM sleep but will rarely report dreaming if awakened from stages 1–4 of regular sleep.

Sleep occurs suddenly in the normal young adult. There is a rapid transition from alpha waves to sleep stage 1, to sleep stages 2–4, and then to REM sleep. After REM sleep, the individual returns to stage 1 and the cycle is repeated. Four or five of these sleep cycles occur each night; each cycle lasts 90–100 minutes and includes 20–30 minutes of REM sleep.

Sleep and REM sleep have been detected in almost all mammals and birds by using the EEG technique. Recorded sleep cycles vary in length from a few minutes in mice and rats to 20–40 minutes in cats to 120 minutes in elephants. REM sleep probably allows a sleeping mammal or bird to return periodically to a state of readiness without actually waking up in case a fight-or-flight response (an alert

EEG and sympathetic activity) is necessary. We do not know whether animals other than humans dream during REM sleep, but it seems likely. Many animals, including cats and dogs, twitch during REM sleep as if in response to some stimulus.

The function of sleep is not known. According to EEG records, fishes and amphibians do not sleep, although they do remain inactive at times. Evidence of whether or not reptiles sleep is conflicting, but they probably do not. Some mammals, such as the antelope, never sleep, and some humans can exist on very little sleep. Whatever occurs during sleep is apparently not essential to vertebrate life. However, if animals that do sleep are deprived of it, they will eventually die, and humans deprived of sleep can develop severe mental disorders.

Perhaps the most widely accepted hypothesis proposed to explain the evolutionary origins of sleep in primitive mammals and birds is that sleep evolved simultaneously with homeostatic temperature regulation. A mammal or bird must take in and utilize an enormous amount of energy to maintain a stable internal temperature. A reduction of activities and a lower body temperature during stages 1–4 of the sleep cycle would conserve considerable energy and reduce stress, thereby prolonging life.

Summary

The fundamental unit of a *neural control system* is the *nerve cell*, or *neuron*. Neurons are specialized for *irritability*, responding to stimuli by generating electrochemical impulses that are rapidly conducted from neuron to neuron within the nervous system of an animal. The electrical component of the impulse is a change in the *electrical potential* (caused by a change in the distribution of ions, between the inside and the outside of the nerve cell membrane). The electrical potential of a cell membrane is called the *membrane potential*. The chemical components of the impulse—the *neurotransmitter molecules*—are released at the *synapse* into the *synaptic cleft*, which is the region between one neuron and the next neuron or *effector cell* in the neural pathway.

A neuron consists of a *soma, dendrites,* and an *axon*. The soma contains most of the cell's organelles and synthesizes neurotransmitters, which are then stored in vesicles and transported to *terminal buttons* at the ends of the axon. The portion of the cell membrane that encloses the soma and the dendrites is specialized for receiving electrochemical impulses; the portion of the cell membrane that encloses the axon is specialized for conducting impulses from the *axon hillock*, where the axon and the soma join, to the terminal buttons.

Electrochemical impulses are conducted by a sequence of four events—the resting membrane potential, the action potential, the propagation of the impulse along the axon, and the transmission of the impulse across the synapse. The *resting membrane potential* of the axon is maintained at −70 millivolts (mV) by the relative concentrations inside and outside the membrane of sodium ions, potassium ions, and large protein molecules with negative charges. These concentrations are maintained by diffusion, electrical attractions and repulsions, active transport by *sodium–potassium pumps*, and the selective permeability of the cell membrane.

The *action potential* consists of a *spike potential* (a rapid change in membrane potential from −70mV to +40mV and back again; also called the *absolute refractory period*) and the *relative refractory period* when the membrane potential gradually returns to its resting state. It begins at the axon hillock and is propagated along the axon to the terminal buttons. If an action potential is generated, it is always propagated in identical form from the axon hillock to the terminal buttons. This is known as the *"all-or-none" law* of neural conduction. The speed of the *propagation of the impulse* increases as the temperature of the axon increases, in axons of larger diameter, and in axons that are coated with myelin. The *myelin sheaths* of vertebrate animals are formed from *oligodendroglial* cells in the brain and spinal cord and from *Schwann cells* elsewhere in the nervous system. These sheaths occur in short segments with exposed areas called *nodes of Ranvier* between them. Action potentials occur only at the nodes of Ranvier, so that the impulse jumps from one node to the next as it is propagated along the axon.

When the action potential reaches the end of the axon, it either travels directly to the membrane of the adjacent cell (an *electrical synapse*) or causes neurotransmitters to be released from the terminal buttons (a *chemical synapse*). All synapses in the human nervous system are chemical synapses. A chemical synapse is composed of the *presynaptic membrane* at the end of the axon, the *postsynaptic membrane* on the adjacent neuron or effector cell, and the *synaptic cleft* (the fluid-filled space between the two membranes). Neurotransmitters released from the terminal buttons become attached to specific receptor molecules on the postsynaptic membrane. There are two categories of chemical synapse (determined by the neurotransmitter and the receptor molecule)—those that produce an *excitatory postsynaptic potential* (EPSP) and those that produce an *inhibitory postsynaptic potential* (IPSP).

The nervous system of an animal makes use of a variety of synapses that differ in terms of the functions they control, the response time, the duration of the effect, and whether the postsynaptic cell membrane is altered directly or indirectly (by initiating the synthesis of cyclic AMP within the postsynaptic cell).

Transduction—the conversion of environmental energy into a neural message—is performed by *sensory cells.* Both the transduction and conveyance of a message to the afferent neural pathway may be carried out by a single sensory cell; the environmental energy initiates a *generator potential* in the dendrite, which may cause an action potential in the axon. In this case, the sensory cell is a *sensory neuron.* Alternatively, many kinds of sensory cells may be organized into a *sensory organ* composed of accessory structures, *receptor cells*, and sensory neurons. The initial change in the membrane potential of a receptor cell is the *receptor potential.* Both the generator potential and the receptor potential are graded responses rather than the all-or-none response of an action potential.

Light energy is transduced by *photoreceptor cells* when pigment molecules in the membranes change molecular form, altering the permeability of the membrane to ions and causing a receptor potential. Photoreceptor cells in the human eye are the *rods* and the *cones.* The rods are located away from the center, or *fovea,* of the retina and are sensitive to dim light, providing visual images in shades of gray. The cones are located in the fovea of the retina and provide detailed color vision. Each cone responds to red, green, or blue light. Rods and cones are connected by synapses to sensory neurons. The human eye contains many accessory structures, including the *iris, cornea,* and *lens.* The arthropod eye is a *compound eye* constructed of many identical subunits, or *ommatidia.* Each ommatidium has a lens, a crystalline cone, pigment cells, and photoreceptor cells. Each photoreceptor cell receives light from a single point in space, and the animal sees a mosaic of separate, nonoverlapping images.

Thermal receptors transduce heat energy. Cold and heat receptors in the human skin respond to skin temperature, and thermal receptors in the hypothalamus respond to changes in blood temperature.

A *mechanoreceptor* transduces distortions of its cell membrane that are caused by sound, touch, pressure, pain, or stretching. Sound is focused, amplified, and converted into vibrations within the human ear by the *external funnel, ear canal,* and *tympanic membrane* of the *outer ear,* the three small bones of the *middle ear,* and the *oval window, basilar membrane,* and *tectorial membrane* of the *cochlea* within the *inner ear.* The vibrations cause membrane distortions in the receptor cells, or *hair cells.* Mechanoreceptors in the *vestibular sacs* and the *semicircular canals* of the inner ear transduce changes in gravity and in the rotation of the head, respectively. The skin contains mechanoreceptors that respond to touch, pressure, and pain. Mechanoreceptors inside the body respond to stretching and pressure in the internal organs and in the skeletal muscles, tendons, and joints.

Chemoreceptors in humans include the *taste receptors* on the tongue, which distinguish between bitter, sour, sweet, and salty, and *olfactory receptors* in the nasal cavity, which discriminate among thousands of different molecules. Other chemoreceptors within the body provide information about the chemical state of the internal fluids and are important in maintaining homeostasis.

Sensory coding is the system that translates the quality, intensity, location, and temporal order of an environmental event into the neural language of action potentials. *Qualitative characteristics* are encoded by the specific sensitivity of each sensory cell; *intensity of the event,* by the frequency of action potentials in the afferent pathways; *location of the event,* by the spatial pattern of afferent pathways with action potentials; and *temporal order of the event,* by the temporal pattern of action potentials in each afferent pathway.

Living representatives of the different forms of animals reflect the major stages in the evolution of nervous systems. In the simplest nervous system, found in the modern coelenterate, the neurons of the radially symmetrical body are arranged in a *nerve net.* More advanced nervous systems occur in animals with bilaterally symmetrical bodies. In the flatworm, the simplest form of this type of animal, the neurons are organized into two parallel *nerve cords* (bundles of neurons that extend the length of the body) and a *ganglion* in the anterior end, or *head.* The consolidation of neurons into cords and ganglia is called *centralization;* the aggregation of neurons and sensory cells in the head is called *cephalization.* Both centralization and cephalization first appeared in the ancestral flatworms and have been major evolutionary trends. A related trend is *cephalic dominance*—the increasing control acquired by the ganglion in the head over most bodily activities. In more complex animals, this control center is called a *brain.* Centralization, cephalization, and cephalic dominance are most pronounced in the hymenopterous insects, the cephalopod mollusks, and the vertebrate animals.

The vertebrate *central nervous system* consists of a hollow, *dorsal nerve cord* (the *spinal cord*) enclosed in the *vertebral column* and a brain enclosed in the *skull.* This system includes the *interneurons,* the somas of *motor neurons,* and parts of the sensory and motor axons. The vertebrate *peripheral nervous system* consists of all neurons and parts of neurons that lie outside the vertebral column and skull, including the somas of sensory neurons, parts of the sensory and motor axons, and the ganglia of the *autonomic nervous system.*

The brain has increased in size and complexity during vertebrate evolution. In the early fishes, the brain consisted of three swellings—the *hindbrain, midbrain,* and *forebrain.*

In modern vertebrates, the hindbrain is differentiated into a medulla oblongata, a cerebellum, and a reticular formation. The *medulla oblongata* controls various feedback systems associated with homeostasis and contains afferent and efferent neural pathways that connect the brain and the spinal cord. The *cerebellum* contributes to the control of body movements. The *reticular formation* is not well developed in the fishes but maintains brain activities in mammals by monitoring the neural messages that enter and leave the brain. Another structure in the mammalian hindbrain—the *pons*—contributes to the control of sleep, breathing, and certain muscular movements.

The midbrain of the less advanced vertebrates integrates sensory information and controls behavior. It has steadily decreased in size during vertebrate evolution, and the forebrain has gradually acquired most of its functions.

The vertebrate forebrain has expanded in size during the evolution from early fishes to modern mammals. The forebrain of fishes, amphibians, and reptiles is composed of the olfactory bulbs, the cerebrum (the two *cerebral hemispheres*), the hypothalamus, and the thalamus. The *olfactory bulbs* receive and organize information about odors and pass it on to the *cerebrum*, where the information is further integrated and then transmitted to the midbrain. This type of cerebrum is called a *smell brain*. The cerebrum of all vertebrates except the fishes is covered by a *cortex*, which consists of *gray matter* (synapses and the somas of neurons). The interior of the cerebrum in these animals consists principally of *white matter* (the axons of neurons). The mammalian cortex is greatly enlarged and is called a *neocortex*. The *hypothalamus* controls the endocrine system and the autonomic nervous system. The *thalamus*—the gateway to the cortex—coordinates afferent messages to the cortex and efferent messages to the spinal cord. The mammalian forebrain also contains a *limbic system*, which deals with olfactory information and emotions, and the *corpus callosum*, which coordinates the activities of the two cerebral hemispheres.

The human nervous system monitors and controls the internal organs through the bloodstream and the spinal and cranial nerves. The spinal cord is the channel of communication between the brain and the spinal nerves. The 31 pairs of *spinal nerves* connected to the spinal cord contain both afferent and efferent neural pathways. Each pair of spinal nerves controls a particular region of the body below the head. The 12 *cranial nerves* are directly connected to the brain; 11 cranial nerves control the head, and one cranial nerve forms a direct communication between the brain and the internal organs.

The human peripheral nervous system has two divisions. The *somatic division* controls voluntary activities and is composed of sensory neurons from the skin, skeletal muscles, and joints and of motor neurons to the skeletal muscles. The *autonomic division* controls involuntary activities and consists of sensory neurons from internal organs and of motor neurons to the smooth muscles of the internal organs, the cardiac muscles of the heart, the exocrine glands, and some of the endocrine glands. The efferent pathways of the autonomic division are organized into two subdivisions. The *sympathetic subdivision* produces the fight-or-flight response during times of stress. The *parasympathetic subdivision* helps to maintain homeostasis and plays an important role in the recovery from stress and during periods of rest.

A *reflex* is a neural pathway that contains only a few synapses between a sensory neuron and motor neuron. A reflex provides fast, involuntary responses to stimuli.

The control center of the autonomic division of the human nervous system—the *hypothalamus*—is influenced by the rest of the brain through connections with the limbic system. The control centers for the somatic division are less discrete and include the neocortex (or *cerebral cortex*) and, to some degree, most of the other brain structures.

A central and a temporal fissure separate the cerebral cortex of each hemisphere into different functional areas—the frontal lobe (motor cortex, premotor cortex, and prefrontal area), the somatosensory cortex, the parietal lobe, the occipital lobe, and the temporal lobe.

The *motor cortex* controls complex patterns of movement. The neurons in each region of the motor cortex control the specific muscles of a particular part of the body. The *visual primary cortex* of the occipital lobes receives visual information from the eyes, and the *auditory primary cortex* of the temporal lobes receives sound information from the ears. The *somatosensory cortex* receives somatic information from the skin, muscles, and joints, as well as information about taste, temperature (via the hypothalamus), and pain (via the reticular formation). The visual primary cortex, the auditory primary cortex, and the somatosensory cortex all contain localized groups of neurons that respond to the stimulation of specific groups of sensory cells. Within each localized group, the neurons are further specialized to respond to distinct aspects of the sensory information.

A large part of the cerebral cortex—the *association cortex*, which is primarily responsible for mental activities—has greatly increased in size and complexity during vertebrate evolution. Information from the environment is linked together, stored, and processed within the association cortex. The storage of information, or *memory*, is an essential component of *learning.* The neural mechanisms of memory are located within the parts of the association cortex and some subcortical structures near the posterior end of the brain. There are two forms of memory. *Short-term memory* probably involves the establishment of temporary neural pathways; *long-term memory* probably involves the establishment of permanent neural pathways due to fundamental changes in the proteins synthesized by the neurons.

The neural mechanisms of *language* are located within two areas of the association cortex. *Broca's area* in the frontal lobe enables thoughts to be spoken; *Wernicke's area* in the temporal lobe enables language to be understood. In most humans, the neural mechanisms of language and of both abstract and logical thought are present only in the left hemisphere of the brain. The association cortex of the right hemisphere deals with nonverbal abilities, such as understanding three-dimensional space, recognizing faces, and perceiving some aspects of music. The discovery that the association cortex differs in the two cerebral hemispheres was confirmed by testing epilepsy patients who had undergone a *split-brain operation* in which the corpus callosum is severed to prevent epileptic seizures in one hemisphere from spreading to the other hemisphere. In normal humans, the intact corpus callosum allows communication between the two hemispheres so that the brain functions as an integrated unit.

States of brain activity, which are an aspect of *consciousness,* are regulated by the reticular formation and the pons of the human brain. Electrical activity in the cortex, which is measured by an *electroencephalogram* (EEG), ranges from a fully awake state to four stages of sleep to REM (rapid eye movement) sleep. Dreams occur most frequently during REM sleep when the pattern of electrical activity is similar to the pattern exhibited during the waking state. The human brain passes through four or five cycles during a night of sleep; each cycle consists of a sequence of the stages of sleep and REM sleep.

Almost all mammals and birds exhibit the same patterns of electrical activity in the brain during sleep, but it is not known whether animals other than humans dream. Fishes and amphibians do not sleep, and reptiles do not appear to sleep. The function of sleep in mammals and birds, which are homeothermic animals, may be to conserve energy and to reduce stress.

The Circulatory System: Lifeline Between the Cells and the Environment

14

The chemical activities within a cell can occur only under certain conditions: structural materials and energy must be present; the temperature, concentration of ions, and pH (ratio of acids to bases) must be within a certain range; and the waste products of earlier reactions must be removed. For a cell to remain healthy and active, these conditions must be maintained at optimal levels and the area surrounding the cell, where materials for molecular construction are acquired and wastes are deposited, must be regularly cleansed and replenished.

Each cell of unicellular and very small multicellular organisms is completely immersed in its external environment, so that the organism or the environment (for example, flowing water) can move from place to place, steadily renewing resources and eliminating wastes. Under extremely unfavorable conditions, these small organisms protect themselves by entering a resting state until optimal conditions prevail once again.

The maintenance of the individual cells in larger multicellular organisms is not as simple, because many cells are situated deep within the body, where they are isolated from the external environment—their source of energy and nutrients and area of waste disposal. Moreover, the cells of multicellular organisms are often highly specialized and consequently can no longer seek out optimal environments or defend themselves against detrimental changes in their immediate environment. Multicellular organisms therefore have special mechanisms that maintain the physical and chemical composition of the area immediately surrounding each cell by transporting materials from the external environment through the body to the cells and then from the cells back out to the external environment.

Multicellular organisms transport materials in several different ways. The body channels of simple oceanic organisms, such as the sponges, are arranged in such a way that seawater passes through the organism and reaches each cell of the body. The seawater moves of its own accord, regularly bringing new materials to the cells and removing waste materials, so that these organisms require no special transport systems. However, the environment does not flow naturally through the bodies of most animals and terrestrial plants. Instead, special structures and fluids in these organisms carry materials from one part of the body to another. These various transport systems will be discussed in this chapter, which places particular emphasis on the human circulatory system.

The Transport of Materials in Terrestrial Plants

Land plants maintain cellular activities by exchanging gases (carbon dioxide and oxygen) with the surrounding air, taking in water and minerals from the soil, and producing materials such as sugars and hormones in one part of the plant for use elsewhere. Gas exchange requires no internal transport structures. Gases are rapidly exchanged during photosynthesis, when the plant uses carbon dioxide and releases oxygen. This chemical activity takes place almost exclusively within the leaves, which are structured in such a way that gases can be moved at a sufficient speed and distance by diffusion alone. Leaves occur in a number of different shapes but are usually relatively small or thin, so that most of the cells are close to the surface. Much of the interior of a leaf is not structured but is simply interconnected air spaces between cells that permit the external environment to enter the leaf through small openings called the *stomata* (Figure 14-1). Carbon dioxide enters the stomata, diffuses throughout the leaf, and enters the interior of each cell by dissolving in a water film that surrounds the cell membrane and diffusing inward. Similarly, oxygen leaves each cell by diffusing across the cell membrane into the thin film of water and moving by diffusion through an air space to the stomata and out into the surrounding air. When the cells are involved primarily with cellular respiration, oxygen diffuses in and carbon dioxide diffuses out in the same fashion.

The film of water around each cell membrane, which is essential for gas diffusion, is maintained by the high (almost 100%) humidity within the air spaces

Figure 14-1
Stomata on the Surface of a Plant Leaf
(a) Magnified photograph of stomata on a leaf (×210). (b) Closer view of a stoma, which is fully open to permit gas exchange (×2,800).

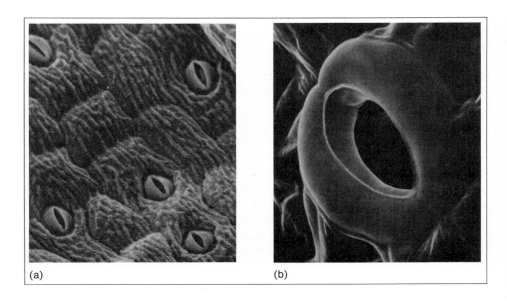

(a)　　　　　　　　　　　　(b)

between the leaf cells. Terrestrial plants are therefore presented with a special transport problem: water tends to move out of the leaf when its stomata are open, but these pores must be open to permit the flow of gases. When external conditions are dry or gas exchange is not essential, the stomata close to prevent water loss. Nevertheless, a substantial amount of water is regularly lost and must be replaced to maintain cellular activities. A single corn plant, for example, requires about 50 gallons of water to grow from a seed to a mature plant. Although some water is used in chemical reactions, about 90% of this water simply passes through the plant and evaporates through the stomata in the leaves.

Most terrestrial plants such as the ferns, gymnosperms, and flowering plants, are *vascular plants,* which have special structures to transport water and minerals and sugars and hormones within the plant. *Nonvascular plants,* including the mosses and most aquatic plants, have no special structures for the internal transport of materials.

The Xylem

The *xylem*—the special structure for the internal transport of water in vascular plants—moves water and minerals from the soil to the leaves to be used in chemical reactions and to replace water that has been lost through the stomata. Unlike animal cells, plant cells have rigid cell walls surrounding the cell membranes. When plant cells inside the stem and roots die, some of these rigid structures form a column or hollow tube, the xylem, shown in Figure 14-2(a). Water and minerals enter the plant through its roots and move via the xylem through the roots and stem to the leaves. In large trees, the xylem may transport water as high as 100 meters above the ground.

Water transport in plants is strictly a one-way system. Water moves into, through, and out of the plant, following a single upward direction; it does not circulate within an enclosed structural system. Plants expend no energy to transport water. The strong attraction of water molecules to one another due to hydrogen bonding moves a thin column of water molecules through each xylem from a root to a leaf. A molecule of water leaves the column through a stoma. The energy from the sun, which converts liquid water to vapor *(evaporation),* exerts a stronger force on the water molecule than the attractive force exerted by the water molecule just below it in the xylem column. The molecule that is evaporating from the leaf exerts an attractive force that pulls up the next water molecule in the column, which pulls up the next, and so on, producing a chain of displacements that extends from the leaf to the root. As the last molecule in the xylem is pulled up, it pulls another water molecule from the soil into the column.

The Phloem

The second transport structure in vascular plants—the *phloem*—moves sugars and other biological molecules in a water solution from one part of the plant to another. Unlike water within the xylem, the solution within the phloem can move either upward or downward. The phloem is constructed of columns of *living* cells that are joined together by small sievelike openings in the cell membranes, as shown in Figure 14-2(b). The exact mechanism that causes this flow is not yet known. It is known that the energy required to move the solution is derived from

Figure 14–2
Transport Structures in Plants

Vascular plants, including the ferns, gymnosperms, and flowering plants, have two types of transport structures—the *xylem* and the *phloem*. In this diagram, small segments of each type of structure have been taken from a tree. (a) Water and minerals are transported upward from the roots of the plant throught the *xylem*—a tubular structure made up of columns of cell walls remaining from plant cells that are no longer living. (b) Sugars and other biological molecules are moved through a plant by the *phloem*, which consists of columns of living cells joined together by sievelike openings in the cell membranes. The associated cells lying near the phloem tissue provide some of the energy required for transport.

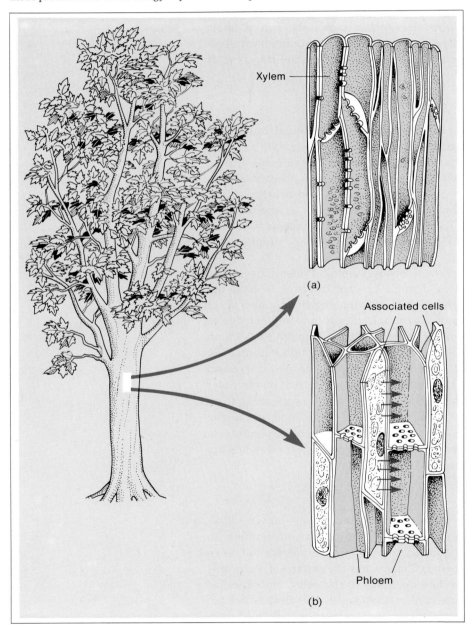

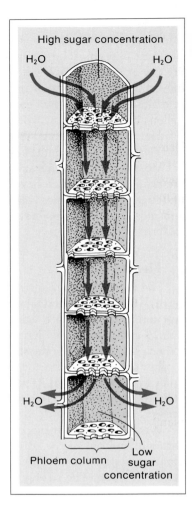

Figure 14-3
The Pressure-Flow Hypothesis of Phloem Transport

Parts of the phloem near the leaves (top of diagram) contain large amounts of sugar, which draw water into the phloem by osmosis. Consequently, the water pressure in the upper portion of the phloem column is high. Parts of the phloem near cells in which sugars are stored or used (bottom of diagram) contain small amounts of sugar, which cause water to leave the phloem by osmosis. The water pressure in the lower portion of the phloem column is low. The difference in water pressure causes the sugar solution to move from one part to another of the phloem column.

the breakdown of sugar molecules in the mitochondria of the phloem tissues and in the associated cells lying next to the phloem.

Although several hypotheses of phloem transport are currently being evaluated, the *pressure-flow hypothesis* (Figure 14-3) appears to be the most probable. According to this hypothesis, phloem cells near the leaves, where photosynthesis occurs, contain high concentrations of sugar molecules. Water molecules move into these cells by osmosis, traveling from areas of high to areas of low water concentration. The influx of water into a cell causes the interior pressure to increase, forcing the solution of water and sugar to move through sievelike openings into the adjacent phloem cells. The water pressure in these cells is therefore higher than it is in each adjacent cell farther along in the column. Conversely, phloem cells near the flowers and roots, where sugars are being used or stored, contain low concentrations of sugar molecules and high concentrations of water, so that water leaves these cells by osmosis. Consequently, the water

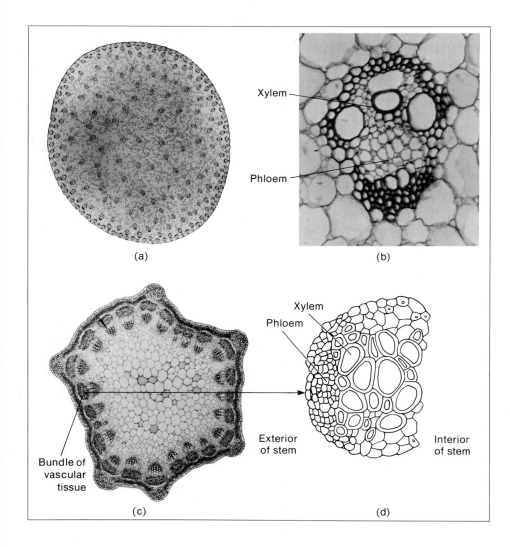

(a)

(b)

Xylem

Phloem

Bundle of vascular tissue

(c)

Xylem

Phloem

Exterior of stem

Interior of stem

(d)

Figure 14-4
Arrangement of the Vascular Tissues in Two Kinds of Flowering Plants

Photographs of cross sections through stems. (a) In the monocotyledon, the xylem and the phloem are mixed together in vascular bundles, which are scattered throughout the stem. (b) A close-up of one of the vascular bundles in a monocotyledon. (c) In the dicotyledon, the bundles of vascular tissue occur only around the periphery of the stem. (d) Within each bundle of a dicotyledon, the phloem columns are always on the outside and the xylem columns are always on the inside.

pressure in these parts of the phloem is low, and material in the phloem tends to move into these areas. The whole process is dependent on the massive uptake of water by cells containing high concentrations of sugars and on the massive loss of water from cells containing low concentrations of sugars.

Evidence suggests that at least some of the molecules in solution are moved through the sievelike openings between phloem cells by active transport and that carrier molecules are involved (see Chapter 4), possibly accounting for the observed energy requirement for phloem transport.

Arrangements of the Two Transport Structures

The xylem and the phloem have the same structures and functions in all vascular plants, but these transport systems are arranged in two, basically different ways in the angiosperms, or flowering plants. In the *monocotyledons,* such as the grasses and lilies, the xylem and the phloem are mixed together in vascular bundles scattered throughout the tissues of the stem, as shown in Figure 14-4(a) and (b). In the *dicotyledons,* which include most trees other than gymnosperms, food plants other than grains, and the majority of garden and wild flowers, the vascular bundles form a ring around the outer part of the stem. Within each bundle in the dicotyledons, the phloem is always positioned on the outside and the xylem on the inside, as shown in Figures 14-4(c) and (d).

A Comparison of Transport Systems in Animals

In simple animals, each cell is close to the surface of the body, so that a direct exchange of materials between the cells and the external environment can occur. Like unicellular and very small multicellular organisms, these simple animals replenish food and oxygen and dispose of waste materials either by moving through their environment or by remaining stationary and permitting the environment to move past them.

Simple animals are either hollow, so that the external environment (air or water) can pass through channels in their bodies, or very thin along one dimension, so that each cell can be near the body surface. One of the larger simple organisms—the *flatworm*—has a flat, long, wide body with almost no depth (Figure 14-5). As in a leaf, most of the cells in a flatworm's body are near the outer surface, so that oxygen and carbon dioxide can be exchanged directly with the external environment. However, because the flatworm consumes particles of food that are too large to be absorbed through the cell membranes, it has a mouth and digestive cavity to engulf and break down food particles into smaller units. The digestive cavity of a flatworm extends throughout the body interior in such a way that the digested contents are brought close enough to all the cells to permit an adequate exchange of food and waste materials (other than carbon dioxide) to occur without a special transport system.

Animals with larger, more three-dimensional bodies have special transport structures that move materials to and from cells buried deep within the body far from the external surface or the digestive cavity. These systems are comprised of a network of vessels containing a fluid—the *blood.* The vessels are not directly con-

Figure 14-5
The Flatworm
The flatworm has a long, wide, flat body with almost no depth, so that all of its body cells are near the surface. Gases can therefore be exchanged directly between the cells and the environment, and no special internal transport system is required. A flatworm is about 1–2 centimeters in length.

nected to the cells, but penetrate the extracellular, or *interstitial, fluid* in which the cells are immersed. The primary function of the blood is to maintain the composition of the interstitial fluid despite the movement of substances into and out of the cells. The blood acquires a supply of nutrients as it passes through the walls of the digestive cavity and a supply of oxygen as it passes through the walls of the lungs or gills. These materials are carried to the interstitial fluid and subsequently pass into the cells. Cellular waste materials pass from the interstitial fluid into the blood, where carbon dioxide is carried to the lungs or gills and other forms of waste are carried to the kidneys and out of the body. A muscular pump—the *heart*—contracts at regular intervals, forcing the blood in a single direction through the vessels.

Invertebrates

Two basic types of transport systems—the open and the closed circulatory systems—occur in the larger invertebrate animals. In the *open circulatory system,* shown in Figure 14-6(a), the blood is not completely enclosed within vessels; in the *closed circulatory system,* shown in Figure 14-6(b), the blood is contained entirely within an unbroken network of vessels. Circulation is slower in an open system because the blood is not enclosed during its entire route and the pumping action of the heart muscles cannot build up enough pressure to make the fluid flow rapidly. In a closed system, the muscular contractions of the heart can produce sufficient pressure to move the blood rapidly.

An open system cannot achieve the high rates of oxygen transport that active animals require. Animals with an open system of circulation are either quite small and sluggish or use the open system only for the transport of food and wastes and a different system for the transport of gases. Insects, for example, have a separate system of vessels—the *tracheal system*—for gas transport. The insect's circulatory system is composed of five muscular hearts, which slowly pump the blood, containing food and wastes (except carbon dioxide), hormones, and other materials,

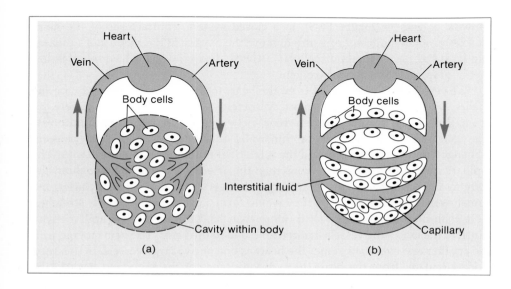

(a) (b)

Figure 14-6
Diagrams of the Two Main Types of Transport Systems in Invertebrate Animals

(a) In the *open circulatory system,* the heart (or hearts) pumps blood through the arteries into a large cavity, where the fluid bathes the cells of the body. The blood is slowly returned to the heart through the veins. (b) In the *closed circulatory system,* the blood remains within a completely enclosed system of vessels and never comes in direct contact with the body cells.

Figure 14-7
The Open Circulatory System
of an Insect

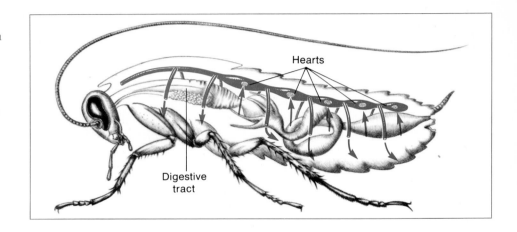

The network of vessels is not completely enclosed. Five separate hearts pump the transport fluid forward and down into open cavities, where the fluid bathes the cells. The blood then moves slowly from these cavities back to the hearts.

through a system of vessels and open cavities (Figure 14-7) in a forward and downward direction. The blood bathes the cells of the body in open cavities below the vessels, providing the necessary materials (except oxygen) for cellular activities and accumulating waste products (except carbon dioxide) from the cells. The blood then moves slowly from these cavities backward and upward to the hearts. Transport is accelerated during physical activity, when the skeletal muscles contract rhythmically, squeezing the cavities and forcing the blood back toward the hearts.

Invertebrate animals that have open circulatory systems include the arthropods (such as insects, spiders, crabs, and lobsters) and most of the mollusks (such as snails, oysters, and clams); invertebrates with a closed circulatory system include the annelids (such as the earthworm) and some mollusks (such as the squid).

Vertebrates

Vertebrate animals have closed circulatory systems with no openings in the network of vessels (Figure 14-8). In a closed system, contractions of the heart muscles produce sufficient pressure to move the oxygen in the blood at a sufficient rate to sustain high levels of cellular activity. The blood of all vertebrates has special properties for carrying oxygen.

The vessels that carry blood from the heart to the tissues of the body are the *arteries.* After leaving the heart, an artery branches and rebranches until a network of small arteries is formed. These arterial networks are connected to minute vessels, called *capillaries,* which have thin walls only one cell thick. Materials are exchanged between the blood and the interstitial fluid only in the capillaries. The capillaries are much more numerous than the arteries, extending throughout the body in a pattern that places a capillary near every cell. If all the capillaries in a human were placed end to end, they would form a continuous line that stretched more than once around the world (more than 25,000 miles)! The capillaries are joined to the *veins*—the vessels that return blood to the heart. The veins merge with one another and eventually enter the heart as one or two large vessels. In this way, blood circulates from the heart through the tissues of the body and back to the heart within an entirely closed system of vessels.

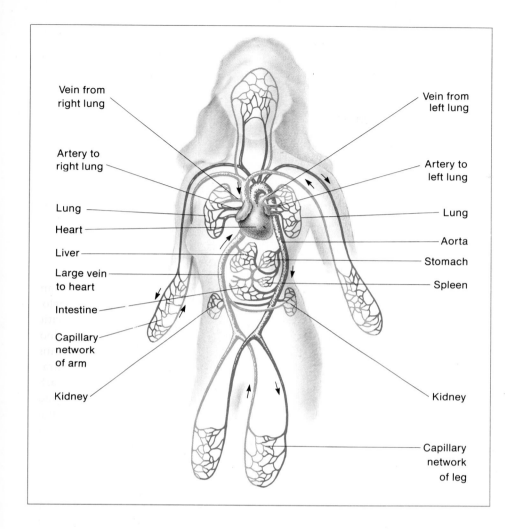

Vein from
right lung

Vein from
left lung

Artery to
right lung

Artery to
left lung

Lung

Heart

Liver

Large vein
to heart

Intestine

Capillary
network
of arm

Kidney

Lung

Aorta

Stomach

Spleen

Kidney

Capillary
network
of leg

Figure 14-8
The Closed Circulatory
System of a Vertebrate—
The Human

Arteries, shown in solid color,
transport blood from the heart to the
tissues. Within the tissues, the arteries
are connected to extensive networks
of *capillaries*. The capillaries are
joined to *veins*, shown in shaded color,
which then merge with each other and
eventually enter the heart as two large
vessels. This diagram is greatly
simplified. In reality, capillary
networks exist everywhere in the
body, and there are many more
arteries and veins.

The primary circulatory system in vertebrates transports blood through the body. A secondary network of vessels—the *lymphatic system* (Figure 14-9)—returns useful materials to the blood from the areas between the cells. A variety of materials, including proteins and fluids, that have left the blood and accumulated in the interstitial fluid are moved into the lymphatic capillaries (Figure 14-10). This fluid, called *lymph,* is transported from the capillaries to larger and larger lymphatic vessels that eventually empty into veins. When these lymphatic vessels are blocked due to injury or disease, fluid accumulates between the cells of the tissues and the area becomes swollen; this condition is called *edema. Lymph nodes* (Figure 14-9) are aggregations of cells at the junction of lymph vessels that play an important role in the body's defense against disease.

A circulatory system must transport materials to and from cells efficiently to maintain the health and vitality of an animal. In humans, the state of the circulatory system—the heart, the vessels, and the blood itself—is a primary determinant of physical health. The amount of exercise that an individual can perform, for exam-

Figure 14-9
The Lymphatic Circulatory System of a Vertebrate— The Human

Protein molecules, fluid, and other materials leave the blood at the capillaries and accumulate in the interstitial fluid around the body cells. These materials are picked up by the vessels of the lymphatic system and returned to the blood via large veins near the heart. The *lymph nodes* are parts of the lymphatic circulatory system that are involved in fighting infectious diseases.

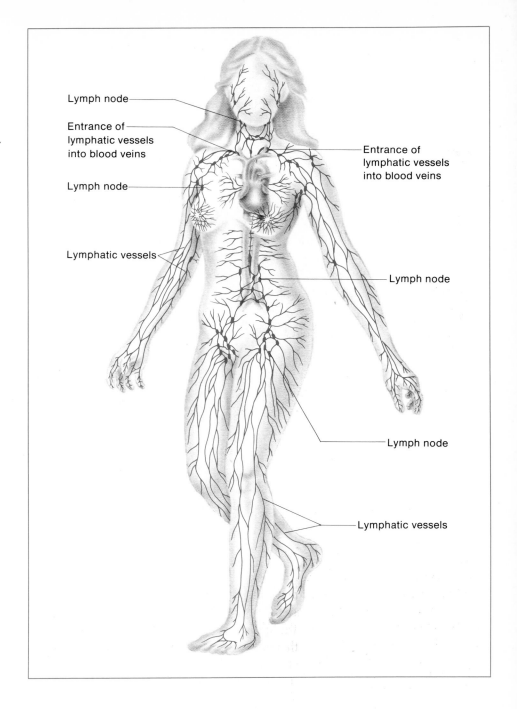

Lymph node

Entrance of lymphatic vessels into blood veins

Entrance of lymphatic vessels into blood veins

Lymph node

Lymphatic vessels

Lymph node

Lymph node

Lymphatic vessels

ple, is determined largely by the circulatory system's capacity to move materials to and from the muscles of the body. Human nerve cells are so dependent on the circulatory system that a cell deprived of blood flow for as short a time as two minutes will die. A defective circulatory system can cause many serious diseases.

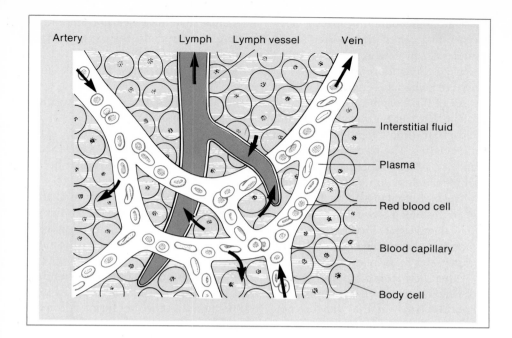

Figure 14-10
The Position of a Lymphatic Vessel Within Body Tissue

Labels on figure: Artery, Lymph, Lymph vessel, Vein, Interstitial fluid, Plasma, Red blood cell, Blood capillary, Body cell

The Characteristics of Human Blood

In the remainder of this chapter, we will emphasize the human circulatory system. We will begin with a discussion of the transport fluid, the blood, and then examine the functions of the heart and the vessels and some of their more common defects and disease conditions.

The chief function of the blood is to aid in maintaining constant physical and chemical states in the cellular environment. In vertebrate animals, including humans, blood carries:

1. Oxygen from the lungs or gills to all parts of the body.
2. Carbon dioxide from the cells to the lungs or gills.
3. Food molecules from the digestive tract to various parts of the body, where they are used or stored.
4. Waste products from the cells to the kidneys.
5. Heat from the deeper parts of the body to the surface, where it can be dissipated.
6. Cells and molecules of the immune system, which fight the invasions of bacteria and other foreign organisms.
7. Materials that repair vessel damage.
8. Special molecules, such as hormones, which are formed in one part and used in other parts of the body.

Most of these transport functions can be achieved by almost any liquid. However, the adequate transport of oxygen requires special materials because only small amounts of oxygen will dissolve in liquid solutions. The transport fluids of most animals contain a special carrier substance that increases the amount of oxygen that can be transported at any one time. In mammals, the oxygen carrier *hemoglobin*, a protein in red blood cells that contains iron, permits the blood to carry about 70 times more oxygen than a water solution could carry. Hemoglobin is bright red when it is combined with oxygen.

All of the oxygen carriers found in animals are combinations of a metal and a protein that can pick up oxygen, hold it during transport, and release it at the appropriate time. Hemoglobin is found in almost all vertebrates and in many invertebrates. The transport fluids of other invertebrates contain the carrier substance *hemocyanin*. The major difference between these two carriers is that hemocyanin contains copper whereas hemoglobin contains iron.

The body contains about 5 liters (1 liter = 1.05 quarts) of blood. Approximately 45% of this blood is composed of cells (primarily red blood cells); the remainder is a fluid called *plasma*. The number of red blood cells varies according to the oxygen content of the air; at high elevations, where less oxygen is available, the body compensates by increasing the number of red blood cells.

When human red blood cells mature, they contain no nucleus, mitochondria, RNA molecules, or Golgi complexes. For this reason, they are often referred to as *corpuscles* to indicate that they are not truly functioning cells. Because a corpuscle has no nucleus, it has no genes to direct the synthesis of new proteins and therefore lives only about four months. Red blood cells are continuously replaced by new cells that are constructed in the bone marrow, particularly in the skull, back, chest, and hip.

Every second of your life, about 2.5 million new red blood cells are made and an equal number of old cells die within your body. To meet this tremendous demand for structural materials, dead red blood cells are broken down in the liver, spleen, and bone marrow and some of the cell components, such as iron and proteins, are moved to the bone marrow to be used again.

Mending Wounds in the Vessels

The blood contains special materials that prevent blood loss by repairing holes in the vessels. As a first defense against blood loss, the injured vessels constrict, temporarily decreasing the flow of blood from the wound. Small spherical structures, called *platelets*, then collect in the vicinity of the wound to plug the opening. Platelets also release substances that cause the nonliquid portions of the blood to condense, or *coagulate*, within the damaged area. The final repair occurs when protein *(fibrin)* strands develop into a fine network around the wounded area, trapping the condensed materials and forming a *blood clot* that seals off the wound (Figure 14-11). The fibrin strands then contract, reducing the clot and pulling the edges of the wound closer together.

After a clot has formed, it shrinks and hardens to form a *scab*, which falls off when the wound has completely healed. The process of clot formation is quite complicated and involves more than 35 special materials.

Some people do not have the appropriate genes to direct the construction of these clotting materials. When these *bleeders* develop a small wound or bruise (capillary breakage), the repair system fails to function and bleeding continues unchecked. The most common condition among bleeders is *hemophilia*. In hemo-

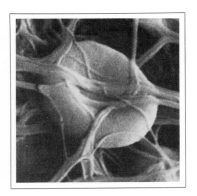

Figure 14-11
Formation of a Blood Clot
Solid materials within the blood, such as this red blood cell, are trapped by *fibrin strands* to form a *blood clot* that seals off the wound.

philiacs, faulty genes are located within one of the sex chromosomes—the X chromosome. The cells of females are rarely defective because they contain two X chromosomes, one of which is almost always normal. However, each male cell contains only one X chromosome; if it is defective, the male will be a hemophiliac. Queen Victoria carried this genetic defect within one of her X chromosomes; although she exhibited no symptoms of hemophilia, Victoria passed the disease on to her son (a male inherits his single X chromosome from the mother). Hemophilia occurred frequently among European royalty during the nineteenth century, because intermarriage was common and the defect was passed on to many royal families.

Sometimes clots develop inappropriately. They may be stimulated to form if the inner walls of the vessels are rough. These clots most often develop in the veins of the leg, where their effects are usually not serious. But if a clot breaks loose, it can lodge elsewhere in the circulatory system. If a dislodged clot blocks a large vessel or interferes with the supply of blood to a vital organ, the consequences can be serious, even fatal.

The Vertebrate Heart and Vessels: Circulation of the Blood

Proper circulation of the blood requires a heart to generate the necessary pressure, a system of vessels to contain the blood and direct its flow, and a mechanism to ensure that the blood flow is one-directional.

The heart is the most muscular part of the network of vessels that make up the vertebrate circulatory system. The arterial vessels that carry blood from the heart to the tissues are also very muscular. The main muscles within the walls of the arteries are circular, enclosing the channel of the vessel. When they contract, the diameter of the vessel becomes smaller; when they relax, the diameter is enlarged. From the arteries, the blood moves into the capillaries, which have very thin walls with no muscles at all. Materials are exchanged between the blood and the interstitial fluid across the walls of the capillaries. The blood moves slowly through the capillaries and enters the veins, which transport it back to the heart. The veins contain some but much less muscle than the arteries. These four functioning parts of the circulatory system—the heart, arteries, capillaries, and veins—form a completely closed system of transport for the blood of all vertebrate animals.

The heart pumps blood throughout the body by rhythmically contracting its strong muscles. The structure of the heart differs, depending on the rate of blood flow required to maintain the activity level of each type of animal. Throughout evolutionary history, as certain kinds of animals became larger and more active, their hearts became proportionately more powerful and efficient pumps to move blood to the tissues.

All vertebrate animals have a single heart, composed of two or more chambers. The simplest *vertebrate heart,* found in the fishes (Figure 14-12), is basically composed of two fairly large, hollow chambers—the *atrium* and the *ventricle.* Blood collects in these chambers until the muscles within the walls are stretched to a certain point. The muscles then contract powerfully to move the blood forward. In fishes, the atrium collects blood that is returning to the heart from the body. On

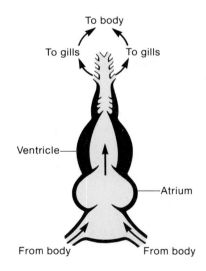

Figure 14-12
The Simplest Vertebrate Heart Is Found in the Fishes
This vertebrate heart is composed of two large muscular chambers—the *atrium* and the *ventricle.* Blood is returned to the heart via the atrium and is then moved into the ventricle, which pumps the blood to the gills. From the gills, blood flows under very little pressure to the rest of the body and eventually returns to the atrium.

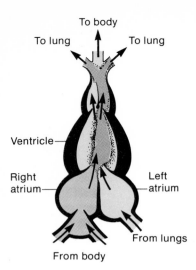

Figure 14-13
The Amphibian Heart
Blood flows from the body into the right atrium, from which it is pumped into a single ventricle and into the lungs. Blood enters the left atrium from the lungs and is pumped into the ventricle again. From the ventricle, blood from the lungs is pumped to all parts of the body and then returned to the right atrium. In this illustration, oxygen-rich blood is brown and oxygen-poor blood is gray.

contraction, blood leaves the atrium and enters the ventricle. When the ventricle contracts, it forces blood out of the heart, through the arteries, and into the capillaries in the gills, where it absorbs oxygen from the surrounding water and releases carbon dioxide from the body. As it passes through the large network of small capillaries in the gills, the blood loses much of its pressure. From the gills, the oxygen-rich blood moves slowly through the capillaries of all the body tissues, exchanging materials as it circulates. The oxygen-poor blood then enters the veins and returns to the atrium of the heart.

The *amphibian heart* is composed of three chambers—two atria and one ventricle (Figure 14-13). Blood that has circulated throughout the body enters the right atrium; blood that has been oxygenated in the lungs enters the left atrium. Both atria empty into the single ventricle, which pumps blood to the lungs and all body tissues. Blood leaving the heart via the single ventricle to circulate throughout the body contains a mixture of oxygen-rich blood (received from the lungs via the left atrium) and oxygen-poor blood (returned from the body via the right atrium). However, the two forms of blood do not become completely mixed, because blood from the right atrium enters the ventricle slightly before blood from the left atrium and therefore lies closer to the exit from the heart. As the ventricle contracts, the oxygen-poor blood from the right atrium leaves the heart first and enters the arteries leading to the lungs. By the time the oxygen-rich blood leaves the ventricle, these arteries are full, so that this blood bypasses them and travels to the rest of the body. Some of the blood that circulates to the body is mixed and has a relatively lower oxygen concentration when it reaches the cells than the unmixed blood from the lungs. However, most amphibians also absorb oxygen through the skin, and oxygen reduction in the blood due to mixing is not a particular disadvantage.

The *reptilian heart* (Figure 14-14) is composed of two atria and a single ventricle, which is partially divided into two chambers (except in the alligators and crocodiles, where division is complete). Blood returning from the body enters the right atrium and is moved primarily into the right portion of the ventricle. The ventricle then pumps the blood to the lungs, where gas exchange occurs. Newly oxygenated blood returns to the heart via the left atrium and is moved primarily into the left portion of the ventricle, from which it is pumped to the body tissues. The partial partition in the ventricle permits only a little of the oxygen-rich blood to mix with the oxygen-poor blood, so that the blood pumped from the reptilian heart to the body has a higher oxygen content than the blood pumped from the amphibian heart. Reptilian blood must contain more oxygen than amphibian blood, because reptiles do not absorb additional oxygen through the skin.

Birds and mammals maintain higher rates of cellular activity than other vertebrates and must therefore transport materials to and from their cells more rapidly than amphibians or reptiles. *Bird and mammal hearts* are similar; both are composed of two atria and two ventricles (Figure 14-15). No mixing of oxygen-rich and oxygen-poor blood occurs within the ventricles because they are two completely separate chambers.

Blood that is low in oxygen and high in carbon dioxide returns to the heart of a bird or mammal by entering the right atrium, where it accumulates before moving to the right ventricle. Although the atrium contracts, this pumping action is not responsible for most of the blood flow from the right atrium to the right ventricle. Instead, the relaxation of the muscles in the right ventricle expands the cavity and draws blood into it. The atrium in birds and mammals is therefore more of an expandable reservoir than a pump. As the right ventricle contracts, blood is pumped to the lungs via the *pulmonary artery,* which branches into smaller arteries and then into capillaries within the lung tissues. Here, gases are exchanged and the blood, which now contains high concentrations of oxygen and low concentrations of carbon dioxide, returns to the heart via *pulmonary veins,* which empty into the left atrium. The left ventricle relaxes and fills with blood from the left atrium. The left ventricle then contracts, forcing the blood under considerable pressure out of the heart through the *aorta* (the largest artery) and, via other arteries, to all parts of the body except the lungs, which are fed by the pulmonary artery. After the arteries branch into capillaries and materials are exchanged between the blood and the interstitial fluids, the blood moves into the veins and returns to the right atrium. The left ventricle is the most muscular chamber of the heart because it must pump blood the greatest distance—throughout the body.

Prevention of Backflow

Four valves within the mammalian heart ensure that blood flows in the proper direction (Figure 14-15). These simple, flaplike structures open and close according to the differences in pressure on the two sides of its structure. Two *atrioventricular valves*—one between the right atrium and ventricle and the other between the left atrium and ventricle—open to allow blood to flow when the atrium contracts (puts pressure on the atrial side of the valve) and the ventricle relaxes. The ventricle contracts after it fills, reducing the size of the chamber and forcing blood back against the valve, which causes the valve to close and prevents blood from returning to the atrium. This simple device ensures that blood flows in only one direction—from atrium to ventricle.

The two *ventricular artery valves* are located between the ventricles and the arteries that lead from them. When the ventricles contract, these valves allow the blood to flow freely into the arteries, as shown in Figure 14-16(a). When the ventricles relax, the pressure of the blood in the arteries causes the valves to snap shut, as Figure 14-16(b) illustrates. There is no backflow of blood from the arteries to the ventricle.

The closures of these four valves are principally responsible for the sounds of the heart. When you listen to the beat of a human heart, you hear a "lub-dub" sound. The first sound ("lub") is the simultaneous closure of the atrioventricular valves on both sides of the heart. The second sound ("dub") is the simultaneous closure of the ventricular artery valves.

A doctor can determine if any of these valves are not functioning properly by listening to the heart. Unusual heart sounds are called *heart murmurs.* If an atrioventricular valve does not close completely, due to the accumulation of scar tissue or the presence of holes, the heart makes a "lub-hiss-dub" sound; the "hiss" is the backflow of blood through a partially open valve. If a ventricular artery valve does not close completely, the heart makes a "lub-dub-hiss" sound, in this case, the "hiss" occurs when blood leaks from the artery back into the ventricle through the incompletely closed valve.

Circulatory problems may develop when a heart valve does not function correctly. These problems are more serious when one of the atrioventricular valves

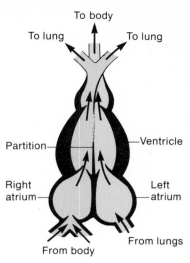

Figure 14-14
The Reptilian Heart
The reptilian heart is similar to the amphibian heart, except that the single ventricle in the reptilian heart is partially separated into two parts. This partition minimizes the mixing of oxygen-poor and oxygen-rich blood from the two atria. This drawing is highly schematic; within the body, the heart is actually inverted so that the atria are positioned on top. In this illustration, oxygen-rich blood is brown and oxygen-poor blood is gray.

Figure 14-15
Schematic Diagram of the Human Heart

The human heart is composed of two completely separate atria and ventricles, so that oxygen-rich blood and oxygen-poor blood cannot mix. In this illustration, oxygen-rich blood is light brown and oxygen-poor blood is gray. Note that the heart is drawn in its actual orientation with the atria positioned on top. Two pairs of valves (indicated by asterisks) prevent backflow of blood in the human heart—the *atrioventricular valves* (between the left atrium and ventricle and between the right atrium and ventricle) and the *ventricular artery valves* (between the right ventricle and the pulmonary artery and between the left ventricle and the aorta). The hearts of birds and other mammals are very similar to the human heart.

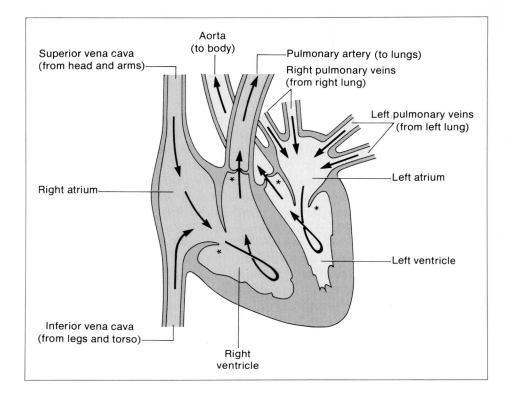

Aorta
(to body)

Superior vena cava
(from head and arms)

Pulmonary artery (to lungs)

Right pulmonary veins
(from right lung)

Left pulmonary veins
(from left lung)

Right atrium

Left atrium

Left ventricle

Inferior vena cava
(from legs and torso)

Right ventricle

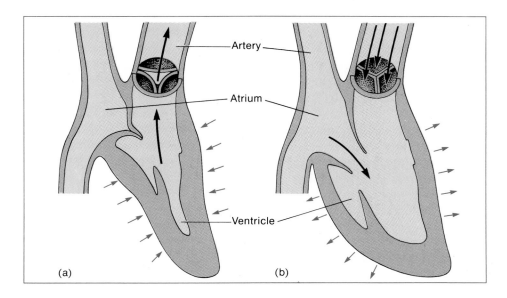

Artery

Atrium

Ventricle

(a)   (b)

Figure 14-16
How the Ventricular Artery Valve Works

(a) The valve opens as the ventricle contracts; blood then flows from the ventricle into the artery. (b) The valve closes when the ventricle relaxes, because blood from the artery flows back toward the ventricle and presses against the flaplike valve.

fails to close completely because blood then accumulates in the atrium and eventually in the vessels of the circulatory system, where the pressure within the engorged capillaries causes plasma to leak out of the vessels. If the valve between the right atrium and ventricle does not close completely, blood backs up in the right atrium and eventually in all organs of the body (except the lungs). The accumulation of fluid and the lack of proper blood flow in these organs interferes with their normal activities. If the valve between the left atrium and ventricle is defective, blood backs up and becomes congested in the lungs. Eventually, so much fluid may leak out of the capillaries and collect in the lungs that the affected person literally drowns.

Defects in the valves between the ventricles and the arteries are usually less serious. In a young person, the only effect may be an increased development of the muscles of the ventricle as it attempts to compensate for the extra work caused by the backflow. The problem is much more serious if a person's heart is already weakened, as is often true of older people. The additional work the heart muscles must perform to compensate for a leaky valve can then lead to heart failure.

Because the heart valves operate according to simple mechanical principles, they are relatively easy to repair or replace. Defective valves are usually replaced by artificial valves made of synthetic materials. Sometimes, however, valves from another mammal—often the pig—are transplanted into human hearts.

Nourishing the Heart Muscles

The heart does an incredible amount of work. For each minute that the body is at rest, the heart pumps 5 liters of blood—an amount equal to the entire volume of blood in the body. Each blood cell travels more than a mile a day. The energy that the heart expends each minute when the body is at rest is equal to the energy that a human must expend to lift a 70-pound weight one foot off the ground. The heart works considerably harder during exercise, pumping up to 37 liters of blood a minute.

Heart muscle cells also require a continuous supply of blood to transport oxygen and nutrients to and remove waste materials from the muscles. Blood within the chambers of the heart does not enter the muscles directly. Instead, blood reaches the muscles from the outer surface of the heart through branches of the main artery (aorta) leaving the left ventricle. These *coronary arteries* (Figure 14-17) divide several times to form numerous smaller arteries and then a network of capillaries that supplies each heart muscle with blood. Blood is collected from the capillaries by the *coronary veins* and returned to the right atrium.

When defects in the coronary arteries prevent blood flow to the heart muscles, a *heart attack* occurs. Some of the main problems that can develop in the coronary arteries are:

1. Blockage by a clot that forms within one of the coronary arteries (a *coronary thrombosis*).
2. Blockage by part of a blood clot that forms elsewhere in the circulatory system and travels to a coronary artery *(coronary embolus)*.
3. A decrease in the diameter of a coronary artery *(coronary occlusion)* due to a build-up of such substances as fats (especially *cholesterol*) or calcium.

Figure 14-17
Vessels That Nourish the Heart
The *coronary arteries* carry oxygen-rich blood to the muscles of the human heart. The two large coronary arteries branch off from the aorta as it leaves the left ventricle. The heart is shown here in its true, somewhat twisted shape in contrast to the diagram in Figure 14-15.

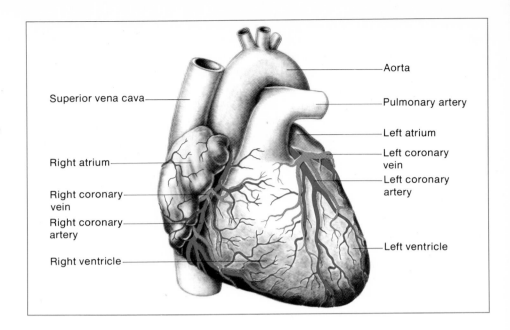

Superior vena cava

Right atrium

Right coronary vein

Right coronary artery

Right ventricle

Aorta

Pulmonary artery

Left atrium

Left coronary vein

Left coronary artery

Left ventricle

In each case, blood is prevented from passing through the vessel to nourish the heart muscles.

When an insufficient amount of blood reaches the heart muscles, they contract in spasms. Substances that stimulate pain receptors then accumulate in the vicinity of the heart and produce a sharp chest pain *(angina pectoris)*. This sensation, which may extend from the chest down the left arm, shoulder, and back, is a strong

Figure 14-18
A Completely Artifical Heart
This artificial heart has been used in humans.

warning that one or more coronary arteries are defective. Medicines that cause the coronary arteries to dilate (expand) temporarily can be used to reduce these spasms, but cannot correct the coronary artery disorder that causes the spasms.

If the blood supply to the heart is blocked for an extended period of time, the heart muscles will eventually die. If a small artery is blocked, then only a few muscle cells will be lost; but if a major artery is blocked, then a large portion of the heart will be destroyed.

Medical scientists are currently experimenting to perfect techniques to overcome these serious heart problems. Areas of experimentation include attempts to develop completely artificial hearts (Figure 14-18) and surgical techniques ranging from replacing coronary arteries with vessels from other parts of the body to transplanting an entire heart from a deceased to a living human.

Control of the Heart Beat

The heart is a remarkable organ. It contracts rhythmically a certain number of times a minute when the body is at rest and increases its beat to higher rates when the body cells require additional oxygen.

The *pulse* is a measure of the impact of blood on the wall of an artery. A pulse can easily be felt at the wrist, where an artery lies just beneath the skin. The impact on the arterial wall is strongest when the left ventricle contracts, so that the frequency of a pulse reflects the frequency of the contractions of the left ventricle. The human *pulse rate,* or normal range of contractions, is 50–100 beats per minute when the body is at rest. Each person has a characteristic heart rate that increases under specific conditions, including exercise, muscular work, high altitude, fever, and certain emotions.

The Pacemaker

The rate at which the heart muscles contract is regulated in several ways. The main control of the heart rate is the *pacemaker*—a small piece of specialized heart muscle tissue located in the wall of the right atrium. Electrical impulses emitted at regular intervals by this tissue stimulate muscle contractions in the four chambers of the heart. The membranes of the heart muscle cells are so close to one another that these electrical impulses travel directly from one cell to another without the use of neurotransmitters. Such connections between cells are called *tight junctions.* Each impulse travels through both atria, causing them to contract almost simultaneously, and on to a second regulator—the *atrial ventricular node*—which transmits the impulse to both ventricles simultaneously. The slight delay in the signal produces a sequence of contractions: first, the two atria; then, the two ventricles.

Irregularities in the heart muscles, such as bumps, scars, or dead muscle tissue from past injuries, can delay the movement of impulses from the pacemaker through the heart muscles. These disturbances, called *heart blocks,* cause the muscles to contract in a less coordinated way. The movement of the electrical impulses from the pacemaker through the heart muscles can be measured by placing special recording instruments on the skin near the heart. Such a measurement, called an *electrocardiogram* (ECG or EKG), detects irregularities in the heart muscles.

Because the pacemaker regulates heart action by transmitting impulses that are similar to electrical discharges, the heart is also vulnerable to interference from electrical activities outside the body. A strong electrical shock may interfere with the activities of the pacemaker to the extent that severe disturbances called *fibrillations*—"quivers" of the heart that result when each muscle contracts and relaxes independently of the others—may disrupt the pumping and circulation of the blood. Because the heart does not beat regularly, the blood is not pumped effectively and circulation fails. Similar fibrillations may occur when a coronary artery is blocked. In either case, uncoordinated contractions, especially when they occur in the ventricles, can cause death within a few minutes.

A defective pacemaker can be replaced by an *artificial pacemaker*—a small battery unit inserted into the body and connected to the heart—which sends out electrical impulses at regular intervals to the heart muscles. An artificial pacemaker can be adjusted to transmit electrical impulses at the rate of the recipient's normal pacemaker.

Control by the Autonomic Nervous System

Although the pacemaker in the right atrium is the primary mechanism governing the heart rate, heart muscle contractions are also regulated by neural and hormonal controls. An area within the *medulla oblongata* of the brain contains cardioinhibitory and cardioaccelerating centers, which form part of the *autonomic nervous system.*

The *cardioinhibitory center* communicates with the pacemaker in the heart via the *vagus nerves,* which contain both afferent and efferent axons. The *afferent nerve axons,* which originate in the pacemaker and terminate in the cardioinhibitory center, provide information about the rate of the heart muscle contractions. The *efferent nerve axons,* which originate in the cardioinhibitory center and extend to the pacemaker, can stimulate the pacemaker to decrease the rate of the heart muscle contractions. The cardioinhibitory center functions to restrain the pacemaker and to hold the heart rate in check.

In addition to feedback from the pacemaker, the cardioinhibitory center receives information from sensory surfaces of the body and from higher brain centers. *Sensory cells* on the internal and external body surfaces transmit information to the cardioinhibitory center about such conditions as severe indigestion, inhalation of irritating fumes, sudden cold temperatures (for example, when the body is plunged into very cold water), and the pressure of blood on the walls of the arteries. When the center receives the information, it stimulates the efferent axons of the vagus nerves, which diminishes the heart rate. Certain emotional states also stimulate the cardioinhibitory center. Many areas of the brain are involved in the regulation of emotion, but the critical pathway that influences the heart rate is from the *limbic system* to the cardioinhibitory center via the *hypothalamus.*

The *cardioaccelerating center* within the medulla oblongata of the brain is stimulated by many factors, including pain sensations from the skin and anticipation of exercise. Efferent neurons from the cardioaccelerating center terminate in the heart muscles themselves, rather than in the pacemaker. When stimulated, these neurons release *norepinephrine* (a neurotransmitter), which increases both the heart rate and the force of the heart's contractions.

The heart rate is also affected by hormonal secretions from the *endocrine glands.* *Thyroxine,* the hormone secreted by the *thyroid gland,* increases the heart rate. *Epinephrine,* a hormone secreted by the *adrenal medullas,* increases both the rate and the strength of the heartbeat.

Blood Pressure

When the left ventricle of the heart contracts, blood is forced into the arteries under considerable pressure. The amount of pressure the blood exerts on the arterial walls depends on the amount of blood that is pumped and on the diameter of each artery. The circular muscles of the arterial walls form the periphery of the channels through which the blood flows. When blood pressure is high, these muscles relax and the diameter of the artery becomes larger; when blood pressure is low, the muscles contract and the diameter of the artery becomes smaller. Because the diameter of the arteries can change, the change in pressure of the blood on the arterial walls is not as great as it would be if the arteries were rigid tubes. The elasticity of the arteries also diminishes the pulsation of the blood from the ventricles to a slow and uniform flow by the time it reaches the capillaries to facilitate the exchange of materials.

Blood pressure is usually measured from the main artery in the upper arm and is expressed as the height in millimeters (mm) that the pressure will raise a column of mercury (Hg). Two measurements are taken: the *systolic pressure,* the maximum pressure during the contraction of the left ventricle, and the *diastolic pressure,* the minimum pressure during the relaxation of the left ventricle. Average blood pressure measurements are about 120 mm Hg for systolic pressure and 80 mm Hg for diastolic pressure in an adult male and 110 and 70, respectively, in an adult female. Extreme deviations from these average values are probable indications of a heart malfunction, unusual amounts of blood in the system, or arterial inelasticity. If the arteries do not expand sufficiently, the systolic pressure will be abnormally high; if the arteries do not constrict sufficiently, the diastolic pressure will be too low.

Blood pressure is highest when it leaves the left ventricle and decreases gradually as blood flows through the vessels. By the time the blood reaches the veins and begins to circulate back to the heart, blood pressure is very low (about 7 mm Hg) and other mechanisms must be relied on to return the blood to the heart. These mechanisms include the intermittent squeezing action caused by movements of the chest during respiration and the contractions of the *skeletal muscles,* which are connected to the bones and are involved in locomotion. Many skeletal muscles undergo rhythmic contractions during normal body movement. As they contract, the skeletal muscles press against the veins and squeeze the blood along (Figure 14-19); as the muscles relax, the veins open up and blood flows into the nearly empty channels. When skeletal muscles are weak or contract infrequently, the blood tends to collect in a specific area of the body—especially in the legs, where it must move upward against gravity to reach the heart. The feeling of faintness you experience when you stand for a long time is caused by a decrease in blood flow back to the heart, which results in insufficient blood flow to the brain.

One-way valves located in the veins permit blood to move back toward the heart (Figure 14-20). These valves are similar to the valves within the heart; they open to allow blood to move in one direction and then snap shut to prevent backflow.

Figure 14-19
Movement of Blood Back
to the Heart
Leg muscles pinch against the veins to
help propel blood from the legs
upward toward the heart.

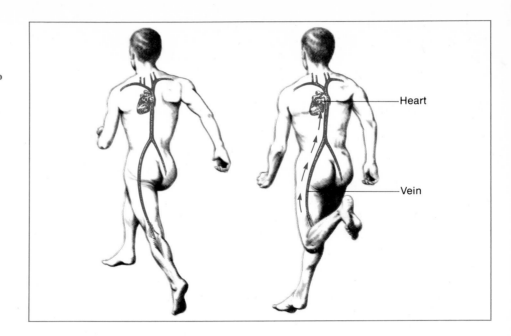

Heart

Vein

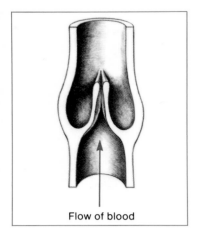

Flow of blood

Figure 14-20
One-Way Valves Within
the Veins
One-way valves are located within the
veins of the body. These valves open
to allow blood to flow toward the
heart and then close to prevent
backflow.

One-way valves in the veins can rupture when they are forced to hold large amounts of blood because contractions of the body muscles are not moving the blood toward the heart at an adequate rate. When the valves fail, the blood is no longer pumped in one direction toward the heart but is sloshed back and forth within the veins. As blood collects in the area where a valve is malfunctioning, the vein becomes engorged with blood, or *varicose*. Varicose veins usually develop in the legs as a result of excessive standing or sitting and little movement of the leg muscles.

The Regulation of Blood Flow

Animals must be able to adjust the rate of blood flow in response to changing conditions. When cellular activity is low, as it is during sleep, the rate of blood flow is lowered to conserve energy. During strenuous activity, the rate of blood flow must be rapid enough to meet the increased demand for the exchange of materials between the blood and the more active cells.

The *cardiac output*—the quantity of blood that the heart pumps per unit of time—is about 5 liters per minute in a resting human. Cardiac output is the product of two factors: (1) the number of contractions per unit of time (heart rate) and (2) the volume of blood ejected from the heart during each contraction. We have already discussed the heart rate, which is controlled primarily by the pacemaker but also by the cardioinhibitory and cardioaccelerating centers within the medulla oblongata of the brain and by hormones secreted by the thyroid and adrenal glands. In this section, we will examine the second factor—the volume of blood ejected at each contraction.

Because the vessels and the heart form a closed circulatory system, the volume of blood expelled from the heart during each contraction can only be increased if the rate at which blood is returned to the heart undergoes a corresponding increase. As the volume of blood returning to the heart increases per unit of time, the muscle contractions that force the blood through and out of the heart become stronger. Blood is returned to the heart more rapidly when the blood pressure is higher, which is caused by a constriction of the arteries.

Regulation by Carbon Dioxide

The autonomic nervous system responds to changes in the carbon dioxide levels within the blood by adjusting the diameters of the arteries to correspond to the level of physical activity. High concentrations of carbon dioxide, a waste product from the breakdown of sugars during cellular respiration, reflect high levels of cellular activity. The amount of carbon dioxide in the blood is detected by nerve cells in two *vasomotor centers,* one on each side of the medulla oblongata, which send electrochemical impulses along efferent nerves (the *vasomotor nerves*) that terminate in the muscles of the arteries. As blood flows through the capillaries in these centers, high levels of carbon dioxide stimulate these nerves and the muscles in the arterial walls respond by contracting. The vessels constrict, increasing blood flow and transporting materials to and from the active cells more rapidly. Carbon dioxide levels also regulate breathing rates, so that the rate at which fresh air is brought into the lungs increases as the rate of blood flow increases. Low levels of carbon dioxide produce the opposite effect: the arteries become dilated, blood pressure drops, and blood flow becomes slower.

Regulation by Information from the Nervous System

Some regulation of blood flow is also maintained by the brain centers that control the emotions, including the cerebral cortex, the hypothalamus, and the limbic system, which emit electrochemical impulses that travel to the vasomotor centers of the medulla oblongata. Certain emotional states can accelerate the heart rate and constrict the arteries; other emotional states can inhibit the heart rate and dilate the arteries to the point that the individual faints. Information is transmitted from the vasomotor centers (via the same vasomotor nerves that respond to carbon dioxide levels in the blood) to the arterial walls, which either constrict or dilate the vessels.

When the body is at rest and physical and emotional activities remain constant, the *parasympathetic division* of the autonomic nervous system acts continuously to maintain blood pressure within the individual's normal range. Special sensory cells called *pressoreceptors,* located within the arterial walls, are sensitive to the amount of pressure exerted by the blood on the vessel walls and relay this information to the vasomotor centers of the medulla oblongata along afferent nerves. The vasomotor nerves from the vasomotor centers carry impulses to the muscles in the arterial walls, causing them to contract or relax. The pathway of nerve stimulation—from pressoreceptors in the arteries to vasomotor centers in the brain to muscles in arteries—keeps the arterial walls in a continual state of partial constriction, maintaining blood pressure within the normal range and allowing the blood to circulate.

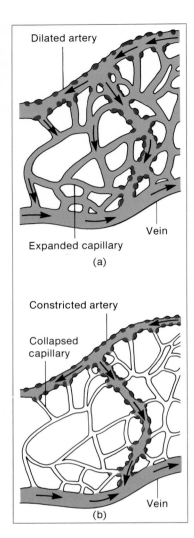

Figure 14-21
The Shunting of Blood from One Area of the Body to Another

(a) Blood flow is increased locally when the arteries feeding the area dilate fully. The capillaries also expand due to the pressure of the blood. (b) Blood flow is decreased locally when the arteries feeding the area constrict. The capillaries then collapse due to the lack of pressure of the blood.

Changes in the Distribution of Blood

Each part of the body requires varying quantities of blood at different times. Because the total volume of blood within the body remains fairly constant, the blood supply can be increased in one area of the body only if the amount of blood is decreased in other areas. Working organs or tissues require more nourishment than resting ones.

Blood is shunted preferentially to the areas of greatest need via the arteries, which dilate in the areas requiring more nourishment and constrict in other areas where blood flow is not as urgent (Figure 14-21). After you eat a heavy meal, blood is shunted from the skeletal muscles to the stomach and intestines to aid in digestion. At this time, you are not inclined to exert yourself and cannot perform physical exercises well.

Changes in the distribution of blood are controlled primarily by *hormones*. Epinephrine, which is released from the adrenal medullas during emergency states, promotes increased blood flow to the skeletal and heart muscles and decreased blood flow to the skin, digestive tract, and kidneys. Other hormones, such as *angiotensin* and *antidiuretic hormones*, produce circulatory effects similar to those of epinephrine. Other hormone-like substances—the *kinins*—act specifically to dilate certain arteries within the body.

Hypotension, or Low Blood Pressure

In some individuals, low blood pressure is normal and is not a symptom of ill health. However, a decrease in blood pressure below the normal level for a particular individual is referred to as *hypotension*—a condition that results from a heart defect or an insufficient volume of blood within the circulatory system. If the heart does not pump blood forcibly because the heart muscles are weak or the heart valves do not close completely, the pressure of the blood leaving the left ventricle will be abnormally low.

A decrease in the volume of blood can be caused by malnutrition, particularly if the individual does not consume enough protein. The concentration of protein molecules in the blood is one determinant of the amount of water in the blood. Water moves into and out of the circulatory system by osmosis from areas of high to areas of low concentration. When the concentration of protein molecules in the blood decreases, the proportion of water in the blood becomes relatively higher; the concentration of water in the blood is then higher than the concentration of water in the interstitial fluid surrounding the capillaries. Consequently, water moves out of the vessels, and the total volume of blood within the circulatory system is reduced. This drop in blood volume results in a decrease in blood pressure.

A decline in blood volume and blood pressure also occurs when blood is lost from an open wound or when a pale yellow fluid called *plasma* (the noncellular part of the blood) leaks from a part of the body that has been burned or crushed. Plasma loss can be as dangerous as blood loss. The amount of blood in the vessels also decreases when an excessive amount of water is lost through perspiration and is not replaced by drinking fluids.

Shock

Shock (usually the result of an injury) occurs when the blood pressure is so low that circulation is sufficiently depressed to interfere with body functions. This condition is extremely dangerous because the blood returns to the heart so slowly that the ventricles do not fill completely before they contract. The pressure developed by the contraction is diminished and the pumping action of the heart eventually fails, resulting in death.

The symptoms of a rapid decrease in blood pressure (shock) are:

1. A faint pulse, because the left ventricle is not receiving enough blood to produce normal pressure.
2. A rapid pulse, because the heart beats more frequently in an attempt to increase pressure.
3. Pale, cool skin, because the arteries in the skin constrict as blood flow is shunted to the more critical organs.
4. Dilated eye pupils—a side effect that is one of the clearest indications of shock.

Most of these symptoms result from the body's attempt to restore normal blood pressure.

First aid for shock includes:

1. Correct the cause of the shock, if possible (for example, control any bleeding).
2. Keep the victim in a prone position.
3. Keep the victim's airway open.
4. Elevate the victim's legs, and keep the head lower than the trunk of the body, if possible.
5. Keep the victim warm.
6. Give the victim fluids (such as water, tea, or coffee) if he or she is able to swallow. Do *not* give the individual alcoholic beverages.

Symptoms of Shock Produced by Emotions

An emotional upset can produce basic symptoms of a rapid drop in blood pressure, except that the pulse is strong rather than weak in the case of *emotional shock.* The hormone epinephrine is released from the adrenal medullas and acts directly on the medulla oblongata in the brain to produce the symptoms. But because no decrease in blood pressure actually occurs, these responses increase blood pressure above the normal level to prepare the body for emergency action—to fight or run away. To keep the lungs, heart, and skeletal muscles well supplied with blood so that the individual can take appropriate action under conditions that produce extreme anger or fright, the heart rate increases and the blood is shunted from the skin and other areas of the body to the skeletal and heart muscles.

If the response to the emotional stress is some physical action, then the increase in blood pressure is only temporary and the individual's normal blood pressure level quickly returns. Due to the social taboos in modern cultures against violent action, however, the physical effects of emotional stress can endure for a long time and may be a major cause of the long-term high blood pressure so prevalent in our society today. If an immediate physical response to stress is impossible, the effects of an emotional upset can be worked off by strenuous physical exercise; running a few miles or playing a set of tennis can return the circulatory system to normal. Apparently, responses of the circulatory system to stress can also be reversed by altered mental states, such as transcendental meditation, although the reduction of stress by employing such mental exercises is not as well understood as the reduction of stress during physical exercise.

Hypertension, or High Blood Pressure

Abnormally high blood pressure, or *hypertension,* develops when the systolic pressure (maximum pressure that occurs when the left ventricle contracts) is greater than 160 mm Hg and the diastolic pressure (minimum pressure that occurs when the left ventricle relaxes) is greater than 95 mm Hg. Like other indicators of body condition, however, whether or not blood pressure is high must be assessed on the basis of the normal blood pressure level for the particular individual. The capacity to increase blood pressure temporarily is adaptive. Long-term high blood pressure is not adaptive and can dangerously strain the entire circulatory system. People with unusually high blood pressure levels are more susceptible to stroke, heart disease, or kidney failure than people with blood pressure levels that fall within the normal range.

Hypertension is actually a complex of diseases. The most common cause of hypertension is the inability of the arteries to dilate in response to the pressure of blood flow when the ventricles contract. *Arteriosclerosis* is the general term for this

Figure 14-22
Advanced Stage of a Plaque Forming Within an Artery
The plaque is covered with muscle cells, some of which are degenerating to form a lesion. The interior of the plaque is filled with fatty debris, which accumulates when muscle cells degenerate and when cholesterol and other fatty materials migrate into the plaque.

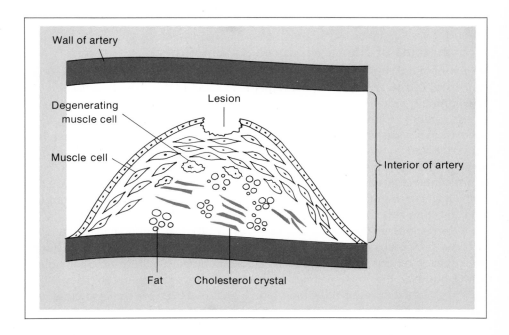

condition. In some cases, the muscles of the arterial walls have lost their elasticity due to lack of exercise; like muscles in other parts of the body, muscles within the arteries must work to remain healthy. In most cases, however, growths or deposits of materials accumulate on the inner lining of the vessels. This "hardening" of the arteries is associated with high blood pressure. One out of every five adult Americans suffers from arteriosclerosis.

Atherosclerosis

The most common condition that prevents the dilation of vessels is the presence of lumpy thicknesses, or *plaques,* on the inner walls of the arteries. The exact cause of this particular form of arteriosclerosis, called *atherosclerosis,* is unknown.

The first thickness that develops on the inner wall of the arteries consists only of muscle cells, which originate from a form of cell division similar to cancer cell formation. As the cells proliferate in the area, some cells die and move inside the plaque. In time, a variety of materials accumulate in the interior of the plaque, including dead cells, fats, and cholesterol crystals, which appear to be attracted to the area (Figure 14-22). This fatty debris suggested the name atherosclerosis from the Greek ("athero" = gruel; "sclerosis" = hardening).

As materials accumulate within the plaque, the vessel becomes smaller in diameter and can no longer fully expand (Figure 14-23). The muscle cells that form the covering of the plaque eventually begin to degenerate, apparently stimulating the development of blood clots that further block the passage of the blood. Finally, calcium is deposited in the area, because it is a part of the normal repair process for lesions. As calcium is added, the arteries begin to resemble corroded pipes. The vessels are unable to expand, and the deposits block normal blood flow entirely.

This process of degenerative change continues until major vessels are completely blocked and the organ nourished by those vessels dies. An area that has died due to vessel blockage is called an *infarct.* When a coronary artery is blocked (a *coronary occlusion*), part or all of the heart dies; this condition is referred to as a *myocardial infarction.* If the affected area is large enough to alter heart function, a heart attack will occur. If one of the vessels to the brain is blocked, the affected region of the brain dies (a *stroke*). For some as yet undetermined reason, methamphetamines ("speed") can also cause strokes. In such cases, tiny vessels in the brain become blocked and the portion of the brain that they nourish dies.

A further danger from atherosclerosis is that the blood pressure may cause the damaged vessel to bulge outward (an *aneurism*) and eventually break. The rupture of a vessel prevents normal blood flow, and the area involved is no longer nourished. Only damaged vessels rupture; healthy vessels never rupture, regardless of blood pressure.

Kidney Diseases

High blood pressure develops not only when vessels do not dilate but also when the kidneys fail to regulate the amount of water retained in the blood, which may result if the kidneys are defective or diseased or if the individual's diet is very high in salt. If the kidneys allow too much water to remain in the blood, the total volume of the blood increases and the blood pressure becomes abnormally high.

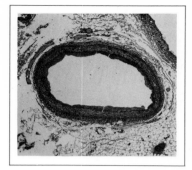

(a)

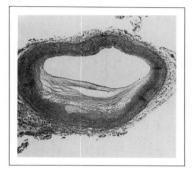

(b)

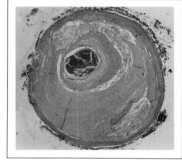

(c)

**Figure 14-23
The Accumulation of Cholesterol and Other Materials on the Inner Wall of an Artery**

(a) Normal artery; (b) an artery partially blocked by deposits; (c) an artery almost completely filled with deposits. A blood clot (dark inner circle) has been trapped within this artery, completely obstructing blood flow.

Preventing High Blood Pressure

Arteriosclerosis, or hardening of the arteries, is the number one cause of death in Americans. It is rare in women before menopause because one of the female hormones, *estrogen,* appears to play a role in preventing the formation of plaques in the arteries. After menopause, however, women are as vulnerable as men to this disease. There also appears to be some genetic predisposition—arteriosclerosis tends to occur more frequently in certain family lines than in others.

The American lifestyle also seems to promote this condition. A particularly disturbing fact is that young men show signs of arteriosclerosis. Autopsies performed on Korean war casualties revealed that over 77% of the Americans, whose average age was 22, already showed symptoms of damaged arteries. Comparable autopsies of Korean soldiers revealed no evidence of such damage. Autopsies on young men killed in Vietnam and in highway accidents showed the same high incidence of diseased arteries among American males.

A number of *risk factors,* or aspects of an individual's body or lifestyle, are correlated with the development of arteriosclerosis. These factors are not necessarily causes of the disease, which remain unknown. Some risk factors are above-normal blood pressure, genetic defects, high-salt diet, cigarette smoking, obesity, emotional stress, diabetes, and a sedentary lifestyle.

For decades, Americans have been advised to maintain low-cholesterol diets, because individuals with high levels of cholesterol in their blood tend to be more susceptible to damaged arteries. The relationship between cholesterol in the diet and in the blood is now less clear than was once believed. The *type* of cholesterol carrier in the blood appears to be more of a factor than the actual amount of cholesterol ingested.

Cholesterol is an animal fat that is carried in the blood in the form of complexes with proteins and other lipids. There are several types of these *lipoprotein complexes,* and the relative amounts of each type seem to vary. Individuals with high concentrations of the low-density lipoproteins (LDL), which are relatively large particles, are more susceptible to artery diseases than individuals with high concentrations of the high-density lipoproteins (HDL), which are relatively small particles. Women in their child-bearing years have higher HDL levels and a lower incidence of arteriosclerosis than men of the same age. It is still too early to tell whether it is possible to alter the relative concentrations of these two forms of carriers and therefore decrease the risk of circulatory problems in men and older women. Recent statistics indicate that middle-aged men who run at least 15 miles a week have significantly lower levels of LDL and higher levels of HDL than sedentary men in the same age category. This observation can be interpreted in two ways. Either men with relatively high HDL become runners, or running has the beneficial effect of raising the HDL level and lowering the LDL level. Long-term studies of men during their transition from sedentary to physically active lifestyles must be conducted to determine whether exercise increases HDL and prevents arteriosclerosis.

Most of the risk factors that have been identified can be altered by a change in lifestyle. The likelihood of arteriosclerosis may be reduced by decreasing the salt content in the diet, maintaining an appropriate weight, not smoking, reducing emotional stress, and increasing physical exercise. As more is learned about the actual causes of the disease, more precise measures to prevent arteriosclerosis will be developed.

Summary

Large multicellular organisms must have some means of transporting materials between their cells and the external environment. Plants have two transport systems: the *xylem*, which transports water and minerals, and the *phloem*, which transports sugars, hormones, and other materials. Most animals have some sort of transport fluid that carries materials through a network of vessels within the body.

Vertebrate animals have two networks of vessels and transport fluids. The primary network is composed of blood vessels, and the transport fluid is *blood.* The secondary network is composed of lymphatic vessels, and the transport fluid is *lymph.* The lymphatic system returns certain materials in the interstitial fluid to the blood.

Vertebrate animals have a *closed circulatory system* of blood vessels and a single heart that pumps blood through the *arteries* to a network of *capillaries*, where materials are exchanged between the blood and the *interstitial fluid* or the external environment. Blood then moves from the capillaries into the veins and back to the heart.

The body contains about 5 liters of blood. The blood of a vertebrate animal transports oxygen, carbon dioxide, food molecules, cellular wastes, heat, cells and molecules of the immune system, materials that repair vessel damage, and hormones. Oxygen is transported in vertebrates by a special carrier molecule in the blood called *hemoglobin*—a protein inside the red blood cells. These special cells are short-lived and must be continuously re-formed. The noncellular portion of the blood is called *plasma.*

The blood contains more than 35 different materials that form *blood clots* during the repair of vessel damage (*coagulation* or *clotting*). An inability to produce any one of these materials results in an excessive loss of blood from injured vessels. *Hemophilia*—a genetic defect that prevents the adequate repair of vessels—is the most common disease among *bleeders.*

The simplest vertebrate heart, found in the fishes, is composed of only one atrium and only one ventricle. Amphibian hearts have two atria and one ventricle. Reptile hearts are similar to amphibian hearts, except that the ventricle is partially divided to form two separate chambers. Birds and mammals have four-chambered hearts that contain two atria and two ventricles.

In a mammal, blood enters the right atrium and moves into the right ventricle, from which it travels to the lungs to pick up oxygen and to release carbon dioxide. The blood then moves from the lungs back to the heart and enters the left atrium. The blood from the left atrium moves into the left ventricle and then leaves the heart via the *aorta.* This large artery branches into smaller arteries, which transport blood to all parts of the body except the lungs. Blood is pumped to the lungs via the *pulmonary artery* and returned to the heart via the *pulmonary veins.* Four valves within the heart prevent the backflow of blood—one *atrioventricular valve* between each atrium and each ventricle, and one *ventricular artery valve* between each ventricle and the artery that leads from it.

The heart muscles themselves are nourished with blood by the *coronary arteries*—branches of the aorta that are connected to a network of capillaries surrounding the outside of the heart. The capillaries are connected to the *coronary veins*, which return blood to the right atrium of the heart. Defects in the coronary arteries cause heart attacks.

The *pulse rate*, or the number of heart contractions per minute, is usually measured by counting the pulsations of blood flow within an artery in the wrist. The heart muscle contractions that pump the blood are primarily controlled by the *pacemaker*—a specialized region of the right atrium that regularly sends out electrical impulses to the heart muscles. These signals are relayed from the atria to the ventricles via the *atrial ventricular node*. A slight delay in each signal causes the two atria to contract just ahead of the two ventricles. The heart rate is also controlled by an area within the *medulla oblongata* of the brain, which increases or decreases the heart rate, depending on the information conveyed by the heart, the brain, or other areas of the body. The heart rate is also affected by hormonal secretions from the *endocrine glands*. Both *thyroxine* (secreted by the *thyroid gland*) and *epinephrine* (secreted by the *adrenal medullas*) increase the rate of heart contractions.

Blood pressure is the force exerted by the blood on the walls of the vessels. This pressure is highest when the blood leaves the left ventricle and diminishes as blood travels through the arteries to the capillaries, where the pressure is much lower. *Systolic pressure* is the maximum force exerted by the blood as the left ventricle contracts; *diastolic pressure* is the minimum force exerted by the blood as the left ventricle relaxes. The elasticity of the arteries diminishes the pulsation of blood flow by dilating during systolic pressure and constricting during diastolic pressure. The blood is returned from the capillaries through veins to the heart by contractions of the surrounding *skeletal muscles*. *One-way valves* ensure that blood moves back to the heart in one direction.

The *cardiac output*—the quantity of blood that the heart pumps per unit of time—depends on the rate of the heart muscle contractions and on the volume of blood moved during each contraction. At rest, cardiac output in a human is about 5 liters of blood per minute. Cardiac output is increased during physical activity and certain types of emotional stress; at such times, the heart rate increases and the arteries constrict, raising the blood pressure. Blood flow is regulated primarily by the carbon dioxide levels within the blood, which are detected by the medulla oblongata, and by the brain centers controlling the emotions. When the body is at rest, blood pressure is regulated by continuous feedback relayed from *pressoreceptors* in the arterial walls to the *vasomotor centers* in the medulla oblongata.

Blood is shunted to the areas of the body that require more nourishment via the arteries, which dilate in areas where the need for blood is greater and contrict elsewhere. Most changes in blood distribution are controlled by *hormones*..

Hypotension, or unusually low blood pressure, may be caused by heart disorders, malnutrition, open wounds, or excessive perspiration. The symptoms of very low blood pressure, or *shock*, are a faint pulse, a rapid pulse, cool skin, and dilated pupils. *Emotional shock* produces similar symptoms, except that blood pressure is raised above the normal level, so that the pulse is stronger than usual.

Chronic *hypertension*, or long-term high blood pressure, usually results from *arteriosclerosis*, or the inability of the arteries to dilate. *Atherosclerosis*, a special form of arteriosclerosis, results when *plaques* develop on the inner walls of the arteries. In the later stages of plaque formation, the artery is blocked by *blood clots* and calcium is deposited around the plaque. Blood flow is obstructed, and the affected area eventually dies. The cause of this disease is not understood, but it may be prevented by decreasing the amount of salt in the diet, maintaining an appropriate weight, not smoking, avoiding emotional stress, and increasing physical exercise.

Respiration: Gas Exchange Between Organisms and Their Environments

15

Cells require a continuous input of energy to maintain their living state. In most organisms, this energy is released when glucose or similar molecules are broken apart in the presence of oxygen. Energy-yielding biological molecules of fat, glycogen, and starch can be stored within an organism and then used to support cellular activities when a plant is unable to undergo photosynthesis or an animal cannot find food. A human with moderate amounts of fat and muscle can live for weeks without eating. Oxygen, by contrast, cannot be stored; animals cannot survive without it for more than a few minutes. The skeletal muscles and heart can temporarily function without oxygen, but food is of no value to certain critical organs, such as the brain and the liver, unless a continuous supply of oxygen is provided.

To maintain life, cells—especially animal cells—must also expel carbon dioxide, a waste product of the breakdown of food molecules. Carbon dioxide combines with water to form carbonic acid, which produces too acidic an environment to support the chemical reactions of cells. To prevent this fatal interference with cell chemistry, carbon dioxide must be removed from the cells of an organism soon after it is formed.

The exchange of oxygen and carbon dioxide between the cells of an organism and its external environment is called *respiration,* or *external respiration.* It must take place continuously if an animal is to survive. This form of respiration should not be confused with *cellular respiration* (the process of breaking down glucose and similar molecules into carbon dioxide, water, and energy that occurs within the mitochondria of cells).

The Exchange of Gases Across Cell Membranes

Carbon dioxide and oxygen pass between cells and their immediate environment by *diffusion*—the movement of molecules from areas of high to areas of low concentration. Gases must be dissolved in water to move across a cell membrane; movement between the interior and the exterior of the cell then occurs freely. Because oxygen is continually used to break down food molecules, it is less concentrated within the cell than it is outside the cell. Molecules of oxygen therefore diffuse *inward,* moving into the cells at about the same rate as they are used as long as the area surrounding the cell is regularly replenished with oxygen. Similarly, because carbon dioxide is formed from the breakdown of food molecules, it is more highly concentrated inside the cell than outside the cell. If this gas is regularly removed from the area surrounding the cell, carbon dioxide tends to diffuse *outward* from the cell.

Gas exchange between the body of an organism and its external environment may occur over the entire outer body surface or over only portions of that surface. The particular arrangement of cells that form a gas exchange surface and the position of these cells on the body vary greatly among different types of organisms. However, any surface that permits the diffusion of gases must meet four basic requirements:

1. The surface area must be relatively large; more diffusion can be achieved over a greater area. Large surface areas usually contain extensive tissue projections and foldings to conserve space.

2. The surface must be moist at all times because oxygen and carbon dioxide must be dissolved in water to diffuse across a cell membrane. Aquatic organisms easily meet this requirement because they are immersed in water. However, it is difficult for organisms living on land to maintain a moist surface because the water tends to evaporate. A number of adaptations in terrestrial organisms minimize but can never totally prevent water loss during respiration.

3. A surface for gas exchange must be close to the active cells of the body or to the fluid that transports gases between these cells and the surface. Diffusion alone cannot carry gases very deeply into the body of an organism.

4. Air or water must be moved over the gas exchange surface. If the environment adjacent to the surface remained motionless, the concentration of oxygen would soon become too low and the concentration of carbon dioxide would soon become too high for diffusion to continue in that area. The body of the organism must move through the environment, or the environment must be moved across the gas exchange surface. The various methods of moving water or air across a gas exchange surface are called *ventilation mechanisms.*

Special Gas Exchange Surfaces of the Body

The cell membranes of unicellular and very small multicellular organisms come into direct contact with the external environment. As long as this environment is replenished frequently by the movement of the organism or its environment, no special gas exchange structures are needed. Larger organisms, however, require special surfaces and/or transport fluids to exchange gases between their cells and the external environment.

The Leaves of Plants

The special surface for gas exchange in most plants is the *leaf,* although gas exchange occurs to some extent in other plant tissues as well. Two important processes involving gases occur in plants. During *photosynthesis,* carbon dioxide is used in the formation of sugar; oxygen is a waste product of this reaction. During *cellular respiration,* oxygen combines with sugar to form carbon dioxide and water and to release the chemical energy in the sugar molecule; carbon dioxide is a waste product of this reaction. At any given time, therefore, the plant may need to acquire oxygen and remove carbon dioxide, or vice versa.

Gases move into and out of a leaf primarily through small openings called *stomata* (see Figure 14-1, p. 356). Most plants have numerous small or flat leaves to provide large surface areas for gas exchange. The interior structure of a leaf consists of cells surrounded by empty, highly humid spaces that cover the cell membranes with a thin film of water to permit gas diffusion. All cells that require high rates of gas exchange are near the stomata on the surface of the leaf. Plants contain no transport fluid for gases. On the outside of the leaves, gases are replenished by the normal movement of the air.

Regular exchange of gases also occurs across the cell membranes of the *root hairs.* Oxygen dissolved in soil moisture diffuses into the root, where it is used in cellular respiration, and carbon dioxide diffuses out of the root and into the soil. Small pores in the outer surface of the stem, consisting of loosely packed cells, also permit gas exchange between the interior of the stem and the atmosphere.

The Skin of Animals

The entire outer surface of the bodies of some animals—most commonly, aquatic invertebrates—is used for the intake of oxygen and the release of carbon dioxide. In small or very flat animals, the movement of gases by diffusion alone is sufficient to maintain cellular activities, because the distance between the external body surface and all of the cells is very short. In larger, more three-dimensional animals, a *circulatory system* ensures that the immediate environment around each cell is high in oxygen and low in carbon dioxide by transporting these gases between the cells and the outer covering, or *skin,* of the animal, where gas exchange can take place.

Two major disadvantages of using the entire body surface for gas exchange are that the skin must be kept moist and that it must be composed of living cells with exposed membranes. Aquatic animals can exchange gases through their entire body surface, but only such terrestrial animals as frogs and some soil-dwelling worms, which live in highly humid environments, are able to respire through their skins. Outer body surfaces that consist of living cells are very fragile and subject to injury. Moreover, when a circulatory system transports gases to and from interior body cells, the gas exchange surface must be abundantly supplied with blood vessels, so that such an outer covering is also vulnerable to excessive blood loss if an injury occurs.

To offset the dangers of moisture loss and injury, the outer surfaces of most animals are modified to carry out functions other than gas exchange. The external covering of an animal is usually composed of dead cells or secretions that protect living cells against predators and invasions by microorganisms. Many animals also have additional structures, such as scales, hair, or feathers, to control water balance and to regulate body temperature.

Gills

Many aquatic organisms exchange gases with a water environment via *gills*—specialized structures that usually project outward from the body (Figure 15–1). The simplest gills are found in some marine worms. Although most worms exchange gases through their entire body surface, some worms protect their delicate bodies from harm by secreting a hard tube around themselves or by living in burrows. Gases cannot then be exchanged across the worm's body surface at a sufficient rate to maintain cellular activities. Because these worms are unable to respire through their skins, gas exchange is accomplished by long filaments of soft gill tissue that extend outward from the tube or burrow into the surrounding water. To

Figure 15-1
The External Gas Exchange Surfaces of a Salamander
The gills of this animal are composed of many projections that provide a large surface area for the diffusion of oxygen and carbon dioxide between the blood and the external environment. A salamander also exchanges gases through its skin.

Gill

protect this soft gill tissue, these worms can retract their gills into their tubes or burrows whenever the water is turbulent.

The gills of some crustaceans, such as fairy and brine shrimp, are finely divided, thin filaments of soft tissue located on each of the animal's many pairs of legs. These tissues float outward from the body and are ventilated as the organism moves through the water. Gills that extend loosely outward from the body are found only on fairly sedentary animals that live in sheltered waters, because they can be easily damaged.

The soft gill tissues of most larger, more active aquatic animals are housed inside a protective flap or cavity across which water is actively pumped. These structures also provide the animal with a more streamlined shape. The gills of some animals, such as clams and snails, that live in turbulent waters are located inside the hard outer shell of their bodies (Figure 15–2). Water is circulated across the gills by the beating of cilia (millions of small, hairlike structures) or by the pumping movements of muscular structures.

The gills of fishes are located on the edges of openings between the *pharynx* (part of the digestive tract just behind the mouth) and the outside of the body. Four or five of these channels, or *gill slits,* usually occur on each side of the pharynx. Primitive fishes used their gill slits to strain food particles out of the water. Later, these gill slits evolved into gas exchange structures when fishes developed protective coverings on their body surfaces and their skin became unsuitable for gas exchange.

Fish gills are covered with thin, delicate *gill filaments* across which oxygen and carbon dioxide diffuse between the organism's blood and the aquatic environment. The total surface area of the filaments is proportional to the activities of the fish and, therefore, to its demand for oxygen. The surface area for gas exchange in the mackerel, a fast swimmer, is about five times as large as the gas exchange surfaces of slow swimmers like the toadfish.

The gills are ventilated when the fish opens its mouth cavity and sucks in water.

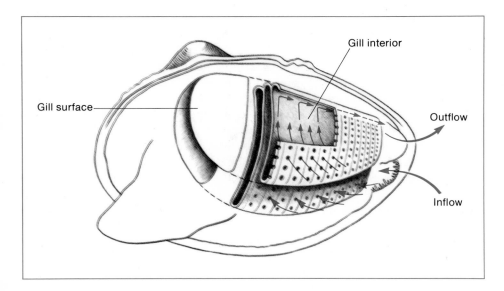

Figure 15-2
The Internal Gas Exchange Surfaces of a Clam

The gills of this aquatic invertebrate are located inside the shell, where they are protected from injury. Parts of the shell and the gills have been cut away here to reveal the pathways of water across the gills.

Figure 15-3
The Flow of Water Across the Gills of a Fish
Water enters through the mouth, flows over the *gill filaments* (where gas exchange occurs), and moves out through the *opercular opening.*

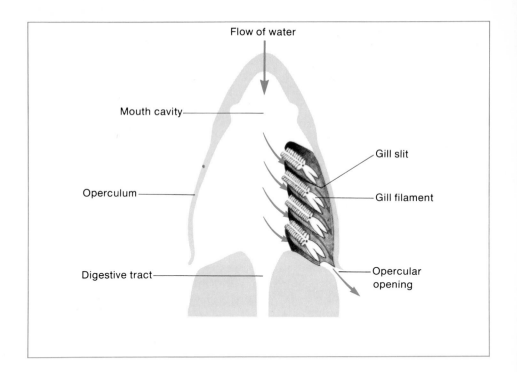

The mouth then closes, forcing water backward through the gill slits. In most fishes, the fragile gill filaments are protected by a stiff covering, the *operculum,* which has an opening to allow water to flow outward (Figure 15–3). Movements of the operculum also help to pump water across the gills. Some fast swimmers, such as the tuna, have very little musculature in their mouths and opercula and are unable to ventilate their gills by pumping movements alone. These fishes swim with their mouths and opercula open, so that a steady stream of water can flow over their gills. Because the need for rapid gas exchange is continuous, these fishes can never stop swimming or they will suffocate.

In the gills of fishes and some mollusks, blood is carried close to the gill surface and flows in the *opposite direction* of the water. This *countercurrent flow* permits gases to be exchanged between blood and water more efficiently (Figure 15–4). Water with a high oxygen content enters the gill and meets blood that has already passed through most of the gill and is almost saturated with oxygen. As the water continues to pass over the gill and lose oxygen, it comes into contact with blood that contains less and less oxygen. At every point on a gill, the oxygen concentration in the water is therefore considerably greater than the oxygen concentration in the blood—a condition that increases the amount of oxygen that diffuses into the blood. Under experimental conditions, scientists have used mechanical pumps to reverse the water flow over the gills of a fish, so that both blood and water moved in a *parallel flow* (in the same direction) rather than in a countercurrent flow. The fish extracted only about 10% of the available oxygen from the water. Under the normal conditions of countercurrent flow, the fish extracted about 80% of the oxygen from the water.

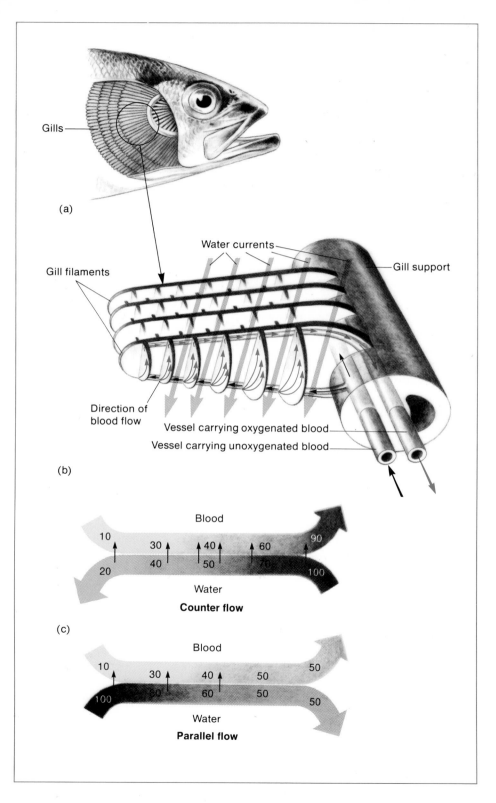

Figure 15-4
Countercurrent Exchange of Oxygen Across the Gills of a Fish
(a) Position of the gills on a fish. The operculum has been cut away here. (b) Greatly enlarged section of the gill, showing the direction of blood flow and water flow through the gill filaments. (c) Diagram of the movement of water and blood in a *countercurrent flow* and a *parallel flow.* The numbers represent relative amounts of oxygen. During countercurrent flow, the amount of oxygen in the water is greater than the amount of oxygen in the blood during the entire period of contact; during parallel flow, the oxygen concentration is greater for only part of the period of contact.

Gas Exchange Surfaces in Terrestrial Animals

Air contains up to 40 times more oxygen than an equal volume of water does. Moreover, oxygen diffuses 300,000 times faster in air than it does in water, although a thin film of water must always be present for oxygen to diffuse through membranes. This tremendous difference in the availability of oxygen may account for the fact that birds and mammals—the animals requiring the greatest rates of gas exchange—evolved on land.

Air is also much lighter than water. Because water is heavy, ventilation in most aquatic animals is a *one-way flow;* water enters through one opening, flows in a single direction over the gills, and moves out through a second opening without changing direction. Ventilation in most land animals is a *two-way flow*—in and out the same opening. The direction of air flow is much easier to change than the direction of water flow.

Very few terrestrial animals have gills, because external gas exchange surfaces are difficult to keep moist in the air. Moreover, soft gill tissues function best when they are supported by water; in air, they tend to collapse and stick together so that their surfaces are not fully exposed to the air. The gas exchange surfaces of most terrestrial animals are located deep within the body—an adaptation that reduces the evaporation of water and protects the delicate tissues from rapid changes in temperature and from injury. Gas exchange surfaces that are folded inward into the body to form hollow cavities are *lungs* or *tracheae.*

Apparently, gas exchange surfaces evolved independently in several different ancestral lines of terrestrial invertebrates, because modern forms of these animals have many different types of structures for breathing air. In contrast, all vertebrates have the same basic type of gas exchange surface, which implies that air-breathing structures evolved only once in the early ancestors of this major group of animals.

The Lungs and Tracheae of Land Invertebrates

Most of the small invertebrate animals that live on land have *diffusion lungs* (Figure 15–5)—hollow body cavities with moist inner surfaces that are folded to increase the total gas exchange surface. These lungs are sometimes called *book lungs,* because their folds resemble the pages in a book. Gases are exchanged between the air in the lung cavities and the circulating blood in the vessels lining the cavity walls. Oxygen and carbon dioxide are transported to and from cells by the blood that moves through these vessels.

Water loss from diffusion lungs is reduced by the presence of very small openings called *spiracles* that connect the lung cavities with the air in the external environment. The spiracles are opened and closed by valves that are operated by muscles. These openings—and therefore gas exchange—are regulated by carbon dioxide levels in the blood. When cellular activity is high, a high carbon dioxide concentration in the blood is detected by the animal's central nervous system, which sends electrochemical impulses along efferent nerves to the spiracles, causing them to open. When cellular activity is low, a low concentration of carbon dioxide in the blood signals the central nervous system to send impulses that close the spiracles. Spiders and terrestrial snails are among the many small animals equipped with simple diffusion lungs.

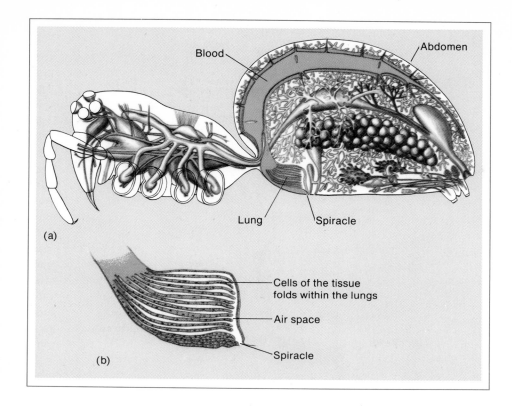

Figure 15-5
The Diffusion Lung
of a Spider

(a) Interior structures of the spider, showing the major blood vessels and the position of the lung on the ventral side of the abdomen. (b) Enlarged drawing of the lung, showing the many folds of the interior surface. Most spiders have at least one pair of lungs, with a single opening, or *spiracle*, to the outside of the body. The spiracle opens and closes in response to carbon dioxide levels in the blood.

Insects and a few other arthropods have a highly specialized type of gas exchange system—a network of hollow, fairly rigid tubes, or tracheae, extending throughout the body (Figure 15–6). A thin layer of water at the ends of the tracheae permits the diffusion of gases. The tracheal system in these arthropods pipes air directly to the fluid surrounding the cells; blood does not transport oxygen and

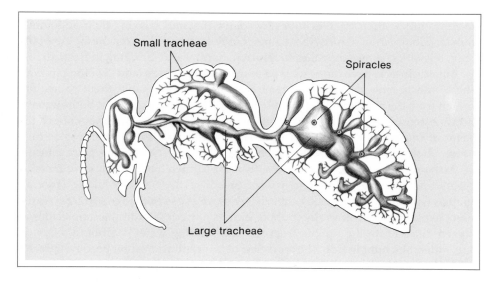

Figure 15-6
The Tracheal Gas Exchange System of a Bee

Tracheal tubes extend throughout the body, allowing gases to be exchanged directly between the air and the interstitial fluid around the cells. The smallest tracheae cannot be shown at this scale.

Figure 15-7
The Tracheal Gas Exchange
System of a Grasshopper
(a) Position of the spiracles on the
outside of the body. (b) Structure of a
spiracle and the connected trachea.

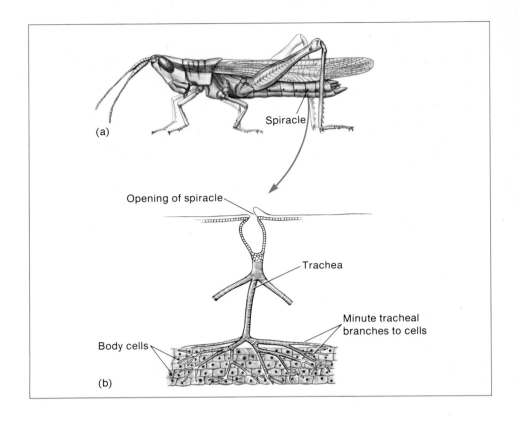

carbon dioxide in these animals. Like diffusion lungs, the tracheae are connected to the outside of the body by small spiracles located on several of the body segments (Figure 15–7).

In smaller, less active insects, the diffusion of gases between the tracheae and the interstitial fluid is sufficient to maintain cellular activities. Larger, more active insects ventilate their tracheal tubes by contracting and relaxing their abdominal muscles. Contraction reduces the diameter of the tubes and forces the air out of the body; relaxation allows the tubes to return to normal size, bringing in fresh air.

Aquatic insects have the same type of tracheal system as land-dwelling insects. These insects have developed ingenious methods that enable them to acquire oxygen from the air or from aquatic plants while remaining in water. Some aquatic insects extend a tube to the surface of the water (Figure 15–8); others pierce the leaves of underwater plants with their spines and draw the oxygen gas formed during photosynthesis into spiracles on their spine tips. One of the more unusual gas exchange surfaces is found in the "water boatmen" and "back swimmers"— insects that are able to live under water much of the time by taking large air bubbles with them. The bubbles themselves serve as gas exchange surfaces. As the insect extracts oxygen from the bubble into its spiracles, additional molecules of oxygen diffuse into the bubble from the surrounding water. Carbon dioxide, in turn, enters the bubble from the spiracles; its concentration soon becomes greater inside the bubble than in the surrounding water and it diffuses out.

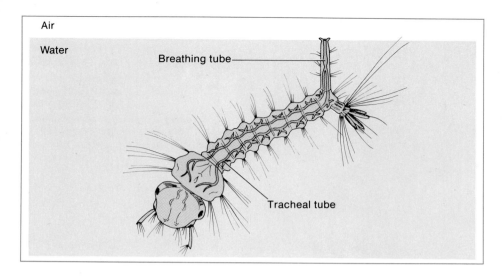

Air

Water

Breathing tube

Tracheal tube

Figure 15-8
Gas Exchange in
an Aquatic Insect
Although this immature stage of an
insect lives and feeds underwater, it
breathes air through a breathing tube
extended to the surface of the water.

The Lungs of Land Vertebrates

The lungs of amphibians, reptiles, birds, and mammals are paired sacs that lie
within the chest cavity. Oxygen taken into the lungs dissolves in a film of moisture
on the lung surface and then diffuses through the thin lung membrane into small
blood capillaries. Once it is in the blood, oxygen is picked up by red blood cells and
carried throughout the body. Carbon dioxide in the blood diffuses from the capillaries to the lung and is then breathed out as a gas.

The lungs of amphibians are simple sacs with no elaborate foldings or subdivisions (Figure 15–9). The lungs are composed of elastic tissues—a common
characteristic of the lungs of all vertebrates. Amphibians fill their lungs by opening
the nostrils and expanding the volume of the mouth cavity, causing air to be sucked
into the mouth. The nostrils then close, the mouth cavity is reduced, and air is
forced into the lungs, causing them to expand. Amphibians exhale by compressing
the body wall while the nostrils are open. As the body cavity becomes smaller, air is
forced out and the lungs contract due to their elasticity.

The lungs of reptiles have more subdivisions and therefore a greater surface
area than the lungs of amphibians. Reptiles can also ventilate their lungs more
effectively than amphibians. The lungs of a reptile are surrounded by and attached
to the wall of the chest, which is supported by many ribs (Figure 15–10). When the
ribs are moved up and apart by muscle contractions, the volume of the chest and
lungs increases, creating a partial vacuum within the lungs, which immediately fill
with fresh air. As the ribs are lowered and brought closer together by muscle
contractions, the volume of the chest decreases, air is forced out, and the lungs
contract by elastic recoil. Air moves in a two-way flow in and out through the
nostrils.

The arrangement of the lungs in birds differs from the position of the lungs in
other vertebrates. Bird lungs are quite small and are directly connected to numerous air sacs, which extend throughout the body—even into the bones (Figure
15–11). Gas exchange occurs only in the lungs, not in the air sacs. This arrangement permits the air to circulate throughout the body rather than back and forth in
a two-way flow, as it does in other vertebrates. In birds, fresh air travels directly to

Figure 15-9
The Lungs of an Amphibian

(a) The salamander is one form of amphibian. (b) Salamander lungs that have been removed from the body and inflated with air. (c) An X-ray photograph of the salamander, showing the position of the lungs in the body.

(a)   (b)   (c)

the *posterior sacs,* through the lungs, into the *anterior sacs,* and out again, so that air that has already given up much of its oxygen and has received carbon dioxide from the blood does not mix with the fresh air in the lungs.

Like mammals, birds must exchange gases rapidly, because their cells must be maintained at a more active level than the cells of other animals. However, mammals—even flying mammals such as bats—function well without the unusually efficient lungs found in birds. The advantage of such efficient lungs is that they enable birds to fly at high altitudes, where less oxygen is available. The air sacs of a bird's respiratory system also fill a large part of the body cavity, substituting air for tissue or fluid in many places and helping to reduce body weight.

Figure 15-10
Ventilation of the Lungs in a Reptile

Movement of the ribs alters the volume of the chest cavity. As the chest cavity expands, the volume of the lungs also increases, drawing in air. As the chest cavity contracts, the volume of the lungs also decreases, forcing out air.

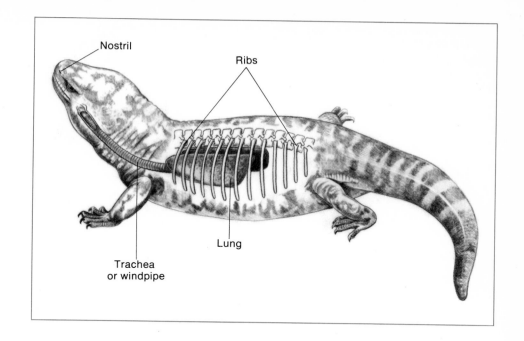

Figure 15-11
Ventilation in a Bird

In birds, air moves through the lungs only as it passes out of the body. In other vertebrate animals, air moves in both directions within the lungs, and does not circulate throughout the rest of the body. (a) Air moving into the bird bypasses the lungs and flows directly into the *posterior sacs* (P). (b) Air moving out of the bird flows through the lungs and the *anterior sacs* (A). Because air flows through the lungs only once, it moves in a single direction. The blood moves in the opposite direction, permitting a countercurrent exchange to take place, and gases are exchanged only across the lung tissues.

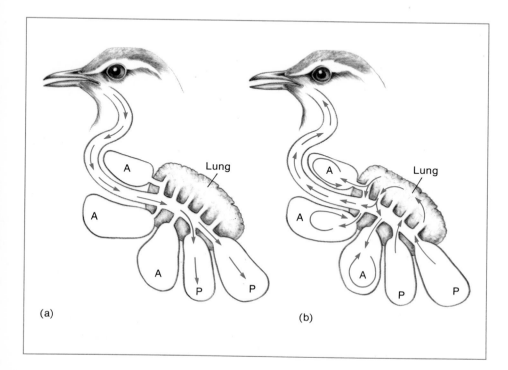

Figure 15-12
The Human
Respiratory System

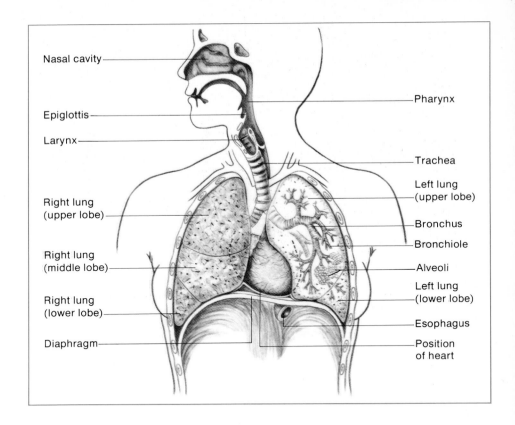

Nasal cavity

Epiglottis

Larynx

Right lung
(upper lobe)

Right lung
(middle lobe)

Right lung
(lower lobe)

Diaphragm

Pharynx

Trachea

Left lung
(upper lobe)

Bronchus

Bronchiole

Alveoli

Left lung
(lower lobe)

Esophagus

Position
of heart

Gas Exchange in Mammals: The Human

The human lungs are two separate hollow cavities composed of elastic, spongelike tissues located on the right and left sides of the body. The right lung is partitioned into three sections, called *lobes;* the left lung, into two lobes (Figure 15–12). The lungs occupy most of the chest cavity and weigh about $2\frac{1}{2}$ pounds, or about 1 kilogram.

The tissues of the lungs are arranged in cup-shaped depressions about 0.2 mm in diameter, called *alveoli* (Figure 15–13). Each alveolus is closely associated with a capillary, through which oxygen and carbon dioxide are exchanged between the blood and the air within the lung cavity. The arrangement of lung tissue into alveoli gives human lungs an enormous gas exchange surface. A pair of human lungs contains about 300 million alveoli, and the total area of these alveoli is about 800 square feet (74 square meters)—the size of a tennis court!

The inner surface of the lung, which is in contact with air, is coated with a thin film of liquid containing a special chemical, a *surfactant,* which prevents the alveoli from sticking together. Due to the property of liquids called *surface tension,* two large surfaces tend to adhere to each other whenever they are covered with a thin layer of moisture, just as two flat pieces of wet glass stick tightly together. A surfactant works like a detergent, decreasing the surface tension of the alveolar liquid.

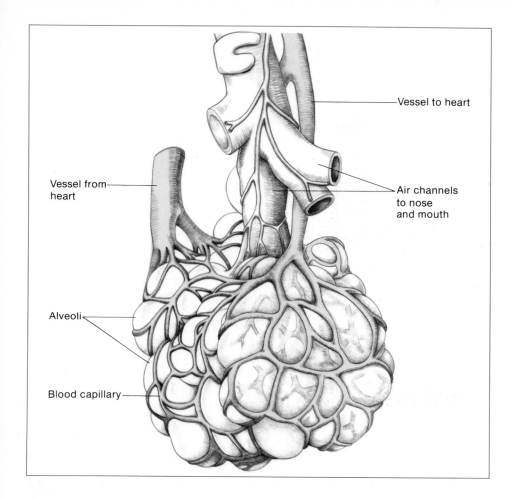

Vessel to heart

Vessel from heart

Air channels to nose and mouth

Alveoli

Blood capillary

Figure 15-13
The Structure of the Human Lung
Section of a human lung, revealing numerous alveoli and the network of blood vessels within the alveoli. Gases diffuse between the interior of the alveoli and the blood within the capillaries.

Occasionally, infants are born without this chemical, a condition called *hyaline membrane disease.* These babies are unable to inflate their lungs completely and usually die shortly after birth.

Lungs are delicate tissues that are vulnerable to injury and disease. Their location deep within the body protects them from most types of damage, but irritating materials and disease organisms are sometimes carried into the lungs with the air during breathing. The incidence of lung disorders has increased as the air has become more polluted in recent years.

Some materials from polluted air collect on the surface of the lungs and form a barrier that impedes the movement of gases. Soot and certain ingredients of tobacco smoke, for example, adhere to the moist surfaces of the alveoli, forming a layer of particles through which oxygen and carbon dioxide cannot diffuse. Other materials in polluted air have a chemical effect on the lungs. *Ozone*, a component of photochemical smog, irritates the alveoli and causes liquid to accumulate in the lungs—a condition similar to drowning. *Nitrogen dioxide,* another chemical component of smog, destroys the surfactant within the fluid of the alveoli, so that they adhere to one another, reducing the gas exchange surface. Many other chemicals in the air irritate the fragile tissues of the alveoli. Continued irritation of the lung tissues often leads to lung cancer—a major cause of death in modern societies.

Pneumonia is a bacterial infection within the lungs. *Simple pneumonia* involves only one lung; *double pneumonia* involves both lungs. As part of its defense against bacteria, the body releases a great quantity of fluid. In the case of pneumonia, this excess fluid accumulates in the lungs, preventing adequate gas exchange and often resulting in death. Pneumonia used to be a common illness, but can now be checked quickly by the administration of *antibiotics*—medicines that act specifically to kill bacteria.

Ventilation

Air enters and leaves the lungs due to changes in air pressure within the lung cavities. The lungs themselves are passive, nonmuscular tissues that adhere to the inner wall of the chest. When the volume of the chest cavity increases, the volume of the lungs also increases. Air then rushes in through the nose or mouth to fill the vacuum that has been created. The lungs and chest do not expand because the air enters; air enters because the lungs and chest expand. When the volume of the chest cavity decreases, air is expelled through the mouth or nose and the lungs contract due to their elasticity.

The contractions of two sets of muscles cause the chest cavity to expand (Figure 15–14). The contraction of one set raises the ribs and moves them outward. The contraction of the other set changes the position of the *diaphragm*—a muscular sheet of tissue that separates the chest cavity from the abdomen. At rest, the diaphragm

Figure 15-14
Breathing Movements in the Human

During *exhalation*, the muscles of the ribs relax, causing them to move downward and inward, and the muscles of the diaphragm also relax, causing it to move upward. These movements decrease the volume of the chest cavity. During *inhalation*, muscular contractions move the ribs upward and outward and pull the diaphragm downward, expanding the chest cavity.

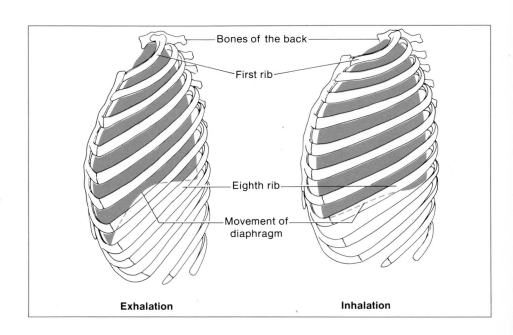

Bones of the back

First rib

Eighth rib

Movement of diaphragm

Exhalation

Inhalation

is a dome-shaped muscle that extends upward into the chest cavity (see Figure 15–12). When this muscle contracts, the dome becomes flatter and lower and the chest cavity becomes larger. The diaphragm is found only in mammals and is a distinguishing characteristic of this group of vertebrates.

Inhalation—the entry of air into the lungs—is the active part of breathing, because it requires energy for muscle contraction. During *exhalation,* or breathing out, the muscles of the ribs and diaphragm relax, causing the ribs to drop downward and inward and the diaphragm to move up to its resting position. Exhalation is a passive activity; the natural elasticity of the lungs returns them to their smaller, resting state.

However, exhalation is not a passive process during vigorous exercise. Air is then forced out of the lungs by strong contractions of the abdominal muscles, which push the diaphragm upward and quickly reduce the volume of the chest cavity. During extremely vigorous exercise, even the neck muscles help to expel air from the lungs.

The lungs expand as the chest does, because they adhere to the chest wall by means of two thin, moist, separate membranes—the *pleura.* One membrane of the pleura covers the inner wall of the chest; the other covers the outer surface of the lungs. Surface tension between the moist surfaces of these two membranes holds them tightly together. The moisture also allows the membranes to slide smoothly over each other without creating a great deal of friction. If these membranes become infected and swollen—a condition known as *pleurisy*—each breath is extremely painful as the inflamed and swollen membranes move across each other.

Surface tension can hold the two pleural membranes together only when they are very close to each other. If air is admitted between the two membranes, as it might be in the case of a wound in the chest or lung, the tension will be broken and the lung will collapse due to its own elasticity, making it useless as a gas exchange surface. After the wound closes, the air between the lung and the chest cavity will be absorbed by the body and the lungs will slowly expand again.

Air may be deliberately introduced between the two pleural membranes to allow a lung to rest, especially after lung surgery or after one lung becomes diseased. Collapsing one lung for a period of time allows it to heal without being continually irritated by the motions of breathing. If the other lung is healthy, an individual can function well with only one lung.

During normal breathing, only about 5% of the total capacity of the lungs is involved in gas exchange. If the abdomen is extended (as it is after a large meal or during pregnancy), so that the diaphragm cannot be fully lowered, total lung capacity may be even further reduced. Girdles and tight belts also prevent normal movement of the diaphragm.

Regulation of Breathing

A resting human inhales and exhales regularly 13–18 times per minute. This rhythm of breathing originates in nerve cells that are localized in two special *respiratory centers* on either side of the medulla oblongata of the brain. These centers are part of the *autonomic nervous system.* Breathing is an *involuntary action*—we do not have to remember to inhale and exhale.

Impulses transmitted to the muscles of the ribs and the diaphragm from the respiratory centers via efferent nerves direct the expansion and constriction of the chest cavity. The depth of quiet breathing is governed by *stretch receptors* located in the lung tissues. The stimulation of these receptors by the expansion of the lungs during inhalation sends impulses to the respiratory centers, which cause inhalation to stop. During quiet breathing, exhalation automatically follows the cessation of inhalation; no special stimulus is needed.

In addition to establishing a basic rhythm of exhalation and inhalation, the respiratory centers adjust the breathing rate in response to bodily reqirements for gas exchange. Breathing slows down during sleep and increases greatly during vigorous exercise, fever, and emotional upset. Breathing can be altered by inputs to the respiratory centers from a variety of chemical and neural sources. Within normal ranges of physical activity, the respiratory centers respond to the concentration of carbon dioxide in the blood. When cells become more active, they break down more sugar. Because carbon dioxide is a waste product of this reaction, it is a sensitive indicator of the body's activity level and, consequently, of the oxygen requirements of the cells. An increase in carbon dioxide in the blood causes an increase in the breathing rate, and vice versa.

Other neural mechanisms are activated during very labored breathing. When the lungs are in a deflated condition, *deflation receptors* in the lungs transmit impulses to the respiratory centers, which initiate a rapid and forced *inspiration* (a gasp for breath). An area within the *pons* of the *hindbrain* also communicates information to the respiratory centers during extremely labored breathing.

Because breathing is controlled by the autonomic nervous system, we cannot hold our breath for very long. When breathing stops, the body continues to function and to produce carbon dioxide, which is not expelled through the lungs. Carbon dioxide accumulates in the blood, transmitting stronger and stronger impulses to the respiratory centers until the stimulus is so strong that it overcomes the individual's will power. If a person is able to stop breathing until consciousness is lost, normal, involuntary breathing will resume.

You have some voluntary control over your breathing; you could not speak if you did not have this control. However, if you concentrate on your breathing, an uncomfortable feeling will develop, because the imposed rhythm will either be slower or faster than your normal breathing rate. When breathing is too slow, the resulting increase in carbon dioxide levels in the blood eventually transmits strong impulses to the respiratory centers, which usually stimulate a gasp. When breathing is too rapid, so much carbon dioxide is removed from the blood that respiration is inhibited. If such breathing, called *hyperventilation,* continues for very long, the individual may faint. Intentional hyperventilation prior to diving into water may cause the diver to faint and then drown. If breathing is too rapid, the blood also becomes more alkaline as the carbon dioxide levels drop (remember that carbon dioxide combines with water to form an acid in the blood). This alkaline condition may cause some kinds of muscle cramps. When swimming, for example, you breathe according to the dictates of the swimming stroke, not your natural breathing rate. When you swim the crawl, you breathe at regular intervals when your head is out of water, and the timing is determined by the rate at which you move your arms. This rate is not necessarily related to your body's need for oxygen. Frequently, the breathing rate is too rapid, and the drop in carbon dioxide levels in the blood that occurs seems to alter the chemistry of some muscles enough to produce cramps.

Similarly, you often experience muscle cramps during the first few days you spend at a high altitude, until your body adjusts the number of red blood cells to coincide with the lower oxygen concentration in the air. In this unusual case, the need for oxygen (detected by special receptors in the arteries) controls the breathing rate and produces very rapid respiration, which removes too much carbon dioxide from the blood.

The chest movements during breathing are so simple and mechanical that they can easily be imitated when natural mechanisms fail. Artificial respiration is an effective means of achieving ventilation when the chest muscles, lungs, or respiratory centers are not functioning properly. Emergency methods include pushing rhythmically on the chest and mouth-to-mouth resuscitation. Mouth-to-mouth artificial respiration is considered more effective; its only danger is that too much air may be forced into the victim's lungs, especially in the case of children. Both methods are performed rhythmically about 10–14 times per minute. More rapid respiration would remove too much carbon dioxide from the blood and the respiratory centers would not be stimulated to initiate breathing.

The Upper Respiratory Tract—Channels to the Lungs

In addition to the lungs and the chest muscles, the breathing apparatus includes channels leading to the lungs that are equipped with special structures for moistening and warming the entering air and keeping it free of foreign particles.

The Nose

On its way to the lungs, air first passes through the nose or the mouth. In a resting state, all mammals (with the exception of some humans) breathe through their noses. The nose has four basic functions; to cleanse, moisten, and warm the incoming air and to detect odors.

The *nostrils* (Figure 15–15) are paired openings to the nose that are lined with hairs to trap particles of dust, small insects, and other foreign materials. Farther inside the nasal cavity, the lining of the nose is covered with a sticky fluid, or *mucus*, which acts like flypaper to trap foreign materials that are not caught in the hairs of the nostrils. This lining also contains tiny *cilia*, which move back and forth in waves, pushing mucus and trapped foreign particles toward the opening of the nose, where they can be blown out. Occasionally, foreign particles are removed by an involuntary, forceful blast of air—a *sneeze*.

Air passing through the nose is moistened by mucus, which drains into the nose from air-filled hollows, or *sinuses*, in the bones of the face. The sinuses reduce the weight of the skull and act as resonators for the voice. Tiny cilia lining the sinuses move the mucus through narrow openings that are connected to the nasal passages. Unfortunately, the structure of the sinuses evolved early in mammals and has not changed significantly during human evolution. These openings provide adequate drainage for animals that walk on four legs. But because humans walk in an upright position, their sinuses frequently become filled with fluid, producing pressure that can result in a painful sinus headache.

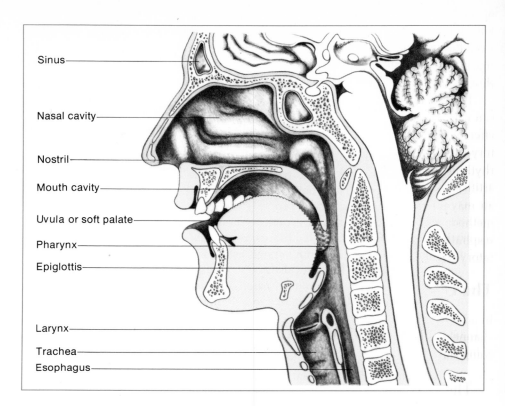

Sinus

Nasal cavity

Nostril

Mouth cavity

Uvula or soft palate

Pharynx

Epiglottis

Larynx

Trachea

Esophagus

Figure 15-15
The Upper Respiratory Tract
of a Human

The nose also warms entering air. Air drawn in through the nasal passages makes contact with the inner walls of the nose, where heat carried in the blood supplied by a rich network of vessels warms the air. A longer nasal passage permits the air to have more contact with these vessels, increasing the moistening and warming effect.

Because the nose is open to the outside environment and is designed to trap foreign materials, it is subject to infections. The nose has a more variable environment (warm, cold, wet, dry) than any other part of the body. Microorganisms, such as bacteria and viruses, can multiply rapidly in the nasal passages, because these variable conditions inhibit the effectiveness of the immune system. When an infection develops, more mucus is produced to prevent microorganisms from moving into the lungs. A runny nose and the movement of mucus from the nose to the throat (post-nasal drip) are common symptoms of this condition.

The Trachea, or Windpipe

Warmed, moistened, and cleansed air is moved from the nose to the *trachea*, or *windpipe* (see Figure 15–15). The entrance to this channel is located in approximately the same position as the entrance to the *esophagus*, through which food moves from the mouth to the stomach. Air from the back of the nose must pass over the opening to the esophagus to enter the trachea, and food from the mouth must pass over the entrance to the trachea to reach the esophagus. To prevent food from entering the trachea and blocking the channels between the nose and the lungs, a flap of tissue (the *epiglottis*) attached to the base of the tongue automatically covers the opening to the trachea during swallowing. An individual cannot inhale and swallow simultaneously.

To prevent food from accidentally entering the nasal passages, another flap of tissue—the *soft palate,* or *uvula*—closes over the respiratory channel during swallowing. This flap of tissue dangles down from the roof of the mouth and often vibrates due to breathing movements during sleep, which causes snoring.

As in the nasal passages, cilia lining the trachea move mucus and foreign particles, including small pieces of food that may slip past the epiglottis, back up to the mouth, where they can be spit out or swallowed. The cough also aids this process. During *coughing,* the epiglottis closes and the chest contracts forcefully. The epiglottis then opens suddenly, and a blast of air passes from the lungs up through the respiratory tract. Frequent coughing usually accompanies a cold, because large accumulations of mucus must be removed from the respiratory tract. Excessive coughing is also symptomatic of heavy smoking. Tobacco smoke destroys the cilia in the trachea, so that material must be coughed up from the respiratory tract.

Part of the trachea—the *larynx* and the *vocal cords* it contains (Figure 15–16)— functions in sound production. During exhalation, the movement of air across the vocal cords when they are contracted produces a sound. The pitch of the sound varies, depending on how much the cords are stretched by the contraction. However, most of the diverse sounds in our speech result from very fine movements of muscles in the tongue and the lips rather than from variations in the length of the vocal cords. The larynx, often referred to as the *voice box* or the *Adam's apple,* can be seen and felt as a bump in the front of the neck.

Below the trachea, the channel divides into two *bronchi* and then branches again into many small *bronchioles* (see Figure 15–12). The walls of both the bronchi and the bronchioles are lined with mucus and cilia, which move foreign materials upward and toward the mouth. At the ends of the bronchioles, the air reaches the actual gas exchange surfaces—the alveoli of the lungs. It is difficult for hayfever sufferers to breathe because their bronchioles constrict and the membranes lining

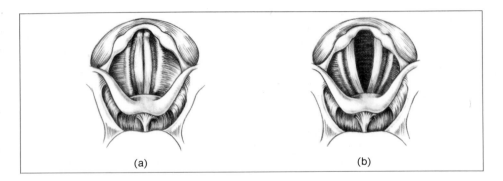

(a) (b)

Figure 15-16
The Vocal Cords
(a) No sound is produced during normal breathing, because the vocal cords in the larynx are relaxed. (b) When the vocal cords are contracted, they produce a sound during exhalation.

the walls of these tubes swell, obstructing the flow of air. This condition is an allergic reaction to foreign proteins, such as those found in pollen, which have been inhaled and identified as foreign matter by the immune system.

Summary

The cells of all organisms must have some means of exchanging gases with their external environment. Animal cells acquire oxygen and release carbon dioxide as a waste product. Plant cells acquire carbon dioxide and give off oxygen during *photosynthesis*, but acquire oxygen and give off carbon dioxide during *cellular respiration*. The exchange of these gases between an organism and its external environment is called *external respiration*.

Carbon dioxide and oxygen move into and out of cells by *diffusion*. The part of an organism that is to be used as a special gas exchange surface must therefore be relatively large, moist, near all body cells or a transport fluid, and able to move the surrounding air or water in relation to its own surface via *ventilation mechanisms*.

The special surfaces for gas exchange in plants are the *leaves*. The interior of a leaf contains cells immersed in highly humid air. Small openings, called the *stomata*, connect the moist interior of the leaf with the external environment. Gases diffuse between the cell interior and the external environment by passing through the humid air and the stomata of the leaf. Plants also have *root hairs* and pores on the stems that serve as regions of gas exchange.

Gas exchange surfaces in aquatic animals include the *skin* and the *gills*. Only relatively small or flat aquatic animals rely entirely on their skin as a gas exchange surface; most aquatic animals have gills that project outward from the body. The gills of active aquatic animals are covered with a flap or are located inside a body cavity to protect them from injury. The gills of fishes are located on the *gill slits* and are usually covered with a protective flap called the *operculum*. Water is moved into the mouth and out through the gill slits, where an exchange of gases takes place. A *countercurrent flow*, in which blood is moved through the gills in the opposite direction to the flow of water, greatly enhances the diffusion of gases between blood and water.

Terrestrial invertebrates have either diffusion lungs or tracheae. *Diffusion lungs* are hollow body cavities with moist inner surfaces lined with blood vessels. Gases move into the lungs and between the blood and lungs by means of diffusion. Evaporation of water from the lungs is reduced by closing the *spiracles*, or external openings to the body cavities, whenever the requirements for gas exchange are low. *Tracheae* are networks of hollow, fairly rigid tubes that carry air to all parts of the body. Gases diffuse across a thin layer of water at the ends of the tracheae into the fluid that surrounds the body cells. The tracheae are connected to the external environment by spiracles, which open and close according to the need for gas exchange.

All terrestrial vertebrates have *lungs*, which are elastic sacs inside the chest. Gases diffuse between the air inside the lungs and the blood capillaries lining the tissues of the lungs. Ventilation occurs when air is moved into and out of the hollow sacs. Amphibian lungs are simple structures with a relatively small surface area. Air is sucked into the mouth and then forced into the lungs; air is forced out of the lungs when the body wall is made smaller. Reptilian lungs are larger and better ventilated than amphibian lungs. The lungs of a reptile are attached to the chest wall. Air enters the lungs when the chest cavity is expanded by lifting the ribs upward and apart; air leaves the lungs when the chest cavity is reduced by lowering the ribs and bringing them closer together. The lungs then become smaller due to their elasticity. Unlike the lungs of other terrestrial vertebrates, bird lungs are connected to many *air sacs*—an arrangement that provides a one-way flow of air through the animal. Fresh air flows from the nose into the *posterior sacs* and then moves through the lungs, where diffusion takes place; from the lungs, air moves through the *anterior sacs* and out the bird's nose. This one-way flow permits a more efficient exchange of gases, because the entering air does not mix with the air being expelled from the respiratory structures.

All mammals have similar respiratory structures. A human has two lungs, which have a total surface area about the size of a tennis court. The right lung is composed of three *lobes*; the left lung, of two lobes. The lung tissue is arranged into millions of cup-shaped depressions—the *alveoli*—which increase the size of the gas exchange surface. A special chemical, a *surfactant*, prevents the moist inner surfaces of the lungs from sticking together. Two moist membranes called the *pleura* (one on the outer surface of the lungs; the other on the inner surface of the chest wall) adhere to each other by surface tension, causing the lungs to adhere to the chest wall. *Inhalation* of air occurs when the volume of the chest cavity is enlarged by raising the ribs and lowering the *diaphragm*. Because the lungs adhere to the chest, they also become larger, and the partial vacuum that is created causes air to rush into the lung cavities. *Exhalation* occurs when the ribs are lowered and the diaphragm is raised again, causing the lungs to become smaller by elastic recoil.

When the body is at rest, the signals that regulate the basic rhythm of breathing originate in the *respiratory centers* of the medulla oblongata in the brain. The signal for inhalation is transmitted via efferent nerves to the muscles of the ribs and the diaphragm. Exhalation occurs when impulses transmitted from *stretch receptors* in the lungs via afferent nerves to the respiratory centers stop these signals. The breathing rate is adjusted according to the body's level of physical activity, primarily in response to altered levels of carbon dioxide in the blood. During extremely labored breathing, *deflation receptors* in the lungs and an area within the pons of the hindbrain also contribute information to the respiratory centers and cause a rapid and deep inhalation (*inspiration*).

In humans, the upper respiratory tract is also involved in the inhalation and the exhalation of air into and out of the lungs during breathing. The *nose* cleans, moistens, and warms the incoming air and also detects odors. The *trachea* carries air between the nose or mouth and the lungs. Its opening at the back of the mouth is protected by a flap of tissue called the *epiglottis*, which prevents food from entering the lungs. A similar tissue called the *soft palate*, or *uvula*, prevents food from entering the nose. *Sinuses* in the bones of the face produce a liquid (*mucus*), which flows into the nasal cavities and moistens the air before it travels to the lungs. Both the nasal passages and the trachea are lined with *cilia* and mucus, which remove particulate matter from the respiratory tract. A special area within the trachea—the *vocal cords* on the *larynx*—functions in sound production. The *bronchi* and *bronchioles* of the lungs are tubes that connect the trachea to the alveoli in the lungs, where the exchange of gases between air and blood takes place.

Energy and Structural Materials From the Environment

16

F*ood* is any substance that an organism can utilize as a source of energy or raw materials for the synthesis of its own biological molecules. The cells of an organism constantly require energy and materials because their molecular components are regularly disassembled and reconstructed. An organism also needs energy and structural materials to grow, to maintain neural and muscular activities, to transport materials across cell membranes, and to manufacture special products such as hormones.

The conversion of the sun's energy into the chemical energy of glucose—a simple molecule consisting of carbon, hydrogen, and oxygen atoms—is the initial biochemical reaction that makes biological molecules available to all forms of life. Organisms that are capable of this biochemical reaction—called *photosynthesis*—are *autotrophs;* multicellular autotrophs are *plants.* The chemical energy released when glucose molecules are broken apart within autotrophic cells is used to support cellular activities. Glucose molecules may also be modified to form amino acids, lipids, nucleotides, and other basic molecules (*monomers*) of life, which are assembled into large *polymers,* such as proteins and nucleic acids, within autotrophic cells.

After the essential monomers of life are synthesized by autotrophs, they can be used as energy sources or as structural materials by animal cells. First, however, the polymers of a food organism must be split into monomers by enzymes during digestion. The released monomers can then be transported across cell membranes and used in various cellular activities. Different kinds of digestive enzymes have evolved through natural selection as genetically determined adaptations in different animal species. An animal's food range is limited by its adaptations for capturing organisms and by the digestive enzymes it can synthesize.

In this chapter, we will discuss the nature of foods, their effects on body tissues, and the adaptations of different organisms for processing food.

Energy

The cells of all organisms require energy to maintain their molecular structures as well as to carry out life functions. The amount of energy the entire organism needs is determined by the number of cells within the organism and the rate at which cellular activities take place.

Energy is initially made available to most forms of life by conversion of the sun's light energy into the chemical energy of glucose. This biological molecule and all forms of the molecules derived from glucose contain chemical energy that can be used to support life.

The main source of energy in most human diets is *carbohydrates*—sugars and starches made of glucose and similar monomers. Synthesized by plants, carbohydrates are abundant in seeds, fruits, and roots. Another source of energy, *fat,* is plentiful in meat, milk products, eggs, and nuts. *Proteins* may also provide energy. However, proteins, especially those obtained from meat, are less available—and consequently more expensive—than carbohydrates and do not form a large portion of the diet of most of the world's population.

Energy Stores

Animals do not have to eat continuously to meet their energy demands because their bodies can store chemical energy in the form of biological molecules. Plants can also store energy and therefore do not have to photosynthesize continuously. The types of molecules used for energy storage vary among different groups of organisms.

Plants store energy for future use primarily in the form of *starches,* which are large carbohydrates constructed from repeating units of smaller glucose molecules. Many plants that live longer than a year store starches in their roots during the winter. The stored starches provide energy reserves for early, rapid growth in the spring. Starch is also stored in the seeds of plants, where it provides energy for the growing embryo (which cannot perform photosynthesis).

Animals store energy reserves in several different types of molecules. *Glycogen,* a carbohydrate similar to starch in plants, is stored in small quantities in the liver and muscles of vertebrates. Glycogen is a short-term storage material; the energy in glycogen reserves can support the body for only a few hours. The long-term storage material in most animals is fat. Fat contains about two times the *calories* (units of energy) per unit weight that carbohydrates contain. Therefore, fat is a lighter storage material for a given amount of energy than starch or glycogen and permits most animals to be mobile while carrying large energy reserves. Many sedentary animals, such as oysters and mussels, that do not have to move their bodies store energy in the form of carbohydrates instead.

Energy Requirements

The amount of energy that must be supplied by an animal's diet is equal to the amount of energy that the animal's body uses. Even the cells of a resting organism must use energy to maintain their structural organization. It is difficult to determine precisely how many calories a human organism needs every day, because the activity levels of people differ greatly. The estimated daily caloric requirements for

Table 16-1
The Number of Calories Required per Day
to Maintain Various Activity Levels

Activity level	Calories
Sedentary	2,500
Exercise	
Light	3,000
Moderate	3,500
Heavy	4,000

Note: These figures are only guidelines to be used in determining the number of calories to include in the diet.

humans in Table 16-1 provide a rough guide to the number of calories that your diet should contain. However, these estimates are based on averages, and everyone's caloric need is unique. The best way to determine how much energy your body requires is to keep a record of your weight. If it increases, you are taking in more energy than you are using and some of the energy is being stored as fat (unless you are building larger muscles). If your weight decreases you are not taking in enough energy to meet your bodily needs and your fat is being used to provide energy.

Starvation results when an amount of food insufficient to sustain the body's energy requirements is taken in for a prolonged period. Once all the fat reserves have been consumed, other biological molecules will be broken down to supply the body with energy. The proteins in the skeletal muscles will be converted into energy first, followed by the muscles of the stomach, intestines, and heart. When the starvation process reaches this point, death is imminent because the muscles of these organs are essential to life.

The diets of many Americans contain too many rather than too few calories. More life spans are shortened by gross overweight than by any other nutritional disorder. Large fat reserves strain the organs of the body, especially the heart and the vessels of the circulatory system.

Structural Materials

Even after an animal reaches adulthood, its body requires materials for the construction of biological molecules. Most cellular structures are torn apart and rebuilt regularly; if the required materials are not present, these molecules cannot be formed. An insufficient intake of structural materials is called *malnourishment*. Although energy can be obtained from a wide variety of foods, materials to meet the body's structural requirements can be found only in certain foods. Despite an abundance of food, many Americans—even those who are overweight—are malnourished.

Nutritional Requirements and Evolution

The diet of each type of animal must include a specific group of minerals, vitamins, amino acids, and fatty acids (components of fats) to provide the raw materials required to synthesize the different kinds of biological molecules the animal needs to sustain life. The specific set of structural materials that cannot be synthesized by the body cells and must be obtained from food differs for each type of organism. Like digestive enzymes, the ability to synthesize particular molecules is a genetic trait acquired through natural selection.

Each animal acquires energy from the biological molecules of a particular array of plants and/or other animals. In general, simpler, more primitive organisms are able to synthesize a greater number of their structural materials in their own cells than more complex animals. The ability to construct many materials was lost during the evolution of more complex animals because these materials became so commonly available in the diet that the enzymes required to synthesize them were no longer necessary and were lost from the genetic repertoire.

For example, most vertebrate animals are able to synthesize vitamin C in their own cells, but this ability has been lost in humans, some other primates, a few types of birds, guinea pigs, and fruit-eating bats. All animals require vitamin C—an essential ingredient that facilitates the construction and maintenance of substances that bind cells together, especially in skin, bone, and muscle—but these few species require vitamin C in their diets because their ancestors regularly consumed an adequate supply and gradually lost the enzymes necessary to manufacture it.

Average minimum daily requirements, which represent the needs of a hypothetical average person for such structural materials as vitamins, minerals, and amino acids, have been estimated. However, individuals of the human population differ in their genetic composition and, consequently, in their nutritional requirements. Individual requirements for structural materials also appear to be developmentally influenced; people who eat smaller amounts of food during infancy and early childhood apparently have lower nutritional requirements when they are adults. Requirements for structural materials, like energy requirements, must be assessed on an individual basis.

Protein Requirements

The enzymes formed from *amino acids* (the components of protein) facilitate the chemical reactions that sustain life. Amino acids also form the basic structures of muscles, skin, hair, nails, red blood cells, and portions of the skeleton. Many of these structures must be replenished frequently so that the amino acids essential to their construction must be available at all times.

Each type of organism and each individual within a population has its own unique set of enzymes, but all enzymes are constructed from only 20 different kinds of amino acids. A deficiency in any one of these amino acids can prevent the construction of the full group of enzymes required by a particular organism. If an enzyme is not present, the chemical reaction it facilitates cannot take place.

Only plants and some unicellular organisms are able to synthesize all of their own amino acids. Multicellular animals must obtain at least some amino acids from their food. Eight essential amino acids cannot be synthesized by humans and must be included in the diet. These eight essential amino acids are then converted by the body into the other molecular forms of amino acids.

Vegetarians who choose not to eat any animal products must be especially careful to include the eight essential amino acids in their diet. Plants are good sources of amino acids, but no single type of plant contains all of the required amino acids, as many animal products, including eggs, milk, and meat, do. A strict vegetarian should eat a variety of plant materials every day.

Most of the protein in the plant material we consume is found in the seeds of *grasses* (wheat, corn, rice, and other cereals) and of *legumes* (beans, lentils, peas, peanuts, and soybeans). Each seed contains an *embryo* (or *germ*), an *endosperm* (a starchy food store for the growing embryo), and a tough outer *seed coat* (or *bran*), as shown in Figure 16-1. Both the embryo and the bran contain proteins; the endosperm does not. Whole-grain flour consists of all three seed components. Processed flour, which is used in making white bread, contains only the endosperm; the bran and embryo, which contain protein, have been removed to make the flour more palatable and to reduce spoilage. Thus, processed flour contains no amino acids.

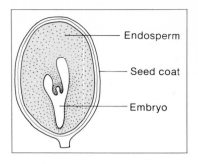

Figure 16-1
The Structure of a Seed

The seeds of legumes provide about twice as much protein as the seeds of grasses. Soybeans contain the highest amount (about 38% protein); peanuts and peanut butter are also excellent sources of protein. Separately, neither the seeds of grasses nor the seeds of legumes provide all of the essential amino acids, but because their amino acids complement each other, a meal consisting of both grass and legume seeds (for example, rice and beans) contains all of the amino acids necessary for the construction of proteins.

It is difficult to determine precisely how much protein an individual needs every day. Children require two or three times more protein per pound than adults because protein is used to build new tissues as well as to maintain existing tissues during growth. The recommended daily amount of protein for an adult is 0.36 g (gram) for every pound of body weight. Thus, a 150-pound person should consume 54 grams (or about 0.12 pound) of protein every day. Meat is not pure protein; it contains fat and other tissues that add to its weight. Beef steak, for example, is about 25% protein by weight.

Many Americans eat more than the recommended daily amount of protein and some actually eat too much protein. When a diet is high in protein and low in fats and carbohydrates, the amino acids of the proteins are broken down and converted into energy. The nitrogen within the protein molecules, which is a waste product of this reaction, must be excreted by the kidneys; a high-protein diet strains the kidneys, requiring them to work much harder to excrete nitrogen.

A swollen abdomen and legs are symptomatic of a severe protein deficiency. In such cases, the blood does not contain enough proteins to hold water in the blood vessels, and water seeps out of the vessels and accumulates in the interstitial fluids between cells. This condition also leads to low blood pressure and the problems associated with poor blood circulation. Another consequence of protein deficiency is enzyme deficiency. When not enough amino acids are available for the construction of enzymes, the chemical reactions these enzymes direct cannot take place, resulting in serious and widespread effects on all body tissues. A remarkable discovery is that conditions of short-term starvation, or *fasting*, in adults do not produce serious enzyme deficiencies; when not enough calories are consumed, the skeletal muscles of the body are broken down and the amino acids from these proteins are used to construct enzymes. Severe enzyme deficiencies occur only when sufficient calories but not enough proteins are present in the diet.

Although adults may be able to survive for limited amounts of time under conditions of very low food intake, the bodies of growing children must be continually supplied with proteins. Without a regular intake of proteins, children develop a condition called *marasmus,* which is common in America and throughout the world. The most severe and long-lasting effect of marasmus is the stunting of the central nervous system. Children suffering from this condition develop fewer and more abnormal nerve cells and exhibit a type of mental retardation that is irreversible because nerve cells develop only during childhood. The effects of protein deficiency on the nervous system during early childhood cannot be corrected by an adequate diet in later life. Marasmus frequently occurs when a child is bottle-fed and the milk is diluted with water.

Vitamin Requirements

A *vitamin* is any biological molecule that the cells of the body require in small amounts but cannot synthesize. A vitamin for one type of animal may not be a vitamin for another. For example, cholesterol is a vitamin for insects because they cannot synthesize this molecule, but it is not a vitamin for humans because we can construct it in our cells. Because the need for vitamins was discovered long before their chemical structures were identified, these molecules were simply referred to by the letters A, B, C, and so forth. The 13 vitamins listed in Table 16-2 are now recognized as essential to human health. Many of these vitamins act in conjunction with enzymes (become part of *coenzymes*) to promote specific chemical reactions.

Animals that live in their natural environments and do not alter their diets substantially rarely suffer vitamin deficiencies. Such disorders are common only in humans and other animals that have departed considerably from their natural diets. For example, vitamin deficiencies are common in domestic and laboratory animals and in zoo animals. Like amino acids, vitamins are biological materials that were so commonly available in the diet during evolution that the genes directing their synthesis were eventually lost from the genetic repertoire.

Much remains to be learned about our precise vitamin requirements and about each vitamin's particular function in the body. Most of our understanding of vitamins is the result of observing certain symptoms of ill health and their disappearance after a particular food is added to the diet. For example, at one time sailors frequently displayed the symptoms of *scurvy*—swollen, bleeding gums and extreme weakness. When citrus fruits were added to the sailors' diets, the symptoms disappeared. Citrus fruits contain a particular type of molecule, now known as vitamin C, that is essential to health; without it, scurvy develops. More vitamins will probably be discovered in the future.

Vitamins occur in both animal and plant materials. Prior to the discovery of vitamins, fruits and vegetables were thought to contain only carbohydrates and water and to be of little nutritional value. We now know they also contain many important vitamins; leafy green vegetables, such as spinach and broccoli, provide 8 of the 13 essential vitamins. Similarly, animal fats have long been considered simple energy foods, and many people have tried to curb their intake of animal fat to guard against obesity and heart disease. However, fatty foods are a rich source of many vitamins. For example, cod liver oil, which is primarily fat, contains large amounts of vitamin D. This vitamin is normally synthesized in the human body through the interaction of sunlight and a *sterole* (lipid) in the skin. During the long winter months in northern climates, there is not enough sunlight to synthesize the amount of vitamin D required by growing children. A lack of vitamin D causes

Table 16-2
The 13 Vitamins Essential to Human Health

Nutrient	Function	Food Source
Vitamin A	Maintains healthy eyes, skin, hair, teeth, gums, and various glands; also involved in the use of fats	Whole milk, butter, eggs, yellow and leafy green vegetables*, liver
Vitamin B_1 (thiamine)	Facilitates energy production by promoting the removal of carbon dioxide from glucose	Whole-grain cereals, legumes, yeast, liver, pork, fish, lean meat, poultry, milk
Vitamin B_2 (riboflavin)	Functions in the release of energy from carbohydrates, proteins, and fats	Milk, whole-grain cereals, liver, lean meat, eggs, leafy green vegetables
Vitamin B_6 (pyridoxine)	Plays an important role in the formation of certain proteins and in the use of fats; also aids in the formation of red blood cells	Lean meats, leafy green vegetables, whole-grain cereals
Vitamin B_{12}	Essential to the construction of *nucleic acids* and in the formation of red blood cells; also aids in the function of the nervous system	Liver, kidney, fish, milk, foods of animal origin in general
Folic acid	Assists in the formation of certain body proteins and nucleic acids; required in the formation of red blood cells	Leafy green vegetables, liver
Pantothenic acid	Involved in the conversion of carbohydrates, fats, and proteins into the molecular forms needed by the body; also required in the formation of certain hormones and neurotransmitters	Eggs, leafy green vegetables, nuts, liver, kidneys
Niacin	Involved in energy-producing reactions in cells	Eggs, meat, liver, whole-grain cereals
Biotin	Involved in the formation of certain fatty acids and in the production of energy from glucose; essential to the function of many chemical systems in the body	Liver, kidney, eggs, leafy green vegetables
Vitamin C	Keeps bones, teeth, and blood vessels healthy; also important in the formation of *collagen*—a protein that helps to support body structures such as the skin, bone, and tendons	Citrus fruits in particular, other fruits, leafy green vegetables, potatoes
Vitamin D	Necessary for the formation of strong teeth and bones; helps the body to use calcium and phosphorus properly	Vitamin D fortified milk, cod liver oil, egg yolk, tuna, salmon
Vitamin E	Aids in the formation of normal red blood cells, muscles, and other tissues; protects fat in body tissues from abnormal breakdown	Vegetable oils, whole-grain cereals
Vitamin K	Essential for normal blood clotting	Leafy green vegetables

Leafy green vegetables include spinach, kale, broccoli, chard, turnip greens, mustard greens, and Brussels sprouts. *Yellow vegetables* include carrots, squash, rutabagas, and sweet potatoes.
Source: Adapted from Don Bingham, "The Nutrition Primer." *Fitness for Living* (September/October 1972), pp. 73–74. Used by permission from Rodale Press, Inc.

rickets (Figure 16-2)—a condition that does not permit the bones to develop correctly. People in northern populations traditionally fed their children cod liver oil during the winter to ensure their normal growth, but this practice is no longer common because vitamin D is now added to most milk that is sold commercially.

Several of the vitamins essential to human health are actually produced by bacteria living in the intestines, but only one vitamin—vitamin K, which is essential for normal blood clotting—is produced by bacteria in a sufficient quantity that it is not required in the diet. Presumably, at some point in our evolutionary history, our ancestors were capable of synthesizing this molecule in their own cells. But once

Figure 16-2
Symptoms of Vitamin D Deficiency
When not enough vitamin D is available in the diet, the bones do not develop correctly in growing children. This condition is known as *rickets*.

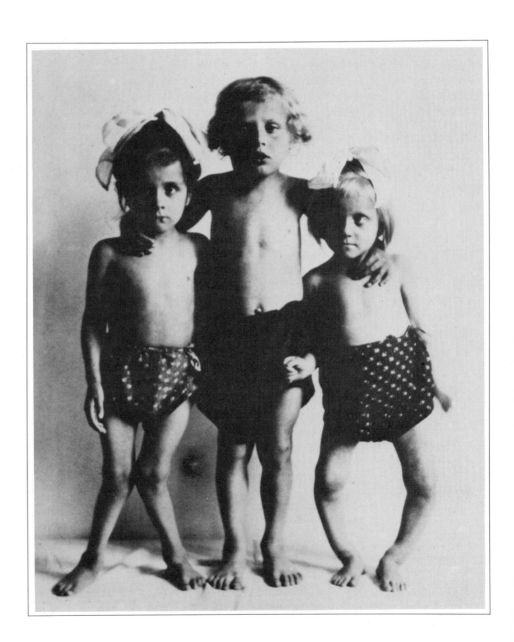

the bacteria that excreted this substance as a waste product became regular inhabitants of the human intestines, human cells lost their ability to manufacture vitamin K. This vitamin is only required in the human diet when prolonged diarrhea or antibiotics such as penicillin and streptomycin disrupt the bacterial population in the intestines.

The need for vitamins varies, depending on the individual's activity and stress levels. Many people take vitamin tablets to ensure that they receive the minimum daily requirements of these special molecules. However, a diverse diet of grains, meat and animal products, fruits, leafy green vegetables, and yellow vegetables usually meets daily vitamin requirements. A diet consisting solely of plant material is lacking in vitamin B_{12}, which should be supplemented by vitamin tablets or by including eggs or milk in the diet.

Vitamins may also help to protect the body against various pollutants found in air, food, and such drugs as tobacco, marijuana, caffeine, and alcohol. If this proves to be true, the minimum daily requirements for vitamins will have to be revised and individually tailored in accordance with the varying levels of pollutants.

Mineral Requirements

All elements that animals utilize in their inorganic state (elements that are not parts of biological molecules) are *minerals.* Carbon, hydrogen, nitrogen, and sulfur are elements, but are not considered minerals because they can be used by an organism only when they are part of a biological molecule. Potassium, calcium, phosphorus, iron, and many other substances are minerals that are usually incorporated into plant tissues from the soil and passed on to animals in the form of food. Some minerals are also ingested by drinking water. Many people supplement their diets with mineral tablets, but any diet that contains a diversity of animal and plant materials should meet daily mineral requirements.

The particular minerals required by an organism provide clues to the past abundance of these materials. The chemistry of primitive plants is based on such unusual minerals as aluminum and silicon—elements required only in trace amounts by more recently evolved plants. Some primitive marine animals require large quantities of vanadium for oxygen transport, whereas vanadium has been replaced by iron in the oxygen-transporting cells of most modern animals. The use of these unusual minerals by primitive organisms suggests that they were once more available for use by living organisms than they are today.

A peculiar requirement of all vertebrates is *iodine*—an element necessary for the construction of *thyroxine,* a hormone produced by the thyroid gland. Iodine is quite rare in many parts of the world, particularly in mountainous areas and in most regions far from present or past oceans. People who live in such areas usually import iodine and add it to their table salt or take it in tablet form. Insufficient iodine in the human diet causes the thyroid gland to enlarge, forming a *goiter* (Figure 16-3). Prior to the discovery that iodine prevents this condition, goiters were fairly commonplace in many parts of the world. The need for iodine may reflect long periods of vertebrate evolution in environments in or near oceans, where this element was plentiful.

Calcium, phosphorus, iron, iodine, sodium, potassium, magnesium, chlorine, bromine, fluorine, cobalt, zinc, vanadium, manganese, and molybdenum are the 15

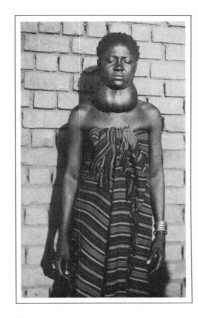

Figure 16-3
Symptoms of Iodine Deficiency

A swollen neck, or *goiter*, results from an enlarged thyroid gland. This condition develops when there is not enough iodine in the diet.

minerals that are essential to human health. Many of the remaining naturally occurring elements are customarily present in the body, but their function and relationships to health are not yet known.

The Processing of Food: Digestive Systems

Because useful energy and structural materials are released and new molecules are built within individual cells, food must be broken down so that it can enter the cells before it can be of use to the body. *Digestion* is the process by which large food molecules are converted into smaller ones. In *hydrolysis*—the biochemical reaction involved in digestion—polymers are split into monomers and a water molecule is added at each location in the large molecule where a chemical bond has been broken (see Chapter 3, page 52). Digestion occurs due to the action of digestive enzymes.

Animals that feed on very small particles do not require elaborate digestive systems to break food down into smaller units. Unicellular organisms engulf small food particles by enfolding them in part of the cell membrane. The food item and the enclosing membrane then separate from the rest of the cell membrane and move into the cell in the form of a *food vacuole* (Figure 16-4). Digestive enzymes enter the vacuole and reduce the food by hydrolysis to small molecules that can be used for energy and structural materials. Some simple multicellular animals, such as sponges, also engulf food particles in this way.

Food items that are too large to fit inside a cell require special processing before they can be used. A few animals, like the spider, digest food externally by secreting enzymes onto the captured prey. After the food has been broken down, the animal sucks the liquid material into its body. However, most multicellular animals consume large food particles by moving them into a muscular cavity within the body, where the food is held while it is broken down by digestive enzymes into smaller particles that can be easily moved into the body cells by a process called *absorption*.

The simplest digestive cavity, which is found in such animals as the starfish,

Figure 16-4
Digestion in a Unicellular Organism—The Amoeba
A food particle is engulfed by the cell membrane, and the food and the enclosing membrane move into the cell, becoming a *food vacuole*. Enzymes then enter the vacuole and break down the food into energy and structural materials.

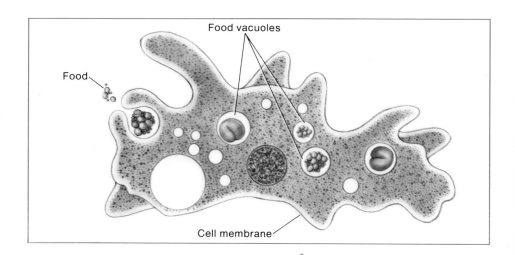

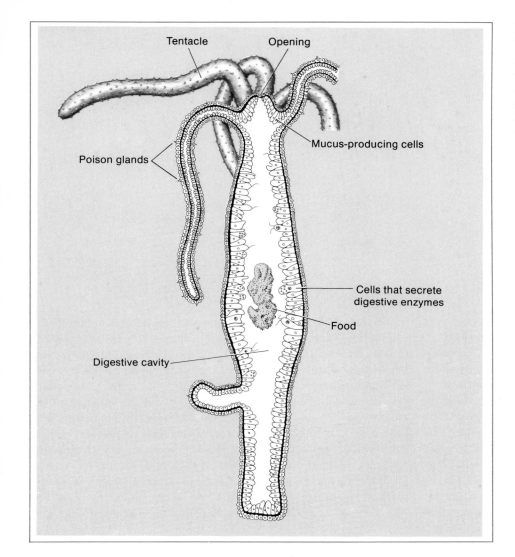

Figure 16-5
The Simple Digestive System of a Hydra
The digestive cavity of the hydra has only one opening to the outside environment. Food enters and feces leave the body cavity through the same opening.

Figure labels: Tentacle, Opening, Mucus-producing cells, Poison glands, Cells that secrete digestive enzymes, Food, Digestive cavity

coral, and hydra (Figure 16-5), has only one opening to the outside environment. Food enters through this opening, which is often fringed with special structures to poison the prey and to keep it from struggling and injuring tissue inside the cavity. The food then moves into the body cavity, where it is digested. The portions of the food that cannot be digested are moved out of the body through the same opening in the form of *feces*. Because the digestive cavity has a single opening, food entering the cavity cannot be separated from the feces leaving the body cavity. To prevent food and feces from mixing, food is not taken in at the same time that feces leave—a restriction that limits the rate of food intake.

A more efficient digestive cavity, found in most animals, is a muscular tube called the *digestive tract* (Figure 16-6). The digestive tract has two openings—one for food intake (the *mouth*) and the other for feces removal (the *anus*). This one-way

Figure 16-6
The One-Way Digestive Tract
Found in Most Animals
In a digestive system with two
openings, food moves from mouth to
anus in a single direction, so that
eating can continue while fecal
material is being removed.

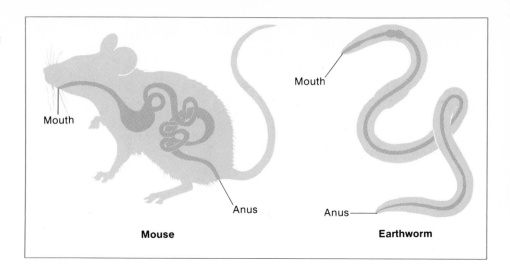

Mouth

Anus

Mouse

Mouth

Anus

Earthworm

system permits food to be taken in and held while fecal material is collected and excreted at the same time. The walls of the digestive tract contain muscles and blood vessels, as well as mucus for lubrication. The food is moved through the tract by rhythmic contractions of the tract walls, which are controlled by the *autonomic nervous system* and hormones produced by the cells lining the walls. Connections between the *efferent nerves* from the *central nervous system* and the muscles of the digestive tract are *tight junctions,* which are electrical *synapses* without *neurotransmitters* (see Chapter 13, page 312). Electrical impulses moving directly between muscle cells produce slow waves of contractions called *peristaltic movements,* which move the food over great distances along the digestive tract in a single direction.

Each segment of the digestive tract is specialized to perform a particular function, so that food passes through an "assembly line" from mouth to anus. The degree of this specialization is greatest in the vertebrate digestive system, which consists of specific regions for different functions: the initial preparation of food, storage, digestion, absorption, and the formation and removal of feces. Each region is separated from the others by special circular muscles which, when contracted, close off that part of the digestive tract.

Preparation of Food

Although some animals, such as *filter feeders* (Figure 16-7) and many fishes and snakes, swallow their food alive and whole, most animals prepare their food in some way before it enters the digestive tract. *Carnivores* (animals that eat other animals) usually immobilize their prey and tear it into small pieces so that it can be swallowed. As long as the meat is shredded into small enough pieces to be swallowed, it does not have to be chewed. Humans are the only animals that chew their meat. Unlike the carnivores, humans have the luxury of time to prolong the taste experience; we chew our meat solely to enhance its flavor.

Plant material requires much more processing than animal meat—a fact that is reflected by contrasting the teeth of plant eaters, or *herbivores,* with the teeth of

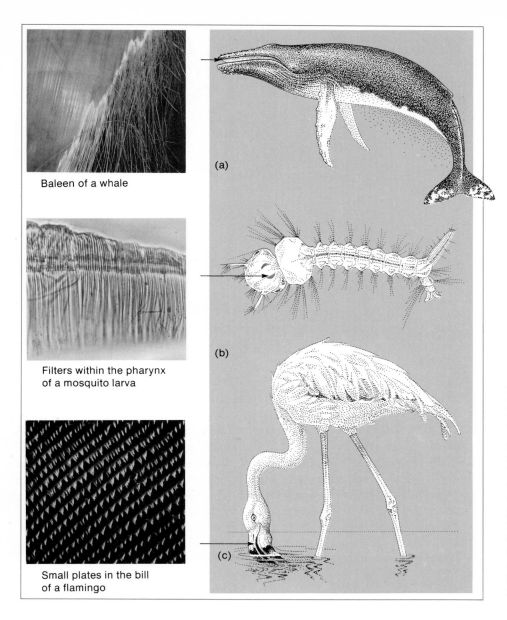

Baleen of a whale

Filters within the pharynx
of a mosquito larva

Small plates in the bill
of a flamingo

Figure 16-7
Filter Feeders Swallow Food
Organisms Whole and Alive
(a) The humpback whale obtains
sufficient food to support its large
body by filtering small organisms
from the water. Its mouth is
constructed of hundreds of small
plates, called the *baleen*. (b) Filters in
the pharynx of the larvae of many
insects sieve bacteria from the
surrounding water. (c) The bill of a
flamingo is covered with small plates
that filter organisms from water and
mud.

carnivores (Figure 16-8). Plant cells are surrounded by tough outer walls of cellulose, and no vertebrates can synthesize the enzymes required to break apart cellulose molecules. Most animals use teeth or some other grinding structure to rupture
the cell walls to obtain nutrients within plant cells. Humans must chew raw plants
extensively to derive nourishment from them. Fruits and vegetables that are carefully cooked or processed in a blender to rupture their cell walls provide more
nutrients than unprocessed forms. Although few nutrients are obtained from raw
plants, they are an important source of roughage in the human diet, which aids in
the formation of feces.

Figure 16-8
Tooth Structure Is Adapted to Food Habits
(a) An *herbivore* has flattened teeth for crushing and grinding plant material.
(b) A *carnivore* has sharp, pointed teeth for killing its prey and tearing it into small pieces that can be easily swallowed.

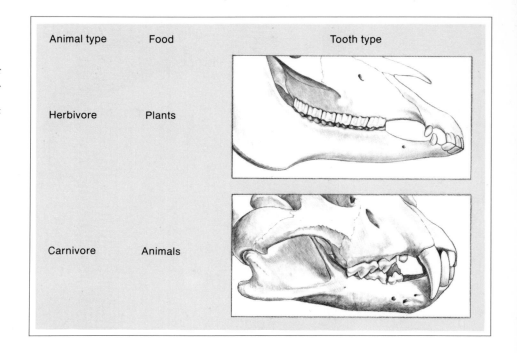

Animal type	Food	Tooth type
Herbivore	Plants	
Carnivore	Animals	

Most mammals process plant materials by chewing, which also serves to mix saliva with the food. *Saliva* is produced by *exocrine glands* near the mouth; it consists primarily of water but also contains some ions (sodium, potassium, iodine, chloride, and bicarbonate), a digestive enzyme, and mucus. The water and mucus lubricate and moisten the food, which holds it together so that it can easily slip down the esophagus to the stomach. Saliva also dissolves some of the materials in the food, which enhances its taste; *taste buds* located on the tongue can detect materials only if they are in a liquid solution. The digestive enzyme in saliva begins the process of carbohydrate digestion by converting starches into sugars.

The production of saliva in humans can be increased merely by the expectation of food. One of the values of cooking food is that the smell and appearance of a well-prepared meal begins the process of saliva production prior to eating. Chewing gum or tobacco also increases saliva production, but excess saliva can be detrimental when not followed by eating, because the ions in the saliva contribute to the formation of tartar on teeth, giving them a yellow appearance and promoting tooth decay.

Seeds require considerable preparation because their nutrients are enclosed within a tough outer coat, the bran. If this seed coat is not ruptured, seeds will travel directly through the digestive tract without releasing any nutrients. Seed-eating birds have a special digestive structure, the *gizzard* (Figure 16-9)—a muscular chamber within which seeds are prepared for digestion. Birds swallow small stones to fill this chamber with abrasive materials. As the muscles of the gizzard move, the small stones grind up the seeds; the crushed material is then moved to the intestines for further processing. The stones in the gizzard become worn from grinding and must be replaced occasionally. Chickens and other birds that live predominantly on seeds can starve to death if they are not provided with abrasive materials.

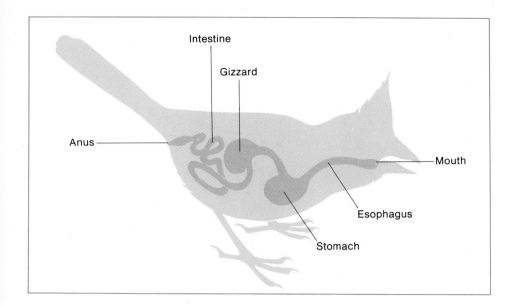

Figure 16-9
The Digestive Tract of a Bird
Seed-eating birds have gizzards,
where seeds are ground up before
digestion.

Seeds form a large part of the human diet. They are usually processed commercially to remove the bran and then ground into flour. Even after processing, however, the starch molecules of flour are difficult for humans to digest. Cooking disrupts the cellulose envelope around the cells within the seed and breaks apart the tight coils of long starch molecules.

Storage

In the digestive tracts of most animals, food is stored prior to digestion in a separate region called the *stomach,* which is connected to the mouth by the *esophagus*—a muscular channel about 25 cm (10 inches) long in the human (Figure 16-10). A large amount of food can be collected in the stomach and released slowly to the digestive areas. Food generally remains in the stomach for several hours.

The stomach is primarily a storage area. (In the earliest vertebrates, this was its only function.) Stomach size varies, depending on the feeding habits of different types of animals. Animals that eat large but infrequent meals have large stomachs. The stomachs of dogs and cats, for example, make up about 70% of their digestive tracts. Most herbivores have smaller stomachs because they eat smaller amounts of food more frequently and have less need for a storage cavity. A horse's stomach comprises only about 8% of its digestive tract.

The stomachs of vertebrates contain a strong acid, which originally evolved in primitive fishes to permit them to kill their prey after swallowing it whole and alive. This function is retained in many modern vertebrates. The acid also prevents bacteria from growing and spoiling the food before it is digested. Food usually contains large numbers of bacteria, which would increase tremendously during the time that the food remained in the stomach if they were not killed.

Some protein digestion also occurs in the stomach. This function probably arose as a secondary adaptation after the stomach evolved in response to the need

Figure 16-10
The Digestive Tract
of a Human

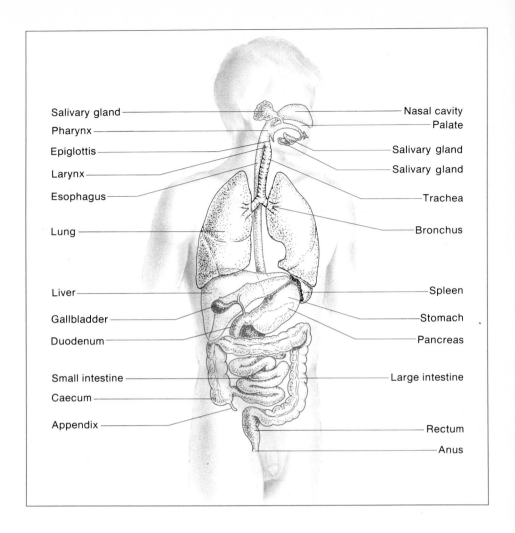

Salivary gland — Nasal cavity
Pharynx — Palate
Epiglottis — Salivary gland
Larynx — Salivary gland
Esophagus — Trachea
Lung — Bronchus
Liver — Spleen
Gallbladder — Stomach
Duodenum — Pancreas
Small intestine — Large intestine
Caecum
Appendix — Rectum
— Anus

for food storage. The acid in the stomach softens protein and dissolves bone. The stomach walls secrete a digestive enzyme that partially breaks down large protein molecules. Another substance, called an *intrinsic factor,* produced by the stomach walls is required for the absorption of vitamin B_{12}. The stomachs of babies and young children also produce an enzyme that aids in the digestion of milk. In adults, milk is digested in the small intestine.

Like the rest of the digestive tract, the stomach walls are lined with muscles that contract to churn the contents of the stomach and eventually move the food into the small intestine. Because these muscles are vulnerable to the protein-digesting contents of the stomach cavity, the stomach walls are normally lined with a slippery film of mucus that cannot be digested by the stomach secretions. Occasionally, however, the walls of the stomach are digested and an *ulcer* forms (Figure 16-11). This painful condition can become serious if the stomach wall is perforated and the contents of the stomach spill into the abdomen. Ulcers usually develop in people who are under psychological stress, when the stomach secretes more acid and enzymes than it requires to digest food. Some drugs and alcohol tend to remove the layer of mucus from the stomach walls, exposing them to digestive secretions. The stomach itself is not essential to food processing and can be removed entirely if it is severely damaged. People who have no stomachs must eat smaller and more frequent meals because they have no capacity to store large quantities of food.

The stomachs of some herbivores are adapted to promote the survival and growth of cellulose-digesting bacteria. In the complex stomachs of these *ruminants* (including cows, sheep, goats, and deer), newly eaten food passes into a compartment called the *rumen.* This part of the stomach may be very large; the rumen of a cow can hold up to 300 liters of fluid. The rumen is filled with a neutral liquid maintained by the production of copious amounts of saliva, an alkaline solution, that provides a favorable environment for the cellulose-digesting bacteria. After the food is initially digested by bacteria in the rumen, the *cud* is returned to the mouth, where it is chewed extensively to break up the plant material and mix it with saliva. The food is then returned to the rumen, where it is further digested by bacteria, and is passed into the intestines.

Some of the contents released from plant cells are immediately available to the ruminant; other plant nutrients are obtained indirectly when the bacteria die and move into the intestines, where they can be digested. These bacteria produce all of the amino acids and many of the vitamins required by the ruminant. These bacteria also convert *urea,* the nitrogen-containing waste material, into amino acids. All urea is removed from the bodies of most animals in the form of *urine.* In ruminants, however, some urea is returned to the rumen for conversion into amino acids.

Digestion

After leaving the stomach, food moves into the *small intestine,* where most digestion and practically all absorption takes place. Here, all of the fats and whatever starches and proteins remain are broken down into their basic units. Digestive materials are released into the small intestine from three areas—the pancreas, the liver, and the walls of the small intestine. The release of digestive materials is controlled by the autonomic nervous system, by hormones secreted by the walls of the digestive tracts, and by materials in the food itself that stimulate specific responses.

The length of the small intestine is determined by the type of food that the animal normally eats. Herbivores have longer small intestines, because plant food requires more processing; carnivores have shorter small intestines because meat requires less processing. Humans and other *omnivores*—species that eat both plants and animals—have small intestines of intermediate length. The small intestine in the human is about 6 meters (20 feet) long and is folded to fit snugly within the abdominal cavity (see Figure 16-10).

The first part of the small intestine is the *duodenum,* which lies adjacent to the stomach and is about 25 cm (10 inches) long in humans. The duodenum receives the highly acidic food from the stomach. An alkaline solution of sodium bicarbonate and water released from the walls of the small intestine and the *pancreas,* which lies just beneath the stomach (see Figure 16-10), is poured into the duodenum to neutralize the food from the stomach immediately. The digestive enzymes secreted by the small intestine cannot function in highly acidic solutions, and the walls of

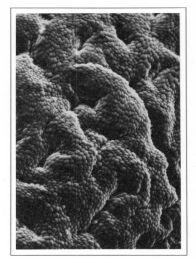

(a)

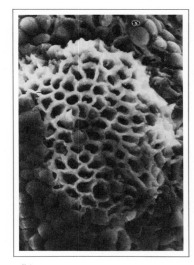

(b)

Figure 16-11
Ulcers—The Digestion of the Walls of the Digestive Tract
(a) The inner wall of a normal stomach. (b) The inner wall of a stomach with an ulcer.

the small intestine are not as well protected from acid as the walls of the stomach. Occasionally, the acid is not neutralized quickly enough, and an ulcer develops in the duodenum. Duodenal ulcers are more common than stomach ulcers.

The pancreas produces several enzymes that break down fats, carbohydrates, and proteins; these secretions enter the small intestine at the duodenum. *Bile*, which is produced by the liver and is an essential material in the breakdown of fatty foods, also enters the small intestine at this point. Bile is not an enzyme; it acts like a detergent, reducing the fat particles to a small enough size to permit them to be mixed with water and broken down by the water-soluble enzymes in the small intestine.

Some animals store bile in a small sac called the *gallbladder*, which lies next to the liver (see Figure 16-10). The gallbladder permits an animal to meet a sudden demand for bile, which occurs when a large amount of fat is eaten. The presence of fat in the small intestine stimulates the production of a hormone that causes the gallbladder to empty. Not all vertebrates have a gallbladder, although all vertebrates produce bile. The human gallbladder is not an essential organ and is frequently removed if it becomes obstructed by cholesterol crystals, or *gallstones*. Bile flows slowly and continuously from the liver into the small intestine, so that digestion remains normal after the gallbladder is removed as long as large amounts of fat are not consumed in a single meal.

The walls of the small intestine secrete additional enzymes that act in conjunction with the enzymes from the pancreas to break down sugars and starches into glucose, proteins into amino acids, and fats into fatty acids and glycerol. All of these biochemical reactions involve hydrolysis—the addition of a water molecule to each pair of monomers as they split from the larger polymer.

Absorption

Amino acids, glucose, and other simple sugars move through the walls of the small intestine into the bloodstream. Vitamins, water, and minerals also enter the blood vessels in the small intestine at this point. This transfer of materials, called *absorption*, is accomplished by simple diffusion and active transport.

Most materials are carried directly from the capillaries of the small intestine to the liver. However, the products of fat digestion—fatty acids and glycerol—move into the cells within the intestinal walls, where most of them re-form into *triglycerides* (three fatty acid molecules combined with one glycerol molecule). These triglycerides combine with protein and cholesterol, which is synthesized in the liver, to form *lipoproteins*, which are transported by the *lymphatic vessels* to special fat cells in certain tissues of the body, where they are stored for later use.

The entire inner surface of the small intestine is covered with small projections—the *villi* and *microvilli* (Figure 16-12)—which greatly increase its surface area and therefore its absorption capacity. Each villus contains a network of blood vessels, which absorbs molecules of food and water.

The Formation of Feces

All food material that is not broken down and absorbed through the walls of the small intestine passes into the large intestine. In humans, the large intestine is about $1\frac{1}{2}$ meters (5 feet) in length and is much larger in diameter than the small

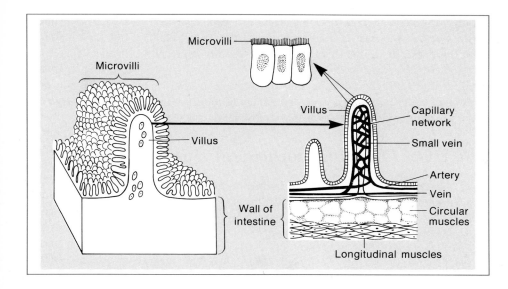

Figure 16-12
The Structure of the Inner Wall of the Small Intestine
The presence of *villi* and *microvilli* greatly increases the surface area—and therefore the absorption capacity—of the small intestine. Each villus contains a network of blood vessels, which absorbs food and water molecules.

intestine. The unabsorbed material consists primarily of undigested plant material, dead bacteria, sloughed-off cells from the lining of the digestive tract, mucus, and bile. The hemoglobin from the red blood cells that have been broken down by the liver and carried to the intestines by the bile gives feces their characteristic reddish-brown color.

As the unused food passes through the large intestine, much of the water it contains is removed and reabsorbed into the bloodstream for reuse. (The digestive process requires a large amount of water.) By the time the fecal material is expelled from the anus, it is a fairly solid material, although it still contains enough moisture to make it pliable.

The consistency of fecal material is important for proper expulsion. Bulky feces stimulate muscular contractions of the intestinal walls, which move the feces through the large intestine fairly rapidly. Because these feces remain in the intestine for only a short time, they retain enough moisture to produce a large, soft stool. A fair amount of indigestible material (roughage) in the diet ensures that the feces will be bulky. Uncooked plant materials and bran are excellent sources of roughage.

When fecal material is too small to stimulate the walls of the large intestine adequately, the feces move slowly and most of the water they contain is absorbed into the bloodstream. Consequently, the feces become smaller, harder, and difficult to expel, causing *constipation*. Small, hard feces also cause the muscles of the large intestine to work harder, creating weak spots that may develop into little pouches, or *diverticula,* that can fill up with fecal material and become infected. About 40% of all Americans develop diverticula at some time during their lives.

In contrast, if the fecal material moves through the large intestine too rapidly, not enough water is removed from the feces, resulting in *diarrhea.* Any irritant to the lining of the large intestine can increase its rate of muscular contraction and consequently the rate at which fecal material is moved. Infections frequently irritate the intestinal lining; some chemicals, such as those in prune juice, are known to produce the same effect. Emotional stress can also cause diarrhea. Diarrhea causes the loss of water that is normally absorbed through the walls of the large intestine and conserved. In adults, this water can be readily replaced by drinking, as thirst dictates. But the need for additional water may not be recognized in babies and small children, and dehydration may become serious. Babies who lose even a small amount of water may lose a large percentage of their total body water; if this water is not replaced rapidly, the water loss can be fatal.

The Liver: Organ of Biochemical Homeostasis

Food molecules that have been absorbed from the digestive tract are carried via the bloodstream to the *liver*—a large, critically important organ that lies on the right side of the abdomen, just beneath the diaphragm (see Figure 16-10). The liver performs four essential functions:

1. Produces bile to be used in the digestion of fats.
2. Converts the end products of digestion into the particular molecular forms required by body cells.
3. Regulates the components of the blood.
4. Converts toxic and foreign substances into compounds that can be excreted from the body.

The liver has the remarkable capacity to convert one type of molecule into another, which permits humans to digest a variety of foods. If the diet includes the essential amino acids, fatty acids, vitamins, and minerals, the liver cells can construct most of the molecules needed to maintain life (Figure 16-13). Amino acids can be converted into glucose for immediate energy, stored as energy in the form of glycogen or fat, converted into nonessential amino acids, or synthesized into the

Figure 16-13
The Conversion of Food Materials by the Liver

Glucose can be converted into almost any of the basic molecules except protein. Essential amino acids can be converted into other kinds of amino acids or glucose, from which a number of conversions are possible. Fats can also be converted into glucose, or their fatty acids can be used for energy or converted into special lipids.

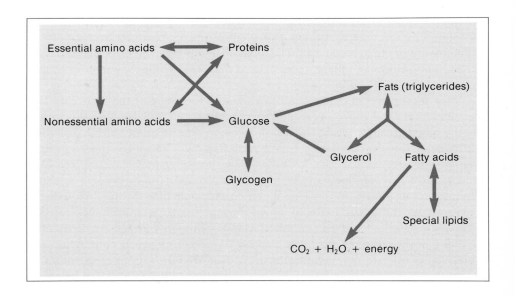

particular proteins carried by the blood that maintain blood pressure. The triglycerides that are stored in fat cells can be used as an immediate energy source or may be transported to the liver for use in the construction of specific lipids, such as cholesterol and phospholipids. The glycerol portion can also be converted into glycogen for energy storage. Glycogen obtained from these sources can be broken down into glucose molecules, which can be released directly into the bloodstream or converted into fat.

Liver cells also render potentially harmful materials nontoxic so they can be removed by the kidneys. The liver's conversion of ammonia—the toxic waste product from the breakdown of amino acids—into its less toxic form, urea, is a critical bodily function. The cells of the liver also regulate the distribution of vitamins and minerals throughout the body.

The feedback between the blood and the liver cells allows the liver to regulate levels of glucose, amino acids, blood proteins, blood-clotting materials, and vitamins in the blood. The most crucial task of the liver is to maintain a fairly constant amount of sugar in the blood. Only the simple sugar glucose is normally used by the central nervous system, and nerve cells may die if sufficient amounts of this energy molecule are not available. When carbohydrates are not included in the diet, the liver cells convert glycogen, amino acids, or glycerol (a component of fats) into glucose.

The pancreas also plays an important role in the utilization of glucose. In addition to producing digestive enzymes, the pancreas synthesizes and releases *insulin.* Without this hormone, glucose cannot cross cell membranes and be converted into energy. Insulin also aids in the conversion of glucose into glycogen—the short-term energy molecule stored in liver and muscle cells. When the pancreas does not produce enough insulin, *diabetes mellitus* develops. Insulin extracted from the pancreas of other animals or synthesized by chemists or bacteria (controlled by the DNA recombination technique) can now be administered to individuals who suffer from this condition.

Regulation of Body Fluids by the Kidneys

The *kidneys* also regulate body chemistry. One of the primary functions of the kidneys is to remove the nitrogen that forms in cells due to the breakdown of amino acids. Carbohydrates and fats are composed of carbon, hydrogen, and oxygen, and the waste products formed from their use are water and carbon dioxide. However, when proteins are formed and broken down, some of the nitrogen contained in the amino acids is not recycled. This is especially true when proteins are used to provide energy, because amino acids are converted into glucose and the nitrogen portion of the molecule is a waste product of the reaction.

Ammonia (NH_3)—the immediate waste product of the breakdown of amino acids—is highly toxic to cells. Only animals that have an abundant supply of water, like the freshwater invertebrates, can dilute ammonia so that it will not be harmful while it is being removed from the body. In most animals, including humans, the liver cells quickly convert ammonia into a less toxic form of nitrogen—urea—which is carried by the bloodstream to the kidneys where it is eliminated in the form of urine. A complete discussion of kidney structure and function is presented in Chapter 17.

Control of Body Weight

Except under conditions of extreme food deprivation, all humans have a layer of fat just beneath the skin and around the organs and membranes of the abdomen. This fat layer serves as an energy reserve and as a cushion to prevent injury and also insulates the body from cold. Long-term *obesity,* or the accumulation of too much fat, is a condition found almost exclusively in humans and their domestic animals. Wild animals eat only when they are hungry, and their food intake is regulated by a sensitive feedback mechanism that prevents overeating. This mechanism is also present in humans, but it is often overruled by emotions. Much of our culture is centered around the social aspects of food consumption. Food is made irresistibly attractive, and we are confronted with social pressures to eat more than the appetite dictates. We also eat to relieve boredom, tension, and frustration.

Appetite is a complex phenomenon, and we do not yet understand all of the factors involved in its regulation. In most animals, appetite is at least partially controlled by the glucose level in the blood—the so-called *blood sugar level.* The liver controls the minimum blood sugar level, preventing it from dropping below a certain point, but the amount of glucose in the blood rises abruptly just after eating (Figure 16-14). This increase is detected by glucose receptors in the hypothalamus of the brain, which cause the individual to feel satiated. Gradually, the excess sugar is removed from the blood and used for energy or converted into glycogen or fat by the liver; the amount of sugar then returns to its minimum level. At this point, the hypothalamus is stimulated and initiates changes in the body that cause the sensation of hunger. This feeling is usually accompanied by stomach pains, which are caused by contractions of the muscles within the stomach walls. After consuming a normal meal containing carbohydrates, fats, and proteins, the period between

Figure 16-14
Change in Blood Sugar Level After Eating a Meal of Carbohydrates, Fats, and Proteins

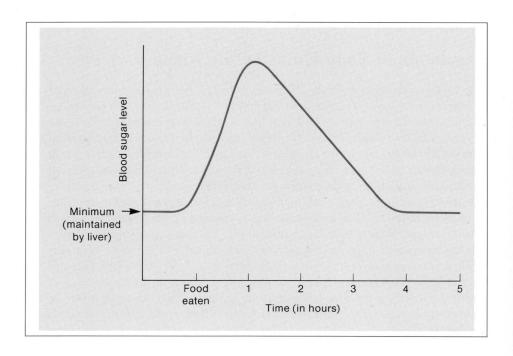

eating and hunger is usually 3–4 hours. For some reason, this control system fails in some people, and they maintain an appetite regardless of their blood sugar level.

A weight loss occurs whenever fewer calories are consumed than the body uses. Weight can be lost by eating less or by exercising more; however, a desired weight loss can usually be achieved more rapidly by dieting than by exercising. The best way to eliminate excess calories is to reduce sugar intake, because sugar does not contribute structural materials to the body and is therefore expendable. Because the body constantly requires amino acids to build enzymes and other structures, proteins should not be excluded from the diet. If the intake of fruits, vegetables, and grains is reduced, the diet should be supplemented with vitamin and mineral tablets. Any reducing diet should include as much salt and water as desired, because these materials are regularly removed from the body through sweat and urine and must be replaced.

Fasting

The quickest way to lose weight is not to eat at all. Very obese people who are more than 100 pounds overweight are now being treated by *fasting,* during which only water, vitamins, and minerals are ingested. Fasting can continue for months as long as the body still contains fat and skeletal muscles that can be broken down into energy and amino acids.

Highly obese people have fasted for more than eight months and lost more than 100 pounds with no undesirable side effects. Such prolonged fasting is always carried out in a hospital under careful medical monitoring. Many people have undergone this extreme treatment, and only one death has been reported. A 20-year-old woman fasted for 210 days, reducing her weight from 260 to 132 pounds. She then began to eat and died shortly thereafter. An autopsy revealed that she had digested all of her fat reserves and about half of her muscles, including some heart muscles. When she became physically active after the fast, the strain on her diminished heart proved fatal.

Long-term fasting is not advisable, although a fast of 7–10 days with a doctor's permission does not appear to be detrimental to overweight individuals. Intermittent fasts of one or two days may even be beneficial; laboratory rats that are occasionally forced to fast for short periods of time live longer than rats that are fed three meals every day.

The chemistry of the body changes during a fast. Very rapid weight loss occurs during the first four days of a fast, because only protein (no fat) is broken down. The proteins of the skeletal muscles are converted into amino acids, and most of these amino acids are converted into glucose, leaving nitrogen as a waste product. The nitrogen is flushed through the kidneys in the form of urea, which requires large quantities of water and temporarily dehydrates the body. In addition, protein contains about 50% fewer calories per unit weight than fat, so that more pounds are lost when the body draws on the calories provided by the muscles than are lost when fat reserves are used.

After the fourth day of a fast, the body begins to make use of the energy stored in fat cells and weight loss takes place more slowly (Figure 16-15). The nervous system continues to use glucose, which is produced by the liver from amino acids, but virtually all other organs and tissues depend primarily on fat as their energy

Figure 16-15
Weight Loss During a Fast
Dashed line indicates how weight
would decrease if there were a
constant loss every day. Solid line
shows how weight actually decreases.

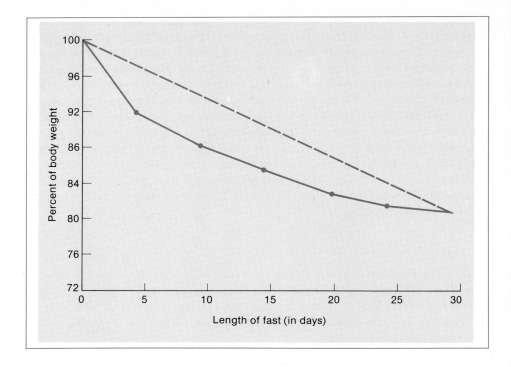

source. Some muscle cells are still broken down to provide amino acids from which glucose, new enzymes, and other essential materials are constructed. After the fifth day of a fast, the nervous system begins to utilize both glucose and *ketones,* which are formed from the fatty acid part of the fat molecule, to conserve protein so that less muscle is broken down. Because the purpose of a short-term fast is usually to lose fat rather than muscle, the fast should continue for several days beyond the fourth day. Children, no matter how obese, should never fast, because they require large amounts of structural materials to continue normal growth.

Food Preferences and New Food Sources

Humans can eat any type of food, but particular populations select only certain plants and animals for consumption. Studies show that immigrants to new countries quite readily change their language and most aspects of their culture before they change their food habits. This aspect of human behavior is deeply rooted.

Despite scientific advances, there are more hungry people in the present century than ever before, primarily due to the vast increase in the world population since 1900 and to the grossly unequal distribution of wealth. Much of the starvation in the world could be relieved if everyone stopped eating meat. Almost 80% of the grain produced in the United States is fed to animals rather than eaten directly by people. (For example, a cow must consume 8 pounds of grain to gain 1 pound.) Although some nutritionists maintain that meat should be included in the human diet, any diet that provides the body with the structural materials it requires is nutritious. It is immaterial from what foods these nutrients are derived.

Although meat animals, including sheep, cattle, pigs, chickens, and fishes, do provide concentrated protein in the human diet, the most efficient biological system for synthesizing protein is not an animal, but the green leaf. Nowhere is protein synthesized more rapidly than in a field of grass during a rapid growth period. Moreover, it is extremely wasteful to harvest only seeds for human consumption because the leaves contain proteins and other nutrients synthesized and accumulated throughout the summer. One aspect of recent biological research has been the development of a practical method of using leaves (even lawn clippings) as a protein source. The technical problem of extracting the nutrients from leaf cells has been mastered, and the protein in leaves has been proved to be valuable to humans. The remaining challenge is to make leaf protein palatable enough for human consumption.

The processing of unicellular organisms, particularly algae and yeast, has also been researched in an attempt to eliminate the use of large animals in the production of attractive protein foods. Algae usually grow abundantly in lakes and ponds and can easily be converted into human food because they do not contain stalks, roots, seed coats, or other structures that require breakdown processing. Like leaves, algae have not yet been made sufficiently palatable to be accepted as a food source.

Yeast can be grown on almost any compound that contains carbon, including molasses, paper waste, garbage, alcohol, and even petroleum wastes from oil refinement. All of these methods must be carefully tested to ensure that the yeast does not absorb toxic materials from the substances on which it is grown. Brewer's yeast—a waste product from the preparation of beer that is commercially available as a nutritional supplement—contains all of the essential amino acids and most of the other nutrients required by humans.

Researchers have also chemically synthesized new foods in laboratories. Vitamin tablets are produced in this way. Amino acids can now be synthesized, so that the missing amino acids can be supplemented in the food of people whose diets consist of a single plant, such as corn or rice. Both sugars and fats are also easy to produce by artificial synthesis. There are, however, two major drawbacks to synthetic food production. First, it requires large amounts of energy to synthesize these biological molecules; second, it removes one of the basic pleasures of human life—dining.

Summary

All organisms must acquire energy and structural materials from the environment to function, grow, and reproduce. *Food* is any substance that an organism can utilize as a source of energy or as a structural material. Plants can obtain food from *inorganic sources.* Except for a few minerals, animals can only obtain food from *biological molecules*, which they acquire by consuming other organisms.

Plants acquire energy from the sun and use it to build *glucose*—through a biochemical reaction called *photosynthesis*. Animals acquire energy from a variety of different biological molecules, depending on the kinds of digestive enzymes they can synthesize. Plants store energy in the form of *starch* (a large *carbohydrate* molecule); animals store energy primarily in the forms of *glycogen* (also a large carbohydrate molecule) and *fat*. The amount of energy an organism requires each day depends on the amount of energy the organism uses. A long-term deficiency of energy intake is called *starvation*.

The structural materials plants require are water, carbon dioxide, oxygen, and minerals. An animal's diet must include all of the structural materials it requires but cannot synthesize. These materials vary, depending on the particular evolutionary history of the animal. Humans require 8 amino acids, at least 13 vitamins, and 15 different minerals in their diet. *Amino acids* are obtained from the *proteins* in food. All animal products, the seeds of *grasses*, and the seeds of *legumes* are good protein sources. Severe protein deficiency causes *marasmus* in children. A *vitamin* is a biological molecule that the cells of the body require in small amounts but cannot synthesize. A natural, balanced diet usually contains sufficient vitamins. *Minerals* include all elements that can be utilized by animals in their inorganic state. A deficiency of structural material is referred to as *malnourishment*.

Digestion is the breakdown of large food items by *digestive enzymes* into particles small enough to be used by the cells of the body. During *hydrolysis*—the chemical reaction facilitated by digestive enzymes—the *monomers* of a *polymer* are separated from one another and a water molecule is added where each chemical bond is broken. In most advanced animals, digestion usually takes place within a digestive cavity or tube, called the *digestive tract*, that passes through the body. This tube has two openings—a *mouth* and an *anus*. Food is moved from mouth to anus by *peristaltic movements* caused by rhythmic contractions of the muscular walls of the digestive tract. Each part of the digestive tract performs a specialized function in food processing. In humans, food is initially prepared in the mouth, where it is chewed and mixed with *saliva*. The components of saliva initiate the digestion of starches, begin to dissolve the food particles (which stimulates the *taste buds*), and moisten the food so that it can move easily down the *esophagus* to the *stomach*. The stomach is primarily a storage cavity, but also initiates protein digestion.

In humans, most digestion and practically all *absorption* occurs in the *small intestine*. Digestive enzymes are released into the *duodenum*—the first segment of the small intestine—from the *pancreas* and the walls of the small intestine, and *bile* is added from the *liver* via the *gallbladder*. Sugars and starches are converted into simple sugars (such as *glucose*), proteins are converted into *amino acids*, and fats are converted into *glycerol* and *fatty acids*.

The basic monomers released from polymers during digestion are absorbed through the walls of the small intestine into the blood vessels and *lymphatic vessels*. Simple sugars and amino acids enter the bloodstream, which transports them to the liver. Glycerol and fatty acids are re-formed into *triglycerides*, which combine with proteins and cholesterol to form *lipoproteins*, and are moved via the lymphatic vessels into fat cells for storage. The inner surface of the small intestine is covered with small projections, called *villi* and *microvilli*, which greatly increase the surface area of the intestine and therefore its capacity for absorption.

Food material that is not digested and absorbed into the blood and lymph moves into the *large intestine* and eventually out the anus in the form of *feces*. Much of the water is removed from the unused food material as it passes through this part of the digestive tract. The water moves across the walls and enters the bloodstream.

The *liver* is essential to the processing and utilization of food. This vital organ (1) produces bile to be used in the digestion of fats, (2) converts the end products of digestion into the particular molecular forms required by body cells, (3) regulates the components of the blood, and (4) converts toxic and foreign substances into compounds that can be excreted from the body by the *kidneys*. The liver regularly converts *ammonia*—a toxic waste product of protein synthesis and breakdown—into *urea*, a nontoxic substance that is excreted by the kidneys. The pancreas secretes a hormone, *insulin*, which is essential to the utilization of sugars.

Body weight—an important factor governing health and longevity—is regulated by diet and exercise. It is difficult for many humans to maintain an ideal weight because they are unable to control their appetites. Appetite is at least partially controlled by changes in the *blood sugar level*. Shortly after eating a meal, the concentration of sugar in the blood is high and appetite is diminished; as the time following a meal passes, the concentration of sugar in the blood decreases and appetite increases. Very obese people have reduced their body weight by *fasting*. This method of weight loss should be undertaken only under the supervision of a medical doctor, however, because serious damage may occur if muscles are extensively broken down.

Food preferences are largely cultural in origin and are usually difficult to alter. Nevertheless, as human populations continue to grow, we may have to make profound changes in our diets. We will probably consume greater and greater amounts of plant material and may eventually acquire our energy and structural materials from unicellular organisms and synthetic foods. In the not so distant future, the pleasures of dining may have to be reserved for special occasions.

Regulation of Water and Temperature

17

Life began in the oceans, and all living things are composed mostly of water. The chemistry of life is therefore based on the specific properties of water. Biological molecules are dissolved or suspended in a watery solution that allows them to interact with each other and to move easily from place to place within the cell or organism. Water also plays an essential role in chemical reactions, because large molecules are constructed and broken apart by the removal and addition of water molecules.

Water is a dominant factor in the regulation of an organism's body temperature. Body temperature must be controlled to some degree because the rate at which chemical reactions occur is temperature-dependent. As the body temperature increases up to about 45°C (114°F), chemical reactions occur more rapidly; beyond this point, the structure of enzymes is destroyed and the organism dies. As the body temperature decreases, chemical reactions occur more and more slowly until water freezes at about 0°C (32°F), at which point the cell structures are damaged by ice crystals.

Three special properties of water facilitate temperature control in living systems. First, water has a *high thermal conductivity,* so that heat generated in one part of the body moves quickly through water solutions to other parts of the body. This property warms the entire body, preventing heat from accumulating in local areas and generating dangerously high temperatures.

Second, water has a *high specific heat,* so that a great deal of heat energy is required to change the temperature of water, compared with other substances. This property helps to maintain a constant body temperature by protecting the tissues, which are predominantly composed of water, from extreme temperature changes.

Third, water has a *high latent heat of vaporization,* so that an unusually large amount of heat energy is required to change water from a liquid to a vapor state (evaporation). Excess heat is removed from organisms primarily by the evaporation of water. In humans and some other mammals, water is removed from the skin in the form of sweat; in many terrestrial vertebrates, most water evaporates from the mouth and tongue. These methods of lowering body temperature are effective but also result in considerable water loss from the body. The amount of water contained in the body and the temperature of the body are therefore closely related states, and their mechanisms of control are interrelated.

In addition to promoting chemical activities and regulating temperature, water is required for the performance of many other bodily functions. Water is essential to digestion, absorption, secretion, excretion, and the transport of materials. Water moistens surfaces involved in gas exchange and cushions the brain and spinal cord to protect them from damage. Moving surfaces, such as joints and lungs, are lubricated by watery solutions. Water also dissolves molecules that enter the nose and mouth, making both smell and taste possible.

In this chapter, we will examine the problems associated with maintaining appropriate amounts of water in the body and the ways in which these problems are overcome by biological adaptations. We will then discuss how body temperature is regulated and the role of water in this homeostatic mechanism.

Osmosis and the Problems of Water Balance

A particular concentration of water must be maintained in relation to the concentrations of other molecules and ions within the tissues. This concentration is usually quite different from the concentration of water in the organism's external environment. Water molecules move passively across cell membranes by *osmosis* from areas of high to areas of low concentration. Whether a cell takes in or loses water depends on the difference between the water concentrations inside and outside the cell (Figure 17-1). If the fluids surrounding a cell have a higher water concentration (for example, pure water) than the fluids inside the cell, water will tend to move into the cell and the cellular materials will become more dilute. If the fluids surrounding a cell have a lower water content (for example, very salty water) than the fluids inside the cell, water will tend to move out of, or dehydrate, the cell.

The most important materials in the process of osmosis are the ions of sodium (Na^+) and chloride (Cl^-). When these ions are not dissolved in water, they combine to form sodium chloride (NaCl), or table salt. Sodium and chloride ions are abundant in the waters of the earth and in the solutions inside organisms and are major factors in determining the direction in which water molecules move across biological membranes. As the concentrations of sodium and chloride ions change

Figure 17-1
Osmosis and the Problems of Water Concentration Within a Cell

Each circle represents a cell; the dots represent dissolved materials (predominantly Na^+ and Cl^-) inside and outside the cell. As the density of the dissolved materials increases, the concentration of water molecules decreases. Arrows indicate the direction in which the water molecules tend to move by osmosis. (a) The environment surrounding this cell has a higher water concentration (contains a smaller amount of dissolved materials) than the cell interior. Water will therefore tend to move into the cell until the concentration of water and ions is the same inside and outside the cell. In the process, the cell will expand and may burst. (b) The environment of this cell has a lower water concentration than the solution inside the cell. Water will therefore tend to move out of the cell, and it will shrivel. (c) The concentration of water molecules is the same inside and outside this cell. About the same amounts of water will therefore tend to move into and out of the cell, and the cell will not be altered by its environment.

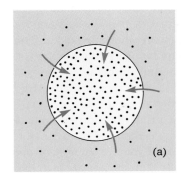 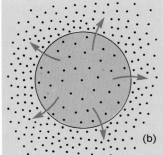 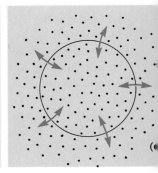

on either side of the membrane, the concentration of water also changes, causing osmosis to occur.

The concentration of materials within the cells of the earliest organisms was approximately the same as the concentration of materials in the oceans; this condition is still true of present-day marine plants and simple marine invertebrate animals. During the course of evolution, some groups of organisms moved from the oceans to freshwater lakes and streams and others moved to dry land. Major groups of fishes have returned to marine waters after evolving for a considerable amount of time in fresh water, where the water content of their cells gradually changed. Organisms that have left their original environments cannot easily maintain appropriate concentrations of water and ions, because their tissues now function optimally at concentrations that differ from the concentrations of water and ions in their present environments.

Freshwater Plants and Animals

The concentration of water within a cell is lower than the concentration of fresh water, so that water tends to move into the cell. If it were allowed to expand indefinitely, the cell would eventually burst. In plants, however, a rigid cell wall restricts cell size as well as the amount of water that can enter the cell and the degree of dilution within the cell (Figure 17-2). Under normal environmental conditions, plants thrive in solutions with high water concentrations.

Animal cells are not protected by cell walls, and fresh water presents a genuine threat to their existence. Excess water is prevented from entering the cells of

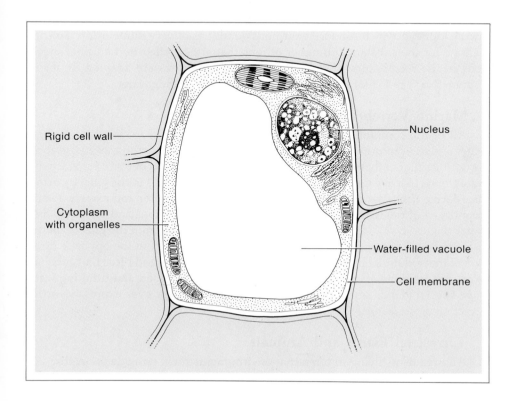

Rigid cell wall

Cytoplasm with organelles

Nucleus

Water-filled vacuole

Cell membrane

Figure 17-2
The Structure of a Plant Cell Allows It to Survive in Pure Water

Most terrestrial and freshwater plants are exposed to solutions that contain a much higher concentration of water than their cells do. (The ion and water concentrations in the cells of marine plants are similar to those in seawater.) A plant cell is protected from the entry of too much water by its *cell wall*, which is rigid and cannot expand. Plant cells in fresh or pure water contain large membrane-enclosed *vacuoles*, or *sacs*, which store excess water to minimize the dilution of materials within the cell. The pressure of the water inside these vacuoles makes the cell rigid and helps the plant to maintain an upright position. When the vacuoles of plant cells are not filled with water, the plant will wilt.

Figure 17-3
The Movement of Water and Ions in a Freshwater Fish

The tissues of freshwater fishes have a higher ion concentration than the surrounding water does. To prevent the accumulation of water and the depletion of ions in their tissues, these fishes avoid swallowing fresh water, have impermeable body surfaces (except for their gills), acquire sodium ions by active transport across their gills, and produce large quantities of dilute urine.

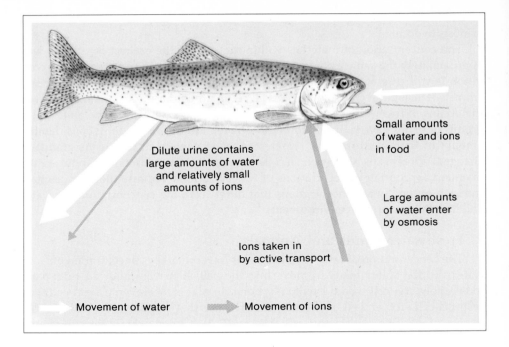

Small amounts of water and ions in food

Dilute urine contains large amounts of water and relatively small amounts of ions

Large amounts of water enter by osmosis

Ions taken in by active transport

Movement of water   Movement of ions

freshwater fishes in three ways (Figure 17-3). First, these fishes avoid swallowing the surrounding water, although they consume some fresh water while feeding. Second, very little water moves through their body surfaces. Third, the large quantities of water that do enter the body across the exposed gill surfaces are removed by the kidneys, which produce enormous quantities of very dilute urine. To compensate for the substantial number of ions that are expelled in the urine, special cells in the membranes of the gills of freshwater fishes are adapted to absorb sodium (Na^+) and chloride (Cl^-) ions from the surrounding water.

Marine Vertebrates

All of the common marine fishes except sharks and skates evolved in fresh water and then returned to the oceans. The ion concentration of seawater is about three times higher than the ion concentration within the cells of these fishes, so that water tends to leave their cells (particularly the exposed cells in the gills) by osmosis. Marine fishes swallow an enormous quantity of seawater and excrete a small quantity of urine, which contains a high concentration of ions and relatively little water (Figure 17-4). In addition, certain specialized cells within the gills of marine fishes excrete sodium and chloride ions by active transport in the opposite direction of that exhibited by freshwater fishes. Marine reptiles and birds that consume excessive amounts of sodium and chloride ions in their food and drinking water have special glands in their noses or the corners of their eyes expressly for the secretion of these ions.

Terrestrial Plants and Animals

All organisms living in terrestrial environments must struggle to acquire and conserve water, which tends to leave the body and evaporate into the dry air. Be-

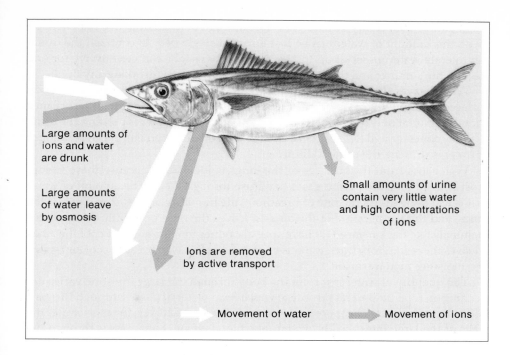

Figure 17-4
The Movement of Water and
Ions in a Marine Fish

The concentration of dissolved
materials is lower in the tissues of
marine fishes than it is in the
surrounding water. To prevent the
loss of water from their tissues and the
accumulation of dissolved materials,
these fishes swallow large quantities
of seawater and excrete most of the
ions they acquire through their gills
and in their urine. All body surfaces
except the gills are impermeable to
water.

Large amounts of
ions and water
are drunk

Large amounts
of water leave
by osmosis

Small amounts of urine
contain very little water
and high concentrations
of ions

Ions are removed
by active transport

Movement of water   Movement of ions

cause the environmental sources of water are usually low in minerals, which form
inorganic ions in water, many terrestrial organisms must also struggle to obtain
and maintain essential ions within their tissues. Some of the adaptations that these
organisms have acquired for regulating the ion and water contents of their cells are
discussed in the following section.

The Achievement of Water Balance
in Terrestrial Vertebrates

A terrestrial animal acquires minerals—and the inorganic ions formed from these
minerals—and water by eating and drinking. Food contains various forms of minerals and water in both free and metabolic forms. *Free water* is already in a liquid
state. Most plants and animals are composed predominantly of water, so that free
water is a component of food when these organisms are eaten. Green vegetables
contain over 90% free water; meat, more than 50%; bread, at least 35%. *Metabolic
water* is formed in the cells, primarily when glucose is broken down during cellular
respiration to form carbon dioxide, water, and energy.

Some animals, such as the kangaroo rat, koala bear, and gerbil, obtain all of
their water from food and do not drink fluids. Several special adaptations enable
these animals to conserve the little water they acquire from their food. However,
most animals drink much of the water they require. *Neurons* within the hypothalamus of the brain are stimulated by low fluid volume and high ion to water ratios in
the blood to cause the sensation of thirst.

The amount of time that an animal can exist without water varies considerably,
depending on the species. Humans become completely incapacitated after only

about 10% of their total body weight is lost through dehydration. In a hot environment, this amount of water can be lost during a single day. In contrast, the camel can tolerate water losses of up to 25% of its body weight and can survive for one week in the summer and three weeks in the winter without drinking any fluids.

Water leaves the body through the kidneys, lungs, large intestine, and body surfaces. The amount of water in the urine is adjusted by the kidneys in accordance with the amount of water in the body. If a low amount of water is present, the urine is highly concentrated (less water compared with other materials); if a large amount of water is present, the urine is dilute.

Water loss from the surfaces of the lungs is inevitable, because these tissues must be kept moist to permit gases to diffuse through them. The amount of water that is lost depends on the rate of breathing, the humidity of the air as it enters the lungs, and the temperature of the air as it leaves the nose or mouth. Small desert mammals like the kangaroo rat have special cavities in their noses to cool the air as it leaves the respiratory tract; water is conserved in this way because cool air retains less moisture than warm air.

The quantity of water lost from the body through the large intestine varies depending on food intake and the rate at which fecal material passes through this part of the digestive tract. The more slowly the fecal material moves, the more water the walls of the large intestine absorb into the bloodstream. Food passes through the large intestine more rapidly when the diet contains a great deal of roughage or when the lining of the intestinal tract is irritated and undergoes rapid muscular contractions. Dirarrhea, which results from such irritation, can eliminate enormous quantities of water from the body.

Water is lost in the form of sweat from the skin of humans and other vertebrates and in the form of saliva from the mouths and tongues of many animals. The evaporation of water from sweat or saliva regulates body temperature by removing heat from the body. As the body becomes warmer, more water is lost in this way. A great deal of water may be removed from the body during strenuous exercise or when the environment is hot and dry.

The Kidneys of Humans and Other Mammals

The kidneys are vitally important in maintaining the correct body chemistry. The liver and the kidneys regulate the composition of the blood and interstitial fluids, which in turn determine the materials within the cells. The kidneys

1. Remove wastes that form during cellular metabolism as well as toxic substances and excess nontoxic materials from the body.
2. Maintain appropriate concentrations of water and ions, especially sodium, potassium, chloride, calcium, magnesium, sulfate, and phosphate ions.
3. Regulate the volume of blood and interstitial fluids by controlling ion excretion.
4. Ensure an appropriate balance of hydrogen ions (H^+) and hydroxyl ions (OH^-), so that the acidity (pH level) of the body fluids is optimal for chemical activities.

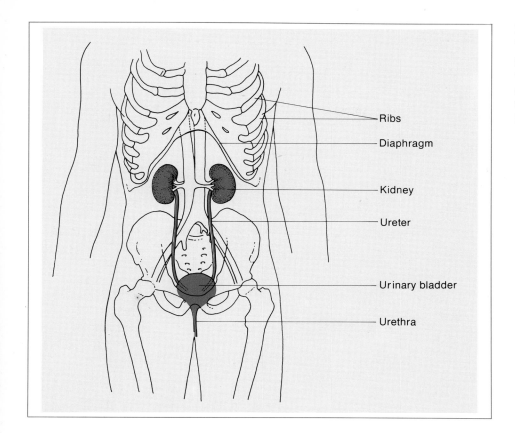

Figure 17-5
The Kidneys and Associated
Structures in a Human
The other organs in the abdominal
cavity, which lie in front of the
kidneys and ureters, are not shown
here.

Labels in figure: Ribs, Diaphragm, Kidney, Ureter, Urinary bladder, Urethra

The Structure of the Kidneys and Associated Organs

The normal individual has two kidneys, one on each side of the back abdominal wall below the diaphragm and behind the liver and stomach. Each kidney is connected by a tube, the *ureter,* to a single urinary bladder (Figure 17-5). Urine is formed in the kidneys and moved to the *bladder*—a muscular reservoir that holds up to 600 ml (about 1 pint) of fluid. When the bladder becomes distended with urine, it empties, usually under voluntary control. Urine moves from the bladder to the outside of the body through a single tube, the *urethra.* In females, the urethra is only about 4 cm ($1\frac{1}{2}$ inches) long and exits the body in front of the vagina. The urethra of a male is 15 cm (6 inches) or more, and exits the body through the penis. In males, the urethra transports both urine and sperm.

Each kidney is composed of more than a million tiny tubes, called *nephrons,* of identical structure and function (Figure 17-6). These nephrons form urine from components of the blood. One end of each nephron is closed and shaped into a cuplike depression called *Bowman's capsule.* A network of capillaries packed within this capsule receives blood from an artery. After leaving the capillaries within Bowman's capsule, the blood travels via another artery (instead of a vein, as is usually the case) to the area surrounding the remaining portions of the nephron

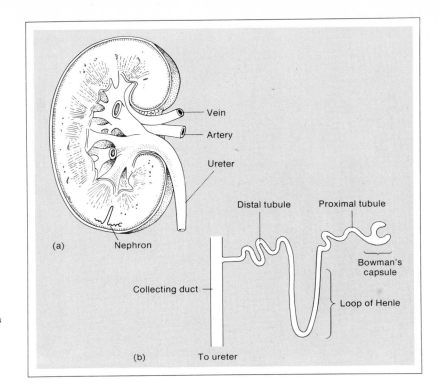

**Figure 17-6
The Nephron**

(a) Position of the nephron inside a kidney. (b) Fluid flows from *Bowman's capsule* of the nephron to the collecting duct as urine is formed. Each kidney contains more than a million nephrons.

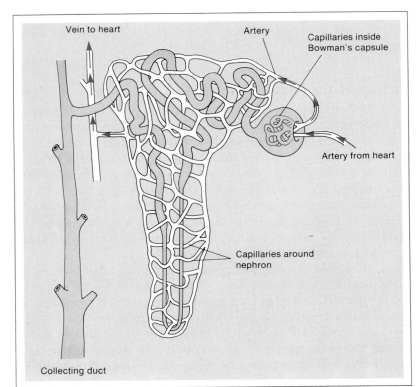

**Figure 17-7
The Circulatory System
Associated with Each
Nephron**

Two capillary networks—one inside Bowman's capsule and the other surrounding the nephrons—exchange materials with the interior of the nephrons during the formation of urine.

(Figure 17-7). There, the artery subdivides again to form a second network of capillaries that nourish the cells of the nephron and exchange materials with the fluid inside the nephron during urine formation. Blood leaving these capillaries moves into veins and is returned to the heart.

Urine Formation: Filtration and Volume Reduction

Components of the blood enter the nephron from the capillary network in Bowman's capsule (Figure 17-8). The blood inside Bowman's capsule is separated from the space within the nephron by a thin barrier of only three tissues—the capillary wall, a membrane, and the outer lining of Bowman's capsule. Blood pressure forces fluid through these tissues and into the nephron. At this point in the circulatory system, the blood pressure is about 90 mm Hg, which is considerably higher than the blood pressure in other capillaries. Most of the components of the blood, except cells and proteins, filter through the barrier into the nephron. These substances include water, glucose, urea and other molecules containing nitrogen, and a variety of ions.

The fluid that enters the nephron at Bowman's capsule is much more dilute

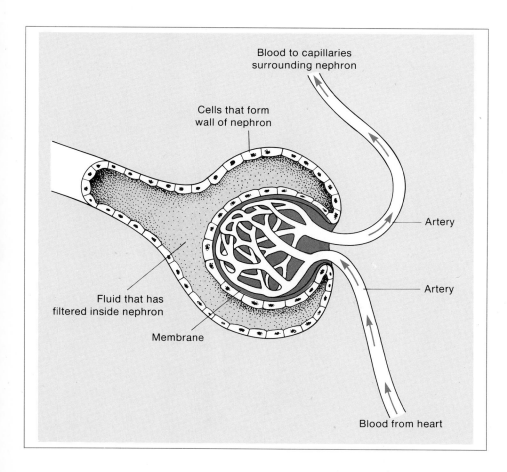

Blood to capillaries
surrounding nephron

Cells that form
wall of nephron

Artery

Artery

Fluid that has
filtered inside nephron

Membrane

Blood from heart

**Figure 17-8
Bowman's Capsule
of the Nephron**

Blood from the heart enters the network of capillaries inside Bowman's capsule. Here, many components of the blood filter through the walls of the capillaries, a membrane, and the wall of Bowman's capsule to enter the interior of the nephron. The components of the blood that remain inside the capillaries move through another artery to capillaries on the outside of the remaining portions of the nephron (see Figure 17-7).

than the concentrated urine that leaves the nephron to enter the urinary bladder. About 180 liters (45 gallons) of fluid enter the nephrons of both kidneys each day, but only about 2 liters ($\frac{1}{2}$ gallon) of urine are excreted. About 80% of the fluid that entered the nephron is removed through the walls of the *proximal tubule* and returned to the blood. Glucose and other nutrients are also actively transported from the nephron back to the blood at this point.

The remaining 20% of the fluid that entered Bowman's capsule proceeds to the next portion of the nephron, the *loop of Henle* (see Figure 17-6), where the urine is further reduced in volume. The loop of Henle is found only in mammals and is an adaptation of the kidney to conserve water. Structurally, the loop is composed of two adjacent limbs—a descending limb and an ascending limb (Figure 17-9). The hairpin turn between these two limbs causes a countercurrent flow, so that the fluid in the descending limb moves in the opposite direction to the fluid in the ascending limb. The *ascending limb* is impermeable to water and actively transports sodium ions out of the nephron and into the surrounding interstitial fluid. A great deal of sodium is pumped out, especially in the area near the hairpin turn. Because the

Figure 17-9
The Loop of Henle Portion of the Nephron

As fluid moves down the *descending limb*, water is drawn out of the nephron due to the high concentration of sodium ions surrounding the limb (proportional to the density of dots in the diagram) and sodium ions move into the nephron. The fluid contains a high concentration of sodium ions and a low concentration of water by the time it reaches the hairpin turn at the bottom of the loop. As the fluid moves up the *ascending limb*, sodium ions are removed by active transport. Thus, as fluid travels through the loop of Henle, water is moved from the nephron into the bloodstream and sodium ions draw the water out and are returned to the nephron.

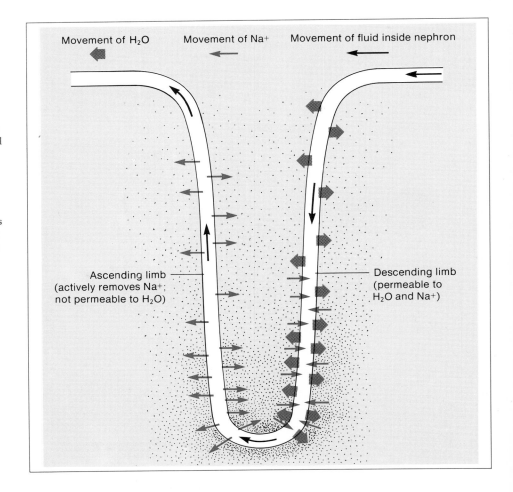

descending limb is permeable to sodium ions and water and is surrounded with high concentrations of sodium ions, which have been pumped out of the ascending limb, water moves out of and most of the sodium ions move into the descending limb. As the fluid moves down the descending limb of the loop of Henle, it is reduced in volume and the concentration of sodium ions is increased.

Once the highly concentrated fluid has moved around the hairpin turn, it enters the ascending limb, where sodium ions are actively transported out of the nephron. At this point, the fluid has been reduced to about 10% of the amount that entered Bowman's capsule. The sodium ions become less and less concentrated as the fluid moves up the ascending limb until approximately the same concentrations of sodium ions and water occur in the blood and interstitial fluids and in the ascending limb.

The amount of water, but not the number of sodium ions, in the nephron fluid is greatly reduced in the loop of Henle. The water is returned to the bloodstream and conserved. The countercurrent flow in the loop of Henle permits the sodium ions to draw water out of the nephron, but also allows most of the sodium ions to recirculate into the fluid to be used again.

Urine Formation: The Adjustment of Ion and Water Concentrations

The reduced volume of fluid moves from the loop of Henle into the *distal tubule* of the nephron, where the ion concentration is adjusted, and on to the *collecting duct*, where the water concentration is adjusted (see Figure 17-6). The fluid from several distal tubules moves into a single collecting duct, so that each kidney contains fewer collecting ducts than nephrons. The presence of two hormones, aldosterone and antidiuretic hormone (ADH), dictates which one of four adjustments—the removal of ions, water, both ions and water, or no ions or water—will be made in the ion and water concentrations in the fluid as it passes through the distal tubules and collecting ducts.

Many different kinds of ions may be pumped out of the distal tubule and returned to the blood. This activity is governed by *aldosterone*—a hormone secreted by the adrenal cortex in response to low blood pressure. When more of this hormone is present, more ions are returned to the blood from the nephron. If the walls of the collecting duct are permeable to water, then water is also drawn out of the nephrons and into the blood, increasing the blood pressure, as illustrated in Figure 17-10(a). The secretion of aldosterone is controlled by the vessels of the kidneys. Specialized cells lining the interior of small arteries around the nephrons produce a hormone, *angiotensin*, which causes the arteries to constrict and stimulates the adrenal cortex to produce aldosterone. As blood volume drops, more angiotensin is produced, the vessels become smaller, and more aldosterone is released. When more ions and water are returned to the blood, the blood volume is restored and the production of angiotensin decreases.

The second hormone, *antidiuretic hormone* (ADH), controls the amount of water in the urine. This hormone, produced in the hypothalamus and stored in the posterior pituitary gland, determines the permeability of the walls of the collecting duct to water. When the water content of the body is high, very little ADH is

Figure 17-10
Activities of the Distal Tubule and Collecting Duct

Fine adjustments in blood volume and the ratio of ions to water in the blood are controlled by these two parts of the nephron. (a) When the blood pressure drops, both aldosterone and ADH are produced, so that ions leave the distal tubule and water leaves the collecting duct. (b) When the blood contains a small amount of water in relation to ions, only ADH is produced and water is removed from the collecting duct and added to the blood. Another adjustment not illustrated here—the production of only aldosterone and the addition of only ions to the blood—occurs when the concentration of water in the blood is high compared with the concentration of ions. When optimal amounts of ions and water are present in the blood, neither hormone is produced, so that ions are not pumped out of the distal tubule and water does not leave the collecting duct.

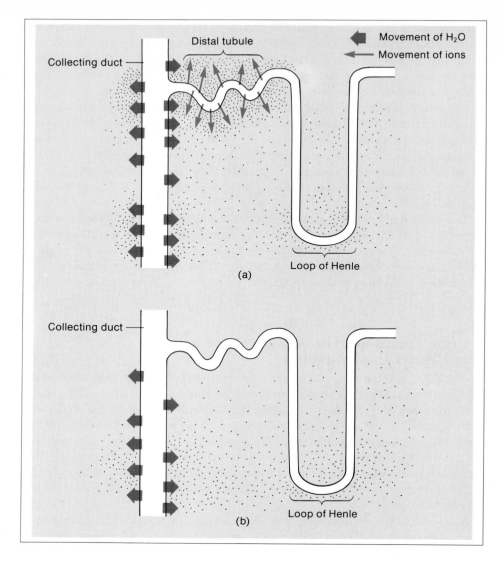

produced, making the walls of the collecting duct impermeable to water, and large quantities of relatively dilute urine are formed. When the body is dehydrated, ADH is secreted, making the walls of the collecting duct permeable to water, which moves out of the nephron by osmosis. At this point, the urine contains relatively small amounts of ions compared with the surrounding area, which has a high ion concentration due to the active transport of sodium ions by the ascending limb of the loop of Henle, as Figure 17-10(b) illustrates. Even more water may be drawn out when aldosterone causes ions to leave the fluid in the distal tubule of the nephron. In the presence of ADH, the urine becomes extremely concentrated and greatly reduced in volume. The production of ADH is increased when blood pressure is low and when a small amount of water in relation to ions is present in the blood. A drop in blood pressure is detected by special receptors in the left atrium of the heart, and this information is relayed by nerve impulses to the hypothalamus. The concentrations of water and ions in the blood are monitored by neurons within the hypothalamus.

Both alcohol and caffeine suppress the production of ADH. When liquids containing either of these materials are consumed, unusually large amounts of dilute urine are formed and the body becomes dehydrated. The production of aldosterone and ADH is correlated, so that the two systems work together to control the balance of water and ions in the blood (Figure 17-10).

Urine Formation: Regulation of Blood pH

The acidity of a liquid is measured by its pH, which ranges between 0 and 14 on a logarithmic scale (Table 17-1). A neutral liquid has a pH of 7, which means an equal number of hydrogen ions (H^+) and hydroxyl ions (OH^-) are present in the solution, as occurs in pure water (H_2O, or H^+OH^-). A drop in pH indicates that the number of hydrogen ions has increased in relation to the number of hydroxyl ions in the solution. As the pH level drops, the solution becomes more *acidic*. A rise in pH indicates that the number of hydroxyl ions has increased in relation to the number of hydrogen ions, and the solution becomes more *alkaline*, or *basic*.

Most chemical reactions are extremely sensitive to the pH level, which is altered in biological systems primarily by changes in the concentration of hydrogen ions. Molecules that contribute hydrogen ions to a solution are acids; a strong acid releases more hydrogen ions than a weak acid. Many strong acids are formed during cellular activities. For example, a high-protein diet results in the production of sulfuric acid. When amino acids are broken apart to be used as a source of energy, the sulfur atoms that form part of the molecular structure of some amino acids are released into the bloodstream in the form of sulfuric acid (H_2SO_4). In solution,

Table 17-1
The pH Scale

pH measure	Ratio of H^+ to OH^-	Example
14	1:10,000,000	Sodium hydroxide (NaOH)
13	1:1,000,000	
12	1:100,000	
11	1:10,000	Milk of magnesia ($Mg(OH)_2$)
10	1:1,000	
9	1:100	
8	1:10	Egg white
7	1:1	Pure water
6	10:1	Milk
5	100:1	
4	1,000:1	
3	10,000:1	Aspirin
2	100,000:1	Lemon juice
1	1,000,000:1	
0	10,000,000:1	Sulfuric acid (H_2SO_4)

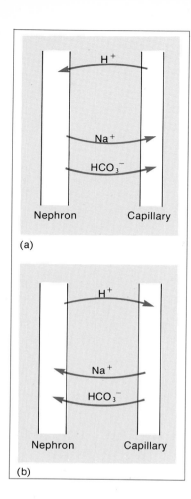

Figure 17-11
Regulation of Blood pH
by the Kidney

(a) When the blood is too *acidic*, hydrogen ions move from the blood into the nephrons. Both sodium ions and bicarbonate ions move from the fluid inside the nephrons into the blood, where they combine with water to form hydroxyl ions and a weak acid, carbonic acid. The sodium ions replace the hydrogen ions in the blood. (b) When the blood is too *alkaline*, hydrogen ions are added to the blood and the sodium and bicarbonate ions that form hydroxyl ions move from the blood into the nephrons.

this molecule dissociates to form two hydrogen ions ($2H^+$) and one sulfate ion ($SO_4^=$). Sulfuric acid is a strong acid because both of its hydrogen atoms form ions in solution. These excess hydrogen ions lower the pH of the blood and must be removed if normal cellular activities are to continue.

Molecules that tend to raise blood pH by yielding hydroxyl ions are called *bases.* In rare instances, too many basic compounds enter the blood, primarily from medicines taken to relieve indigestion (antacids) or constipation (laxatives). Prolonged vomiting can also increase blood pH because it removes acids from the body.

The pH of human blood is maintained at the optimal level of 7.4 by the movement of ions between the fluid inside the nephrons and the blood in the surrounding capillaries. Virtually all excess hydrogen ions in the blood enter the urine by secretion (not filtration)—by active transport across the walls of the capillaries into the portion of the nephron beyond Bowman's capsule (see Figure 17-7). At the same time, sodium and bicarbonate ions (HCO_3^-) move out of the nephron and into the blood, as Figure 17-11(a) illustrates. Sodium ions promote the formation of hydroxyl ions from water, increasing blood pH and making the blood more alkaline. The reaction is

$$Na^+ + H_2O \longrightarrow Na^+OH^- + H^+$$

The hydrogen ions that this reaction yields combine with the bicarbonate ions to form carbonic acid (H_2CO_3). This acid is very weak because the hydrogen ions tend to be held within the molecule and do not contribute greatly to the number of hydrogen ions in the blood. The sodium ions replace the hydrogen ions that moved from the blood into the nephrons, so that the overall electrical charge of the blood is conserved. Thus, blood pH increases due to the removal of hydrogen ions and the formation of hydroxyl ions.

The blood can also become too alkaline, although this condition is less common. When the blood contains too many hydroxyl ions in relation to hydrogen ions, hydrogen ions are transported from the nephrons into the blood. Both sodium and bicarbonate ions are then removed from the blood and added to the fluid in the nephrons, as shown in Figure 17-11(b), to reduce the formation of hydroxyl ions in the blood. Thus, the number of hydrogen ions is increased and the number of hydroxyl ions is decreased; the blood becomes more acidic, and blood pH is restored.

The Regulation of Body Temperature

Most chemical reactions are extremely sensitive to temperature. In general, the rate of a biological reaction approximately doubles with each 10°C increase in temperature up to about 45°C (114°F), at which point enzymes tend to break apart.

All animals must find external environments with or maintain internal environments at temperatures that will support their cellular activities. This problem is more acute for terrestrial animals because air has a much lower specific heat than water. Air temperature is altered by the addition or removal of relatively small amounts of heat energy and fluctuates considerably. Terrestrial animals are also exposed to more heat from solar radiation, which penetrates air much more readily than water. Temperatures on land often exceed the range within which most living things can remain active; this condition rarely occurs in large bodies of water.

Animals can be divided into two groups according to their method of temperature control. A *poikilothermic* ("poikilo" = varied; "thermal" = heat) *animal* has a relatively variable body temperature that is determined largely by the heat acquired from its external environment (Figure 17-12). A poikilotherm is unable to generate sufficient heat internally to maintain high body temperatures. An animal that regulates its temperature in this way is also called *cold-blooded* or *ectothermic* ("ecto" = outside). All animals except birds and mammals are poikilotherms. A *homeothermic* ("homeo" = similar) *animal* maintains a relatively constant internal body temperature independently of the temperature of its external environment (Figure 17-12). Homeotherms can generate enough internal heat from chemical reactions to maintain a high body temperature. Homeotherms are also called *warm-blooded* or *endothermic* ("endo" = inside). All birds and mammals are homeotherms.

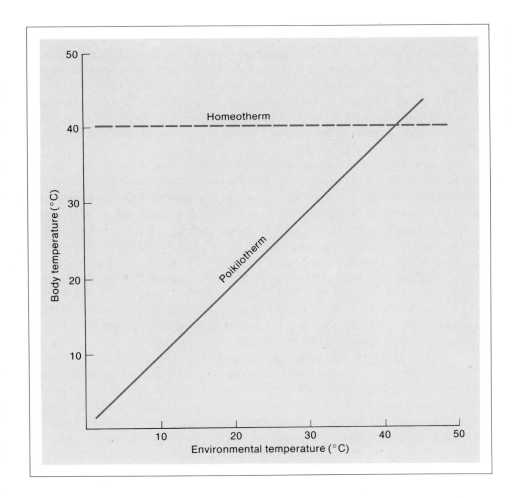

**Figure 17-12
The Basic Difference
Between a Poikilotherm
and a Homeotherm**
The body temperature of a poikilotherm is closely correlated with the temperature of its environment. The body temperature of a homeotherm remains almost constant over a broad range of environmental temperatures. In this example, the homeotherm maintains a body temperature of approximately 40°C.

An organism acquires heat from virtually all chemical reactions that occur in the body, but most body heat is produced by *cellular respiration.* Cellular respiration is usually an *aerobic reaction* (involving oxygen) during which glucose combines with oxygen to form carbon dioxide, water, and energy. Some of the energy produced by this reaction becomes part of ATP molecules, but most of this energy becomes heat. Cellular respiration may also be an *anaerobic reaction* (involving no oxygen); this form of respiration occurs primarily in the skeletal muscles of animals during rapid movement and also produces heat. In addition to the heat produced within the body, an organism acquires heat from the external environment when the temperature outside the body exceeds the temperature inside the body. The body can also be warmed by solar radiation; heat can be accumulated on a cool day by basking in the sun.

An organism loses body heat when the temperature of the body is higher than the temperature of the external environment. Terrestrial organisms also lose body heat when water evaporates from their body surfaces; aquatic organisms do not lose water through evaporation because they are immersed in water.

Direction of sun's rays

(a)

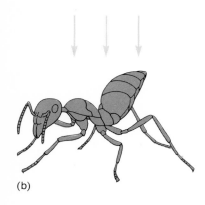

(b)

**Figure 17-13
Posture Influences the Amount of Heat an Animal Absorbs from the Sun**

(a) Some ants hold their entire body at right angles to the direction of the sun's rays when they need to acquire heat. This posture exposes the maximum body area to solar radiation. (b) When the ant needs to reduce heat input, it holds its abdomen upright. This posture reduces the body area that is exposed to solar radiation.

Poikilothermic Animals

Because a poikilotherm's body temperature is highly dependent on environmental heat sources, its internal temperature is most effectively controlled by the selection of a suitable environment. When external temperatures are cold, a poikilotherm cannot increase its body temperature by generating heat because the rate at which its chemical activities take place is directly proportional to its body temperature. A decrease in temperature lowers the rate of chemical activity and therefore the rate of heat production.

The body temperature of aquatic poikilotherms is closely correlated with the temperature of the surrounding water. Aquatic animals therefore seek waters with preferred temperatures. Some aquatic poikilotherms apparently are equipped with sensitive temperature receptors that enable them to detect appropriate water masses. Terrestrial animals can usually choose among a greater variety of environmental temperatures (warm sun, cool shade, and so on). They also receive solar radiation by basking in the sun; the color of an animal and its posture can greatly influence the amount of heat acquired in this manner (Figure 17-13). Terrestrial vertebrates decrease their body temperatures by moving into cooler areas and by evaporative cooling from body surfaces. Reptiles make rapid breathing movements similar to panting to circulate air over the moist surfaces of their mouths when their bodies become too warm.

Most poikilotherms enter some form of resting state when environmental temperatures become extreme. Adult animals seek safe shelter from predators during winter and allow their body temperatures and cellular activities to decrease. As long as their body fluids do not freeze, poikilotherms can remain in this state for long periods without eating because very little energy is required to maintain tissues at cold temperatures. The pupal stage in the development of insects and the egg stage in the development of many animals are adapted for long-term survival under biologically inhospitable conditions.

Homeothermic Animals: Physical Adaptations

Birds and mammals have acquired physical adaptations for insulating the body from the environment and elaborate feedback systems of heat production and evaporative cooling that enable them to maintain constant, relatively high body temperatures. Each species of bird and mammal maintains its own particular body temperature, which is usually around 40°C (103°F); the average temperature in the human is 37°C (98.6°F).

Physical adaptations of the body facilitate heat retention or loss, depending on the climatic conditions under which the species has evolved. Birds and mammals in cold environments have thick, fluffy outer coverings of feathers or fur and thick layers of fat beneath the skin, which insulate the body and prevent the loss of body heat. The extremities of mammals living in cold climates are often smaller, thus minimizing heat loss from these surfaces (Figure 17-14). The extremities of many homeotherms can function at temperatures considerably below the temperature of the body core. The considerable heat loss that can occur when warm blood moves into these cooler extremities is minimized by shunting blood flow from peripheral veins to the larger interior veins of these body parts (Figure 17-15). The major artery and vein in an extremity lie adjacent to each other, causing a countercurrent flow; warm blood in the artery flows from the heart to the extremities, and cool blood in the vein flows in the opposite direction. Most of the heat in the arterial blood moves into the venous blood and is carried back into the body core instead of being transported to the extremity, where it would be lost to the environment.

Most homeotherms can tolerate only slight increases in body temperature, because their optimal temperature is already near the lethal point. Homeotherms living in hot environments have acquired adaptations that either promote heat loss or enable the animal to avoid the heat. Because heat tends to move from areas of

Figure 17-14
Size of Ears Reflects Climate
(a) The ears of an artic fox are very short, which reduces the amount of heat loss when the blood circulates through them. (b) The ears of a desert fox are much longer because this animal must dispel excess body heat.

(a) (b)

Figure 17-15
Conservation of Heat by Countercurrent Blood Flow

(a) Under warm conditions, when heat loss can be tolerated or is necessary, blood returning from the leg of a bird flows through superficial veins which lie near the surface of the leg. Heat carried in the blood from the body core is lost to the environment as the blood flows through these veins. (b) Under cold conditions, blood returns through the large central vein instead of the superficial veins, thereby reducing heat loss. The central artery and vein lie close to each other, so that the artery contains more heat than the vein along the entire length of the leg. Heat always moves from the artery to the central vein, so that most of it is returned to the body instead of being lost to the environment from the lower parts of the leg.

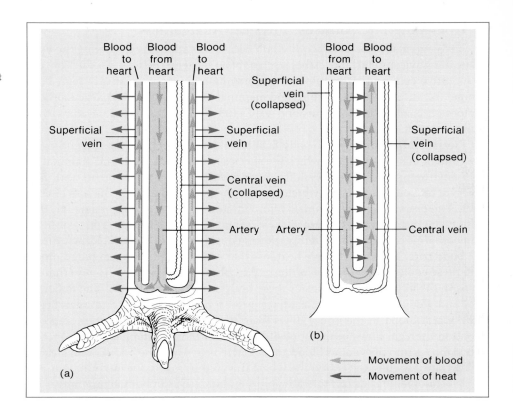

higher concentration to areas of lower concentration, heat tends to move into the animal when the temperature of the environment is higher than the body temperature, and adequate heat loss can only be accomplished by evaporation of water. This solution is not useful for most desert animals because evaporative cooling requires large quantities of water. These animals simply avoid the heat as much as possible by dwelling in holes, shady crevices, or caves during the day and becoming active during the cooler night hours. Desert animals have also acquired many physical adaptations for conserving water, so that some water is available for evaporative cooling when necessary.

Most larger desert mammals, such as sheep and camels, can travel far enough to obtain water regularly. These animals are also covered with sleek, light-colored hair that reflects the sunlight and insulates their bodies against the heat. A few large desert mammals, notably the dromedary camel, accumulate large amounts of heat during the day and then lose heat during the night when environmental temperatures are cooler than the animal's internal body temperature. The camel's body temperature can drop as low as 34.5°C during the night, so the animal can begin the day with a below-normal body temperature; the camel then accumulates heat during the day until its body temperature reaches about 40.5°C, at which point sweating begins. This unusually broad range of body temperatures for a homeotherm allows the camel to conserve water that would otherwise be lost through evaporative cooling.

Homeothermic Animals: Feedback Mechanisms

Sensitive feedback systems maintain the internal temperatures of birds and mammals within a few degrees of their optimal body temperatures. The mechanisms involved in temperature regulation are integrated by two neural centers within the hypothalamus of the brain: one initiates heat loss from the body; the other governs the production and conservation of heat. These feedback mechanisms vary slightly in birds and mammals; the discussion here pertains primarily to mammals.

Under normal conditions, the centers in the hypothalamus are sensitive to small deviations in the temperature of the blood flowing through the brain and respond by causing slight changes in the amount of blood flow to the body surfaces. These dilations and constrictions of small arteries in the skin are mediated by the autonomic nervous system. When the body temperature is slightly elevated, the arteries increase in diameter and more blood reaches the body surface, where the heat it carries is lost to the environment. When the temperature of the body drops slightly, the arteries become smaller and less heat is lost to the environment.

Adjustments to more extreme changes in body temperature are also controlled by the hypothalamus. If a dramatic rise in blood temperature is detected by the hypothalamus, it coordinates not only an extreme dilation of vessels in the skin (as much as 12% of total cardiac output) but also the release of water to the body surfaces. In humans, water for cooling the body is released through about 2.5 million *sweat glands* distributed over the surface of the body, which can produce up to 4 liters of sweat per hour. *Sweat* contains water, ions (primarily Na^+ and Cl^-), and a chemical that promotes the dilation of the small arteries in the skin. Heat from the blood moves into the sweat, and the water evaporates into the environment, taking the heat with it. As long as the lost water is completely replenished by drinking fluids, sweat effectively prevents an increase in body temperature. Other mammals may produce sweat in more localized areas, such as the pads of the feet, or copious amounts of saliva, which evaporates from the mouth and tongue, especially when the animal pants (rapid, shallow breathing). Some animals also spread saliva over their body surfaces by licking, and its evaporation removes heat from the body.

If a pronounced drop in body temperature occurs, the blood and special receptors on the skin relay the information to the hypothalamus, which initiates the constriction of blood vessels in the skin, increases heat production by increasing muscle tension, and, if the body temperature is more than 2.5°F below normal, induces shivering. Because the rapid muscle contractions of shivering produce no work, most of the energy from cellular respiration in these tissues becomes heat. The rate at which heat is produced may be increased four times by shivering.

Like the poikilotherms, some homeotherms can withstand very low temperatures. Some mammals undergo major changes in temperature regulation during the winter when their body temperatures drop significantly, and these lower temperatures are maintained by feedback mechanisms stimulated by the hypothalamus. This *hibernation*, or *winter dormancy*, conserves the animal's energy. Humans have no neural mechanisms that reset the body temperature to a lower level. A drop in the body temperature of a human causes progressive loss of brain and muscle function. The hypothalamus fails to regulate body temperature after it falls below 32°C (90°F); below this point, body temperature is correlated with the temperature of the external environment.

The hormones of at least two glands are also involved in temperature regulation. *Thyroxine,* a hormone produced by the thyroid gland, increases cellular respiration throughout the body except in the brain. When temperatures are cold, the thyroid gland secretes greater amounts of this hormone. The body responds slowly to thyroxine, however, and this hormone probably only adjusts the rate of heat production to seasons of the year. *Epinephrine,* a hormone secreted by the adrenal medulla, is released during stress and increases the rate of cellular respiration. Epinephrine stimulates the breakdown and use of glycogen and fats and also causes the blood vessels in the skin to constrict. Both reactions increase body temperature during stress.

Disturbances in the Control of Body Temperature

A prolonged increase in body temperature may result from unusually high levels of cellular activity or from the failure of the mechanisms that promote heat loss. Increased cellular activity usually does not raise body temperature significantly, although internal temperature may increase a few degrees during strenuous exercise or during a fever. Most frequently, body temperature increases due to an inability to lose heat. A significant drop in body temperature occurs when the rate of heat loss exceeds the body's ability to produce heat.

When the body cannot dispel excess heat adequately, the temperature can quickly rise to a dangerous level and *heat stroke* or *sunstroke* may result. When the environmental temperature and the humidity are high, heat energy cannot be transferred passively from the body by moving from areas of higher concentration to areas of lower concentration, and very little heat is lost through sweating because the environment is already highly saturated. Heat stroke may also occur during strenuous exercise if not enough fluids are consumed to replace the water lost in sweat. The symptoms of heat stroke are a hot skin and a sharp rise in body temperature. The victim often becomes unconscious and may be delirious or experience convulsions (involuntary muscle spasms). The affected person should be given a cold bath or applications of ice packs; fluids should be administered if the victim is conscious, especially if the heat stroke is due to prolonged sweating.

When an individual suffers from *heat exhaustion,* or *heat prostration,* the body temperature remains approximately normal because blood is shunted to the skin and sweat is released. However, the rest of the circulatory system is unable to adjust to the redistribution of blood flow. A drop in blood pressure occurs and causes the problems associated with shock. Complete rest in a cooler environment is usually sufficient for recovery.

Heat cramps—exceedingly painful cramps in the skeletal muscles of the body—are also associated with temperature regulation. This condition results when considerable amounts of ions (especially Na^+ and Cl^-) are lost from the tissues in sweat and are not replaced. The temperature-control mechanisms function, and the body temperature is normal. Muscle cramps that are caused by persistent sweating can be prevented by drinking a 0.2% salt solution (about one-half teaspoon of table salt per liter of water) instead of pure water.

A *fever* usually accompanies an infection. Temperature regulation does not fail during a fever; the body temperature rises several degrees and is maintained at this new level by feedback mechanisms. The initial temperature increase is caused by a decrease in blood flow to the skin, which conserves the heat being produced and creates the sensation of being chilled. Shivering may occur, which rapidly increases heat production. The body does not sweat during the development of a fever.

A fever may help to counteract infectious microorganisms. Many fevers are initiated by the release of materials (probably prostaglandins) from the white blood cells of the immune system. Aspirin is often effective in reducing a fever, and aspirin is known to suppress the formation of prostaglandins. When a fever subsides, the blood vessels in the skin dilate and sweating occurs, causing the body temperature to return to normal.

Another disturbance in temperature control occurs during prolonged exposure to cold. Long-term shivering will eventually exhaust the glycogen reserves stored in the muscles and liver. Consequently, the body will no longer be able to generate much heat and its temperature will drop. When the body temperature reaches 35°C (95°F), the brain and muscles no longer function at optimal levels and the person becomes disoriented, irritable, and clumsy in movement. When the body temperature reaches 90°F, collapse occurs and death is probable; even the intake of energy foods will not raise the body temperature at this point, because the hypothalamus is no longer controlling the internal temperature. The affected person must be revived by administering warmth from an external source, which is most safely and effectively accomplished by placing the victim and a healthy person together inside a sleeping bag or under blankets. Even after normal body temperature is restored, the individual should be hospitalized, because circulatory, respiratory, or kidney failure may occur after the body temperature has dropped below normal (*hypothermy*).

Summary

Life originated in the oceans, and the chemistry of cells is based on the properties of water. The properties of water that are relevant to temperature regulation include (1) high thermal conductivity, (2) high specific heat, and (3) high latent heat of vaporization.

The appropriate concentration of water must be maintained in the cells of an organism for it to survive. Because water tends to move from areas of high to areas of low concentration by osmosis, too much water will move into or out of the cells unless the concentration of water in the external environment and within the cells is the same. The concentrations of sodium and chloride ions are particularly important determinants of the concentration of water within cells and in their environment.

The plant cell wall—an adaptation that limits the amount of water that can move into a cell by osmosis—permits plants to survive in fresh water. Freshwater fishes prevent water from diluting their cells by (1) not drinking, (2) having impermeable body surfaces, except for their gills, (3) producing large quantities of dilute urine, and (4) transporting sodium and chloride ions into the body across their gill surfaces.

Marine plants have about the same water concentration as their environment and require no special adaptations. However, the cells of marine fishes have a higher water concentration than seawater. These fishes prevent the desiccation of their cells by (1) drinking enormous quantities of seawater and producing small amounts of urine with very high ion concentrations, (2) removing ions from the body by active transport across their gills, and (3) having impermeable body surfaces, except for their gills.

A variety of adaptations for acquiring and conserving water and ions have evolved in terrestrial plants and animals. Terrestrial animals acquire both *free* and *metabolic water* by eating and drinking; they lose water through the kidneys, lungs, large intestine, and body surfaces.

The human kidneys (1) remove wastes that form during cellular metabolism as well as toxic substances or excess nontoxic materials from the body, (2) maintain appropriate concentrations of water and ions, (3) regulate the volume of blood and interstitial fluids by controlling ion excretion, and (4) ensure an appropriate balance of hydrogen and hydroxyl ions, so that the *pH* (acidity) level of the body fluids remains optimal for chemical activities (about 7.4).

Humans have two kidneys, each of which is connected to the *urinary bladder* by a *ureter*. Urine moves from the bladder to the outside of the body through the *urethra*. The basic structural and functional unit of a kidney is the *nephron*; there are more than 1 million in each kidney. A nephron is a long tube comprised of five functionally different regions: (1) *Bowman's capsule*, which filters fluid from the blood into the nephron; (2) a *proximal tubule*, where important nutrients and about 80% of the water and ions in the urine are returned to the blood; (3) the *loop of Henle*, where water is moved from the nephron into the blood; (4) a *distal tubule*, which transports ions back into the blood if necessary; and (5) a *collecting duct*, which moves water from the nephron into the blood if necessary. All parts of the nephron except Bowman's capsule play a role in the regulation of blood pH by appropriately transporting hydrogen ions (H^+) and molecules that promote the formation of hydroxyl ions (OH^-) into and out of the nephron.

A large volume of fluid from the blood is converted into urine as it passes through the nephrons. The volume of the fluid is greatly reduced by the active transport of sodium ions out of the nephron and the passive movement of water from the nephron by osmosis due to the high concentration of sodium ions in the fluid surrounding each nephron. The composition of the fluid is altered by the active transport of glucose and other nutrients out of the nephrons and into the blood, the adjustment of water and ion concentrations in response to feedback mechanisms from the body, the adjustment of *acids* (molecules that form hydrogen ions) in relation to *bases* (molecules that form hydroxyl ions), and the passive movement of urea and other wastes through the nephron. All these activities are precisely regulated so that the composition of the body fluids remains constant, regardless of what foods or liquids are ingested or what cellular activities occur. Many nephrons in each kidney empty urine into the same collecting duct. From this duct, the urine is moved into the ureter and then into the urinary bladder, where it is held until it is expelled from the body during urination.

Most chemical reactions are extremely sensitive to changes in body temperature. The rate of a biological reaction approximately doubles with each 10°C increase in temperature. Above 45°C, however, enzymes are irreversibly altered, which leads to death.

The *poikilotherms* (all animals except birds and mammals) have no internal mechanisms for regulating their body heat; these animals adjust their body temperatures primarily by acquiring heat from their external environments. Birds and mammals are *homeotherms*, which have internal mechanisms for maintaining constant, optimal body temperatures.

Adaptations for temperature regulation in homeotherms include physical characteristics, such as fat and feathers or fur, and internal feedback mechanisms that are integrated by the hypothalamus of the brain. Slight deviations in body temperature detected by the hypothalamus are corrected by the reduction or expansion of the diameters of the arterial vessels in the skin; in addition, large deviations are corrected by the increased production of *sweat* or *saliva* (if the body temperature is too high) or increased muscle tension or shivering (if the internal temperature is too low). Two hormones— *thyroxine* and *epinephrine*—also play roles in raising body temperature. Some homeothermic animals maintain significantly lower but constant body temperatures during winter *hibernation*.

Occasionally, the temperature-control mechanisms in humans fail or lead to circulatory problems. *Heat stroke* results when the body temperature rises above normal due to insufficient sweating. *Heat exhaustion* results when blood is shunted to the skin and the remainder of the circulatory system is unable to adjust to this change, so that the blood pressure decreases. *Heat cramps* in skeletal muscles result from excessive sweating with no ion replacement. During a *fever*, the temperature-control mechanisms of the body continue to function, but the body temperature rises several degrees and is maintained at this new level. Under conditions of *hypothermy*, the body may lose the ability to raise its internal temperature because shivering has exhausted the body's glycogen stores or because the hypothalamus has become too cold to carry out its regulatory function.

Infectious Diseases: Invasion of the Body by Foreign Organisms

Microorganisms, or *microbes,* are small life forms that cannot be seen without the aid of a microscope. These small organisms exist almost everywhere on earth and greatly outnumber the larger organisms; the total weight of microbial life in the world is about 20 times heavier than the weight of visible animal life.

Highly exacting physical, chemical, and biological conditions must be met for each species of microbe to survive and grow. Each environment supports a unique assemblage of microscopic life forms. Many microbes grow and reproduce in or on the human body. There are millions of invisible organisms on your skin, in your nose, mouth, and throat, and on the lining of your digestive tract. Everything you touch and everyone you meet is covered with this unseen form of life. Whenever you exchange the dollar bill in your wallet, shake hands, or kiss, you pass on thousands of microorganisms to others and receive thousands more in return.

Most of these microorganisms are harmless, and some microbes are even beneficial to humans. However, some microorganisms interfere with normal functions of the human body. These disease organisms and their interactions with the defense systems of the human body are the subjects of this chapter.

Parasitic Microbes: Evolution of the Parasite–Host Relationship

The association between microbes and certain diseases was not discovered until the latter part of the nineteenth century. The first germ theory of diseases was proposed by Louis Pasteur (1822–1895)—a French chemist who also demonstrated that microorganisms are reproduced by other microorganisms rather than formed by spontaneous generation. Pasteur's germ theory was later confirmed when a

German country doctor, Robert Koch (1843–1910), demonstrated that *anthrax,* a disease affecting sheep and cattle, was caused by a microbe.

Microorganisms that live in or on another organism, or *host,* and benefit at the expense of the host are called *parasites.* When the host organism is harmed by this association, the condition is called a *disease.* An *infectious disease* results when a parasitic microbe moves from one host individual to another. A single parasitic species can usually multiply in only one or two closely related species and may even be restricted to specific types of tissues within their hosts.

Most existing parasite–host relationships are not fatal to either species. If a parasitic species killed its host population or if the host population evolved a defense system that prevented its parasites from surviving, the parasitic species would have no place to live and would become extinct. Most parasite–host relationships have existed over a long period of time and have produced evolutionary adaptations that allow both host and parasite to survive and reproduce.

Serious Parasitic Damage

Microorganisms can cause serious illnesses in their hosts when the environment of the host is altered or when a parasitic species invades a new type of host organism. The most common environmental factor that increases the vulnerability of the host to the parasite is *malnutrition.* A delicate balance exists between the activities of a parasite and the defenses of its host. Serious illness can result when the defenses of a host population are reduced due to inadequate nutrition. Famine and disease have usually occurred together throughout human history.

The disease *poliomyelitis* (polio) provides an example of the interference of cultural changes with the natural interaction between parasite and host. Paradoxically, polio only became a serious disease after the role of microbes in disease was understood and measures had been taken to improve standards of hygiene. A clear relationship exists between the occurrence of paralysis from polio and the standard of living of the host population. In communities with advanced standards of hygiene, occurrences of polio are more serious and affect an older age group; in communities with low standards of hygiene, very young children contract polio, which manifests itself as a relatively harmless infection of the intestinal tract and protects the child from further infection by the polio microbe. In areas where hygiene is poor, polio remains a common childhood disease but the development of paralysis is rare. In areas where hygiene is advanced, young children do not contract the disease and are therefore vulnerable to infection at a later age. Resistance to the disease apparently declines with increasing age. Paralysis develops most frequently in older individuals, and its incidence is directly related to the higher standards of living in many areas of the world today. (Polio is now largely controlled by the administration of a vaccine, which will be discussed later in the chapter.)

Most diseases that are fatal to humans are caused by microorganisms that normally infect other animals. Although microbes are usually specific to their hosts, they will invade a new species if the host with which they have established a harmless equilibrium becomes so abundant that large populations of the microbes exist. Large populations of organisms generally support a large variety of genetic types. A particular combination of genes may allow some forms of microbes to invade a new host species. The new host is vulnerable to serious illness because it has no specific defenses against the invading parasites. The *black plague*—a disease that has killed millions of people throughout history—is caused by a microbe that is normally parasitic to rats. In crowded and unsanitary urban areas, where rat populations were large and humans were often weakened by stress, the disease frequently spread from rats to fleas to humans. Similarly, the microbe that causes *encephalomyelitis* (inflammation of the brain and spinal cord) normally resides harmlessly in birds that evolved in North America. House sparrows and pheasants—the only birds on this continent that become ill from this microbe—evolved on other continents and were introduced into North America only recently. In the past few decades, the encephalomyelitis microbe has produced severe disease symptoms in two new host species—horses and humans—reaffirming the importance of establishing a balance between host and parasite over a long period of adjustment so that neither becomes seriously injured.

When two populations of the same species are isolated from each other for a long period of time, one population may establish a relatively harmless relationship with a particular microbe and the other population may never be exposed to that microbe. Migrations between these two populations can then cause a severe outbreak of disease in the previously unexposed population because the genetic and developmental adaptations required for the survival of host and parasite are not operative, as is true when a microbe moves from one host species to another. For example, *measles* is a relatively minor childhood illness in European populations, where the microbe has existed for centuries. However, the first European explorers carried the measles microbe to previously unexposed human populations that had no specific defense against the disease; measles was a serious and often fatal disease in these new host populations.

Many current diseases were not common in primitive human populations. There is little evidence of smallpox before the first century A.D., of measles before the sixth century, or of cholera before the sixteenth century. It is doubtful that these diseases, as well as mumps, influenza, and poliomyelitis, were prevalent throughout most of human history, because the population density in any one area was probably so low that the parasites rapidly exhausted their sources of new hosts and became extinct.

Disease Microorganisms

Microorganisms were first seen by Anton van Leeuwenhoek (1632–1723), a Dutch naturalist who ground lenses and constructed microscopes as a hobby. The four major groups of disease microbes are the protozoa, bacteria, fungi (yeasts and molds), and viruses. The organisms in each of these groups have an almost unlimited variety of shapes, structures, behaviors, and environmental requirements.

With the exception of viruses, disease microbes perform all the biological functions observed in other heterotrophic life forms. They take in food and convert it into energy and structural materials, which the microorganism then uses for maintenance, growth, and reproduction. Disease microbes are also composed of the same kinds of atoms as other organisms, and these atoms combine to form proteins, lipids, carbohydrates, and nucleic acids that are basically similar to these materials in all other life forms. In contrast, viruses exhibit only the ability to reproduce, which they accomplish by manipulating the cellular activities of other living organisms.

Protozoa

The *protozoa* are members of the kingdom Protista. These unicellular organisms, which are among the largest microbes, are eukaryotic cells containing a nucleus, chromosomes, mitochondria, and the various other intracellular structures characteristic of this class of complex cells. (See Chapters 5 and 11 for a review of cell structures and types.) Protozoa are highly mobile and are capable of seeking out food or more favorable environments. Most of the more than 15,000 species of protozoa feed on biological molecules that have already been formed. Practically any kind of living or dead biological material is food for some type of protozoan.

Protozoa are found in virtually every environment that contains both water and biological molecules. You are probably the host of at least a dozen different species. One species lives in your mouth, where it eats the bacteria and food particles that are lodged between your teeth; several species live in your intestines, where they feed on large populations of bacteria. Most of the protozoan inhabitants of the human body are harmless; a few species even consume potentially harmful bacteria. However, some protozoa feed directly on human tissue (Figure 18-1). For example, a protozoan causes *amoebic dysentery*—an intestinal disorder that results when a particular type of amoeba feeds on the cells lining the intestine.

Four different species of protozoa (genus *Plasmodium*) cause *malaria*—a disease

Figure 18-1
Protozoa That Cause Human Diseases
(a) This amoeba causes amoebic dysentery when it infects the intestine. (b) This protozoan, *Trypanosoma*, shown here with human red blood cells, causes African sleeping sickness.

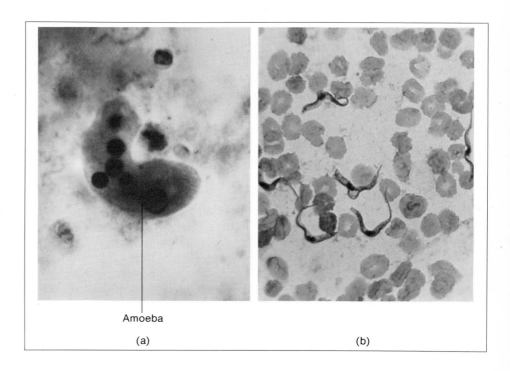

Amoeba

(a)

(b)

that has ravaged millions of people over centuries. The bite of an *Anopheles* mosquito injects these protozoa into the body. Two weeks later, as many as a billion protozoa may be living in the victim's body. At periodic intervals, these protozoa invade the red blood cells, where they multiply until they eventually break the cell membranes. The released protozoa invade more cells. The protozoa population within a single host acts in *synchrony*—invading cells at the same time and breaking out of cells at the same time. Each time the invaded red blood cells burst open, the victim suffers chills and a high fever. Malaria has largely been controlled by the eradication of the *Anopheles* mosquito.

Fungi

Yeasts and molds belong to the kingdom Fungi. Like the protozoa, the *fungi* are composed of eukaryotic cells with characteristic intracellular structures (although the number of nuclei may vary due to developmental events). All fungi obtain food by absorbing biological molecules across their cell membranes. They generally release digestive enzymes into the food mass outside the cell and then absorb the liquified material. Some microscopic forms of yeasts and molds can cause infectious diseases (Figure 18-2).

Figure 18-2
Disease-Causing Fungi

(a) A fungus that grows on moldy peanuts and rice; when eaten, it can cause cancer of the liver ($\times 2,170$). (b) A fungus that causes skin lesions ($\times 105$). (c) A fungus that causes athlete's foot and ringworm of the scalp ($\times 2,000$). (d) A fungus that causes hair infections ($\times 670$).

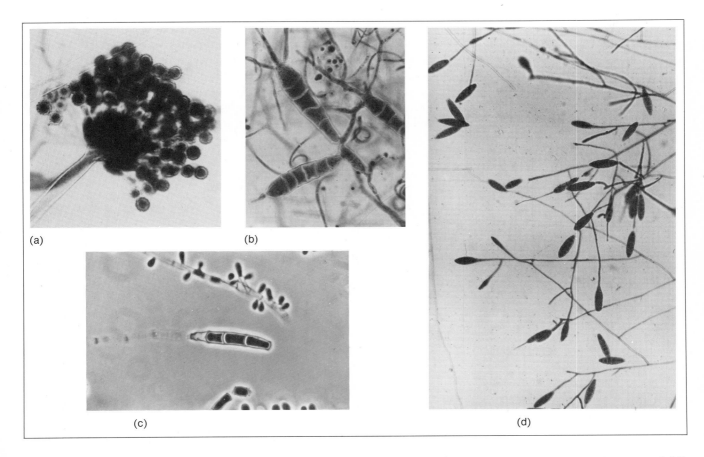

(a)

(b)

(c)

(d)

Many fungi are useful to humans. Certain fungi are important in the preparation of foods, such as bread and wine. *Penicillin,* a medicine used to treat bacterial infections, is the product of a fungus. Together with bacteria, the fungi break down the biological molecules of dead organisms and therefore play a critical role in recycling materials. They can also be a nuisance; certain fungi cause food to become moldy and clothing to mildew.

The human body provides a favorable environment for numerous types of fungi, which are able to live on the scalp, hair, fingernails, most skin surfaces, and inside the mouth, throat, lungs, and intestines. Considering their prevalence, however, fungi are responsible for only a small percentage of human diseases, including athlete's foot and ringworm. Serious fungal infections, especially in the lungs, can be fatal because few effective treatments for fungal infections have been found.

Bacteria

Bacteria belong to the kingdom Monera and are probably the oldest forms of life. These organisms outnumber all other living things; they can and do exist practically everywhere. All bacteria are simple prokaryotic cells without complex intracellular structures (see Chapters 4 and 11).

Most bacteria do more good than harm. Bacteria are the most active microorganisms in breaking down the complex molecules of dead plants and animals into simpler chemicals that can be reused by living organisms. Without bacteria, the materials required to sustain life would be trapped in dead organisms and all life would end. Only about 150 of the several thousand known species of bacteria cause human diseases; a few of these are shown in Figure 18-3.

Many types of bacteria can survive in hostile environments for long periods of time by forming a capsule around their cells to protect the genetic material. These capsules can withstand extreme temperatures, ranging from freezing to boiling. Bacteria embedded for millions of years in the ice of Antarctica were revived when

**Figure 18-3
Bacteria That Cause
Human Diseases**

These microbes cause cholera,
(b) local skin infections, (c) syphilis,
and (d) typhus.

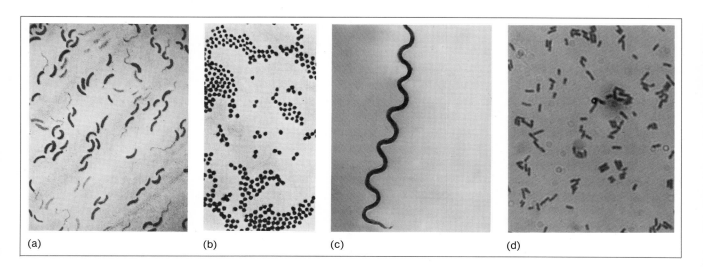

(a)   (b)   (c)   (d)

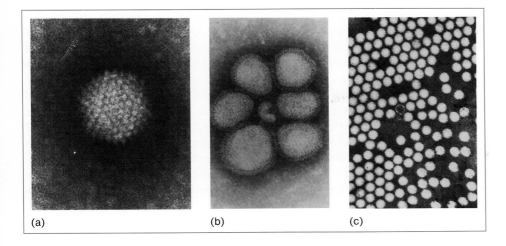

Figure 18-4
Viruses That Damage Human Cells

These viruses cause (a) the common cold, (b) influenza, and (c) polio.

warmed and were able to reproduce. A few of the bacteria that cause human diseases can enter such highly resistant states, and they are some of the more persistent and deadly killers in the microbial world.

Some diseases are not caused by the bacteria themselves but by biological poisons, or *toxins,* that are produced and released by bacterial cells. Diphtheria, cholera, scarlet fever, and tetanus are among the many diseases caused by bacterial toxins. Each toxin has a unique effect on the human body. The *diphtheria toxin,* for example, is an enzyme that prevents amino acids from linking together to form proteins. In the presence of this material, protein synthesis cannot occur in the infected cells. Recent studies indicate that the genes that direct the synthesis of these toxins are the genes of viruses residing within the bacteria and are not included in the bacterial genetic codes.

Bacteria are responsible for the majority of human infectious diseases, but *antibiotics* such as penicillin, streptomycin, and tetracycline interfere with the bacterial synthesis of certain molecules, such as proteins, and stop the growth and reproduction of bacteria. The use of antibiotics has reduced once major, lethal bacterial infections to much less serious diseases.

Viruses

Viruses (Figure 18-4), the smallest microbes, were not actually observed until the development of very powerful microscopes during the past few decades. These microorganisms are not cells but consist only of genetic information in the form of DNA or RNA molecules enclosed in a protein envelope (Figure 18-5). Viruses are not usually placed in any of the five kingdoms. A virus functions by invading a cell and taking over its metabolic activities by shutting off the synthesis of the cellular materials in some way and instructing the cell to manufacture the structures and enzymes required to produce more viruses.

There are many kinds of viruses, and some type of virus can invade the cells of any organism—plants, animals, fungi, protozoa, algae, bacteria, methanogens, or blue-greens. The simplest viruses have an outer envelope of a few proteins; the

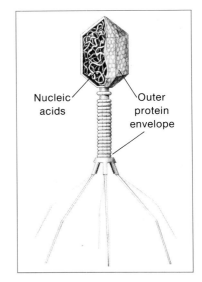

Nucleic acids

Outer protein envelope

Figure 18-5
The Structure of a Virus

A virus is composed only of an outer protein envelope and nucleic acids that contain genes. It is not a cell and contains no structures for building its own proteins.

envelope of more complex viruses also contains some lipids and carbohydrates. Frequently, the genetic code for these materials is not contained in the genetic material within the virus but is derived from the cells in which the virus multiplies. In exceptional situations, some of the host cell's DNA may even be incorporated into the structure of a virus.

A virus invades a cell by attaching itself to the cell membrane. The proteins that make up the outer envelope of each kind of virus have a special affinity for the outer surfaces of their particular host cells. Viruses that invade bacterial cells inject their nucleic acid molecules into the interior of the cell, leaving the protein envelope outside. The protein envelope of animal viruses either fuses with the cell membrane or is removed after the virus is inside the cell.

Once the protective envelope has been removed, the nucleic acid of the virus moves into the nucleus of an animal host cell. There, the genetic code is transcribed into messenger RNA molecules, which are moved to the ribosomes for translation into enzymes. Once an adequate amount of these enzymes has been synthesized, the viral nucleic acid makes many copies of itself and the proteins of the outer envelope are synthesized. These materials are then assembled into individual viruses, which are released from the host cell one at a time over a period of many hours (Figure 18-6). Each released virus then goes on to attack another cell.

Figure 18-6
Polio Viruses Inside a Cell
Polio virus particles in a cytoplasmic fragment from an infected tissue culture cell (× 300,000).

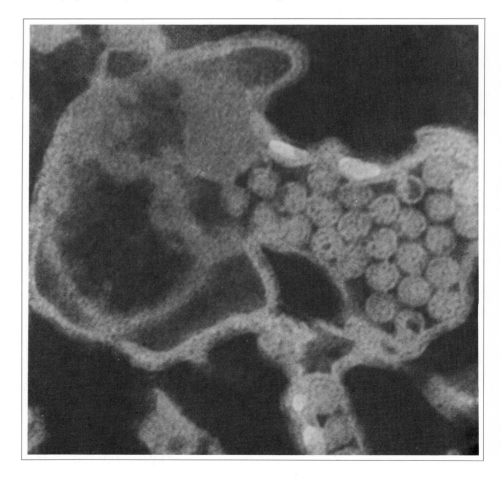

Most of the viruses that cause disease kill their host cells. Viruses often shut down their host's protein and RNA synthesis, so that it cannot survive. Many viruses produce toxins that poison the cell. Infected cells also swell up, changing the permeability of the cell membrane and interfering with the normal exchange of materials between the cell and its external environment. Eventually, the affected cell's own lysosomes (organelles in eukaryotic cells) release enzymes that digest the cell. Some viruses, such as the viruses of herpes and German measles, also damage the cell's chromosomes.

About 100 different viruses are known to cause human diseases. The viruses that cause smallpox, measles, and chicken pox spread throughout the body and do not concentrate in any particular type of tissue. The polio virus may strike the cells of the nervous system; mumps occur in the salivary glands; influenza and the common cold localize in the respiratory tract.

Potentially harmful viruses can remain dormant in the human body without causing an immediate and dramatic illness until they are activated by certain environmental conditions. The herpes virus that produces cold sores or fever blisters may be dormant within its host for years. The viruses lie inactive within nerve cells and produce no symptoms of disease. If the host is exposed to unusually intense sunlight or some other type of stress, the viruses are often activated and migrate to the skin where they invade cells and cause small skin eruptions. These sores are the sites of intense viral reproduction.

The Body's Defenses Against Infectious Diseases

Many infectious diseases appear to result from unsanitary conditions, and most human cultures have developed measures to protect the human body from microorganisms. Food is washed and cooked, water is inspected, and personal belongings are kept as clean as possible. Humans regularly scrub their skin, brush their teeth, and even rinse their mouths with antiseptic to minimize the number of microorganisms on the body.

Although these measures are certainly of some value, if we relied only on the defenses that have been devised culturally to protect us from the great number of microbes on the human body, we would probably be dead within a week. Natural defenses that have evolved over a long history of human contact with various types of damaging microorganisms provide the most effective protection against infectious diseases.

Preventing Invasion

The skin and linings of the nose, lungs, mouth, and intestinal tract are the first line of defense against disease microbes. The infectious microbes that occur on the body surface rarely penetrate these tough outer layers. The body is vulnerable to invasion by these routes only if these tissues are broken, as they would be by a cut or an insect bite.

Individual microorganisms that could cause disease do not live long on unbroken skin or internal linings because the microbes that normally inhabit these areas create a hostile enviroment. Some beneficial microorganisms consume the invading microbes directly; others produce chemicals that interfere with their activities.

The body itself produces many materials that prevent growth of foreign microbial populations. The salts in sweat and natural antibiotics in the fluid covering the eyes and in saliva and ear wax curtail the activities of foreign microbes. The membranes lining the respiratory, digestive, urinary, and genital tracts are covered with hairlike projections (cilia) and a layer of sticky fluid (mucus), which work together to trap microbes and move them out of the body. The stomach and vagina are also highly acidic, and most microbes cannot survive in such an environment.

Fighting an Invasion—The Local Response

Microbes occasionally penetrate the body's outer defenses and may even reach tissues that support their growth and multiplication. The presence of foreign microbes or their toxins stimulates a complex sequence of bodily defenses that kill the invaders and keep the infection from spreading. How quickly and efficiently the defense system detects and fights the invading microbes determines how close the individual comes to serious illness or death.

Several substances released from cells that have already been damaged by infectious microbes initiate the body's internal defense system. Within 24 hours of invasion, *interferons*—molecules constructed of proteins and carbohydrates—stimulate other cells to manufacture materials that inhibit the reproduction of all types of viruses. An individual who is already infected with a relatively harmless virus is protected to some extent against the effects of more damaging microbes as long as the infected cells continue to release interferons.

Unfortunately, interferon molecules cannot be obtained from other animals (interferon from one species is not active in another species), and these molecules cannot be artificially synthesized in laboratories because their structure is not yet known. A recent breakthrough has been the use of a recombinant DNA technique in interferon production: genes containing instructions for synthesizing human interferons are placed inside bacteria, which then manufacture the interferon molecules. Interferons will probably soon be a widely available method of increasing the body's natural defenses against viruses.

Other substances released from damaged cells trigger an inflammatory response, which manifests itself as redness, swelling, heat, and pain in the infected region. Some chemicals that are released cause the veins in the area to constrict and the arteries to dilate; others increase the permeability of the capillaries, so that fluid leaks out of the blood and into the infected areas. One chemical causes the materials in the infected area to clot, thereby walling off the infection so that the disease microbes cannot flow into uninfected areas. The chemicals released from damaged cells and the pressure of the accumulating fluid stimulate the nerves to produce the sensation of pain, which serves to protect the infected area; the individual is reluctant to touch or move a swollen and painful area of infection. At this stage, the symptoms of the disease primarily result from the body's defense system rather than from the cell damage caused by foreign microbes.

Three materials that leak out of the capillaries help to contain the infection— white blood cells, antibodies, and complement. *White blood cells* are always present

Table 18-1
Human White Blood Cells and Their Functions

Type of white cell	Function
Neutrophil	Engulfs and digests foreign material and cellular debris
Monocyte	Engulfs and digests foreign material and cellular debris
Eosinophil	Detoxifies foreign proteins
Basophil	Produces a substance that prevents the blood from clotting
Lymphocytes	
T cell	Destroys foreign material, releases toxic substances, attracts other white cells, and regulates B cells
B cell	Produces antibodies

in the blood and act to combat foreign substances that enter the body. These cells move continuously throughout the body and are capable of directed locomotion toward the site of a microbial invasion. Five different categories of white blood cells have been described (Table 18-1). Although much remains to be learned about the specific properties of each category, we now have a general idea of how each type of white blood cell functions in destroying foreign materials.

White blood cells usually reach the infected area first, where they form a barricade to prevent the microbes from invading other parts of the body. The white cells then begin to engulf and digest the invading microbes (Figure 18-7), which usually destroy many of the white cells in return. The dead white cells, microbes, and fragments of damaged body cells that accumulate during this process form the whitish fluid, *pus,* that oozes from the infected area somewhat later.

Antibodies are protein molecules formed by *lymphocytes* (special white cells located primarily in the lymph nodes and the spleen). These molecules usually travel through the bloodstream to the infected area within a few hours after the body has been penetrated by a foreign microbe. Antibodies attach themselves to specific foreign microbes and clump them together (Figure 18-8), so that the other white cells can find and destroy them. Lymphocytes can produce antibodies that are specific to any kind of protein or sugar that is foreign to the body. Each type of microbe has a slightly different outer membrane that a particular type of antibody recognizes.

The first time that a particular foreign microbe invades the body, antibodies form slowly; production does not peak until the second week after the onset of an infection. Antibodies are most valuable in fighting infection when the same type of microbe invades the body a second time. The number of antibodies in the blood then increases rapidly, reaching a peak in only a few days. This much faster and more effective response, called *immunity,* will be discussed in the next section. How quickly antibodies reach the infected area is critical to the defense of the body; a second infection by the same type of microbe may produce no symptoms at all.

Complement is a substance that is always present in the blood. It is constructed

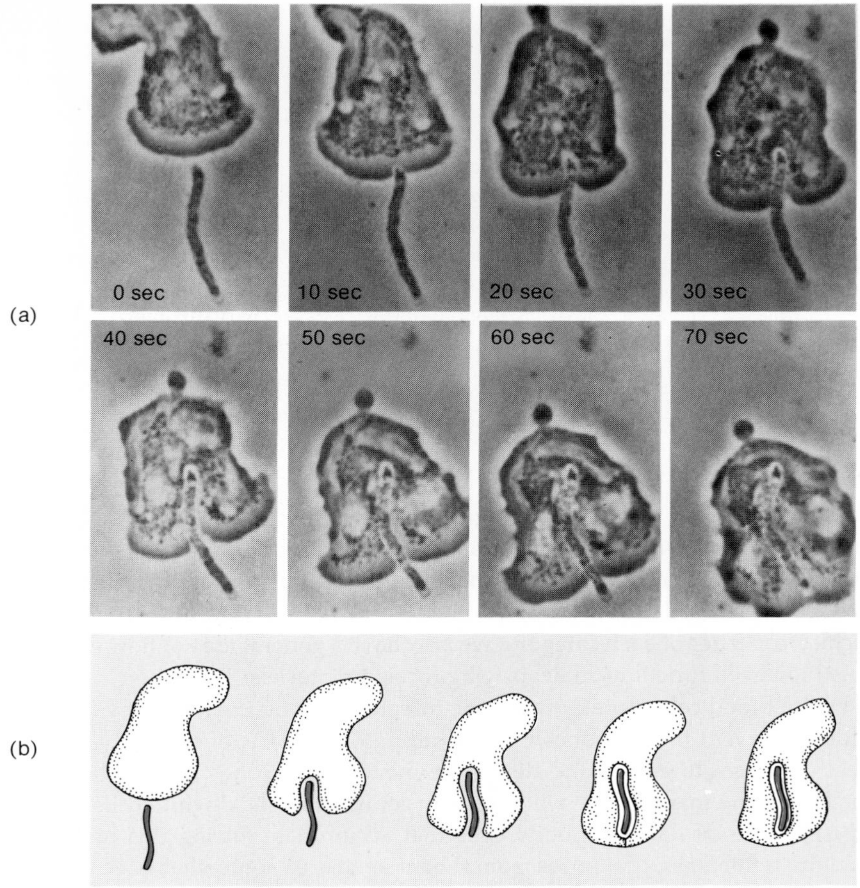

**Figure 18-7
A White Blood Cell
Engulfing and Consuming
a Bacterium**

(a) Photographs taken during the
engulfing and consuming process.
The large structure is a human white
blood cell; the rod-shaped structure is
a bacterium. (b) An illustration of the
same process.

of a complex system of 11 enzymes that act in sequence to aid in the body's defense against infection. Once an antibody has detected and attached itself to a foreign microbe, complement combines with the antibody. One of the complement enzymes then makes a small hole in the outer membrane of the microbe, causing it to rupture and die.

Other enzymes in complement promote the synthesis of several materials that help the white blood cells to find and ingest microbes. One material released in the infected area guides the white cells directly to the microbes; another material, *histamine,* increases the permeability of blood capillaries to white cells, permitting them to move rapidly from the blood into the infected area. A third substance helps the white blood cells to adhere to microbes.

Although antibodies and complement perform important functions independently, their most valuable role in fighting an infection is the assistance they give to the white blood cells. The effect of complement on the efficiency of white cells is especially important in a first infection when a relatively long time elapses before the body can produce large numbers of antibodies specific to the new microbe.

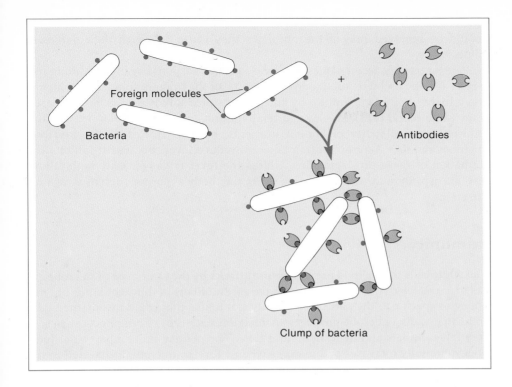

Figure 18-8
How Antibodies Function
Antibodies attach themselves to certain molecules on the outer membranes of foreign microbes, causing them to clump together.

Labels within figure: Foreign molecules, Bacteria, Antibodies, +, Clump of bacteria

Fighting an Invasion—The General Response

If the local defense system does not successfully contain the infection, the invading microorganisms enter the circulatory system and spread throughout the body. A second line of defense is then activated by the body.

First, microbes are removed from the blood and ingested by white cells that are located in tissues throughout the body. These cells are particularly concentrated on the inner linings of small blood vessels in the liver, spleen, bone marrow, and lymph nodes. The microbes that escape the white cells in the blood vessels are moved by the lymphatic vessels to the *lymph nodes*—areas of tissue located at frequent intervals throughout the lymphatic system (see Figure 14-9, page 364). As the lymph fluid flows into these coarse, spongy meshes of tissue, the microbes are filtered out and ingested by white cells packed within the nodes.

Lymph nodes and lymphatic vessels are especially plentiful in the respiratory passages and the digestive tract—the areas of the body that microbes are most likely to invade. As lymph accumulates in the infected area, the nodes swell and may become painful, particularly near the original site of invasion. The development of "swollen glands" (a misnomer because nodes are not glands) at the angle of the jaw, in the armpit, or in the groin indicates that an infection is nearby.

Lymphatic tissue guards the throat and nose. The tonsils and adenoids, which act as filters to trap and localize microbes, often become inflamed and swollen. In extreme cases, their protective function is overwhelmed and they harbor microbes rather than destroy them. The adenoids or tonsils may then cause persistent, long-term respiratory infections and may have to be surgically removed. However, the tonsils and the adenoids are important parts of the body's defense system and should be removed only if the problems they cause outweigh their defensive values.

A fever usually accompanies the body's general response against invading microbes. During an infection, the hypothalamus of the brain is stimulated to maintain a higher body temperature by substances—probably *prostaglandins*—released from white blood cells. An elevated temperature stimulates cellular activity and reduces the availability of certain nutrients, such as iron, that the invading microbes require. Aspirin is often taken to reduce fever, because it inhibits the formation of these prostaglandins. Unless the fever is dangerously high (above 104°F), aspirin may only interfere with the body's normal defenses against infections.

Immunity

The white cells that help to prevent future attacks by the same type of microbe are called *lymphocytes.* An individual may have vast numbers of different kinds of lymphocytes specific to the particular microbes or toxins that could invade the body. Until an invading microbe produces an infection, each type of lymphocyte exists in very low numbers and in an immature state. The population of a particular lymphocyte rapidly increases and matures when a foreign microbe is detected in the body. In the course of a lifetime, an individual is exposed to many different kinds of microbes and forms large populations of the lymphocytes specific to each microbe, which continue to exist in the body after the infection is gone. Although their numbers decline slowly with time, the "memory cells" that remain can divide rapidly to defend the body against future attacks by the same microbe.

The Development of the Immune System

The source of all blood cells, including white cells, is the *bone marrow.* However, the bone marrow produces only primitive, immature lymphocyte cells that migrate to other parts of the body, where they develop into mature, functional cells. Most immature lymphocytes are formed during fetal life (prior to birth).

About 50% of the lymphocytes that leave the bone marrow pass through the *thymus*—a mass of tissue located within the chest cavity just below the neck. The thymus is quite large during infancy but shrinks to a very small size in early childhood and remains small throughout life. Cells that pass through this organ before entering the circulatory system develop into a particular kind of lymphocyte, called *T cells.* The outer membranes of these cells have special sites that recognize foreign cells or substances. The T cells help to defend the body against an infection by (1) attaching themselves to the invader and destroying it directly, (2) releasing materials that are toxic to the invader, (3) producing a substance that attracts the white cells that specialize in ingesting foreign materials, and (4) regulating the activities of a second type of lymphocyte, the B cells. The T cells are especially effective in detecting and destroying cells that have been infected by viruses. They are also responsible for the rejection of tissue transplants.

Immature lymphocytes that do not pass through the thymus—the *B cells*—probably enter the blood directly from the bone marrow. The B cells synthesize antibodies, which aggregate foreign microbes.

It is essential for the lymphocytes to come into contact with the invading microbes, so that both the T and B cells can enlarge and multiply. Each newly formed B cell then constructs specific antibodies to combat the particular intruder. A single B cell can synthesize and release about 2,000 identical antibody molecules a second. From this point on, these B cells manufacture only that specific type of antibody.

After the foreign microbes have been neutralized and digested, the activated lymphocytes return to their former inactive states. The T and B cells specific to this particular type of microbe are now present in large numbers to protect the body from a second invasion by the same kind of microbe.

The Ability to Distinguish Self from Nonself

Of course, it is crucial that the body's defense system be able to distinguish between the body's own molecules (*self*) and the molecules of foreign microbes (*nonself*), so that it does not act on its own cells or materials. Apparently, the lymphocytes can differentiate between the multitudes of possible foreign invaders and the body's materials, although how this recognition is achieved is not yet known.

The current theory explaining the ability to distinguish between self and nonself, the *clonal selection theory,* suggests that at some point in early embryonic development, lymphocytes are formed that can act specifically against all kinds of proteins and complex sugars, including those manufactured by the body itself. Later in development, the particular lymphocytes that could act against the body's molecules are destroyed, leaving only cells that are designed to recognize foreign molecules. After infancy, one or a few lymphocytes remain that are specific to each molecule that is not part of the body. When a foreign microbe is recognized, both the T and B cells specific to that microbe divide rapidly and the newly formed B cells begin to manufacture antibodies.

A fetus has no defense system against foreign materials because the immune system does not fully develop until several months after birth. However, any microbes that could enter the fetus must first pass through the mother's circulatory system and are usually eliminated there by her defense mechanisms. Occasionally, however, the mother's defense system fails and the fetus becomes infected. For example, German measles often infects both mother and fetus and may severely damage the developing tissues of the unborn child. The virus that causes German measles usually affects the chromosomes, and the tissues cannot develop properly if the genetic instructions contained within the chromosomes are altered.

Antibodies and white blood cells from the mother are present in the infant's bloodstream when it is born and continue to protect it for about six weeks after birth. Antibodies are also passed from mother to child through the mother's milk if the baby is breast-fed.

Artificial Immunity

The control of infectious diseases by *artificial immunity* was a great medical triumph. There are two basic forms of immunization. *Active immunization* is the stimulation of an individual's own immune system by exposure to the disease microbe or its toxin. *Passive immunization* is the process by which antibodies from a human or animal known to be immune to a disease are transferred to another individual.

Active artificial mobilization of the immune system against a particular microbe is accomplished by *vaccination,* which introduces just enough foreign material—a microbe or its toxin—into the body to cause the number of lymphocytes and antibodies to increase. The first vaccination, a dangerous experiment, was performed by Edward Jenner in 1796. Jenner recognized that people who contracted *cowpox,* a disease of cattle, generally did not contract the more serious human disease *smallpox.* He withdrew liquid from a cowpox sore and injected it into the body of a young boy. The experiment was successful: the boy did not develop smallpox later when he was injected with material from a smallpox sore. Smallpox and cowpox viruses are so similar that the boy's immune system did not distinguish between them; the lymphocytes his body developed to combat cowpox protected him from contracting smallpox.

Sometimes the injection of killed microbes provides sufficient immunization against a disease because the body's defense system recognizes the foreign molecules that are still present on the dead microbes. Lymphocytes and antibodies specific to the foreign molecules are then produced and remain in the lymph nodes and the bloodstream for a few years to protect against further invasions. Live microbes are more effective in stimulating the immune system but also have the potential to cause disease. Most of the live vaccines used today are composed of laboratory-grown microbes that have been altered to reduce their damaging effects. Vaccines that contain live but altered microbes include the smallpox, measles, mumps, and oral polio vaccines.

Passive immunization provides protection in the absence of the actual invasion of a particular disease microbe. *Gamma globulin*—the part of the blood that contains antibodies—is extracted from an animal that had previously contracted the disease and injected into an individual who has not yet had the disease. The recipient gains passive immunity because his or her own immune system does not play an active role in the production of antibodies. Passive immunity is effective for a much shorter time (a few months) than immunity from a vaccination because the cells that actually produce the antibodies (the B cells) are not transferred. Passive immunity is applied when exposure to a microbe is unexpected—for example, to combat the venom from a spider or a snake bite or to protect against tetanus when there is an open wound.

Most infectious diseases are now quite rare due to the effectiveness of vaccination programs. However, the microbes that cause these diseases still exist, and humans who have not been properly vaccinated against them are still vulnerable to infection.

Undesirable Side Effects of the Immune Response

The immune system also responds to a variety of noninfectious foreign materials, including pollen, drugs, certain foods, bee or wasp stings, the chemicals in such plants as poison ivy and poison oak, and the tissues from other individuals.

Allergies   An *allergy* (literally "altered reaction") is the result of an inappropriate reaction of the immune system. Most of the symptoms of an allergy are caused by the excessive production and release of histamine and other materials, which is an appropriate response only if the body is invaded by a foreign microbe. Histamine dilates the arteries and increases the permeability of the capillaries in the infected area, causing the affected tissues to swell.

Tissue Transplants   Except in the case of identical twins, each individual carries a unique genetic code which directs the cells to synthesize a unique collection of molecules. The tissues of one individual contain molecules that are recognized as foreign by the immune system of another individual. In particular, the T cells of the immune system of one individual recognize the molecules on the surfaces of the cells from another individual as foreign. When tissues or organs are transferred from one person to another, the immune response may therefore cause the rejection of the tissue transplant.

A Mother's Immune Response to Her Fetus   The particular proteins and sugars found on the cells of a fetus differ to some degree from the proteins and sugars found on the cells of the mother. The fetus inherits 50% of its genes from the mother and 50% of its genes from the father. Although different genes can produce different molecules, in most instances the mother and the fetus exist together in harmony for approximately nine months with no evidence of an immune reaction. The immune system of the fetus is not sufficiently developed to form specific lymphocytes that could destroy the cells of the mother, but the mother's immune system can and does form lymphocytes that are capable of destroying certain cells in the fetus. For some unknown reason, the immune defenses of the mother do not usually harm the developing child.

Under certain circumstances, however, the mother's immune system does harm the fetus. A familiar example is *erythroblastosis fetalis,* which occurs when a mother who does not have the Rh molecule on her red blood cells (is Rh-negative) carries a child who has inherited this molecule from its father (is Rh-positive). As some of the fetal red blood cells leak into the mother's circulatory system, her immune system develops antibodies to the Rh molecule. If large numbers of these antibodies move into the fetus, they destroy its Rh-positive red blood cells, producing a potentially fatal condition.

For unknown reasons, the mother forms antibodies against the Rh molecule so slowly that sufficient quantities to harm her first child are usually not generated, but enough antibodies are present to destroy the red blood cells of the fetus during the second or third pregnancy. Now that this disease is understood, effective treatment can prevent the mother's immune system from seriously damaging the child she is carrying.

Autoimmunity: Fighting the Body's Own Cells   *Autoimmune diseases* are the result of attacks of the immune system on the body's own cells and molecules. The factors that stimulate this abnormal behavior of the immune system are not known. Several severely debilitating diseases, including *rheumatoid arthritis* and *lupus (systemic lupus erythematosus)*, are apparently caused by such attacks of the immune system. In rheumatoid arthritis, the joints are the targets of this attack; they become inflamed and eventually immobilized. Lupus is a multiple-organ disease in which the immune system may attack any part of the body. Autoimmune diseases produce gradual, long-term degenerative changes in the body.

Some Unconquered Enemies

Most of the major infectious diseases have been conquered, primarily due to improved nutrition and the development of effective vaccination procedures. Antibiotics have also played a large role in the control of diseases that are caused by bacteria. A few diseases have been virtually eliminated by sanitation improvements; for example, cholera, an intestinal disease, is prevented largely by the isolation of sewage from drinking water.

A few diseases are still uncontrolled because we do not know enough about their effects on human tissues or even, in some cases, their causative factors. Two such diseases—influenza and cancer—are considered here in some detail.

Influenza

One of the more infectious diseases common in modern times is *influenza,* or the *flu*—the only viral disease that continues to produce worldwide epidemics and one of the few illnesses that can recur again and again. The illness begins abruptly with familiar symptoms—aching muscles, chills, a fever, a headache, and extreme fatigue, followed by a runny nose and a sore throat. A cough almost always develops and may be severe. The disease usually runs its course in 3–7 days, although the fatigue may last much longer.

Influenza itself is rarely fatal. Most deaths associated with the disease are the consequence of *bacterial pneumonia,* which often develops after the individual has already been weakened by influenza. These deaths are now largely prevented by treating the bacterial infections with antibiotics.

The influenza virus was one of the first viruses to be identified. It is a medium-sized virus (Figure 18-9) with genetic material arranged in eight separate coils of RNA, each of which directs the synthesis of a different protein. The outer envelope of the virus is composed of six of these proteins, which are recognized by the immune system. Two of these occur as spikes extending from the outer envelope. The *hemagglutinin,* or *H protein,* binds the virus to the target cell. If the activity of this protein is inhibited by an antibody, the virus cannot infect a cell. The *neuraminidase,* or *N protein,* frees the newly formed viruses from the host cell. Antibodies that become attached to the N protein cannot save the infected cell from damage but can prevent the newly formed viruses from escaping from the cell and spreading to new cells.

The formation of antibodies to combat these two outer proteins is the body's main defense against an influenza attack. The natural immune system apparently fails to prevent repeated attacks of influenza because the virus undergoes genetic changes so frequently that the proteins on its outer envelope are unrecognizable to the system. The antibodies against one genetic strain of the influenza virus therefore provide little or no protection against the new mutant forms. An effective vaccine would have to be developed as soon as each new form of virus was detected. Because it is difficult to anticipate a new strain of virus and to develop large quantities of vaccine in a short period of time, an epidemic is usually well underway before the protective vaccine is available.

The influenza virus undergoes two types of protein changes. Small mutations in

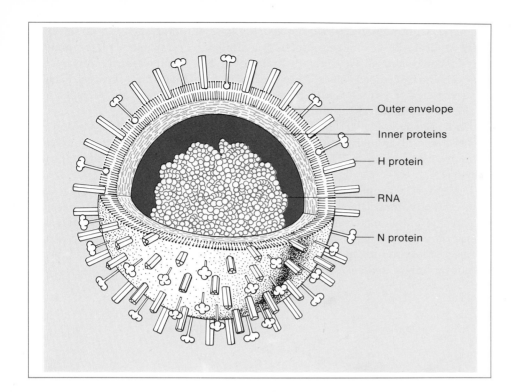

Figure 18-9
Diagram of the Influenza Virus

The spikes on the outer envelope of the virus are made of two different kinds of proteins—the *hemagglutinin,* or *H protein,* and the *neuraminidase,* or *N protein.* The H protein binds the virus to a target cell; the N protein frees the newly formed viruses from an infected cell. Antibodies interfere with the functions of both H and N proteins.

Labels on figure:
Outer envelope
Inner proteins
H protein
RNA
N protein

the RNA coils occur almost every year and create minor protein changes in the new strain. Immunity to a previous, similar strain of virus usually provides some protection against these new forms of viruses. Major changes in the viral RNA that cause extensive restructuring of at least one of the outer proteins occur every 10–12 years. Immunity to a previous strain of virus usually provides no protection against these new and very different forms of viruses.

The proteins of the influenza virus were first described in 1933. The first major change of these proteins—a restructuring of the H protein, which binds the virus to a target cell—was observed in 1947. Immunities developed from previous influenza infections were completely ineffective against this new viral strain and the disease became widespread. In 1957, both the H and the N proteins were altered and an enormous influenza epidemic resulted. In 1968, the H protein changed again and many more people contracted the disease, which was referred to as "Hong Kong flu." A worldwide epidemic in 1978 of the "Russian flu" resulted from several minor protein changes rather than a major restructuring of the virus. This variant of the influenza virus had less impact in the United States, where it was called the "Texas flu."

Major changes in the protein structure of the influenza virus occur when one or more of the eight RNA coils are entirely replaced. The only known mechanism that could initiate such an abrupt change in the virus is *independent assortment*—the rearrangement of chromosomes that takes place during sexual reproduction (see

Chapter 7). Although viruses do not actually have chromosomes and do not normally replicate by sexual reproduction, a similar rearrangement of the genetic material could occur if two different strains of viruses simultaneously infected the same cell and exchanged coils of RNA during the formation of new viruses. Each of the new types of viruses would have an altered but complete set of genes and would inherit characteristics from both of the original viral strains.

Evidence suggests that viral RNA is not restructured in human cells but in the cells of domestic animals, where simultaneous infection with more than one viral strain frequently occurs (Figure 18-10). The restructured viruses then invade human cells.

Once they are formed, the new viral strains completely replace the original viral strains due to natural selection. The original viruses are unable to multiply because most of the individuals they invade have developed effective immune defenses against them from previous infections. However, the new strains, although few in number, can reproduce successfully because individuals have not yet developed a specific immunity to them.

Figure 18-10
The Proposed Origin of New Strains of Influenza Viruses
Here, influenza viruses from a human and from a pig simultaneously infect another pig. The disease localizes in the lungs, where viruses from both strains invade the same cells. Each lung cell produces many new viruses—some identical to the human form, some identical to the swine form, and some with both characteristics. The newly formed viruses with characteristics of both the human and the swine viruses contain RNA coils from both of the original strains.

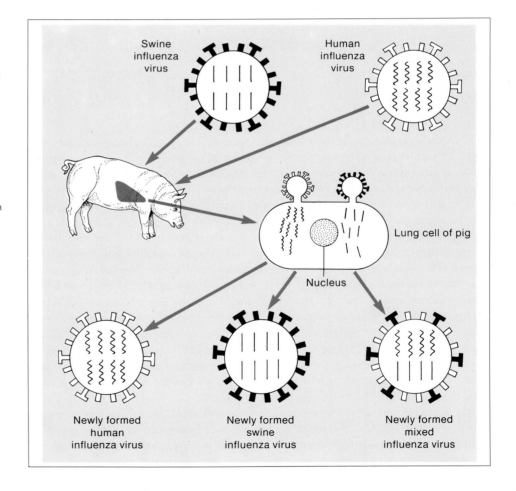

Cancer

Cancer is a disease of all advanced forms of animals and plants. Unlike most other illnesses, cancer is not a single disease caused by a single type of microbe. It occurs in more than 100 different forms and is probably caused by many different factors. One common property shared by all of these disorders is the apparent malfunction of the system that regulates cell division; when a normal cell becomes a cancer cell, it continues to divide uncontrollably. Another unusual characteristic of *malignant* cancers is that the abnormal cells *metastasize,* or move from one part of the body to another (Figure 18-11). Most cancer patients are not killed by their primary tumor. Instead, they succumb to multiple, widespread tumor colonies established by cancer cells that detach themselves from the original tumor and travel through the body, often to distant sites.

At least 4 million cells divide every second in an adult human. Every time a normal cell passes through a cycle of division, it has the potential to lose control—to become a cancer. However, most people do not develop a fatal cancer. Clearly, the body must possess remarkable control systems that prevent aberrant cells from arising or progressing to the cancer state.

Cancer statistics are deceptive because they appear to indicate that few, if any, control systems are at work. About a million new cases of cancer are diagnosed every year in the United States. In each person who develops cancer, the tumor probably arose from a single abnormal cell. Since each person generates billions of new cells every year, the occurrence of deviant cells is actually very small.

The body may produce tens to hundreds of genetically different and potentially cancerous cells each day. Most cancer experts now think that the immune system recognizes cancer cells as foreign and destroys them before they can increase in number. When the immune system becomes weakened or ineffective, the cancer cells multiply and invade normal tissues. If the cancer is not treated, it may eventually proliferate and destroy the body by disrupting the activities of vital organs.

There is no direct evidence to support the theory that a defective immune system permits cancer to develop. However, the immune system does deteriorate slowly with increasing age, and the incidence of cancer is much higher in the elderly. Also, up to 10% of the individuals who have certain immune-deficiency diseases are afflicted with cancer. Many of these people are young, and their incidence of cancer is far greater than the incidence of cancer in normal individuals of the same age. Moreover, patients who have undergone therapy to suppress their immune system (a frequent treatment for individuals who receive tissue transplants) develop cancer more often than individuals with normally functioning immune systems. Patients who have had kidney transplants, for example, are 100 times more likely to develop cancer than individuals in the same age group of the general population. Stress may also contribute to the development of cancer because the body's reaction to stress interferes with the functioning of the immune system.

It is also possible that cancer does develop in individuals with normal immune defenses but that the aberrant cells evade destruction in some way. Some abnormal cells may fail to produce molecules that identify them as foreign, so that they are not recognized by the immune system. Alternatively, cancer cells may release large quantities of foreign molecules that saturate and overwhelm the immune system.

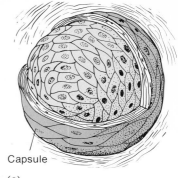

Capsule

(a)

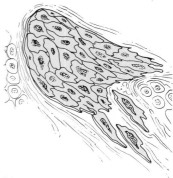

(b)

Figure 18-11
Two Kinds of Cancers:
Benign and Malignant

Cancer is uncontrolled cell division in some part of the body. (a) A benign cancer is contained within a capsule. It grows but does not spread to other parts of the body. (b) A malignant cancer is not contained within a capsule. It grows faster than the benign form and invades surrounding tissues. Cells detach from a malignant cancer and establish new cancers elsewhere in the body.

Several different factors probably contribute to the formation of unusual cells. A genetic change resulting from a mutation or from the presence of a virus in the cell could alter the cell so that it divides uncontrollably. Some types of cancer appear to be inherited. If defective DNA is passed from parent to offspring, a mutation has probably occurred. However, the DNA of some types of viruses becomes an integral part of the cellular DNA; other viruses that contain RNA can induce the infected cell to make DNA copies of their RNA molecules. In both cases, the viral DNA could be passed from parent to offspring just as if it were a mutation. It has been shown that viruses do cause cancer in other animals.

Environmental factors such as radiation, food additives, pesticides, drugs, cigarette smoke, and industrial wastes have been linked to the incidence of cancer. Studies have shown that many of these factors damage the DNA of other animals and probably cause mutations of the DNA in humans. As much as 80% of all human cancers are caused by physical and chemical agents in the environment.

On the basis of indirect evidence that cancer occurs because the immune system fails to monitor the body properly, many new treatments are being developed and tested to stimulate the immune system to higher levels of defense. Tuberculosis bacteria produce a powerful generalized immune response. When these microbes (the BCG vaccine) are injected into tumors, they stimulate the immune system to send white cells and antibodies to the area. This treatment has been successful in destroying cancer cells in some patients.

Another new cancer treatment is the administration of *interferon*—a cellular molecule that stimulates the production of a substance that prevents viral reproduction. Interferon also slows cell division and regulates the activities of the immune system. The cancerous tumors of some patients who have received injections of interferon have regressed significantly. It is still too early to determine whether increasing the amount of this natural body product will play an important role in controlling human cancers.

Treating cancer by manipulating the immune system is still in the experimental stage. Each year, however, researchers acquire more basic knowledge about the nature of normal and cancer cells, producing new insights into the cause or causes of cancer and developing new methods of fighting this tragic disease. Many treatments have proved to be effective in controlling the spread of cancer within the body, and a few treatments actually appear to have cured certain kinds of cancer. These treatments include surgery, the application of chemicals (*chemotherapy*), and radiation therapy.

Summary

Infectious diseases are caused by *parasitic microorganisms*—invisible life forms that live at the expense of their *host organism*. Most parasite—host relationships are not fatal to either organism, because the two species have evolved together over a long period of time. Severe damage to the host occurs when this relationship is altered; the host may be malnourished or weakened in some way or a new host population or species may not have the necessary defenses to combat a particular parasitic species.

There are four major groups of *disease microbes:* protozoa, fungi, bacteria, and viruses. Only a small number of microorganisms in each group cause human illness. *Protozoa, fungi,* and *bacteria* are heterotrophic cells. A *virus,* on the other hand, is not a cell; it is simply nucleic acid enclosed within a protein envelope. A virus can reproduce only by invading a cell and altering cellular activities so that viral proteins and nucleic acids are constructed from the viral genes. The cell is damaged in the process.

The human body's defenses against infectious diseases include its outer fortifications, a local response, and a general response. The outer fortifications are the skin, the linings of the digestive and respiratory tracts, beneficial microbes, and antimicrobial substances released onto exposed body surfaces. The local response prevents the disease microbes from spreading to other areas of the body. This response, which is initiated by damaged body cells, includes the release of *interferons* (if the microbe is a virus), inflammation, and destruction of the microbes by the coordinated activities of *antibodies, complement,* and *white blood cells.* If the disease microbes evade destruction by the local response, the body initiates a general response, which includes the ingestion of the microbes by white cells in the blood vessels and the *lymph nodes* and the development of a fever.

Immunity is the body's resistance to a certain type of microbe. Lymphocyte populations of T and B cells develop during the first infection by a microbe. The *T cells* fight the microbe directly; the *B cells* manufacture microbe-specific antibodies that aggregate the microbes and deactivate them in other ways. The body is capable of developing lymphocytes to combat all forms of foreign proteins and carbohydrates. The *clonal selection theory* explains how the immune system develops the ability to distinguish foreign molecules from normal molecules of the body.

Artificial immunity can be acquired without contracting the disease. *Active immunization* results from the deliberate exposure to attenuated or dead forms of the microbe (a *vaccination*). *Passive immunization* results from the transference of antibodies from an immune animal to an unexposed individual.

The immune system is adapted to recognize foreign materials, and it sometimes responds inappropriately to materials that are foreign but not infectious. Some undesirable side effects of the immune defense system are *allergies,* rejections of *tissue transplants, erythroblastosis fetalis,* and various *autoimmune diseases.*

Despite the immune response that usually develops after a first infection by a particular microbe, humans are susceptible to *influenza* year after year. The influenza virus regularly undergoes genetic changes that alter its surface proteins. The specific lymphocytes developed against the virus in one year are therefore ineffective against later genetic strains.

Cancer occurs when certain body cells continue to divide uncontrollably. A new approach toward understanding and treating cancer involves the immune system. Researchers believe that cancerous cells may normally be destroyed by the immune system as soon as they appear; cancerous tumors continue to grow because the immune system fails to recognize and/or destroy the cancer cells. Some new cancer treatments include stimulation of the immune system to higher levels of activity and the administration of large doses of interferon.

Movement and Exercise

19

Movement is one of the most striking characteristics of life. All organisms have the ability to move at least part of their bodies in a *nonrandom* direction. The speed of this movement varies greatly, from the slow bending of a plant to the rapid sprint of a cheetah.

All cells are capable of locomotion, but only multicellular animals have evolved a truly effective means of rapid, long-distance movement. Some early multicellular organisms developed cells specialized for movement, which allowed these animals to seek the slow-moving unicellular organisms on which they fed. Later, when multicellular animals began to eat one another, a very strong natural selection for rapid locomotion occurred—the predator that moved the fastest caught the most food; the prey that escaped its predators survived to reproduce.

This selection for rapid locomotion has continued throughout the history of most animal groups. Different animals move in different ways. Depending largely on the type of environment in which they live, animals crawl, swim, fly, burrow, paddle, or run. Some animals have no legs and move by other means; other animals have as many as 400 legs. Most adaptations for movement can be attributed to the need to capture food or the need to avoid being eaten. As Table 19-1 shows, many carnivores and the herbivores on which they prey are fast runners.

The muscular, circulatory, respiratory, digestive, and temperature-regulatory systems—and the nervous system, which coordinates these diverse systems—all participate in locomotion. In this chapter, we will discuss how each of these systems is involved in promoting efficient body movement.

Table 19-1
Maximum Speeds Recorded for Various Animals
Speeds are given in miles per hour (1 mile = 1.6 kilometers).

Animal	Speed (in mph)	Diet
Cheetah	70	Carnivore
Pronghorn antelope	61	Herbivore
Lion	50	Carnivore
Thomson's gazelle	50	Herbivore
Quarter horse	48	Herbivore
Elk	45	Herbivore
Coyote	43	Carnivore
Gray fox	42	Carnivore
Rabbit (domestic)	35	Herbivore
Mule deer	35	Herbivore
White-tailed deer	30	Herbivore
Cat (domestic)	30	Carnivore
Human	28	Omnivore
Squirrel	12	Herbivore
Pig (domestic)	11	Herbivore
Chicken	9	Herbivore
Spider	1	Carnivore
Garden snail	0.03	Herbivore

Movement Without Muscles

The muscles involved in locomotion are specialized tissues for rapid movement; interactions between nerves and muscles permit an animal to respond quickly to outside stimuli. Many organisms, however, derive no advantage from moving quickly and can function well without special locomotive tissues. Movement without muscles can be achieved in various ways.

Plants

Plants move either by the *differential growth* of certain structures or by *changes in water pressure.* The slow bending of stems in the direction of light and of roots in the direction of water is accomplished by changes in the lengths of the cells on one side of the plant. *Auxin,* a plant growth hormone, is transported to one side of the stem or root, where it increases the ion concentration and elasticity of the cells so that water enters by osmosis. The cells elongate, causing an increase in the length of only one side of the plant structure, which then bends in the opposite direction to the elongated side. A stem bends toward light when auxin is moved to the shaded side of the stem, causing it to bend away from the shade.

Some plants move more rapidly in response to light or touch by changing their water pressure. Flowers open and close by this method. When cavities in the base of the petals fill with water, the area expands and stiffens, lifting the petals and

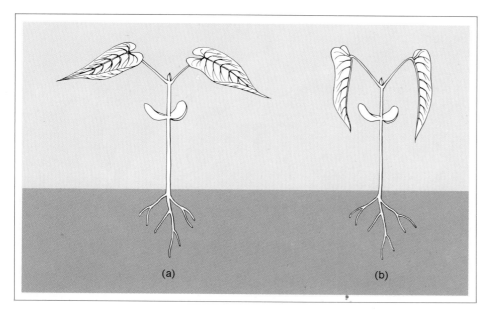

Figure 19-1
Leaf Movements in a Bean Plant
(a) The leaves are raised during the day when the base of each leaf is filled with water. (b) The leaves are lowered during the night when water is removed from the base of the leaves.

pulling them open. When these cavities are emptied, the petals grow limp and collapse into a closed position. The leaves of many plants are also raised and lowered in daily cycles by changes in water pressure (Figure 19-1). Lowering or closing the leaves at night decreases the loss of water when photosynthesis is not taking place.

Single Cells

The nonrandom movement of materials occurs within all types of cells, whether they are free-living unicellular organisms or parts of a multicellular plant or animal. This definite pattern of movement within a cell ensures that nutrients are transported to areas where they will be utilized, waste materials are removed, and organelles are well distributed. Precisely controlled movements also occur during cell division when the paired chromosomes separate and move to opposite sides of the dividing cell. All of these movements and strong contractions of muscle cells involve the same kinds of special protein molecules.

Many cells are able to move through space in a nonrandom direction. The amoeba, a unicellular organism, moves by forming blunt projections of the cell, called *pseudopodia* ("false feet"), into which the rest of the cell flows; the cell reforms as a rounded structure in this new position (Figure 19-2). Another pseudopodium then forms, and the process is repeated. A series of these small movements enables the entire cell to travel a few centimeters in an hour. White blood cells in the human body use the same method of creeping locomotion to track

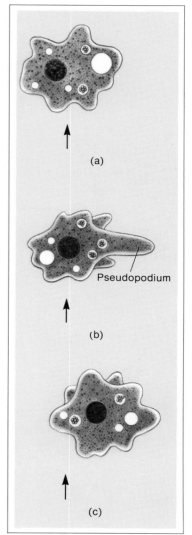

Figure 19-2
Movement in an Amoeba
(a) Position and shape prior to movement. (b) A pseudopodium is formed. (c) After movement of the entire cell to a new position. (Arrows indicate position prior to movement.)

Figure 19-3
Cilia on a Unicellular
Organism (×2,430)

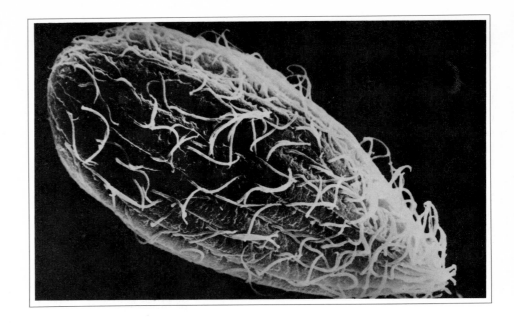

down invading microbes. These cells are capable of moving through the blood vessels to particular destinations independently of the direction of blood flow. Cancer cells also move through the body independently.

Most unicellular organisms move more rapidly than cancer or white blood cells or the amoeba. The outer surfaces of these organisms are covered with thousands of small, hairlike structures called *cilia* (Figure 19-3), which sweep back and forth like the oars on a boat. During the backward sweep (away from the direction of movement), the cilia are stiff and friction created in the surrounding fluid propels the cell forward (Figure 19-4). The cilia then become limp and curled as they move forward and reposition themselves for rowing action. Cilia beat 10—40 times a second, moving the cell at a speed of about 5 meters an hour.

Very small multicellular animals, such as the larvae of aquatic invertebrates and the flatworm, also propel themselves by cilia. Cilia serve as sense organs in some unicellular organisms and direct the movement of food into the bodies of filter feeders. The human respiratory tract is also lined with cilia, which form currents of mucus to move foreign particles away from the lungs.

Instead of thousands of cilia, some unicellular organisms have just a few long structures, called *flagella,* which produce successive waves to propel the organism forward. The sperm of animals and some plants also swim toward the egg by means of flagella (Figure 19-5).

Muscular Movement

The movement of all larger animals is regulated by powerful muscle contractions that convert chemical energy into mechanical work. A *muscle* consists of a bundle of elongated cells, usually arranged in parallel succession, that can contract or shorten rapidly, resulting in a pull or a squeeze.

Figure 19-4
Locomotion by Means of Cilia
Ciliary movement consists of a stiff backward sweep that propels the organism forward (direction of large arrow) followed by a limp, curled return of the cilia to their original position.

Locomotion in the earthworm provides a good example of the squeezing and pulling capacity of muscles. The earthworm is constructed of a series of body segments, each containing two sets of muscles—one extending the length of the segment and the other encircling the segment, as shown in Figure 19-6(a) and (b). When the circular muscles of the segments contract, they squeeze the front of the body into a longer, thinner shape. Body fluids are pumped into the front segments of the body, so that they become stiff and extend forward. Hairs along the front segments then grasp the ground, and the rest of the body is pulled forward as the longitudinal muscles of the segments contract to shorten the body, as shown in Figure 19-6(c).

The muscles of more complex animals, such as mammals, are involved in many bodily functions, including skeletal movements, heart contractions, blood vessel constrictions, and rhythmic squeezing movements of the digestive tract. Three types of muscles are responsible for these different activities—smooth, cardiac, and skeletal muscles.

Smooth Muscles

Smooth muscles line the digestive tract, blood vessels, urinary bladder, reproductive tract, skin, and bronchioles in the lungs. Smooth muscle cells, which are elongated and arranged in parallel succession, do not exhibit the banded patterns observed in cardiac and skeletal muscles. Smooth muscle contractions are controlled by the *autonomic nervous system,* and smooth muscles are categorized as *involuntary muscles* because their contractions are not consciously controlled by the brain.

The *rate* at which smooth muscle cells contract is governed by the autonomic nervous system, but each group of smooth muscles is stimulated in a different way, depending on which organ system of the body is involved. In some muscle systems, such as the digestive tract, the muscle cells are *self-stimulating;* they require no neural impulse from the nervous system to contract. These muscle cells are directly bound together so that an electrical impulse generated in one muscle cell quickly passes to all other cells in the muscle mass. Muscle contractions occur slowly so that the muscle cells do not become fatigued. The stimulation of one muscle cell, usually by stretching, causes all the muscle cells to which it is connected to contract rhythmically. In contrast, the smooth muscles that line the arteries of the circulatory system are not self-stimulating but receive impulses from motor neurons and are controlled entirely by the autonomic nervous system. The urinary bladder contains both types of muscles—self-stimulating muscles and muscles that are stimulated by motor neurons.

Cardiac Muscles

Cardiac muscles—the muscle cells of the heart—form a closely interconnected network; they do not lie in parallel succession. Cardiac muscles exhibit a banding pattern similar to the one observed in skeletal muscles but function in a similar manner to the smooth muscles of the digestive tract. Each muscle cell is capable of self-stimulation; once one of the cells is stimulated (normally within the *pacemaker*), the impulse travels to all cells in the muscle mass. The rate of contraction is governed by the autonomic nervous system.

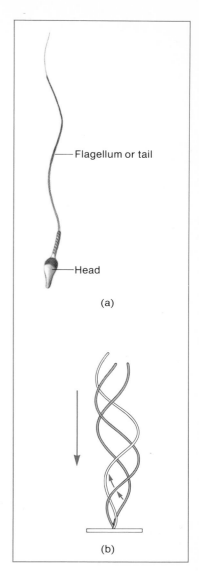

Flagellum or tail

Head

(a)

(b)

Figure 19-5
Locomotion by
Means of Flagella

(a) The flagellum of an animal sperm.
(b) A sperm is propelled forward
(direction of large arrow) by
successive waves of the flagellum.

Figure 19-6
Muscles and Locomotion in a Worm

(a) The earthworm is composed of a series of segments, which respond in an integrated way during locomotion. (b) Cross section through an earthworm, showing the positions of the longitudinal and circular muscles. (c) The earthworm moves by using the longitudinal and circular muscles in its segments. Hairs hold portions of the body in place while other portions are moved. The front portion of the body is extended forward by contractions of the circular muscles (1 and 2). The longitudinal muscles then contract to draw the rest of the body forward (2 and 3). The front portion of the body is extended forward again by contractions of the circular muscles (4), and the rest of the body is drawn forward again by contractions of the longitudinal muscles (5). (Arrow indicates direction of movement.)

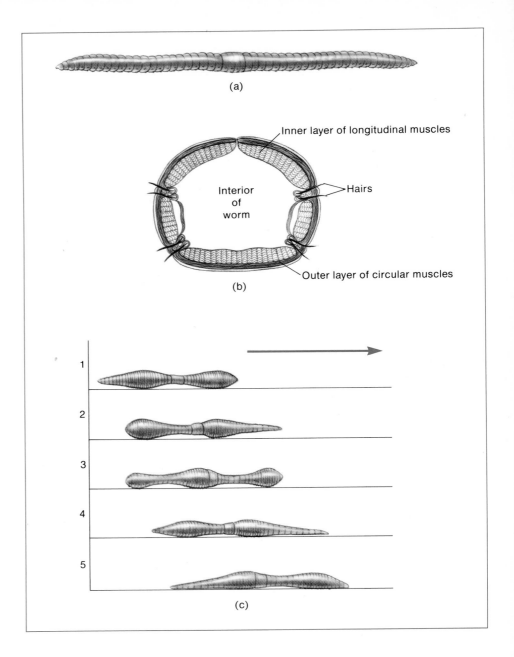

(a)

Inner layer of longitudinal muscles

Interior of worm

Hairs

Outer layer of circular muscles

(b)

1

2

3

4

5

(c)

Skeletal Muscles

The *skeletal muscles*, which regulate movements of the skeleton, are referred to as *voluntary muscles* because they are largely consciously controlled. Contractions of the skeletal muscles enable an animal to walk, move skillfully, position its body, vocalize, breathe, eat, move its eyes, make facial expressions, and perform other physical functions. Compared to smooth and cardiac muscles, skeletal muscles contract quickly, are controlled voluntarily, and become fatigued easily. The

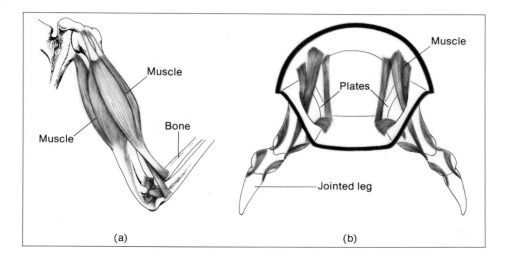

Figure 19-7
Muscles Cause Skeletal
Movements in Most Animals
(a) Arrangement of muscles and bones
in the human arm. (b) Arrangement of
muscles and rigid plates in an
invertebrate.

mechanism of contraction is well understood only in the skeletal muscles, which differ from the other two muscle types in structure.

In addition to muscles, most animals have some form of rigid material, or *skeleton*, that aids their movement by supporting the softer body tissues. Support materials may be internal or external to the skeletal muscles. In vertebrates, the support material, *bone*, is internal to the skeletal muscles, as shown in Figure 19-7(a). In invertebrates, the support material usually lies external to the skeletal muscles in the form of rigid plates, as shown in Figure 19-7(b). Muscles are connected to the bones or rigid plates in such a way that when the muscles contract, the bones or plates of the skeleton move and locomotion results.

Bone—A Dynamic Tissue

Bones not only provide support for locomotion but also protect the body from injury. In vertebrates, all bones were initially on the outside of the body (Figure 19-8), where they provided protection against predators and the physical environment. An internal skeleton developed only in the more advanced fishes. A remnant of the primitive external skeleton remains today in the form of the skull, which surrounds and protects the vulnerable soft tissues of the brain.

We tend to think of bones as dry, inert, and brittle objects because that is how we usually see them, whether we see the skull of a cow in a field or the skeleton of a dinosaur in a museum. But living bones are dynamic, resilient structures in which a variety of important chemical activities takes place.

Several types of materials are stored in the bones. *Calcium* and *phosphorus*, which are crucial to body chemistry, are stored in bones when excess amounts are ingested in the diet. When insufficient amounts of these minerals are available, the stores in the bones are released to be used in cellular activities. Some *fat* is also

Figure 19-8
Early Fishes with
an Exoskeleton

All or most of the bones in these early
vertebrates were external to the softer
body tissues. This hard external
covering probably provided
protection against predators and the
physical environment.

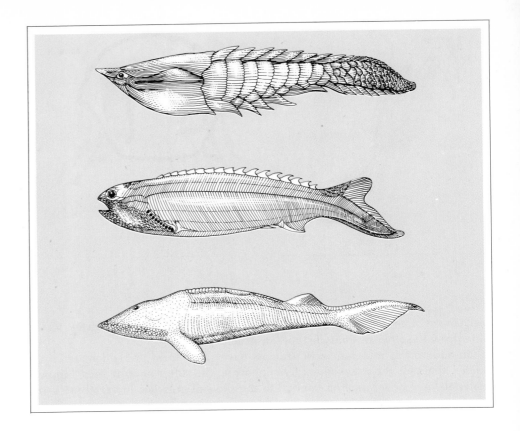

stored in the bones and can be used as a source of energy. Another vital function
carried out by the bone tissue is the production of blood cells, which takes place in
the interior (*marrow*) of some of the longer bones.

Structure

Bones are constructed of hard material interspersed with many fibers and small
canals. Blood vessels and living bone cells located within these canals exchange
materials regularly with each other and with the hard material surrounding them
(Figure 19-9). Mineral deposits, primarily calcium and phosphorus, secreted by the
bone cells into the hard material surrounding the canals permit the skeleton to be
both strong and lightweight.

Bones are continually restructured throughout life in response to the body's
needs for support. Minerals are added or removed according to the amount of
pressure placed on the bones during locomotion. The mineral content of the bones
of athletes is much greater than it is in less active people of the same age. Weight
lifters and discus throwers have the highest mineral levels in their bones, followed
by runners and football players. Increased pressure on these tissues makes the
bones stronger. Remaining in bed for a few weeks causes an appreciable loss of
minerals, although they are rapidly restored after the individual returns to normal

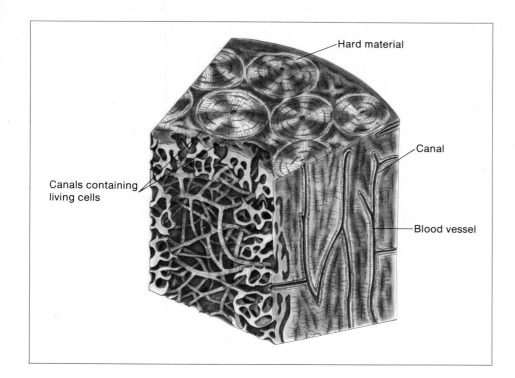

Hard material

Canal

Canals containing living cells

Blood vessel

Figure 19-9
The Structure of a Bone
Living cells in small canals within the bone secrete the hard material of the bone. Blood vessels supply nutrients to the cells.

activity. The continual stress of physical activity is essential to the maintenance of healthy bones that are resistant to fractures.

Astronauts, who must endure extended periods of lower gravity and little physical activity, experience drastically reduced mineral levels in their skeletons. On the Gemini V flight, which lasted eight days, losses reached more than 20% in some bones, due primarily to the lack of weight on the astronauts' skeletons.

Repair

Because bones contain living cells, they are capable of repairing themselves when they are damaged. After a minor fracture or after the bones of a major fracture have been realigned, the first step in the repair process is the formation of a *blood clot,* which seals off the bleeding vessels and holds the broken ends of the bone together. During the first few weeks after a break, the blood clot is replaced by *cartilage*—a more flexible structural material than bone. New bone tissue then begins to spread over the cartilage and slowly invades and replaces it.

After two or three months, the site of the fracture is covered with a mass of new bone tissue, which bulges over the broken ends. In the final stages of healing, this excess material is removed until the repaired bone eventually conforms to the outline of the original bone. Both the resistance to fractures and the ability to mend fractures decrease progressively with age; the elderly are more likely to break bones, and it takes much longer for them to heal.

Figure 19-10
The Knee Joint

Where two or more bones meet, a *joint* is formed. The bones are held together by *ligaments*; friction is reduced by a film of liquid held in the joint by a membrane. The ends of the bones are constructed of *cartilage*—a material similar to bone but softer and more flexible.

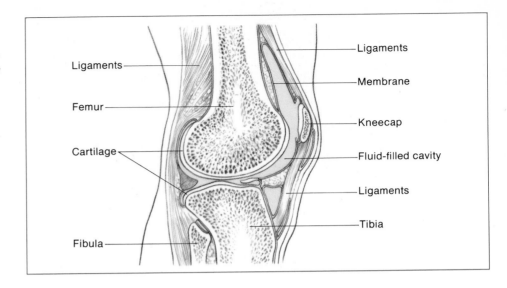

Ligaments

Femur

Cartilage

Fibula

Ligaments

Membrane

Kneecap

Fluid-filled cavity

Ligaments

Tibia

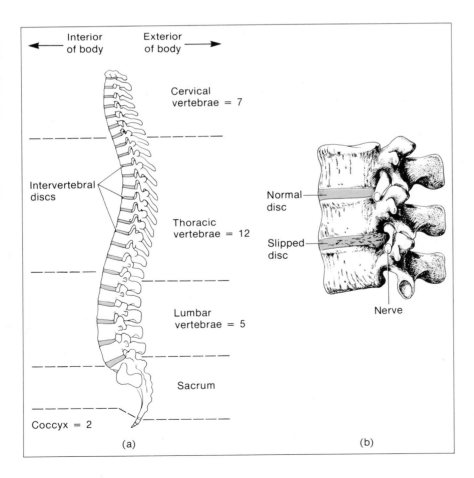

Interior of body

Exterior of body

Cervical vertebrae = 7

Intervertebral discs

Thoracic vertebrae = 12

Lumbar vertebrae = 5

Sacrum

Coccyx = 2

(a)

Normal disc

Slipped disc

Nerve

(b)

Figure 19-11
Arrangement of Bones and Discs in the Human Backbone

(a) The *vertebrae* (the bones in the back) are separated by discs. (b) A *slipped disc* occurs when some of the disc material bulges out and presses on a nerve.

Joining Bones Together

Despite their great differences in size, about the same number of bones form the skeletons of all mammals. Even the necks of giraffes and whales have the same number of bones, although they are quite different in length. Most human skeletons are constructed of 206 bones, but minor variations can occur; for example, one out of every 20 people has 13 rather than 12 pairs of ribs.

Most of the bones forming the skeleton are movable. Where two or more bones meet, a loose connection, or *joint*, is formed. The bones forming a joint are held together by tissues called *ligaments* (Figure 19-10). Although these tissues are very strong, ligaments are the weakest elements in the joint. Violent movements can stretch or tear the ligaments around the joint; this occurs when you receive a *sprain*. Severe stress on a joint may stretch the ligaments until one bone is moved out of alignment with another. This painful condition, called a *dislocation*, usually happens at the shoulder or the hip, although an elbow or a knee is occasionally dislocated. If too much stress is repeatedly applied to a joint, the blood vessels in the ligament may dilate, permitting fluid from the blood to leak into the tissues and cause the area to swell. A familiar example of this condition is *tennis elbow*. The susceptibility to stress injuries can be minimized by a gradual program of physical training during which the ligaments are slowly strengthened.

The *spine*, or *backbone*, is made up of 26 separate bones, or *vertebrae*, as shown in Figure 19-11(a). Each vertebra is separated from adjacent vertebrae by spongy, cushion-like pads, or *discs*, that function like shock absorbers. A *slipped disc* occurs when one of these structures bulges out and presses against a sensitive nerve ending, as shown in Figure 19-11(b). Slipped discs and other back problems usually result if the muscles that hold the structures of the backbone in their proper positions are not sufficiently developed. Lifting heavy objects involves the chest and abdominal muscles as well as the back muscles. If all of these muscles are not strong, the tissues surrounding a disc may receive more pressure than they can withstand. These tissues tear, and the disc becomes squeezed between two vertebrae.

All of the joints in the body are lubricated to prevent friction when the bones move against each other. A film of liquid is present between the bones of a joint that has a wide range of movement, and the cavity at the joint is often surrounded by a membrane that helps to hold the bones loosely together (see Figure 19-10).

The Movement of Bones

Every skeletal muscle is attached to at least two different bones in such a way that changes in the length of the muscle alter the positions of the bones relative to one another. Muscles are attached to bones by *tendons* (Figure 19-12), which are tough cords similar to ligaments. Of the four structures involved in movement—bones, muscles, ligaments, and tendons—the tendons are the strongest and the least likely to break. However, severe stress may rip the tendon from its attachment to a bone. Tendons seldom tear at the points of attachment to the muscle itself.

Some tendons are very short; others are quite long. Long tendons allow the muscles to be located quite far from the bones they move. For example, the muscles

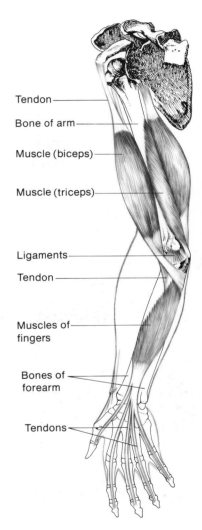

Tendon

Bone of arm

Muscle (biceps)

Muscle (triceps)

Ligaments

Tendon

Muscles of fingers

Bones of forearm

Tendons

**Figure 19-12
The Four Types of Tissues Involved in Movement**

Bones are moved by the contraction of *muscles*. Bones are joined to other bones by *ligaments*, and muscles are connected to bones by *tendons*. Long tendons permit muscles to be located quite far from the bones they move; the fingers are moved by muscles located in the forearm and the palm of the hand.

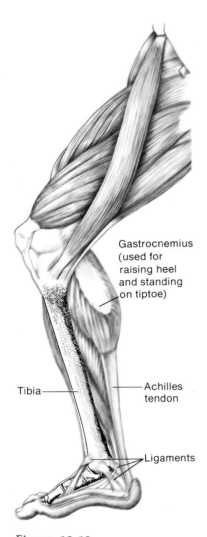

Gastrocnemius (used for raising heel and standing on tiptoe)

Tibia

Achilles tendon

Ligaments

**Figure 19-13
The Achilles Tendon
in a Human**

The ankle is free of muscles and light in weight because its movement is regulated by this very long tendon.

that lower the fingers are located in the palm of the hand and in the arm; the muscles that raise the fingers are located in the forearm and are connected to the bones of the fingers by long tendons that lie along the top of the hand and the fingers (Figure 19-12). These tendons are easy to see and feel on top of the knuckles, especially when the hand is clenched. If these muscles were near the bones they move, the fingers and the top of the hand would be considerably more bulky and less useful in grasping and manipulating objects.

A particularly long tendon, commonly referred to as the *Achilles tendon,* connects the calf muscle to the bone of the heel (Figure 19-13). This long connection between muscle and bone frees the ankle of muscle so that it is light in weight. In some swift-running mammals, such as deer and horses, this tendon extends from the heel all the way up to the back of the thigh, eliminating the calf muscle completely; the legs of these animals are slim and light so that they can run faster for longer periods of time.

Skeletal muscles are voluntary muscles because their activities can be consciously controlled by the individual. Not all skeletal muscle movement is voluntary, however. Most muscles regularly undergo small contractions that are regulated by neurons in the spinal cord. These small movements help to keep the muscles healthy. Involuntary control of skeletal muscles is also essential for maintaining posture. Information about the state of all the muscles and the position of the body is transmitted by sensory neurons from the muscles, joints, and tendons as well as from the semicircular canals and the eyes to the central nervous system. Small deviations in body posture are corrected by involuntary muscle contractions directed by motor systems of the brain.

The Structure of Muscles

Although the number of bones in the mammal skeleton has not changed markedly during vertebrate evolution, distribution of the muscles has been greatly altered to produce such different movements as swimming, flying, and running. Muscles never work by pushing; they can only pull or squeeze. Skeletal muscles perform mechanical work by shortening, or contracting, their length and pulling bones into new positions. To allow a bone to move in two directions, skeletal muscles must occur as *antagonistic pairs*—one muscle pulls the bone in one direction; the other pulls the bone in the opposite direction. Most of the 650 skeletal muscles in the human body (Figure 19-14) occur in antagonistic pairs. As one muscle shortens, the other muscle is usually stretched.

A single muscle, such as the *biceps* of the arm (Figure 19-14), is made up of hundreds of thousands of long, cylindrical *fibers.* Each fiber is a single muscle cell and the fibers of a skeletal muscle are arranged parallel to one another in bundles, as shown in Figure 19-15(a) and (b). Each skeletal muscle cell, or fiber, contains several nuclei. During embryonic development, several muscle cells fuse together, but their nuclei remain intact. The total number of muscle cells in an adult is determined before birth and no further cell division occurs after that point. Of course, muscles increase in size when they are exercised regularly, but this occurs because the size of the individual muscle cells increases, not because the number of muscle cells increases.

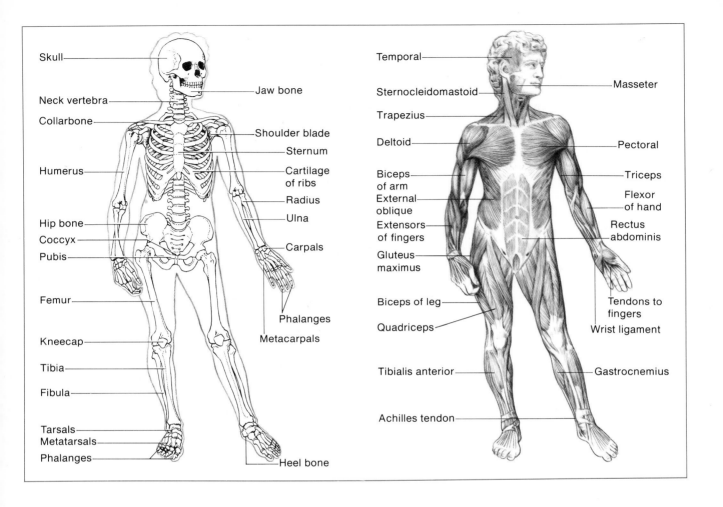

Skull

Jaw bone

Neck vertebra

Collarbone

Shoulder blade

Sternum

Humerus

Cartilage of ribs

Radius

Ulna

Hip bone

Coccyx

Pubis

Carpals

Femur

Phalanges

Metacarpals

Kneecap

Tibia

Fibula

Tarsals

Metatarsals

Phalanges

Heel bone

Temporal

Masseter

Sternocleidomastoid

Trapezius

Deltoid

Pectoral

Biceps of arm

Triceps

External oblique

Flexor of hand

Extensors of fingers

Rectus abdominis

Gluteus maximus

Biceps of leg

Quadriceps

Tendons to fingers

Wrist ligament

Tibialis anterior

Gastrocnemius

Achilles tendon

Each fiber is made up of hundreds of smaller *fibrils,* shown in Figure 19-15(b) and (c), which are also long, cylindrical, and arranged parallel to one another. When a muscle fiber contracts, all of its fibrils contract at once; each fiber either remains inactive or exerts its full strength when it shortens in what is characterized as an "all-or-none response."

A muscle can produce different amounts of power by regulating the number of fibers that contract. Much less power is required to lift a pencil than a 50-pound weight. During a slight muscle contraction, only a few fibers contract (although all of the fibrils in each fiber contract). During a stronger muscle contraction, more fibers are recruited. The number of fibers involved in a muscular movement is controlled by the motor cortex and associated areas within the brain.

Skeletal muscles have a regular pattern of dark bands, shown in Figure 19-15(c), that smooth muscles do not have. (Cardiac muscles are also banded.) Researchers observing these bands through high-powered microscopes have only recently discovered that these structures actually cause muscular contractions. The bands are located in the fibrils, as shown in Figure 19-15(d); the darker shade of the

**Figure 19-14
The Major Bones and
Skeletal Muscles of
the Human Body**

Figure 19-15
Structure of a Skeletal Muscle

Each muscle is made of bundles of *fibers* (a and b), which are composed of hundreds of *fibrils* (b and c). The banding pattern of the fibrils (c and d) is produced by the overlapping of two protein molecules, *actin* and *myosin* (e).

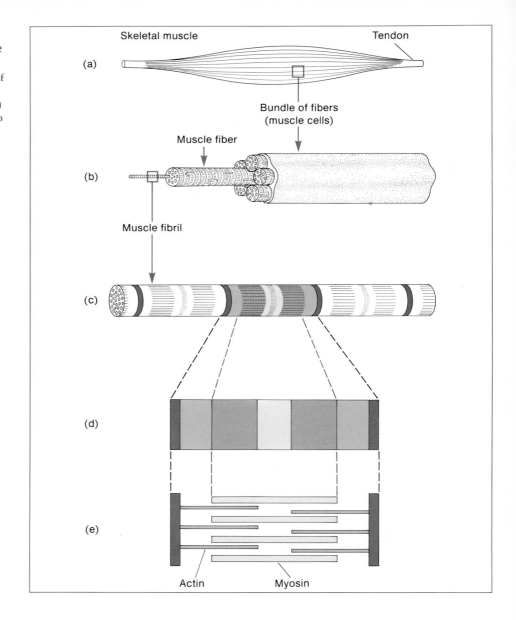

bands is caused by the overlapping of two long protein molecules, *actin* and *myosin*, as shown in Figure 19-15(e).

A fibril shortens during a muscle contraction because the myosin molecule, which remains stationary, pulls the actin molecule across its surface. Small projections, called *cross bridges*, on the myosin molecule repeatedly attach themselves to the actin molecule and, by pulling and letting go, move the actin molecule across the myosin (Figure 19-16). The attachment appears to be a chemical one during

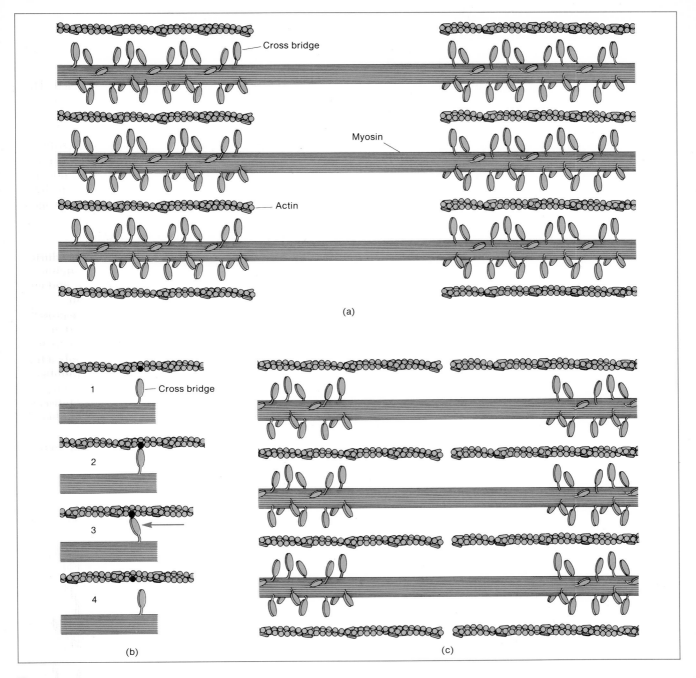

Figure 19-16
The Mechanics of Muscle Contraction

(a) The arrangement of actin and myosin molecules during muscle relaxation. Molecules (not shown here) between the actin and the myosin molecules prevent them from interacting chemically. (b) The muscle contracts when the actin molecule is accessible to the myosin molecule; *cross bridges* on the myosin molecule become attached to the actin molecule (2) and pull it along the myosin molecule (3). Contact between the cross bridges and the actin molecule is then broken (4) and the sequence is repeated (1). (c) The arrangement of actin and myosin molecules during a complete muscular contraction.

which the actin molecule is briefly bound to the myosin and then released. This description of muscle contraction is called the *sliding filament theory*. The mechanics of muscle contraction are now well established, but the chemical nature of the reaction is not completely understood.

The immediate source of energy for muscle contraction is adenosine triphosphate (ATP) which is formed from adenosine diphosphate (ADP) and phosphate when glucose molecules are disassembled inside a mitochondrion. It is not surprising, therefore, that these very active cells contain large numbers of mitochondria. Glucose is stored in muscle cells in the form of the large molecule, glycogen.

Stimulation of Muscle Contraction

Even the simplest movement of the body is a highly complex process during which many muscles contract, relax, or maintain a steady tension. The activities of these muscle groups must be coordinated if movement is to be smooth and efficient. This important function is performed by the central nervous system.

Motor neurons carry the stimulus from the central nervous system to the muscles (Figure 19-17). Each motor neuron is connected by *synapses* to more than one muscle fiber and, when stimulated, causes all of the fibers connected to it to contract. Some neurons are connected to hundreds of fibers; others, to only a few. The dexterity or precision of the muscular movement diminishes as the number of fibers that a neuron stimulates increases. One neuron is connected to only five muscle fibers in the eye because eye movement requires highly sensitive muscular adjustments, whereas one neuron may control several thousand different fibers in an arm or leg muscle.

Neurotransmitter molecules (acetylcholine) released by the motor neuron travel across the synaptic cleft to the muscle cell. There, they generate electrical

**Figure 19-17
The Structure of
a Motor Neuron**

The electrochemical impulse received by the *dendrites* or the *soma* (the body of the cell) travels down the *axon* (direction indicated by arrows). The message is then transmitted from the axon to the muscle via neurotransmitters released at the *synapse*.

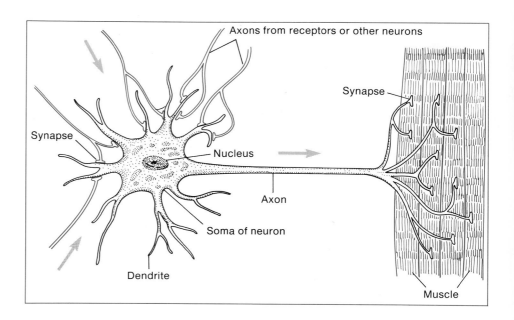

Axons from receptors or other neurons

Synapse

Synapse

Nucleus

Axon

Soma of neuron

Dendrite

Muscle

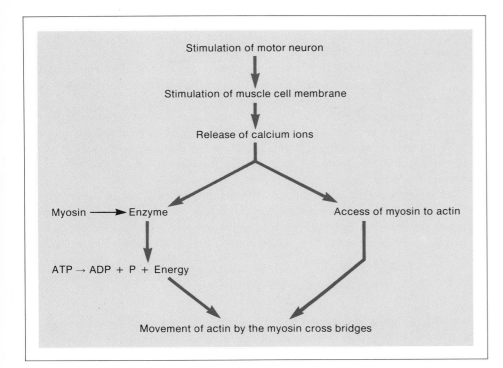

Figure 19-18
The Sequence of Events
in the Transfer of an
Electrochemical Impulse
to a Muscle Contraction

impulses that travel along the outer membrane of the muscle cell, or fiber. The impulses affect all of the fibrils inside, stimulating the release of calcium ions (Ca^{++}) stored in packets within a membrane network surrounding each fibril. The release of calcium ions (1) temporarily converts myosin into an enzyme that causes the breakdown of ATP into ADP, phosphate, and energy and (2) allows the myosin molecules to interact with the actin molecules so that contact between the cross bridges and the actin can be made. This sequence of events is diagrammed in Figure 19-18. After the fibrils contract, the calcium ions are returned to the packets to be used again. The movement of calcium ions is accomplished by active transport and therefore requires energy, which is released from ATP molecules.

The energy within the ATP molecules that propels the reaction is derived primarily from the breakdown of muscle glycogen into glucose and the conversion of glucose into carbon dioxide, water, and energy. In addition to ATP molecules, muscle cells also contain a second form of high-energy molecule called *creatine phosphate.* Some of the energy obtained from the glucose molecules may be temporarily stored in creatine phosphate molecules and transferred to ATP molecules later.

Energy Use During Exercise

Muscles are beautifully adapted to perform all types of work, ranging from simple adjustments in body posture to rapid flight. The rate of cellular activity varies more in muscles than in any other body tissue; muscular activity can be increased as much as 50 times above the resting level. However, extremely rapid rates of muscle contraction cannot be sustained for very long. Except in a short sprint, the speed that an animal can maintain depends much more on the rate at which the heart and lungs can supply the muscles with oxygen than on the activities of the muscle cells themselves.

Energy for muscle contraction can be derived from glucose either with the use of oxygen (aerobic respiration) or without the use of oxygen (anaerobic respiration). Which system of energy release is implemented depends on the rate at which energy is required for muscular activity and the rate at which oxygen can be delivered to the cells.

Aerobic exercise occurs when the rate at which muscle cells utilize oxygen in the breakdown of glucose is no greater than the rate at which oxygen can be transported to the muscles. First, glycogen is converted into glucose. Glucose is then broken down into carbon dioxide, water, and energy by the chemical reaction

$$\underset{\text{glucose}}{C_6H_{12}O_6} + \underset{\text{oxygen}}{6O_2} + \underset{\text{water}}{6H_2O} \longrightarrow \underset{\substack{\text{carbon} \\ \text{dioxide}}}{6CO_2} + \underset{\text{water}}{12H_2O} + \text{Energy}$$

About 40% of the energy is transferred to ATP molecules (ADP + phosphate + energy $\longrightarrow$ ATP + H_2O); the remainder is released as heat. An animal can continue aerobic exercise for long periods of time.

Anaerobic exercise permits more vigorous exercise for shorter periods of time than aerobic exercise. In the anaerobic reaction, glycogen is also converted into glucose, but the glucose is only partially broken apart in the absence of oxygen:

$$\underset{\text{glucose}}{C_6H_{12}O_6} \longrightarrow \underset{\text{lactic acid}}{2C_3H_6O_3} + \text{Energy}$$

Only about 2% of the energy originally present in the glucose molecule is transferred to ATP molecules. Some of the energy remains in the lactic acid molecules, and some is released as heat. Further breakdown of the lactic acid molecules cannot take place in the absence of oxygen. The amount of energy released from glucose during the anaerobic reaction is only 5% of the amount of energy released during the aerobic reaction.

The rate at which the muscles can contract during the anaerobic reaction is not limited by the rate at which oxygen can be supplied to the muscle cells, as is true during aerobic exercise. For example, it is possible to complete a 100-yard dash within 10 seconds. About 7 liters of oxygen would be required if the energy for muscle contraction during this dash were derived from the aerobic breakdown of glucose, but the lungs can only supply about 0.5 liter of oxygen in 10 seconds. The body is able to function without oxygen intake during such a rapid movement. In fact, the runner does not need to breathe at all, since oxygen is not used and carbon dioxide is not produced.

Anaerobic exercise cannot be continued for long periods of time because (1) lactic acid accumulates in the muscles and gradually lowers the pH (increases acidity), causing pain and interfering with normal chemical activities, and (2) the inefficient use of glycogen causes the energy reserves in the muscles to be used up rapidly. Aerobic exercise can utilize the energy stored in fats; anaerobic exercise requires glucose. When all of the energy for muscular activity is provided by the anaerobic reaction, the body's glycogen reserves are depleted in less than a minute and the muscles can no longer function. Of course, exercise levels between the entirely anaerobic type, such as sprinting, and the entirely aerobic type, such as slow walking, employ a combination of anaerobic and aerobic energy release. The amount of time that an exercise can be continued depends largely on the relative proportions of these two types of muscle activity.

Certain muscles are adapted to long-term aerobic exercise and are not easily fatigued. The fibers of such muscles contain *myoglobin*—a pigment similar to the hemoglobin of red blood cells—that temporarily stores oxygen and gives the muscles a uniformly darker color (not to be confused with the dark bands present in all skeletal muscles). Fibers containing myoglobin are referred to as *slow-twitch fibers;* they contract more slowly and undergo fatigue less readily than fibers without myoglobin. For example, the leg muscles of birds, which are used predominantly for standing and walking, contain this pigment and are dark in color. Other muscle fibers, called *fast-twitch fibers,* are adapted for short-term anaerobic exercise and contain little or no myoglobin. The breast muscles of birds, for example, contain very little myoglobin and are light in color. These muscles are used for rapid flight; the energy for more rapid muscle contractions is derived anaerobically and does not rely on oxygen stored in myoglobin. Similarly, rabbits use their skeletal muscles primarily to sprint to the safety of their burrows, and their muscles are generally light in color. Hares, by contrast, are endurance runners and tend to have darker muscles.

The specific types of muscles in humans do not differ as markedly. Most muscles contain a mixture of both kinds of fibers. In general, however, muscles involved in maintaining posture (such as back muscles) are darker in color than muscles involved in lifting (such as arm muscles). Slight differences also exist among humans. Some individuals have a higher percentage of darker fibers; others, a higher percentage of lighter fibers. These differences are often important factors in determining the top athletes in certain competitive events. World-class marathoners, for example, tend to have a higher proportion of dark fibers in their leg muscles than sprinters do.

The Body's Response to Aerobic Exercise

Regular involvement in some form of intense, aerobic exercise is an effective way to maintain good health. Muscular work affects practically all organs of the body profoundly and beneficially. The greatest health benefits are achieved when large muscle groups are worked to the degree that heart and respiratory rates are elevated and the level of activity can be maintained for at least 30 minutes. Physiologically, this is the level of exercise at which the lungs and circulatory system can just meet the demands of the muscle cells.

In this section, we will examine the response of the entire body to this level of exercise. We will attempt to understand how exercise promotes good health and how all of the body systems, considered individually in earlier chapters, function in an integrated way to achieve maximum physical performance.

The Warm-up Stage

In the initial stage of aerobic exercise, the body goes through a short, partially anaerobic phase called the *warm-up* that lasts from two to six minutes, depending on the severity of the exercise and the physical condition of the individual. During this often uncomfortable period, the muscles are working at a greater rate than when they are at rest, so that more oxygen is required. At first, oxygen transport is inadequate to meet the body's needs because the respiratory and circulatory systems adjust slowly to new levels of activity. The muscles temporarily operate in a partially anaerobic manner, and some glucose is converted into lactic acid and energy. The presence of lactic acid in the muscles causes them to feel heavy and somewhat painful.

After a few minutes of exercise, the carbon dioxide produced from the aerobic breakdown of glucose accumulates in the bloodstream and stimulates the respiratory system. The hypothalamus, through the autonomic nervous system, causes the adrenal medulla to release epinephrine. These activities produce an increase in the breathing rate and the heart rate as well as in the volume of blood that is pumped through the heart at each contraction. The time lag between the onset of activity and the stimulation of the lungs and heart causes lactic acid to accumulate in the body. The only way to reduce the level of lactic acid is to convert it into the other molecules, but such chemical reactions only proceed in the presence of oxygen. All available oxygen is used for muscle contraction as rapidly as it is transported to the muscles during the remainder of the exercise period. The lactic acid produced during the warm-up stage therefore remains in the body until after exercise is completed and excess oxygen is once again available.

After this brief period of mixed anaerobic and aerobic exercise, the heart and lungs function rapidly enough to maintain sufficient oxygen flow to the muscles and allow completely aerobic exercise to begin. The movements of the muscles themselves improve the speed of blood flow throughout the entire circulatory system. Muscular contractions around the veins increase the rate at which blood is returned to the heart. When large muscle groups are being exercised, they require greater blood flow. The circulatory system responds to this need for increased blood flow by favoring the exercising muscles, including the heart, the diaphragm, and the muscles that move the ribs, at the expense of other body tissues. Blood vessels in the working muscles dilate to increase blood flow. As blood is shunted to these muscles, its flow is decreased elsewhere in the body by the constriction of vessels, including the vessels that supply the skin, which becomes pale and cool. Heat energy released from the breakdown of glucose in the muscle cells is carried by the blood and normally leaves the body through the skin by the evaporation of sweat. Because the vessels in the skin constrict during the early stages of exercise, however, most of the additional heat produced by the working muscles does not reach the skin and is not removed from the body. The retention of heat allows the muscles to be warmed during the warm-up process. Normally, the temperature of skeletal muscles is slightly below 98.6°F. The optimal temperature of these muscles during contractions is two or three degrees above 98.6°F; this temperature is quickly reached during the first few minutes of aerobic exercise.

The joints of the moving bones also readjust at this time. The volume of the spongy cartilage covering the ends of bones increases about 12% during the first 10 minutes of exercise, which helps to cushion the joints and protect them against injury. This increase in volume is apparently due to the addition of water to the cartilage.

The Sustained Phase

After the warm-up, all body systems stabilize at a higher rate of muscular activity and the individual experiences a "second wind"—the feeling that the whole body has shifted into the correct gear for the level of exercise being undertaken. Breathing becomes regular, although more air is inhaled and exhaled during each breath than when the body is at rest. The heart and lungs are now adjusted to working at an increased but steady rate and volume; the amount of oxygen transported to the muscles is equal to the amount of oxygen required for aerobic exercise.

The heart pumps about 5 liters of blood per minute when the body is at rest, but a much greater volume of blood—up to 30 liters per minute—is pumped through the heart during the sustained phase of aerobic exercise. The rate of heart contraction, which is measured by the pulse rate, may be as high as 215 beats a minute and the systolic blood pressure may exceed 175. The coronary arteries, which carry blood to the heart, dilate to permit the heart to contract more rapidly.

The rate and volume of breathing are also much greater during exercise than when the body is at rest. Normal breathing occurs 10–20 times a minute, but may occur up to 45 times a minute during exercise. The volume of air that is moved into and out of the lungs increases from about 6 liters per minute at rest to as much as 200 liters per minute during exercise. The increased activity of the respiratory muscles themselves may utilize 10% of the total amount of oxygen that is consumed. Individuals who are not accustomed to such heavy breathing may find it distressing initially but exhilarating after training.

Body temperature is also readjusted during the sustained phase of aerobic exercise. The removal of excess heat from the body is critically important; an increase in body temperature of about 15°F above normal is fatal. Heat dissipation therefore takes precedence over muscular activity. Because the amount of blood in the body is limited, as the need for heat dissipation increases and blood is shunted to the skin, the amount of blood flow to the muscles must decrease. More blood is shunted to the skin if heat or humidity levels in the environment are high because the body cannot release heat through the skin as efficiently under these conditions as it can when the environment is cool and dry. As more blood is diverted from the working muscles, exercise is more difficult and performance is relatively poor.

A significant amount of the 40 liters of water contained in the human body—sometimes as much as 2 liters per hour—may be removed from the body in the form of sweat. If only water were lost in sweat, the concentration of salt in the body fluids would increase and thirst receptors would stimulate the individual to drink until the lost water was replaced. But sweat contains salt as well as water, so that the concentration of salt in the body does not increase as much as it would if only water were lost. The body's full need for fluids is not usually detected by the thirst receptors, and the desire for liquids diminishes after about 50% of the lost water is replaced.

Thirst can be maintained until the lost water is completely replaced by drinking a 2% salt solution (about one teaspoon of salt per quart of water) instead of pure water. One advantage of physical conditioning is that the composition of sweat becomes altered, so that very little salt is lost in sweat and drinking liquids (as dictated by thirst) more closely compensates for the water loss.

Many people drink sweetened beverages to quench their thirst, but the water in sugar solutions is not absorbed into the bloodstream from the stomach as rapidly as the water in salt solutions. Sugar solutions tend to accumulate in the stomach until the individual simply feels too full to drink. The intake of sugar in solution or food also decreases the body's ability to produce sweat.

If more than 2% of the body weight is lost in the form of water, *dehydration* results. The blood pressure drops because blood volume is reduced, and the body temperature may rise rapidly because too much water has been lost for the body to produce sweat. The body cannot endure for long in a dehydrated state. It is important to drink water before and after exercising or even during exercise if possible. Sweat also contains salt (sodium chloride), potassium, and small quantities of vitamin C, and extra amounts of these important nutrients should be taken when a large amount of sweat is released from the body.

Fatigue

Even moderate exercise cannot be continued indefinitely. At some point, the body experiences *fatigue*—the feeling that it simply cannot continue to move. The symptoms of fatigue generally reflect a variety of causes.

One of the first symptoms, mental fatigue, is protective in that it forces the individual to stop exercising before the body is severely strained. Both sensory neurons in the muscles and lactic acid levels in the blood send information to the cerebral cortex and associated areas of the brain concerning the condition of the muscles. Only mild muscle fatigue and small amounts of lactic acid in the blood produce this effect in the unconditioned person. Through physical training, however, the brain either becomes less sensitive to this information or the muscles become more tolerant of their fatigued state. Drugs that remove this mental inhibition can be administered to improve performance; however, the use of such drugs is very dangerous. If the brain does not stop the body from exercising before it becomes completely fatigued, injury or even death can result from overexertion.

When the synapses between the motor neurons and the muscles become fatigued, coordination is disturbed and reaction time is slowed. The synapses appear to be more vulnerable to fatigue than the neurons themselves. Exactly what happens during synapse fatigue is not yet known.

Loss of muscle power is primarily due to the depletion of glycogen in the muscles. The initial glycogen content of the muscles and liver is of decisive importance in determining the length of time that exercise can continue. The amount of these glycogen stores and, consequently, the endurance period during long-term exercise can be increased by certain diets. The effects of various diets on endurance are given in Table 19-2. Clearly, a high-carbohydrate diet for three days prior to strenuous exercise can greatly increase the length of exercise time before fatigue occurs. This diet is even more effective if the glycogen reserves in the muscles are emptied by strenuously exercising four days prior to the athletic event. If little exercise is performed during the three days of the high-carbohydrate diet, the glycogen reserves will be built up to the maximum level.

Table 19-2
The Effects of Various Diets on Glycogen Stores and Endurance

Diet	Grams of glycogen per 100 grams of muscle	Time to fatigue* (in minutes)
Mixed or "balanced"	1.75	115
Fat and protein only	0.60	60
Carbohydrates only	3.50	170
Carbohydrates only for three days prior to exercise; before that, a low-carbohydrate diet	4.0	More than 240

*The time to fatigue is measured at an exercise intensity of 75% maximum.

The practice of this dietary regime, called *carbohydrate loading,* is controversial but its effect on long-term endurance is impressive. However, frequent use of this diet (more than once a month) may result in malnutrition, especially protein deficiency. The diet cannot be used by athletes during the competitive season if it is necessary to reach peak performance levels every week. Another disadvantage of carbohydrate loading is that the body retains more water when it stores glycogen, so that the athlete's weight increases slightly. A weight increase would impair performance in sports such as gymnastics and jumping where the body must be lifted.

Aerobic exercise can continue after the glycogen reserves in the muscles are exhausted, although it is often painful and performance is poor. When this occurs, fatty acids from fat reserves are broken down to produce carbon dioxide, water, and energy. Although fat contains more calories per gram than glycogen, more oxygen is required to release the calories in fat and they are used less efficiently. A well-trained marathon runner frequently uses up the glycogen reserves in the muscles after running 18–20 miles, a point in the race often referred to as "the wall." The energy for muscle contractions is then primarily provided by fatty acids from fat reserves. Evidence suggests that women draw more regularly on their fat reserves during endurance events than men do. Some energy can also be obtained from body protein, although this amount is usually less than 2% of the energy used during exercise.

When preparing for activities in which endurance is not a major factor, diets other than a high-carbohydrate one are recommended. If strenuous exercise is to last less than an hour, the normal glycogen reserves will be sufficient and diet is less important. The most effective preparation for such an activity is to eat a light meal that is not particularly high in carbohydrates about $2\frac{1}{2}$ hours prior to the event. A large intake of sugar tends to decrease the capacity for short-term strenuous activity.

People are often forced to stop exercising due to severe abdominal pains. These pains, commonly called *side ache* or *stitch,* most frequently occur just below the ribs. Their cause is not known; they may result from cramps in the muscles of the diaphragm or, if the pain is lower, from straining the membranes that support the abdominal organs. Side aches can be minimized by a gradual training program that includes sit-ups to strengthen the abdominal muscles, so that the organs are held more tightly in place within the abdominal cavity.

Individuals who are not involved in a rigorous training program or competitive sport tend to become fatigued due to a lack of motivation. The tendency to stop exercising prematurely must be overcome before a truly effective conditioning program can begin.

Recovery from Exercise

After a strenuous workout, the body still contains lactic acid from the initial anaerobic breakdown of glucose and very little ATP remains in the muscle cells. The lactic acid must be removed and the ATP level must be restored before the body can function normally again. Both of these processes must take place in the presence of oxygen, since the energy needed to drive them is acquired through aerobic reactions. Oxygen is also required to replenish the red blood cells and the myoglobin in the muscles. The oxygen for all of these recovery needs is obtained from heavy breathing and an accelerated heart rate after exercise has ceased, which is often referred to as "paying off the oxygen debt."

Most of the lactic acid produced during the anaerobic phase of the warm-up is converted back into glycogen, primarily in the liver. The energy to drive this reaction is derived from breaking some of the lactic acid molecules apart in the presence of oxygen to produce carbon dioxide, water, and energy. About 15% of the accumulated lactic acid is broken down to release energy to convert the remaining 85% of the lactic acid into glycogen, which is stored again in the muscles.

It takes the body at least one hour to return to its previous resting state after a strenuous aerobic workout. This recovery time can be reduced by massaging the muscles and lightly exercising muscles not involved in the major exercise, thereby increasing blood flow to the muscles so that lactic acid is transported more rapidly to the liver for conversion into glycogen. Light exercise also helps to restore the excitability of the fatigued synapses.

Some soreness from small injuries to some of the muscle tissues and tendon attachments usually develops about 12 hours after a beneficial exercise period and may last for several days. When the tissues mend, they are stronger than they were before the exercise. The muscles may also swell slightly, contributing to the soreness and stiffness.

Muscle cramps—strong, involuntary, sustained contractions of a muscle—frequently occur after exercising. The cause of these cramps is not known. Muscle cramps can usually be prevented by taking a warm bath after exercising and by drinking a 2% salt solution if a great deal of water and salt has been lost in sweat. Once a cramp develops, the muscle should be massaged and gently stretched.

The Effects of Physical Conditioning

Noticeable improvements in physical and mental health result when a period of aerobic exercise is included in an individual's daily activities. The entire skeletal system is strengthened and becomes less vulnerable to injury. The tendons and ligaments and the ends of bones where they meet to form a joint become thicker.

The number of fibrils within the muscle cells increases, so that stronger contractions are possible. Whole new networks of capillaries develop within the muscles and joints, enhancing the blood flow and improving endurance. Well-exercised muscles develop less lactic acid and store more glycogen than muscles that are seldom used.

Muscular movements become more precise and the muscles are used more efficiently after physical training. New connections between the motor neurons and the muscles are not made, but the central nervous system exerts finer control over the activities of existing connections.

Like any other muscle, the heart increases in size when it is exercised regularly. Stronger heart muscles pump more blood with each contraction, which conserves energy—the heart can beat at a slower rate and still move the same amount of blood through the circulatory system. The muscles in the blood vessels are strengthened by dilating and constricting during physical exercise. An individual with more muscular vessels is less likely to develop long-term high blood pressure. The number of small arteries and capillaries that form branches from the coronary arteries also increases, providing more routes to the heart muscles, so that if a heart attack occurs, its effects are less severe.

More and more people are turning to exercise to prevent heart diseases and other conditions that cause premature aging and death. Exercise further improves the quality of everyday life by increasing our mental alertness and endurance, and by helping us to maintain an ideal weight and develop attractive, supple bodies. These desirable states of health were natural by-products of the great physical effort our ancestors had to make just to stay alive. Although our environment no longer demands it, exercise is still a vital component of a healthy and vigorous life.

Summary

Nonrandom movement is characteristic of life; it occurs in all cells and all multicellular organisms. In some cases, the entire organism moves from place to place in a nonrandom direction. Movement of the entire organism, or *locomotion*, has evolved primarily in response to predator–prey interactions.

Plants move parts of their bodies in response to environmental conditions. This movement is achieved by two mechanisms—*differential growth* and *changes in water pressure*. Unicellular and small multicellular organisms exhibit a variety of methods of locomotion. Some cells, like the amoeba, change their shape by forming *pseudopodia*; others move more rapidly by means of *cilia* or *flagella*.

Locomotion in larger animals is accomplished by powerful contractions of *muscle cells*. Animals have three types of muscles—*smooth*, *cardiac*, and *skeletal*. The *skeletal muscles* are involved in locomotion. Each skeletal muscle is attached to two or more *bones* in vertebrates and to two or more external rigid plates in most invertebrates.

The bones of a vertebrate animal are living tissues. Two or more bones are held together in a *joint* by *ligaments*, and *muscles* are attached to bones by *tendons*. The *vertebrae* of the spine are separated from one another by spongy *discs*. Bones do not just provide skeletal support. Minerals and fats are stored in the bones, and blood cells are formed in the bone *marrow*.

Skeletal muscles perform mechanical work by shortening their length, pulling bones into new positions. Most muscles occur in *antagonistic pairs*—one muscle pulls a bone in one direction; the other muscle pulls the bone in the opposite direction. Skeletal muscles are *voluntary muscles;* their contractions can be consciously controlled.

Each skeletal muscle is made up of muscle cells, or *fibers*, and each fiber contains hundreds of *fibrils*. All of the fibrils in each fiber either remain inactive or exert their full strength when they contract in what is characterized as an "all-or-none response." The working parts of a muscle are the *actin* and *myosin* molecules within each fibril. According to the *sliding filament theory*, a fibril contracts when the myosin molecules become attached to and slide the actin molecules across their surfaces. The fibril shortens as the actin molecules overlap the myosin molecules. This movement occurs when an electrical impulse travels along the membrane of the fiber and causes the release of calcium ions (Ca^{++}). These ions (1) temporarily convert myosin into an enzyme that releases energy from adenosine triphosphate (ATP) and (2) allow contact to be made between the *cross bridges* on the myosin molecules and the actin.

The energy that drives the muscle contraction is derived primarily from ATP molecules, which can be formed by either aerobic or anaerobic cellular respiration. *Aerobic exercise* (glucose + oxygen + water $\longrightarrow$ carbon dioxide + water + energy) occurs when oxygen is available; *anaerobic exercise* (glucose $\longrightarrow$ lactic acid + energy) occurs when the circulatory system is unable to transport oxygen fast enough to meet the oxygen requirements of the muscle cells. The aerobic reaction yields about 20 times more ATP per glucose molecule than the anaerobic reaction. Some muscle fibers, the *slow-twitch fibers*, are adapted for aerobic exercise and contain the darkly pigmented *myoglobin* molecules, which temporarily store oxygen. *Fast-twitch fibers* do not contain myoglobin and are adapted for anaerobic exercise. When the level of physical activity increases, the muscles shift from the aerobic to the anaerobic reaction. Anaerobic forms of exercise cannot be continued for very long due to the build-up of lactic acid in the blood and the depletion of glycogen in the muscles.

Strenuous, aerobic exercise is the optimal form of physical conditioning. The initial stage of aerobic exercise, or the *warm-up*, is a combination of anaerobic and aerobic exercise, because the respiratory and circulatory systems, which determine rate of oxygen transport, are slow to respond to the new activity level. During the warm-up, blood is shunted from the skin to the working muscles, and the layers of *cartilage* on the ends of bones in the working areas of the body become enlarged.

During the *sustained phase* of aerobic exercise, the respiratory and circulatory systems transport a sufficient amount of oxygen to the muscles to facilitate the aerobic reaction. Heat generated by this reaction is removed from the body by the dilation of the blood vessels in the skin and the production of sweat.

Fatigue from aerobic exercise may result from mental fatigue, *synapse* fatigue, or the depletion of glycogen reserves in the muscles. Glycogen reserves can be increased by a form of diet called *carbohydrate loading*. Immediately after exercise ceases, the body's "oxygen debt" must be paid off. Heavy breathing and an accelerated heart rate while the body is at rest produce excess oxygen, which is used to convert lactic acid into glycogen and to replenish the red blood cells and myoglobin.

Physical exercise strengthens the bones, muscles, ligaments, and tendons; improves neuromuscular coordination; increases the number of capillaries that feed the muscles; and strengthens the heart muscles and vessels. All of these beneficial changes in the body add to the quality of life and prevent premature aging.

Animal Behavior

20

The most striking characteristic of life is activity, or *behavior.* Virtually all of the behavior that an organism exhibits in its natural environment contributes to the individual's survival and reproduction. Adaptive behaviors promote internal homeostasis, food acquisition, predator evasion, mate attraction, survival of the young, and other important functions of life. All organisms exhibit some degree of behavior. Even plants move and orient parts of their bodies in response to environmental changes. In this chapter, however, we will deal primarily with the most highly developed form of activity exhibited by life forms—animal behavior.

Animal behavior is produced by the nervous system and usually can be described in terms of a stimulus and a response. A *stimulus* is a signal that influences an animal's behavior through its sensory receptors. The signal may originate in the external environment or within the animal's body. Once the signal has been received, nerve impulses travel from the sensory receptors to the central nervous system, where they are evaluated through complex interactions among neurons, which initiate command messages to specific *effectors* (muscles or glands) to produce a behavioral response. The particular behavior, or *response,* that results from the original stimulus depends on the pattern of the interactions that occur among the neurons in the animal's nervous system. This pattern is influenced by both genetic and environmental factors.

Some aspects of an animal's behavior are largely inherited; genes govern the organization of neurons during development in such a way that when a certain stimulus is perceived, the animal behaves in a predictable manner. All individuals of the same species respond similarly. *Inherited behaviors* are modified by *natural selection*—the persistence of the genes that cause adaptive behaviors and the removal of the genes that produce behaviors that reduce reproductive success. Inherited behaviors are sometimes called *innate* or *instinctive behaviors.*

In addition to inherited behaviors, most animals develop learned behavior patterns from interacting with their environments. *Learning* is an adaptive modification of behavior toward a stimulus that can be traced to a specific experience with that or a similar stimulus in the animal's lifetime. During learning, new relationships develop among neurons in the nervous system, so that the animal acquires new behaviors and modifies inherited responses.

The relative contributions of the genotype and the environment to behavior vary considerably from one group of animals to another and among the many behaviors exhibited by each animal. Most behaviors are partially inherited and partially learned.

Orientation Movements

Animals are able to move themselves or at least position parts of their bodies toward or away from certain objects or local environments. Behaviors of this type are largely inherited and are called *orientation movements.* The two types of orientation movements are kinesis and taxis.

In the simpler orientation movement, *kinesis,* the animal moves in a *nondirected* manner that eventually brings it toward or away from an environmental stimulus. The animal's response to the stimulus—a change in the speed of movement or the rate of turning during locomotion—alters its position in relation to the source of the stimulus. An example is the humidity kinesis of wood lice. In an environment that has both humid and dry areas, the lice move faster when they are in dry areas and slower or not at all when they are in humid areas, thereby permitting them to spend longer amounts of time in the more favorable, humid areas. However, the lice turn in a random manner, so that their movement is not actually directed toward a humid area. Their sensory receptors detect only the humidity, and the animal responds to this information by varying its rate of movement.

The more complex form of orientation movement, *taxis,* is movement in a definite, *directed* path toward or away from a stimulus. Sensory receptors detect not only the intensity of the stimulus but also its location. For example, some butterflies fly directly toward the sun when chased by a predator. Certain aquatic snails that feed on oysters move directly toward their prey by following a concentration gradient of molecules the oysters release as waste products. The snails move along a path of increasing molecular concentration until the oyster is located.

Fixed Action Patterns of Behavior

A newly hatched chicken walks about, scratches the ground, and pecks at seeds. A newly hatched duckling enters a pond, where it swims, dives, and feeds underwater. These stereotyped behavioral patterns, or *fixed action patterns,* are inherited; they require no previous experience and are characteristic of the species. All young chickens show the same behaviors, as do all young ducks. Even when ducklings are raised by a chicken mother or chicks are raised by a duck mother, the behaviors of the young chicks and ducks are typical of their own species and not of their foster mothers.

In contrast to kinesis and taxis, which always occur in response to a perceived stimulus, fixed action patterns usually occur in response to a stimulus only when the animal is in a specific physiological state, such as hunger. Many aspects of mating behavior, for example, depend on the concentration of sex hormones in the body. When these hormone levels are low, even a very strong stimulus will not elicit mating behavior; when hormone levels are high, however, even a weak stimulus will trigger mating behavior. Fixed action patterns also differ from orientation movements in that a single stimulus often elicits a sequence of different behaviors; once the first behavior in the fixed action sequence is performed, the others follow automatically. In kinesis and taxis, there is only a single type of behavioral response to the stimulus.

An example of a fixed sequence of behaviors occurs during the construction of a cocoon by a type of tropical female wandering spider. The stimulus is the location of a suitable site for the cocoon, and the response occurs only when the female is in a particular physiological state—ready to lay her eggs. She responds to the stimulus by spinning silken threads to make a floor to the cocoon and then surrounds it with a rim. The eggs are deposited on the floor and the opening of the cocoon is closed. If she is interrupted—for example, if the floor and the eggs are removed—the spider does not adjust her construction to compensate for the change; she simply continues the behavioral sequence. Her construction behavior includes approximately 6,400 movements; once the first occurs, the others follow automatically regardless of any further feedback from the environment.

The stimulus that elicits the fixed action pattern is often very specific and only a small part of the phenomenon to which the animal is actually responding. For instance, a female turkey displays parental care behavior only in response to the *call* of her chick; if presented with any object fitted with a tape recording of a newborn turkey, she will exhibit this behavior. In the schooling fish *Pristella riddlei*, only the sight of a conspicuous black mark on the dorsal fin of another fish (Figure 20-1) elicits schooling behavior; in the absence of this mark, the fish will not form a school.

The genetic basis of fixed action behaviors has been experimentally demonstrated in some species. These demonstrations often involve mating individuals from one species with those of another species to determine what behavior traits appear in the offspring. Such matings rarely occur in natural populations, but can be conducted under laboratory conditions. For example, two species of African parrots differ in the way individuals transport nesting materials. In one species, strips of leaves are tucked under the rump feathers, which have small hooks to hold the leaves in place. In other species, nesting material is carried in the bill. The offspring of matings between these two species exhibit a mixture of the two transport methods—usually attempting to tuck the leaves under their tail feathers but failing to release them from the bill. The offspring are unable to transport nesting materials and therefore to reproduce successfully because they have inherited components of different behaviors from two species.

Two advantages of inherited behavior are that the response to a stimulus is immediate and that the sequence of movements is complete and functional the first time it occurs. These aspects of inherited behavior are especially important when a wrong response could be fatal. Learning the correct response by trial and error is inappropriate behavior when being attacked by a predator; a dead animal cannot

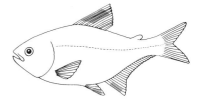

Figure 20-1
Schooling Behavior in This Fish Is Elicited by the Sight of a Black Mark on the Dorsal Fin of Another Fish

Figure 20-2
A Moth Evades Predatory Bats by Means of Inherited Behavioral Responses

In these photographs, a moth responds to an ultrasonic sound similar to the squeak produced by a hunting bat. The moth alters its flight pattern suddenly when it detects the sound.

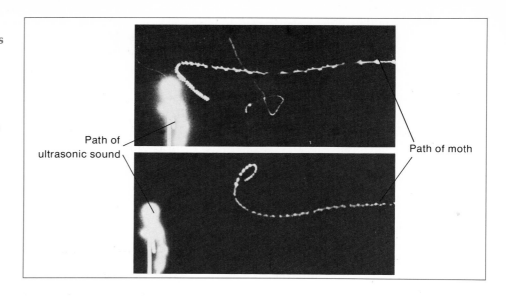

Path of ultrasonic sound

Path of moth

learn from its mistakes. Bats hunt at night, locating their prey by sending out beams of ultrasonic squeaks and following the returning echoes that bounce off flying insects. Moths respond to the squeak of a bat by exhibiting inherited anti-predator behaviors that help them avoid capture. Some moths fold their wings and fall to the ground; others reverse their direction of flight (Figure 20-2). In general, whenever the environment remains highly predictable throughout generations of a species so that the best response to a stimulus is always the same, inherited behavioral responses are more advantageous than learned responses. Fixed action behaviors are also valuable when there is little parental care, as is true of the majority of invertebrates, fishes, amphibians, and reptiles. The young of these species simply would not survive if they had to learn what foods to eat, where to find shelter, and which animals were dangerous predators.

Courtship Behavior

An important category of fixed action behavior is *courtship* and *mating behavior.* When mating occurs between members of different species, the offspring are usually unhealthy and incapable of reproducing and may not develop to adulthood at all. Thus, there is a strong selection for recognizing potential mates of the same species. Each reproducing animal must be capable of signaling its species identity and recognizing the signals of the opposite sex. Courtship behavior, which precedes mating, is unambiguous, elaborate, highly stereotyped, and unique to each species. To a large extent, courtship behaviors are governed by genetic programming.

Courtship behavior in the three-spined stickleback (Figure 20-3), a small freshwater fish, has been extensively studied by the European biologist Nikolaas Tinbergen and his colleagues. Outside the breeding season, these fish are silvery gray in color and move in schools. In the springtime, rising temperatures trigger

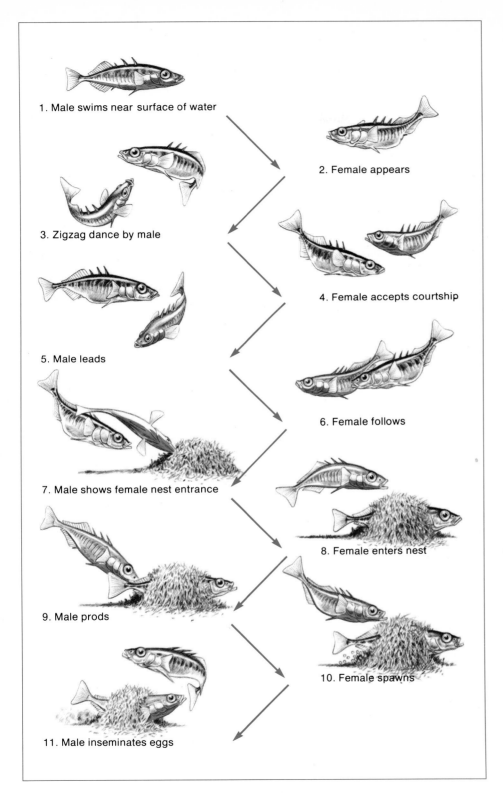

**Figure 20-3
Courtship in the
Stickleback Fish**

The sequence of courtship behaviors in the male and female are largely inherited and proceed in a stereotyped manner. Each action by the male is the stimulus for a particular response on the part of the female, and vice versa. During the breeding season, the female stickleback is uniformly silver in color with a large, swollen belly containing up to 100 eggs. Attracted by the swimming movements of the male (1), a female approaches (2), swimming in a head-up posture to display her enlarged belly, which triggers courtship behavior in the male. He performs a zigzag dance, moving between his nest and the female (3). She accepts courtship (4) and follows him to the nest (5 and 6), where the male makes a series of quick thrusts with his nose to indicate the entrance (7). The female enters the nest (8), and the male prods his head against her tail and quivers (9), causing the female to release her eggs (10). She then leaves and the male enters the nest to release his sperm onto the eggs (11).

1. Male swims near surface of water

2. Female appears

3. Zigzag dance by male

4. Female accepts courtship

5. Male leads

6. Female follows

7. Male shows female nest entrance

8. Female enters nest

9. Male prods

10. Female spawns

11. Male inseminates eggs

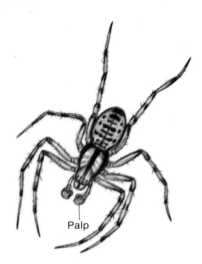

Figure 20-4
The Enlarged Palps of a Male Spider

Palps resemble antennae and extend from the spider's mouth region. Each palp of the male spider can be filled with semen and function as a penis.

solitary behavior, and the males search for appropriate sandy areas on which they build nests of water plants. During the breeding season, male sticklebacks have bright red bellies—a signal that provokes aggression between males. (A male stickleback kept in a tank near a window in Tinbergen's laboratory regularly attempted to "attack" a red mail truck as it passed in the street outside.) Each male defends the area surrounding its nest, driving other males away. Courtship behavior begins when the male advertises that he is ready to mate by swimming in a particular pattern near the surface of the water, and the courtship sequence continues as shown in Figure 20-3. After mating, the male chases the female away, attracts another mate in the same manner, and the courtship sequence is repeated. The male stickleback guards the fertilized eggs in his nest until they hatch and keeps the brood together for a day or so afterward, carrying the young that wander off back to the nest in his mouth.

Many birds combine both songs and stereotyped movements in their courtship. The green heron is a North American bird that nests in tall trees along streams or in marshes. Courtship begins when the male advertises himself by posturing with up-pointed bill on a treetop and giving a call that sounds like *skow*. Females are attracted to the male's call, and one soon perches nearby and answers with a slightly different call, *skeow*. The duet continues for some time, until the male allows the female to approach. Both birds then perform an elaborate aerial dance above the nest, displaying the unique colors of the plumage of their species to one another. The pair later returns to the treetop, where the male enters the nest, displays his brightly colored feathers, and sways from side to side, pointing his bill up and away from the female and uttering a soft *aaroo—aaroo*. The male then points his bill down, snaps it sharply, and bobs and bows. The female enters the nest, and the birds stroke their bills together and preen one another's neck feathers. These behaviors signal a readiness to accept physical contact, and copulation takes place.

Insects and spiders also exhibit elaborate courtship behaviors. After locating a female spider on her own web, the male spider of a web-building species spins a web for himself. He deposits a large drop of *semen* (sperm and fluid) on the top of his web and then crawls beneath it to draw the semen through the web into his enlarged *palps*—short, hollow, appendages hanging down beside the mouth parts in front of the first pair of legs (Figure 20-4). Each palp functions as a penis. The male approaches the female in a dancelike movement, waving his palps back and forth, and further courts her by setting up a specific pattern of vibrations on her web. In time, she allows him to insert first one palp and then the other into her reproductive tract and her eggs are fertilized.

Courtship behaviors: (1) guarantee that the mates are of the same species; (2) reduce aggression that might otherwise occur between the mating individuals; and (3) tend to synchronize the physiologies of the pair, so that the sperm and eggs are released at the same time.

Learned Behaviors

Learning is a behavioral change that results from experience. A learned behavior is acquired during the lifetime of an individual and is not transmitted to offspring

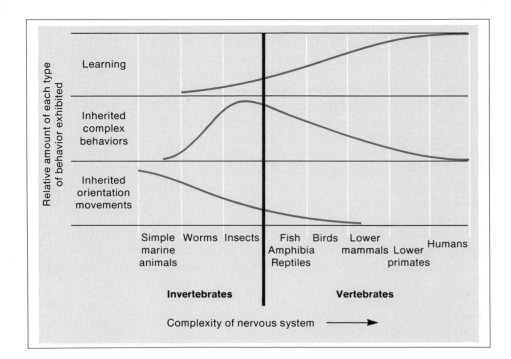

Figure 20-5
**Ways in Which Behaviors
Are Acquired in the Major
Animal Groupings**
Animals with more complex nervous
systems exhibit greater learning
capacities than animals with simpler
nervous systems.

through genes. The capacity to modify behavior is regulated by special features of the nervous system that permit the incorporation of new information, so that new associations can be formed or previous associations broken. To be able to learn, an animal must be able to *remember*—to store new information so that the animal can retain the effects of past experiences and recall them at the appropriate time. The animal must also be able to distinguish in its memory between rewarding and unpleasant experiences. Animals with simpler nervous systems exhibit less capacity for learning than animals with more complex nervous systems (Figure 20-5).

It is not yet known precisely what happens within the nervous system during learning. There are probably both structural and chemical changes in neurons that modify their relationships with one another. New synapses may form, creating new pathways for electrochemical impulses or strengthening already existing ones. Some evidence suggests that the affected neurons may synthesize new forms of molecules by activating previously repressed genes. These new molecules could be involved in information transmission or in memory storage. It is generally agreed that learning requires the integrated activity of many cells and that these cells must be capable of modifying their relationships with one another.

In all learned behaviors, certain associations and movements are more easily learned than others. Each animal inherits tendencies to learn only certain things. Herring gulls, for example, readily learn to recognize their own chicks—an adaptation to nesting in colonies on the ground, where the young are apt to wander from the nest and mingle with other chicks. The closely related kittiwake gulls, by contrast, cannot discriminate between their own nestlings and the nestlings of neighbors or of other species. Any chick placed in a kittiwake's nest is accepted and fed. Kittiwakes build their nests on the narrow ledges of cliffs where there is no room for the young to wander, so that they remain where they hatch. Under normal conditions, the chicks in the nest of a kittiwake are always its own, and the parents only learn to recognize the location of their nest, not to recognize their chicks.

Another example of the inherited tendency to learn certain things is the ability to solve detour problems. Faced with a detour, an animal must move farther away from its goal in order to eventually reach it. Dogs do not solve detour problems as readily as tree squirrels (Figure 20-6), because a large animal that travels on the ground usually does not encounter such a situation. Tree squirrels, however, must frequently solve detour problems in their natural settings—usually by descending from one tree before moving up another.

Learning is therefore dependent on two inherited factors. First, the capacity to learn is directly related to the complexity of the nervous system, which is similar for all animals within a taxonomic group. (For example, fish have less complex

Figure 20-6
A Detour Problem
A dog has great difficulty solving this problem; a tree squirrel quickly learns to go around the stake to reach the food.

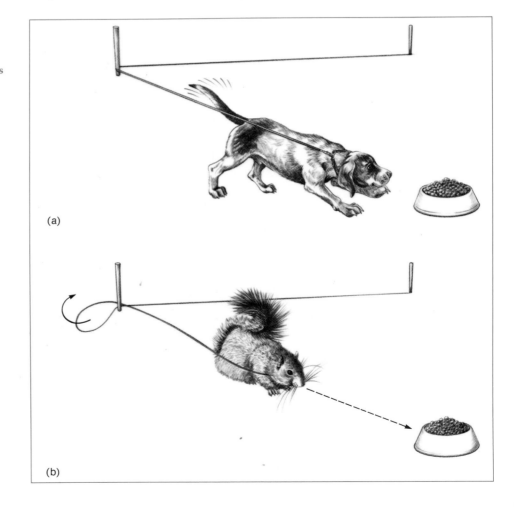

(a)

(b)

nervous systems than mammals; within the mammalian group, mice have less complex nervous systems than primates.) Second, the ability to learn is influenced by inherited tendencies, which result from natural selection and may be quite different for closely related species. When learning a task provides no advantage, an animal exhibits little or no tendency to master that particular task.

Habituation

The simplest form of learning is *habituation.* The animal stops responding to stimuli that are biologically meaningless—that are not rewarding or unpleasant. Birds soon learn to ignore a scarecrow, and young mammals learn not to flee from falling leaves. A previous association between stimulus and response is broken; the stimulus is perceived, but the response disappears. In all species, habituation is adaptive because it prevents the animal from wasting time and energy on responses to stimuli that are of no particular significance.

Classical Conditioning

Another form of simple learning, *classical conditioning,* results when an association is formed between a previously neutral stimulus and a physiological response controlled by the autonomic nervous system. In early studies by the Russian biologist Ivan Pavlov, a dog was trained to salivate in response to the sound of a bell. Before this conditioning, the dog salivated in response to the presence of food in its mouth. However, after Pavlov repeatedly sounded a bell just before placing food in the dog's mouth, the dog began to salivate whenever the bell was sounded, even if no food was present (Figure 20-7). The response became associated with a neutral stimulus that was regularly linked with the natural stimulus.

Figure 20-7
The Training Procedure Used by Ivan Pavlov in His Studies of Classical Conditioning
(a) Prior to training, the presence of food causes a reflex action; the brain receives sensory information indicating that food is present and stimulates the salivary glands.
(b) During training, a bell is sounded just before food is presented; these two stimuli eventually become associated in the dog's brain. (c) After training, the sound of the bell alone causes the dog to salivate.

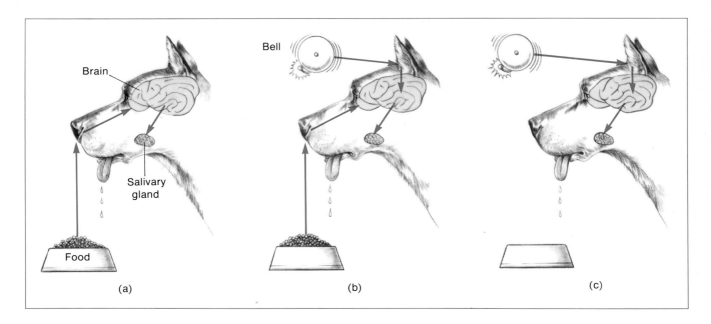

(a)   (b)   (c)

Disruptive emotions such as unreasonable fears (*phobias*) and some forms of psychosomatic illnesses develop in humans due to classical conditioning. A neutral or harmless stimulus becomes associated with something that is truly threatening to the individual; thereafter, the emotional or physiological responses occur whenever the neutral stimulus is perceived. Such maladaptive associations can be treated by behavioral therapy—repeatedly linking the harmless stimulus with something pleasant so that its previous association is broken.

Operant Conditioning

Most animal behaviors are modified by *operant conditioning*—the formation of a new association between a voluntary behavior and its consequences. If a new behavior produces beneficial results, the animal will continue to modify its behavior in this way; if the behavior produces unpleasant results, the animal will not repeat it. The nervous system also forms associations in the absence of immediate physical rewards or punishments, so that not all forms of complex learning involve operant conditioning. However, when learning is studied under laboratory conditions, the animal must be rewarded or punished for its behavior to control the associations that are made so that they can be measured. Therefore, although operant conditioning does not encompass all of the conditions under which an animal forms associations, this form of complex learning has been studied more thoroughly than any other.

In one laboratory example of operant conditioning, a cat is placed in a cage that can be opened from the inside only by pressing a lever. At first, the cat moves around restlessly and tries to escape; after a while, it steps on the lever by chance and the door opens. If the cat is placed in the cage repeatedly, the second and third trials may be repetitions of the first. In time, however, the cat begins to concentrate on the lever; soon it moves directly across the cage and presses the lever immediately after being placed in the cage. The cat has learned to eliminate the behaviors that produce no reward and to increase the frequency of the behavior that is rewarded, even though the animal's activities during the early stages of learning were not systematic.

Evidence suggests that all the major groups of vertebrate animals and most invertebrates are capable of this form of complex learning. Operant conditioning is highly useful to animals in their natural settings. If an animal repeatedly finds food in the same place, its searching behavior will become more localized and the animal will begin to look for food in places where it appeared regularly in the past. Animals that feed on a large variety of foods learn to concentrate on edible items and to avoid noxious ones by operant conditioning.

Navigational Learning

Some of the more remarkable examples of learning occur during navigation when an animal travels from one particular site (usually a nest) to another and then back again. Some animals leave scent marks as guides for their return trip. Others learn the position of environmental cues, such as rocks, shrubs, or the sun; on their outward journey, these animals note and remember cues in the appropriate sequence and then recall them in reverse during their return trip. Seed-eating ants

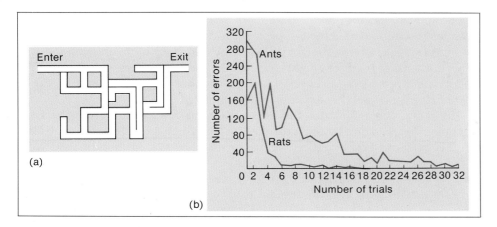

(a)

(b)

Figure 20-8
Navigational Learning Under Laboratory Conditions
(a) A *multiple-choice maze* consists of a sequence of *choice points*. An animal must learn the correct way to turn at each choice point as well as the correct sequence of turns. (b) A comparison of maze learning by ants and rats. Each animal ran the maze repeatedly until it made few or no errors per run. When the number of errors decreases as experience (the number of trials or runs) increases, learning has occurred. The rats learned more rapidly than the ants, although both the rats and ants eventually mastered the task.

travel in an erratic searching pattern during their journey away from the nest as they look for seeds. Once a seed is picked up, the ant travels in a straight line directly back to the nest. Somehow, the ant keeps track of its position in relation to the nest during the outward trip. Many ants use the sun as a navigational cue and are able to make adjustments to compensate for the change in its position as time passes during the foraging journey.

Navigational learning is often studied under laboratory conditions by using a *multiple-choice maze* similar to the one shown in Figure 20-8(a), which can be solved only by learning from experience. In a typical maze, the animal may be provided with several cues, such as colored markings or the position of lights and other objects, at each *choice point*. To travel through the maze repeatedly and make few errors, the animal must master each choice point and also remember the correct sequence of turns. Most animals can learn to run a maze. The results of a study comparing the maze-learning abilities of ants and rats are provided in Figure 20-8(b).

Honeybees not only navigate between their hive and daily food sources but also communicate the location of rich food sources (patches of flowers containing pollen and nectar) to one another. This communication, first described by the Austrian biologist Karl von Frisch, is in the form of two dances performed in the hive by returning foragers. The "round dance"—movement along a path in the shape of a figure eight, as shown in Figure 20-9(a)—is performed when the forager has found a rich source of food within 80 meters of the hive. The dance communicates that food is nearby; the odor of the dancer communicates the type of food.

The "waggle dance," shown in Figure 20-9(b), is performed if the forager has found a rich patch of flowers more than 80 meters from the hive.

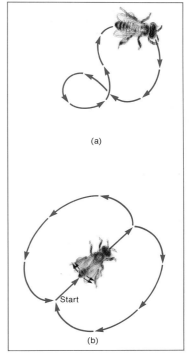

(a)

Start

(b)

Figure 20-9
Dances of the Honeybee
(a) The "round dance" is performed in the hive by a returning forager when she has found a rich patch of flowers less than 80 meters from the hive. (b) The "waggle dance" is performed when the food source is farther than 80 meters from the hive. The bee begins this dance by running along a straight path, wagging her abdomen back and forth, and producing a buzzing sound by vibrating her wings. She then turns right and returns to the start of the path without wagging her abdomen. The bee repeats the waggle portion of the dance along the same straight path but this time turns left to return to the point of origin. The dancer performs these movements many times, alternating right and left turns at the end of the straight path.

Top of hive

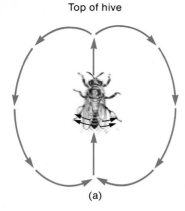

(a)

Top of hive

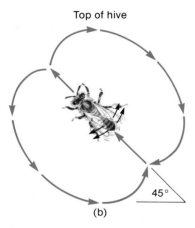

45°

(b)

**Figure 20-10
The Waggle Dance
Communicates the Location
of Food with Reference to the
Position of the Sun**

(a) In this particular dance, the forager
moves vertically upward along the
wall of the hive to indicate that
the bees should fly directly toward the
sun to reach the food. (b) Here, the
dance communicates that food will be
found if the bees fly along a path that
is 45° to the left of the sun.

The amount of energy required (the distance the bees must travel) to reach the blossoms is communicated by the speed of the dance as well as by the number of abdominal waggles on the straight path and the frequency of the buzzing sounds. When the distance is great or flight is against the wind, the dance is slower and the rate of sound production is lower; when the distance is short or flight is with a tailwind, the dance and the sounds are more animated.

The forager communicates the direction of the food source from the hive during the waggle dance by using the position of the sun as a reference point. The dance is performed on a vertical surface of the hive; the position of food with regard to the position of the sun is given by the angle of the straight path of the dance with respect to gravity. If the flight to the food is in the same direction as flight toward the sun, then the straight path of the waggle dance is vertical and upward, as shown in Figure 20-10(a). If the flight to the food is directly away from the sun, the straight path of the dance is vertical and downward. If the food is located 45° to the left of the sun, the path of the waggle dance is 45° to the left of a vertical and upward path, as shown in Figure 20-10(b). Like ants, bees can adjust their navigation and waggle dance to compensate for the sun's movements during the day.

Generalizations and Insightful Learning

A surprisingly large number of animal species possess the ability to make *generalizations.* In experimental situations, most birds and mammals can learn to choose the wider of two stripes, regardless of absolute width, or the darker of two colors, regardless of actual intensity. Animals generalize from experience about a variety of specific objects or problems.

The *oddity problem* is one of the more difficult tests of the ability to generalize. Several objects of two types, such as triangles and circles, are presented to the subject. In each trial, a reward is covered by the only odd object. A monkey quickly learns to choose the triangle when presented with a triangle and two circles. When the same type of test is then carried out with a different set of objects (say, disks of different colors), the monkey chooses the odd color. It has learned the principle underlying different but related problems. Progressive improvement in an animal's ability to solve a sequence of similar problems is called the *formation of learning sets,* or learning to learn.

The ability to make generalizations varies among mammals (Figure 20-11). Clearly, learning can be achieved more swiftly when the animal can form generalizations based on previous experiences with similar problems. The ease with which an animal learns a task therefore depends not only on the type of animal but also on its previous experience with similar problems.

Many problems designed to test higher learning abilities require the formulation of an abstract concept. Chimpanzees and monkeys are able to learn the symbolic values of chips of various colors and sizes. They learn that certain chips can be exchanged for food, others for being released from the cage, and others for playing with the keeper.

In *insightful learning,* or *reasoning,* problems are solved by thinking; in other words, the process of learning is internalized. Insightful learning requires an understanding of relationships between things and is most often observed in the use

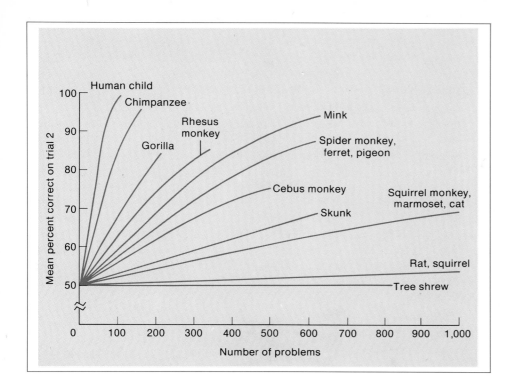

Figure 20-11
The Ability of Various Types of Mammals to Make Generalizations

Each curve depicts a *learning set*, which measures the animal's improvement in performance when it is given 100–1,000 similar types of problems, such as oddity problems, to solve. The measure of performance (indicated on the vertical axis) is the mean percent of correct responses that the animal achieves on the second trial of a large group of problems (50–100 in a sequence). As the animal becomes more experienced with the principle underlying the solution to all the related problems, it makes the correct response more frequently on the second trial. The ability to generalize is much more advanced in human children and chimpanzees than in other animals. The tree shrew essentially has no ability to make generalizations.

of tools by animals. Chimpanzees figure out that they can reach a bunch of bananas suspended from the ceiling by piling boxes on top of one another and climbing up to the bananas. Before taking any action, they look at the boxes, at the bananas, and at the place beneath the bananas, apparently thinking out how different solutions would work. This is planning with *foresight,* a capability once believed possible only in humans.

Exploration, Curiosity, and Play

Most animals are curious and will *explore* new objects and surroundings by sniffing, chewing, feeling, or climbing. As is true of complex learning, the degree of curiosity reflects an animal's evolutionary ancestry and the complexity of its nervous system (Figure 20-12). Unlike most types of complex learning, exploratory behavior is not associated with any immediate physical reward or punishment. Primates, for example, will learn to solve a puzzle without any incentive except the performance of the task itself (Figure 20-13). Exploration and curiosity are adaptive behaviors which provide an animal with information about its environment. A rabbit that knows the area around its burrow is able to reach safety much faster than one that has not explored its environment.

Play behavior is widespread among the young of vertebrate animals. Play is apparently related to learning and is a way in which the young acquire knowledge about their own physical abilities, the social behaviors of their group, and their

Figure 20-12
Responses of Animals to Novel Objects

The response score reflects the time spent investigating new objects. Primates (baboons, monkeys, and lemurs) and carnivores (tigers, foxes, and raccoons) spent more time exploring new objects than rodents (prairie dogs, porcupines, and kangaroo rats) and reptiles (snakes, lizards, crocodiles, and turtles). All curves decline with time after initial exposure to the novel objects, because the animals become habituated and gradually lose interest.

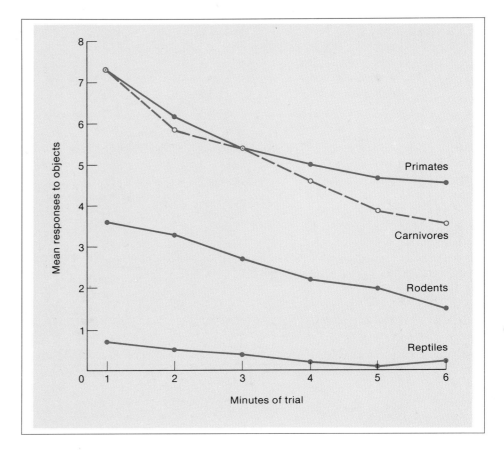

Figure 20-13
Curiosity in Primates

Both humans and monkeys exhibit a great deal of curiosity, especially the young. In both photographs, puzzles are being solved in the absence of any expectation of physical reward.

physical environment. Species that exhibit the greatest curiosity also exhibit the greatest amount of play.

Play activities usually involve both fixed action and learned components of behavior. Some forms of play improve physical coordination and provide opportunities to practice adult behaviors. In predatory species, play incorporates such aspects of hunting as chasing, stalking, biting, and shaking. In play fighting, puppies and kittens often grab each other by the throat in just the way they will kill their prey when they are older. Animals that fight with other members of their own species as adults practice the actions during play, although serious damage is inhibited. Play behaviors develop gradually in each species according to a defined pattern, so that the first appearance of each behavior is related to the age of the animal.

Imprinting

The learning of some kinds of behavior is genetically programmed to take place during specific sensitive periods of only a few hours or days during the animal's life. This type of learning, called *imprinting,* occurs rapidly and its effects are difficult to reverse. Imprinting, which has been observed in many ground-nesting birds, was first identified by the German biologist Konrad Lorenz, who showed that less than a day after hatching, young geese form a strong attachment to the first slowly moving object they see, which is normally their mother, and follow it thereafter. The goslings "become imprinted" on the object. When raised in the absence of a mother, goslings will imprint on a variety of objects, even a box pulled by a string. Goslings easily become imprinted on a human and will attempt to follow the person everywhere (Figure 20-14). Once a gosling has imprinted on a human, it cannot be induced to follow its own mother later because the sensitive period for acquiring this association has passed.

Imprinting has been observed in other vertebrates as well. Hoofed mammals, such as horses, sheep, bison, and zebras, imprint on their mothers during a sensitive period shortly after birth that lasts for only a day or two. These animals can also become imprinted on a human. Domestic dogs form lasting social attachments to other dogs or humans during the fourth to sixth week after birth.

Many animals acquire knowledge about the appearance and behavior of potential mates through imprinting. An animal should learn to identify mates early in life when it is associating with its own parents instead of when it reaches sexual maturity and begins to mingle with other species. The jackdaw, a crowlike bird, learns what its mate should look like early in life by imprinting on its parent. If a jackdaw is raised by humans, it will join a flock of other jackdaws when it is old enough to fly but will court humans rather than jackdaws when it reaches reproductive maturity. Imprinting for mate choice occurs much earlier in the development of jackdaws than the appearance of sexual behavior. The hand-reared bird regards humans as sexual companions because it was exposed only to humans during a sensitive period when it was young. Hand-raised turkeys and other domestic fowl may also become sexually imprinted on humans.

When the recognition of mates is not strictly inherited, rapid evolutionary change in the population is facilitated. If the recognition of features that signal

Figure 20-14
Konrad Lorenz with Goslings That Have Imprinted on Him Rather Than on Their Mother

Figure 20-15
An Infant Monkey with Its Cloth and Wire Surrogate Mothers

Due to the deprivation of social contact with other monkeys when young, this animal will not be able to learn normal adult behavior as it grows older. Infant monkeys preferred the cloth models even when they were fed from bottles attached only to the wire models.

species identity were strictly inherited, as is true of animals that do not interact with their parents, the young would reject mates with adaptive modifications of appearance, song, or behavior and accept only mates that exhibited ancestral traits. Individuals with such rigid mate recognition would tend not to reproduce when evolutionary changes in these traits were occurring in the population.

Primates learn many aspects of adult social and parental behavior during a certain part of their infancy. If they are deprived of learning during this sensitive period, adult behavior will be abnormal. The American psychologist Harry Harlow studied the effects of social deprivation by raising infant monkeys in separate cages without mothers or companions. The monkeys were provided with surrogate mothers—wire models which, in some cases, were covered with terry cloth (Figure 20-15)—and were fed from bottles attached to these models. As adults, these animals were unable to develop normal social behaviors. In groups, they failed to form a stable social system and an exceptional amount of conflict and deaths resulted. Sexual behavior was affected even more profoundly. The males attempted to mate but were unable to orient themselves to achieve copulation. The females were able to reproduce when mated with exceptionally gentle, patient, and skillful males but were indifferent or hostile to their babies.

Human infants are also influenced by their early environment. Prolonged separation from the mother during the first year of life can result in serious retardation and even death. Humans are especially sensitive to such deprivations between the ages of six months and one year, when babies normally develop close bonds with their mothers (or foster mothers). If this relationship is disrupted, a basic attitude of distrust develops. Other sensitive periods in human development include:

1. The ages of *two and three years.* A child who is not allowed to explore the environment during this period tends to be dependent on others and insecure about his or her own ideas as an adult.

2. Approximately *five years of age,* when a child is sensitive to sex-specific behavior and identifies with adults of its own sex if an acceptable role model is available.

3. During *puberty,* when a child is sensitive to new values and the attitudes of peers.

The learning that takes place during these sensitive periods has a profound and lasting effect on adult behavior.

Imprinting is both an inherited and a learned phenomenon. Genetic factors control the capacity to become imprinted at a specific time and the behavior that is released in the imprinted individual. But the object toward which these stereotyped behaviors are directed is learned as a result of experiences during the sensitive period. Imprinting is known to influence the associations of young with their parents, mate selection, choice of habitat, and social behaviors, including parental care.

Behavioral Ecology

Behavior is an integral part of ecology that enables an animal to recognize and actively seek out the resources it requires and to evade predators. Within a group of animals with nervous systems of similar complexities, the degree of learning involved in the ecology of these species is related to the variability of each dimension of the niche. In environments where a resource is always plentiful and never varies, animals exhibit a greater degree of inherited behavior because there is no advantage in learning to use new types of resources. In environments where preferred resources are not predictably available, it is advantageous for the animal to be opportunistic and to modify its behavior to obtain other accessible resources. The advantage of learning may only apply to the dimensions of the animal's niche that are variable. For example, many crowlike birds, such as ravens, build their nests in a characteristic manner according to inherited fixed action patterns. However, these birds learn to choose their nesting materials because the types of materials that are available vary from place to place and the birds must be flexible with regard to this dimension of their niche.

Habitat Selection

A *habitat*—the particular type of environment in which individuals of a species live—usually provides the food, shelter, nesting sites, and specific conditions of temperature and moisture suited to the animal's needs. Because an animal has a better chance of surviving and reproducing in its appropriate habitat, most animals inherit the capacity to recognize and move into these special places.

Most animals have sensory receptors that enable them to detect and move directly toward suitable habitats from a distance. Some flies detect the warm air emitted from buildings during autumn and move into these favorable habitats. Hermit crabs live inside the shells of other marine animals; their young are able to choose shells of the appropriate size and location without previous experience. Two different populations of the same species of deer mice live in different habitats. In laboratory experiments, when each type of mouse was given its choice between a meadow or woodland environment, it chose the habitat in which it was normally found. Habitat selection in deer mice has been shown to be an inherited trait that is strengthened by early experience; the inborn preference is stronger when the mice are born and raised in their preferred habitat.

Two common species of house rats also exhibit inherited habitat preferences. The black rat is a descendent of wild populations that nested in trees and prefers to live in the upper floors of buildings. The brown (or Norway) rat is a descendent of wild populations that lived near water and prefers cellars and sewage systems.

Some inherited habitat preferences may be changed if the animal is exposed to other environments at an early age. Adult chipping sparrows prefer to live in pine trees rather than broad-leaved trees. Hand-reared 2–4-month-old chipping sparrows that have had no access to any type of vegetation exhibit the same preference for pine trees, suggesting that this habitat preference is inherited. However, young chipping sparrows that have been raised with oak leaves and twigs exhibit a preference for oak trees at the ages of 2–4 months; the inherited behavior of these birds was modified by early experience, probably by some form of imprinting.

Migration—Change of Habitat

Some animals change habitats seasonally, often traveling great distances between their summer and winter residences. These movements are called *migrations* because they occur between two defined sites, rather than *dispersal*, which is simply

Figure 20-16
The Migratory Paths of Freshwater Eels

(a) The eels live in freshwater lakes and rivers in eastern North America and western Europe but migrate to the Sargasso Sea to breed. The adult eels die; the young eels return to the freshwater habitats of their parents in Europe or North America. (b) One explanation for this peculiar migratory behavior is that it evolved millions of years ago when Europe and North America were closer together and to the Sargasso Sea. The positions of the continents 135 million years ago appear above; the positions of the continents 65 million years ago appear below.

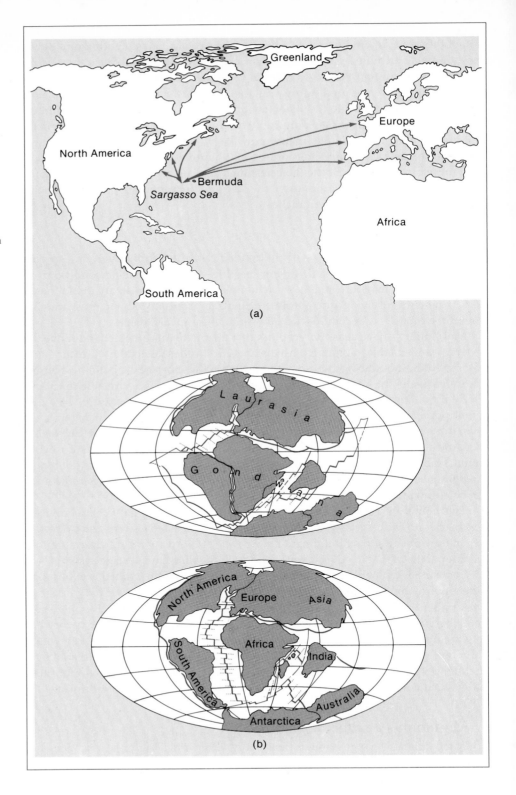

(a)

(b)

the movement away from a site. Seasonal migrations are more common among birds than any other group of animals because long distances can be covered more rapidly by flight than by movement on the ground or in water.

Birds migrate because food is scarce in northern areas during the winter. Seed-eating birds can find food in winter and usually do not migrate; birds that eat fish, insects, or nectar must fly south to warmer climates that provide adequate food supplies during the winter. Migratory birds fly back north in the spring when food supplies are again abundant and longer days provide more time for foraging. These birds travel great distances, making use of such navigational landmarks as coastlines, mountain ranges, and rivers as well as the position of the sun, star patterns, and the magnetic north pole.

Some of the more spectacular migration journeys are undertaken by fish. These journeys are usually to and from breeding areas and coincide with the age of the animal rather than with the seasons of the year. The migration of freshwater eels provides a remarkable example. These eels grow to reproductive maturity in lakes and rivers throughout western Europe and eastern North America. At about 9 years of age, they move downstream to the Atlantic Ocean and travel to a particular site—the Sargasso Sea—southwest of Bermuda, shown in Figure 20-16(a), where they spawn in the ocean depths. The adults presumably die after reproducing. The young hatch, leave the Sargasso Sea, and travel back through the Atlantic toward the fresh waters of either Europe or North America, depending on the origin of their parents. The young are less than 1 centimeter long when they begin this journey and about 8 centimeters long and three years old when they reach their destination. The ability of these young eels to migrate across the ocean to the habitats of their parents must be inherited and is believed to be regulated by chemical cues as well as oceanic currents. The ability of the adult eels to migrate to the Sargasso Sea is probably also inherited, although the eels could have learned the way when they made the reverse journey shortly after they were born. One explanation for the bizarre migration of the freshwater eel is that Europe and North America were once close together and the Atlantic coasts of both continents were not far from the Sargasso Sea, as Figure 20-16(b) illustrates. When the continents began to drift apart, the eels retained their original breeding grounds despite the ever-increasing distances they had to travel to reach them.

Some insects also migrate. The monarch butterflies of North America evolved in tropical climates and cannot survive prolonged freezing temperatures at any stage of their life cycle. Although reproduction of this species actually takes place in northern climates, young monarchs emerge from their cocoons in late summer and migrate to southern California, central Mexico, and Florida (Figure 20-17) for the winter months. In the spring, they return north to their breeding habitats where a series of short-lived generations grow to maturity and reproduce during the summer months. The adult monarch butterflies die after reproducing. The young that emerge from cocoons in later summer fly south, and the migratory cycle is completed. The annual return of these young monarchs to southern areas must be governed by inherited navigational skills, because these butterflies are several generations removed from the ancestors that flew north the previous spring.

(a)

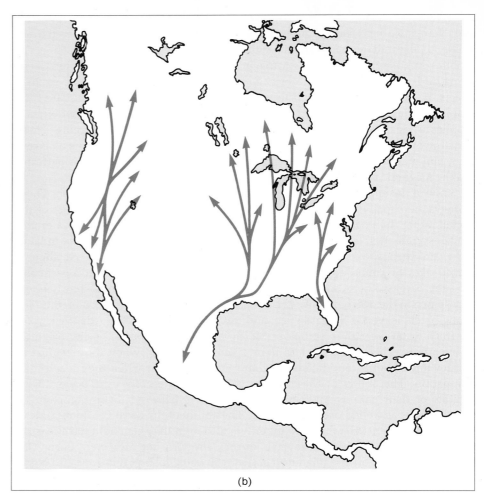

(b)

Figure 20-17
The Migration of Monarch Butterflies
(a) In their winter habitats, these butterflies often form dense clusters on trees. (b) The known migratory paths of North American monarch butterflies. In autumn, the butterflies travel south in large groups; in spring, each butterfly makes the trip north alone.

Feeding Behavior

Most animals are highly restricted to specific types of food. If an animal fed on many types of food, it would be unable to locate, catch, or handle each type efficiently. Every prey species exhibits a different antipredator behavior, and it is difficult for a predator to adapt to a large variety of prey behaviors. The more specialized the diet, the more efficiently an animal obtains and processes its food. Similarly each plant species employs an antiherbivore defense that includes specific toxic chemicals; it would be difficult and energetically costly for an herbivore to have mechanisms for detoxifying a large variety of chemicals.

Many insect herbivores feed on a single plant species. These insects produce enzymes that detoxify the specific plant toxins and may even use these toxins to identify edible plants. A few species of closely related beetles are specialized to feed on a single plant species, the St. John's wort. This small shrub produces a chemical toxin that the beetles render harmless. These beetles explore their environment in search of food but feed only when they detect this particular chemical. The plant's chemical defense is the stimulus that initiates feeding behavior in the beetles.

Behavior is an important factor in predator–prey interaction. An increase in predator efficiency (the number of prey captured by a single predator within a given time period) due to natural selection in predator populations has led to the evolution of such traits as an acute sense of hearing, smelling, or seeing the prey, running and sprinting abilities, and various attack behaviors. Selection in prey populations has produced traits that decrease predator efficiency, including camouflage, escape behaviors, and acute senses.

The most common antipredator adaptation, *camouflage,* includes physical and behavioral traits that make the prey difficult for the predator to locate. Prey that resemble their background environment exhibit *associated behaviors,* such as selecting a matching background, assuming an appropriate posture, and remaining motionless when a predator is detected (Figure 20-18). Many insects and fish have false eyespots—color patterns resembling eyes located somewhere on the body other than the head to deceive the predator about the direction of flight (Figure 20-19).

Escape behaviors are usually quite elaborate. To make pursuit difficult, the pattern of movement is often erratic, such as the unpredictable direction of turns exhibited by a sprinting rabbit. Prey that move in groups, such as fish schools, bird flocks, and antelope herds, may evade predators by rapidly moving away from one another in many directions. This type of escape behavior is distracting to the predator and interferes with its ability to concentrate on and capture a single individual. Birds within a flock may move closer together instead of apart when the predator is another bird. An eagle or hawk risks injuring itself if it dives into a tightly packed flock at a speed of about 300 kilometers an hour and is reluctant to attack under these conditions.

Figure 20-18
Camouflage Must Be Accompanied by Associated Behaviors

The spider is both predator and prey. This spider resembles the bark of a pine tree. For its camouflage to be effective, the spider (outlined in brown) must assume an appropriate position on the bark and remain motionless.

Access to Resources

An animal usually searches for food and other resources in a small, restricted area. If this area is used predominantly by a single animal or group of animals but is not defended, it is a *home range.* If the feeding area is defended by a resident animal or group of animals that have a monopoly on its resources, it is a *territory.* In some cases, the individuals of a population are aggregated within a small area and *dominance hierarchies* develop in which the more dominant animals have first access to the resources.

The home range is often the area immediately surrounding the animal's nest, place of shelter, or sleeping site. An animal is less vulnerable to predators when it forages in familiar surroundings with known escape routes and sheltered areas. Social factors may also play a role in restricting an animal's activities to a defined area. Members of a population are less likely to encounter one another when each individual has its own home range, thereby reducing the time and energy expended in and the risk of injury from aggressive interactions. Home ranges are essential when an adult is feeding young in the nest, because food must be brought to the young regularly and the young must not be left unprotected for long periods of time. Most mammals have home ranges.

An animal, mating pair, or group of animals defends a territory against close competitors (usually other members of the same species) to ensure exclusive use of

Figure 20-19
Predator Evasion by Deception

The lantern fly has a false head on its tail (shown here on the left) formed by modifications of its wing tips to resemble eyes, antennae, and a beak. The real head is shown here on the right, pressed close to the branch. When it is startled, the insect jumps forward suddenly in the opposite direction from the direction in which its false head is pointing.

Figure 20-20
Advertisement of Territorial Ownership

Perched on a cattail, a male redwing blackbird advertises ownership of his territory by singing and displaying his brightly colored wings. The bird is defending a circular area of 1,000–10,000 square feet of marshland. A male must secure a territory in order to attract mates.

the resources within the area, which may include food, shelter, nesting sites, mates, or space for courtship displays. Territorial behavior is exhibited by almost every major group of animals. The defense of a territory generally follows a sequence of behaviors—advertisement of ownership (Figure 20-20), then threat, and finally fighting. Advertisement or threat usually resolves the issue and reduces the need for actual combat. Such diverse forms of animals as dragonflies, fish, and birds perform similar sequences of these three defense behaviors.

An *advertisement* is a signal indicating that the area is owned and will be defended. Bird songs, cricket chirps, and the odor (urination and defecation) of mammals are all advertisements. Many insects emit olfactory advertisements. The apple maggot fly maintains the exclusive use of an apple for its young by laying her eggs inside and then covering the surface of the fruit with a scented secretion that deters other flies from laying their eggs in the same apple.

Threat often involves a particular posture or stance and the raising of hair, feathers, or fins, so that the animal looks larger than its normal size. This action may be accompanied by vocalizations and a movement toward the intruder. If advertisements and threats fail to intimidate the intruder, *fighting* ensues. The original owner of the territory is rarely defeated in a fight. The losing contender displays submissive signals, such as exposing vulnerable parts of its body to the winner or exhibiting juvenile behavior as a gesture of appeasement, that permit it to leave the territory without incurring additional physical injury. Territorial fighting rarely ends in death.

In a dominance hierarchy, a set of aggressive–submissive relationships is established among the members of a group of animals of the same species in such a way that that each adult is socially ranked relative to every other adult of the same or both sexes. Such hierarchies are found throughout the animal kingdom and are as prevalent among invertebrates as vertebrates. When individuals of a population

Figure 20-21
Behavior in Accordance with Social Rank

Aggression between two female wolves is avoided by submissive behavior. Here, the individual with lower status in the hierarchy licks the muzzle of the individual with higher status in the hierarchy.

are grouped together in a small area, as they often are when resources are highly localized, the more dominant animals have first access to mates, nesting sites, and/or food. Dominance hierarchies are common in seabird colonies along coastlines where suitable nesting sites are scarce. Many birds nest closely together, and the more dominant birds occupy the better sites.

Hierarchies are formed through repeated threats and fighting among animals in a group. After the dominance ranking is established, encounters cease to be aggressive and each individual is submissive to its superiors (Figure 20-21). Factors that determine who dominates whom include size, age, strength, general health, previous experience, and status of parents. The formation of a hierarchy provides animals with the benefits of group living without the problems of frequent fighting.

Sociobiology

Sociobiology is a relatively new area of study dealing with the origin, evolution, and adaptive nature of social behaviors. Many invertebrates and vertebrates form *societies* in which animals of the same species live together cooperatively. The advantages of socialization include organized defense against predators, group construction of nests and shelters, communication of the location of rich food sources, cooperative hunting, and division of labor. Social behavior in some animals, including humans, apes, elephants, and dolphins, even extends to the aid of disabled or endangered members of the group.

Based on the principles of natural selection, animals should behave selfishly, channeling all of their time and energy into their own survival and reproduction. But some animals exhibit social behaviors that seem to contradict evolutionary theory in that the behavior of one animal enhances the survival and reproductive success of other members of its group.

Pair Bonds and Parental Care

A *pair bond* is a social commitment formed between a mating male and female. The pair cooperate in building a nest and caring for the young and sometimes even bring food to each other. A pair bond is developed and reinforced by greeting rituals, such as the touching of bills by birds (Figure 20-22) or sniffing and licking by mammals, which promote continuance of the relationship and suppress aggressive tendencies.

A pair bond develops after courtship if the animals are *monogamous.* The primary reason for monogamy—the system of mating in which each female mates with only one male and each male mates with only one female—is the mutual care and feeding of the young by both parents. Monogamy is common in animals whose young are helpless at birth and have a high energy requirement. Most songbirds are monogamous, although they may change partners from year to year. Swans, geese, and eagles retain the same mates for life. Monogamy is less common among mammals because the young suckle milk from their mother and the father does not participate in the feeding.

Parental care by one or both parents includes feeding, sheltering, and grooming the young and protecting them from predators. All aspects of parental care require

Figure 20-22
Mutual Displays by a Male and Female who Have Formed a Pair Bond
Whenever mates meet, they reinforce their social bond and suppress aggression by performing a ritualized greeting. (a) Gannets, (b) Adelie penguins, (c) wandering albatrosses, and (d) grebes.

(a)　(b)

(c)　(d)

the sacrifice of time and energy that would otherwise contribute to the parent's survival; in many cases, the parent actually risks its life to protect its young. Some animals display distractive behavior in the presence of predators. Many ground-nesting birds lure predators away from their young by pretending to be injured, displaying a "broken" wing and staggering or limping from the nest area. In terms of natural selection, the parent that makes these sacrifices to ensure the survival of its offspring will leave more genes in the next generation than the parent that makes no sacrifices and loses all or most of its offspring. In a sexually reproducing species, each offspring inherits copies of one-half of each parent's genes. Therefore, if a parent has more than two offspring, selection favors traits that ensure the survival of the young rather than the parent.

Altruistic Behavior and Kin Selection

An *altruistic behavior* reduces the chances of the performer's survival or reproduction and increases the chances of another animal's survival or reproduction. When altruistic behavior is directed toward relatives, it is similar to parental care in that it is brought about by natural selection.

Any altruistic act that increases reproduction in a relative increases the number of copies of the performer's genes in subsequent generations. Animals that are related have copies of many of the same genes; on the average, about one-half of the genes of brothers and sisters are identical because these genes are inherited from the same parents. As the closeness of the relatives decreases, the number of genes that the individuals have in common also decreases.

There is no indication that an animal evaluates the degree of relatedness of each animal it encounters and adjusts its behavior accordingly. But animals that live in closely interrelated groups tend to exhibit cooperative and altruistic behavior. The extension of evolutionary theory to include selection for traits that increase the survival and reproduction of relatives is called *kin selection.*

Care of the young by nonparental members of a group is an example of altruistic behavior. When an individual sacrifices its own reproductive opportunities to enhance the reproductive success of others, the behavior is called *alloparental* ("allo" = different) *care.* This behavior is most highly developed in ants, bees, wasps, and termites. These insects live in colonies in which only one or a few females reproduce; the other members of the colony are closely related to the reproducing females and help them feed and care for the young. Alloparental care has also been observed in the young adults of at least 60 species of birds, including certain woodpeckers, jays, and wrens, and among mammals, including porpoises, elephants, and such primates as lemurs, monkeys, chimpanzees, and humans. This behavior can be explained by natural selection and is expected among relatives, especially when the caring individual has little opportunity to reproduce due to age, lack of a nesting site, or other factors.

Many unanswered questions related to altruistic behavior remain. The use of *warning signals* provides an example. A bird screeches when it detects a predator, and other birds nearby seek safety; a prairie dog cries out when a predator approaches. Is the animal that gives the warning signal more or less vulnerable to capture than the animals that receive the signal? Are the individuals that benefit from the warning closely related to the animal that gives the warning?

Sociobiologists believe certain aspects of human social behavior may have originated in genetically determined behavior, and prevalent cooperative and altruistic behaviors could have resulted from kin selection. However, it is well established that most human social behaviors have been developed or learned from cultural experiences with parents, siblings, peers, and others.

Summary

The behavior of an animal reflects the responses of its nervous system to a *stimulus*—a signal that originates in the environment or within the animal's body. Most behaviors are partially inherited and partially learned, indicating that the activity of the nervous system is influenced by both the animal's genes and its experience.

The advantages of an *inherited behavior* are that the response is immediate, functional, and adaptive the first time it occurs. The simplest form of inherited or *instinctive behavior* is an *orientation movement. Kinesis* is *nondirected* movement toward or away from an environmental stimulus; *taxis* is *directed* movement toward or away from an environmental stimulus.

Fixed action patterns often involve a specific sequence of stereotyped behaviors that are largely programmed in the animal's genes. Usually the animal must be in a particular physiological state to exhibit this behavior. In many cases, the stimulus that elicits a fixed action pattern is only a small component of the phenomenon to which the animal is actually responding.

In one type of fixed action pattern, *courtship behavior*—a series of elaborate, highly stereotyped, specific behaviors—is performed by a male and female prior to mating. Courtship behaviors ensure that the mates are of the same species, reduce aggression between the male and female, and synchronize the physiologies of the pair so that the eggs and sperm are released simultaneously.

Learning is a behavioral change that results from experience acquired during the individual's lifetime. When learning occurs, structural and chemical changes in the nervous system create new relationships among the neurons. Whether or not an animal is able to learn a new behavior depends on two inherited characteristics: (1) the complexity of an animal's nervous system, and (2) an animal's tendencies to learn only certain things.

The simplest forms of learning are habituation and classical conditioning. *Habituation* is the gradual elimination of a behavioral response to a stimulus that is of no biological significance to the animal. In *classical conditioning,* an association is formed between a previously neutral stimulus and a physiological response controlled by the autonomic nervous system.

More complex forms of learning include operant conditioning, navigational learning, generalization, and insight. In *operant conditioning,* an animal's behavior is modified by increasing the frequency of acts that result in a reward and decreasing the frequency of acts that produce unpleasant results. *Navigational learning* enables an animal to move through its environment and find its way back to its nest or shelter. Honeybees not only have superb navigational abilities but also are able to communicate the route to a rich food source to other bees by means of two dances, the "round dance" and the "waggle dance." Many kinds of animals are able to make *generalizations*—to use the knowledge they have acquired about one kind of problem to solve similar learning problems. *Insightful learning* is internalized; the individual develops solutions to problems by thinking.

An animal acquires information about itself and its social and physical environment through *exploration, curiosity,* and *play,* which are *adaptive behaviors.* This information may be useful at some later time.

Imprinting—rapid learning in response to a particular object or environment—occurs only during a sensitive period in the animal's lifetime. The age at which the animal is sensitive and the exhibited behaviors are inherited; the object or environment toward which the response is directed is learned. Through imprinting, some animals learn to associate with their parents, choose an appropriate mate, develop adult social and parental behaviors, and select a *habitat* (a suitable environment in which to live).

The *behavioral ecology* of an animal reflects both inherited and learned behaviors. In general, a greater degree of learning is associated with resource acquisition when the resource is more variable.

Animals inherit the capacity to recognize and move into suitable habitats. In some cases, the type of habitat selected is also influenced by early experience and is therefore a form of imprinting. Some animals change habitats seasonally or as a function of age. This type of movement from one specific site to another is called *migration.* Migratory animals have remarkable navigational abilities.

Behavior is a critical factor in determining food choice. The coevolution of predator and prey behaviors makes it difficult for a predator to feed successfully on a large variety of prey species because it exhibits only a limited number of specialized behaviors for finding and capturing food. The more common antipredator adaptations exhibited by prey are *camouflage* and *escape* behaviors.

An animal acquires resources within a restricted area. If this area is not defended against other members of the species, it is a *home range;* if the area is defended, it is a *territory.* Territorial defense involves a sequence of three behaviors—*advertisement, threat,* and *fighting.* Most territorial disputes are resolved without physical conflict. Some animals live together in groups or *societies* with other members of their species. In *dominance hierarchies*, a social ranking is established for each member of a group of animals of the same species. The more dominant animals have first access to resources, including mates, nesting sites, and food. An animal of lower rank exhibits submissive behaviors to an animal of higher rank. Dominance hierarchies permit animals to live in close proximity without continual fighting.

Sociobiology is the study of the origins, evolution, and adaptive nature of social behaviors, including cooperation and altruism. The most common social behaviors promote *pair bonds* between mates and *parental care* by parents for their offspring. Selection favors these behaviors because they increase the reproductive success (the number of copies of genes) of the animal exhibiting the behavior. An *altruistic behavior* decreases the chances of the performer's survival or reproduction while it increases the chances of another animal's survival or reproduction. When directed toward relatives, natural selection may favor altruistic behaviors because animals that are related possess many of the same genes; an increase in the number of relatives therefore increases the number of copies of an animal's genes. The extension of evolutionary theory to include selection for traits that increase the survival and reproduction of relatives is called *kin selection.* Examples of altruistic behaviors include *alloparental care* and *warning signals.* Humans exhibit both cooperative and altruistic behaviors, but these behaviors are probably the result of cultural training rather than genetic programming.

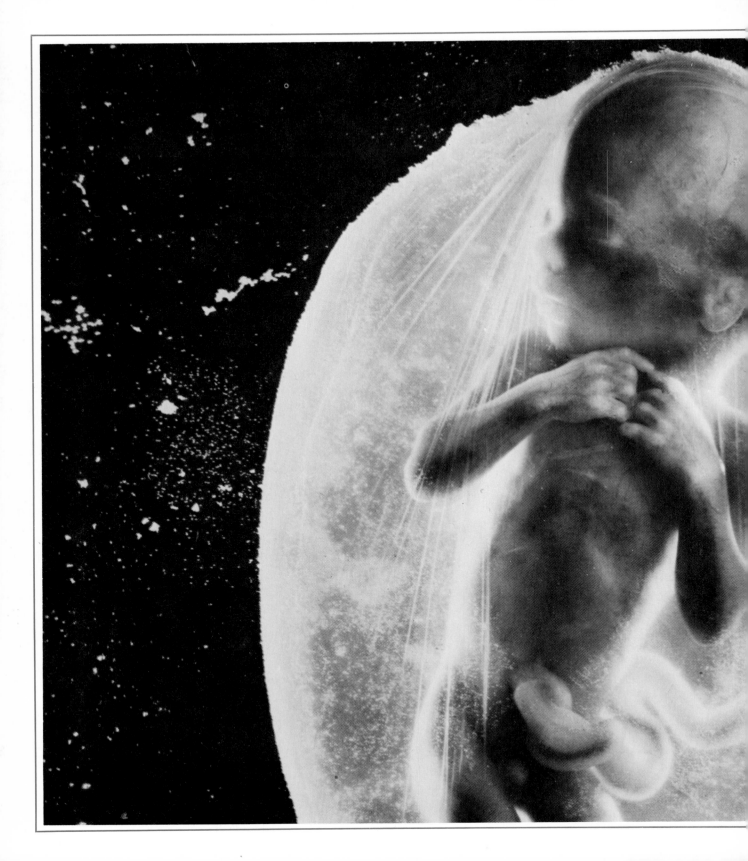

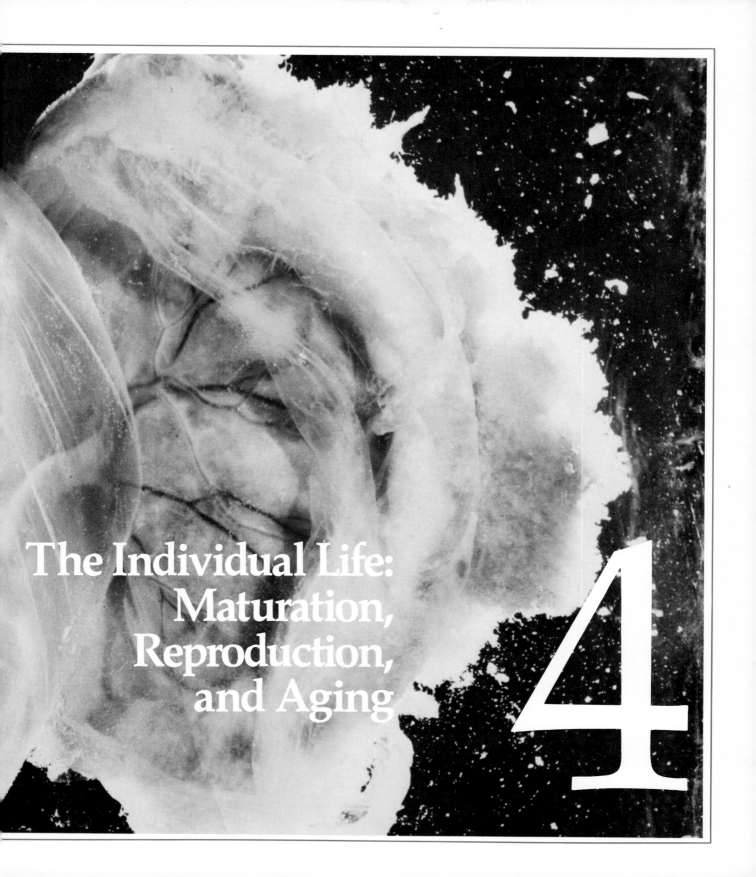

The Individual Life:
Maturation,
Reproduction,
and Aging

4

Patterns of Life History

21

A characteristic property of living material is its capacity to change. Each level of biological organization—an individual molecule, a cell, an organism, or an entire population of organisms—is continually in transition from one state to another. Some biological changes take place within a few seconds; other changes may not be evident for decades, or even centuries. In Part 2, we considered the long-term genetic changes that occur in populations as a result of natural selection. Because some individuals in a population leave more offspring than others, certain genes tend to increase and other genes tend to decrease in frequency, so that the genetic composition of the population changes from generation to generation. In Part 3, we examined the short-term physiological and behavioral adjustments critical to the survival of the individual—changes that have evolved in response to fluctuations in the particular environment of the organism. In Part 4, we will consider another time period during which biological changes occur—the lifetime of the individual.

Maturation, Reproduction, and Aging

An individual lifetime can be divided into three phases: maturation, reproduction, and aging. During *maturation*, the individual undergoes profound changes as the structures of its body develop and begin to perform specific functions. For multicellular organisms, a single cell is converted into a mature individual with diverse tissues and organs. After maturation is achieved, the anatomy, physiology, and behavior of an individual adjust to carry out *reproduction* of offspring. *Aging*—the final phase of a lifetime—usually begins shortly after the period of peak reproduction. This period of gradual deterioration of the body is the least understood of the three phases of life.

The development, reproduction, and continued survival of an organism throughout all stages of life require energy and nutrients, but only limited amounts of these resources are available in the environment. The way in which resources are allocated to the three stages of life is an inherited adaptation of the individual to its environment. The ultimate evolutionary test of how well genes regulate resource allocation is the number of surviving offspring, or *progeny,* an individual produces during its lifetime.

The lifestyle that maximizes reproductive success differs for each set of environmental conditions. In some environments, more total progeny are produced when adult organisms utilize large portions of their resources to guarantee their own survival. These organisms are usually large and have fairly elaborate defenses and mechanisms of homeostasis so that they are relatively invulnerable to most environmental dangers. Due to their large size, these organisms have a long period of growth and development prior to reproductive maturity; once they reach maturity, they regularly allocate a small amount of their resources to reproduction, leaving a few offspring at a time. Considerable amounts of energy and nutrients are reserved for the continued maintenance of their own bodies. The total number of offspring left during their lifetimes may be relatively small, but each offspring will probably survive to reproduce many times. In other environments, a greater number of surviving offspring are left when adult organisms reserve only a few resources for their own survival and allocate most available resources to reproduction. These organisms are usually small, have a short developmental period, and produce a large number of small offspring. They seldom live long enough to reproduce more than once.

The particular sequence of events that occurs during an individual's life span is called a *life history pattern.* The total life span and the number of times reproduction occurs are aspects of an individual's life history pattern. Selection favors events in an organism's life on the basis of whether or not they increase reproductive success.

In this chapter, we will consider the broad patterns of individual life histories and the effects of several important variables on the reproductive output and continued survival of the parent. In Chapters 22 and 23, we will examine the specific events and biological mechanisms involved in each of the three phases of life.

Inheritance: Two Parents Versus One

The offspring produced by *sexual reproduction* carry various combinations of genes from two parents; the offspring produced by *asexual reproduction* inherit genes from only one parent. Selection favors asexual or sexual reproduction in a particular environment on the basis of which system produces more surviving offspring.

Asexual Reproduction

Two different forms of asexual reproduction are known to occur in multicellular organisms—vegetative reproduction and parthenogenesis. In *vegetative reproduction,* some part of the adult body other than an egg or sperm develops into another, complete individual. The strawberry plant provides a familiar example. After this plant is mature, a long, narrow segment, or *runner,* grows outward away from the parent plant and eventually becomes a complete individual with its own roots, stem, and leaves. Offspring produced by vegetative reproduction are genetically identical to one another and to the parent.

In asexual reproduction by *parthenogenesis,* a female produces a diploid egg, containing two sets of chromosomes, that develops without being fertilized by a sperm. In one form of parthenogenesis, the single set of chromosomes within the haploid egg replicates to form a second, identical set of chromosomes prior to development; in another form of parthenogenesis, meiosis is terminated after the first meiotic division (Figure 21-1). When parthenogenesis has been the mode of reproduction for more than one generation, virtually no variability is exhibited among the offspring of each female because the DNA in the homologous chromosome that each female carries has been formed by replication. (Note how the homologous chromosomes are formed in the diploid eggs shown in Figure 21-1.) Offspring from parthenogenesis may exhibit a somewhat greater degree of genetic variation than offspring from vegetative reproduction but are much less genetically varied than offspring from sexual reproduction.

An advantage of asexual reproduction is that the offspring receive most, if not all, of a set of genes carried by an individual that survives and reproduces successfully. If the offspring are exposed to the same environmental conditions in which

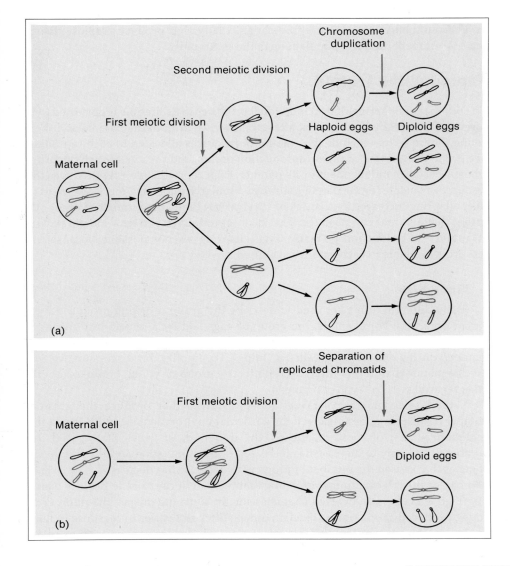

(a)

(b)

Figure 21-1
The Formation of Diploid Eggs Prior to Parthenogenesis
(a) In some parthenogenic organisms, the chromosomes are duplicated after meiosis is complete. (b) In other parthenogenic organisms, meiosis terminates after the first meiotic division. The replicated chromatids then separate, and each chromatid becomes a chromosome. In both (a) and (b), it is assumed that the homologous chromosomes in the maternal cell contain somewhat different genes. The homologous chromosomes in each daughter cell (diploid egg), however, are always identical.

the parent survived and reproduced, they will also reproduce successfully. Another advantage is that a population is able to increase more rapidly through asexual reproduction than through sexual reproduction, because all the members of an asexual population are females capable of producing young.

Sexual Reproduction

The sets of genes carried by offspring produced by sexual reproduction differ from each other and from the sets carried by either parent. This genetic variation increases the chance that some progeny will inherit a combination of genes that will allow them to survive and reproduce even if the environment and other organisms in it change over time. If the environment does not change, however, then the complete set of genes that ensured the parents' survival and reproductive success will not be passed on to the offspring. The prevalence of sexual reproduction among plants and animals may be primarily an adaptation to rapid changes in coexisting populations. As long as some populations are capable of undergoing rapid evolutionary changes by reproducing sexually, they produce a rapidly changing environment for other populations in the community.

Reproductive Output

The number of offspring that an individual produces in its lifetime is governed by a variety of interrelated factors that affect the survival of both the parents and their young. These factors include the amount of resources allocated to each reproductive event, the number of times reproduction occurs, and the size of each offspring when it becomes independent of its parents. Each reproductive event reduces the resources available for the parent's survival. Similarly, allocating more resources to each offspring increases its chances of survival but limits the number of other offspring that a parent can produce. The ideal reproductive pattern—indefinite survival and the production of many large offspring—is not possible because the resources available to each parent are limited.

Litter Size

Many different terms are used to describe the group of young produced by a parent organism. For example, the group of eggs laid by a female bird during a reproductive season is called a "clutch," but the group of young born to a female mammal during one season is called a "litter." To simplify this discussion, we will use the mammalian term "litter" to refer to any group of young produced at one time, regardless of the type of organism.

Natural selection always favors the individuals that produce the greatest number of offspring possible given the particular environmental conditions. In the evolutionary past of these organisms, the frequency of genes that caused the number of progeny to increase has increased in the population and the frequency of genes that caused the number of progeny to decrease has decreased. The *optimal litter size*, or the largest number of young that the parent can raise to healthy maturity (Figure 21-2), is an evolved characteristic. In some organisms, the number of young produced at one time is fixed; in others, litter size varies in response to fluctuations in the amounts of available resources, including energy, nutrients, water, shelter, and nesting sites. The parents' ability to acquire the necessary resources for their offspring and themselves is an important factor in the evolution of an optimal litter size.

Litter size is also influenced by the number of times that the parent can reproduce. In organisms that can produce several litters in a lifetime, selection favors small litter size so that the parent can survive to reproduce again, increasing the accumulated number of offspring from all litters. In contrast, organisms such as the Pacific salmon, that typically have only one litter in a lifetime, produce as many offspring at one time as possible, often using up their own body tissues in the effort; selection does not favor conserving resources to ensure the survival of these parents.

Number of Litters

Many organisms, including annual plants and most insects, reproduce only once and then die; other organisms, including mammals and birds, reproduce several times during their lifetime. The probability of the parent's survival from year to year largely determines whether an organism reproduces once or several times. In environments where the organisms have little chance of living for another year, selection favors the reproduction of as many offspring in a litter as possible.

Adult survival is poor in temporary and highly seasonal environments. Organisms that inhabit *temporary environments* (such as ponds that form after a rain or forest lands after a fire) are called *colonizing species.* Most colonizing organisms produce all of their young in a single litter. The offspring are usually adapted for dispersal to new, more favorable areas, whereas the adults are either immobile or are unable to travel long distances. Once an adult becomes established in a temporary environment, it has little chance of surviving and uses all of its resources to produce offspring. In a *highly seasonal environment,* conditions fluctuate throughout the year and the adult organism reproduces and dies during a single season. The offspring are resistant to the unfavorable environmental conditions and remain dormant (inactive) until the next favorable season when they develop into adults and reproduce.

In contrast, when the probability of the adult's survival from one year to the next is high, as it is for most large plants and animals, fewer offspring are produced per litter and resources are conserved to permit the adult to survive until the next reproductive season. When environmental conditions affecting the survival of the young change from year to year, the reproduction of several small litters in different years increases the probability that some offspring will be born at a favorable time. If all of the young were produced in a single, large litter and environmental conditions were extremely poor, all of the offspring would die.

Age of First Reproduction

If reproduction begins at an early age, more descendants (children, grandchildren, and so on) can be produced over a given time. The offspring are born sooner and, in turn, produce their own offspring sooner. In populations that are increasing in size, such as human populations, the age of first reproduction may be the most influential factor affecting population growth. Selection favors reproductive maturity at an early age unless specific factors oppose it.

(a)

(b)

Figure 21-2
Optimal Number of Young in Two Types of Birds
The number of eggs produced by a female bird during a single reproductive season is determined by the interaction between the individual's genes and its environment. (a) Nest containing 11 pheasant eggs. (b) Nest containing 5 eastern meadowlark eggs.

The need for a long developmental period is the most important factor opposing early reproductive maturity. Large, complex bodies often provide a competitive edge in acquiring resources, capturing prey, or escaping predators. If selection favors larger body sizes, reproduction must be delayed until a later age to allow more time for development.

Some plants also have long developmental periods to allow them to compete successfully for resources. Trees must compete for sunlight to obtain enough energy to reproduce. Trees that have channeled their resources into growth for several years have greater access to sunlight. The age of the first reproduction in these plants is delayed until they are large enough to acquire sufficient resources to reproduce.

Season of Reproduction

Because large amounts of energy and nutrients are needed to form and nourish offspring, organisms typically bear their young during the season of the year when food sources are most abundant. When the seasonal availability of resources determines the time of reproduction, mating is also seasonal since a specific time interval elapses between mating and the appearance of young in most organisms. In temperate environments, most plants grow and reproduce only between late spring and early fall. Consequently, animals that feed on plant material bear their young in the spring, when vegetation is plentiful, and rear their offspring during the summer. Predators of plant-eating animals also produce their young in the spring and early summer so that large prey populations will be available when their offspring are growing.

Competition from other organisms also affects resource availability, particularly in tropical environments where physical conditions do not change dramatically during the year. Tropical animal species that compete for food and tropical plant species that compete for animal pollinators usually reproduce at different times of the year, so that the demand for resources is staggered. Even in more northern environments, competition for resources determines the time of reproduction. Many small forest plants grow and flower in early spring (Figure 21-3) before the leaves of the trees emerge to shade the forest floor.

When mates are difficult to find, organisms must breed at specific times. Mating in some ants, for example, occurs only at certain times of the day on a few consecutive days of the year. During these few hours, each colony releases winged males and females which form huge mating swarms. Even the place of mating may be specific; many mating swarms form only above the highest point in the area. Each colony in the population must recognize and respond to the same environmental cues for mating between colonies to occur.

An even more remarkable example of mating synchrony is provided by the periodical cicadas. The adult forms of these insects appear and mate above the ground only during a few weeks every 13 or 17 years, depending on the species. This extreme reproductive cycle is believed to be a means of avoiding heavy predation. Because these insects appear so infrequently, their potential predators are unable to rely on them as a food source and do not evolve adaptations for capturing them. Only a small percentage of the adult population is lost to predators with

Figure 21-3
A Plant That Reproduces Early in Spring
To compete for sunlight, this lady's slipper grows and flowers before the leaves appear on the trees and shade the forest floor.

generalized feeding habits, because the cicadas appear suddenly in very large numbers. The relatively low densities of predators that are present when the adults appear are simply unable to eat the cicadas fast enough to have a significant effect on the insect population.

Maximum Life Span

Maximum life span is the length of life possible for individuals living under ideal environmental conditions. Some organisms live for a very short time because the adult form is not adapted to survive through different seasons of the year. The adult reproduces during one season and then dies. Just prior to death, it leaves offspring in the form of eggs or seeds, usually with low rates of cellular activity, enclosed within layers of material that are resistant to unfavorable environmental conditions. When the favorable season returns, the young grow rapidly into adults and the cycle repeats itself. Desert annual plants, for example, germinate and grow into adult forms only during the short rainy season of each year (Figure 21-4). The adult desert annual does not have the deep roots, waxy leaves, or water-storage organs that would allow it to survive the dry season of each year. It reproduces when conditions are wet, and its drought-resistant seeds survive in the soil until the next rainy season.

It is easy to understand why this type of organism has such a short life span. It is more difficult to understand why organisms that are able to survive throughout all seasons of the year do not live forever. Most organisms seem to have a built-in timer that dictates their maximum life span. For these species, death from old age is predictable and as much a part of their biology as growth and reproduction.

Figure 21-4
The Appearance of Desert Annual Plants After a Rain
Only the seeds of these plants are present during the dry season. After the rainy season, the seeds germinate, grow, and reproduce in a short period of time. The offspring then remain dormant in the soil in the form of seeds until the next rainy season.

Table 21-1 Comparative Ages of Maturity, Average Age at Death, and Maximum Life Span of Various Mammals			
Species (all males)	Average age at maturity (years)	Average age at death (years)	Maximum life span (years)
Human	21	70	115+
Elephant (Asiatic)	25	60*	77+
Camel	8	31*	40+
Horse	5	22	40+
Bull	4	19	30+
Dog	1.5	11.5	34
Cat	1	10	35*
Sheep	1.5	9	15+
Mouse	66–67 days	2.2	6*

*Approximate

The fact that the ages of the oldest individuals of a population are about the same for all populations of a species but vary greatly among different species suggests that maximum life span is genetically controlled. All individuals of a species appear to carry genetic information that determines the maximum possible length of time they can live. In humans, the average life span varies as a function of different environmental conditions, but the maximum life span of 115–120 years is similar for all populations and apparently has not changed throughout recorded history. The maximum life spans for several species of mammals are listed in Table 21-1.

One factor that seems to influence maximum life span is the number of litters that an individual is capable of producing during its lifetime. Once all reproductive functions have ceased, natural selection cannot evolve adaptations for prolonging life. If certain genes cause a fatal disease in an individual after its reproductive period is over, those genes cannot be selectively removed from the population. Selection may even favor terminating life after reproductive capabilities cease. In some cases, the removal of a post-reproductive individual from the population ensures that more resources will be available for its progeny.

Thus, life spans are limited by the lack of selection for survival after reproduction ceases. But why does the reproductive period of an organism terminate? The ability of most organisms to reproduce declines with increasing age. In some organisms, the decline is sudden; in others, it is gradual. The rather sudden decline in the reproductive capacity of human females at about 45 years appears to be genetically programmed. At present, the reasons why individuals cannot continue to reproduce indefinitely or live forever cannot be logically explained.

Average Life Span

In general, death occurs at a predictable age and is preceded by a gradual aging process. Most organisms that are kept in captivity exhibit a pattern of progressive physical deterioration with increasing age, but only a few animals live long enough in the wild for the aging process to become evident. The causes of premature death, such as predation and starvation, vary with each species, and each cause may affect a particular age group more than others.

Because acts of predation are seldom observed and dead organisms are rapidly consumed by scavengers and decomposers, the causes of death in a wild population are difficult to determine. To determine the number of deaths in wild populations that can be attributed to the aging process, biologists have developed a tool they call a *survivorship curve*, which graphs the percentage of the population still alive at successive ages of life and provides an indirect means of examining the causes of mortality. To construct a survivorship curve for a captive or a wild population, a large group of individuals born at the same time are marked so that they can be identified in the future. The number of individuals still alive at different time periods is then recorded until all of the marked individuals have died.

Four basic types of survivorship curves are shown in Figure 21-5. In curve I all deaths are associated with the aging process; in this type of population, mortality occurs only when the maximum physiological life span has been reached. Survivorship curve I is rarely representative of natural populations but is typical of organisms kept in captivity in zoos, greenhouses, or laboratories. The survivorship curves of technically advanced human populations that benefit from high levels of sanitation and medical care are similar to the type I curve.

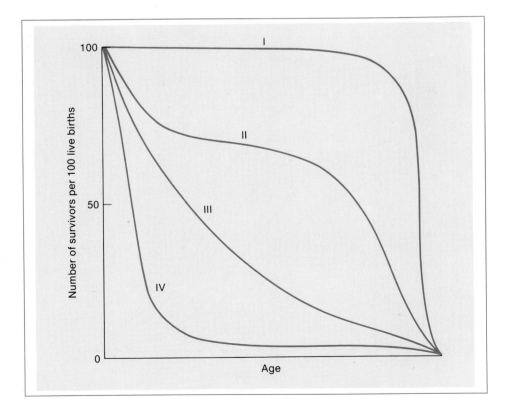

Figure 21-5
The Four Basic Types of
Survivorship Curves
In curve I, all death is attributed to old age. Curve II indicates high mortality in the young and the old but relatively good survival in between. In curve III, a constant percentage of the population dies at each age; in curve IV, almost all mortality occurs in the very young.

Survivorship curve II indicates a relatively high death rate in the young and the old, but very little mortality during the intermediate periods. This type of survivorship is typical of mammals; it reflects the effectiveness of adaptations in large and complex life forms to promote survival during the reproductive period, including all mechanisms of homeostasis and such antipredator strategies as learning and rapid locomotion. The relatively high rate of mortality in younger individuals demonstrates their vulnerability to predation and disease.

Survivorship curve III is representative of many natural populations, especially plant, bird, and reptile populations. This curve reflects the death of a constant percentage of the population at each age: as the total number of individuals declines, fewer individuals die at each age. The probability that a particular individual will die is the same for all ages; as an individual grows older, it does not become more or less vulnerable to death. This type of survivorship occurs in populations where most deaths result from predation and the ability of an individual to avoid capture does not improve or deteriorate with age. Such conditions can develop in a population that is vulnerable to a large number of different types of predators. For example, each type of predatory bird feeds at a specific time and employs a particular attack pattern. After several encounters with one species of predatory birds, a prey individual may learn not to be active at certain times or to utilize certain flight patterns to escape that particular predator. However, this experience will be of no advantage in avoiding predators that hunt at different times and employ different attack patterns.

Figure 21-6
The Effects of Cultural Changes on Human Survivorship Curves

Improvements in medicine, sanitation, and nutrition have reduced the causes of death other than senility to the point that the human survivorship curve is approaching a type I curve. (All data are from Sweden.)

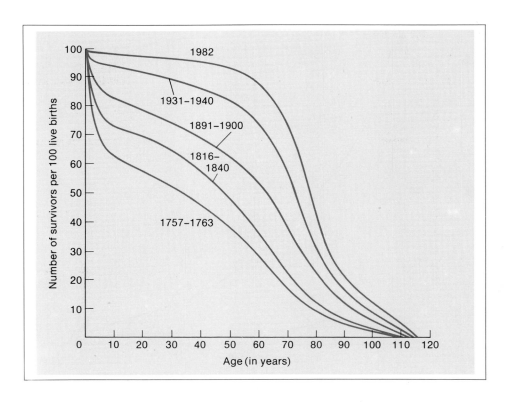

Survivorship curve IV indicates a high death rate in the young but a good chance of survival thereafter. This is the most common type of survivorship; it occurs in populations composed of individuals that produce a large number of relatively small, helpless offspring and provide no parental care. Many fishes, for example, release their eggs and sperm into the water, and the fertilized eggs develop far away from their parents. These small, slow-moving offspring are equipped with few adaptations for survival and are vulnerable to all types of mortality, particularly predation. However, these fishes apparently outgrow the causes of mortality in their environment and become relatively invulnerable to death after they reach a certain size. Trees and most marine organisms also exhibit this type of survival.

If a population were able to evolve adaptations that made it invulnerable to the causes of mortality in its environment, its survivorship curve would approach the type I curve and all death would eventually result from old age—a progressive trend exhibited by human populations in technically advanced cultures for the past 200 years (Figure 21-6). Changes in the survival rates of most organisms reflect genetic changes. However, increases in the survival rates of human populations in the recent past reflect cultural changes—improved standards of living and technology that make these populations less vulnerable to the environment. The most dramatic change has been a reduction in juvenile mortality through improvements in sanitation, nutrition, and the conquest of major childhood infectious diseases. The maximum life span has not been significantly altered by these cultural improvements, however, and death from old age remains a predictable phenomenon in human populations.

Summary

An individual lifetime is divided into three phases: *maturation, reproduction,* and *aging.* The amount of time that an organism spends in each phase and the types of activities that occur during these phases are adaptations that have been selected for because they have a strong influence on the reproductive success of the individual.

One characteristic of an individual's *life history pattern* is the mode of reproduction. In *asexual reproduction,* the offspring are produced by only one parent and are genetically similar to one another and to their parent. Organisms that inhabit unchanging environments tend to reproduce asexually. In *sexual reproduction,* the offspring are produced by two parents and differ genetically from one another and from either parent. This genetic variation increases the chance that some offspring will inherit the appropriate set of genes to survive and reproduce in a changing environment. An important aspect of environmental change is the evolution of other populations of organisms.

The reproductive success of an individual depends not only on its mode of reproduction but also on the number of offspring produced at one time (*litter size*), the number of litters produced in a lifetime, and the age at which reproduction begins. These aspects of life history are selected for in each individual to maximize reproductive success under the particular environmental conditions experienced by its ancestors. Mating and reproduction are usually highly seasonal activities, because resources are seasonally abundant due to a limited growing period during the year or to the activities of competing species.

Evidence suggests that length of life may be genetically programmed—that each kind of organism has a *maximum life span*. The onset of old age occurs after the time of peak reproductive output because natural selection does not effectively promote survival after reproduction has ceased. Why reproduction declines or ceases after a certain age is not yet understood.

Survivorship curves, graphs that indicate what percentage of a population is still alive at successive ages of life, provide an indirect means of examining the causes of mortality. Four basic types of survivorship curves are recognized: (I) all mortality occurs from physiological aging; (II) high mortality occurs in the very young and the old but little mortality occurs during the reproductive ages; (III) mortality occurs at the same rate for all ages; and (IV) very high mortality occurs in the young with a good chance of survival thereafter. Human populations are moving from a type II to a type I survivorship curve in many technically advanced cultures.

Reproduction of the Individual: From Cell to Organism

22

A new organism is formed by the process of *reproduction.* The ability of existing individuals to develop new individuals is a fundamental property of all life forms, but methods of reproduction vary greatly. The process of reproduction may include mating behavior, fertilization of an egg by a sperm, embryonic development, and parental care of the young. The simplest form of reproduction, which occurs in unicellular organisms, is the division of one cell to form two cells. The most complex form of reproduction, which occurs in humans, begins with the mating between a man and woman and continues until parental care is no longer essential to the child's survival.

The function of sexual intercourse in human reproduction was not known until the eighteenth century. Earlier studies of other mammals had shown that eggs originate in the ovaries of females and eventually develop into offspring. The role of the male in reproduction was not revealed until after the microscope was invented; sperm were observed for the first time around 1700.

When a sperm fertilizes an egg to form a single cell, the *zygote,* two sets of genes—one from the father and one from the mother—combine to provide a blueprint for development of the zygote into an individual. The sequence of changes that occur during the conversion of genetic instructions into body structure has been observed in a large variety of organisms. The mechanisms that cause these changes are still not well understood. After the union of a sperm and an egg in multicellular organisms, the zygote undergoes mitosis to form two cells, each containing a copy of the genetic material carried in the chromosomes. Every successive division of these cells doubles the number of cells in the developing embryo (from 4, to 8, to 16, etc.) until it contains millions of cells. During the development of the individual, the cells of the original zygote somehow become specialized in form and function into the cells of an animal (nerve, muscle, skin, etc.) or the cells of a plant (root, stem, leaf, etc.). The development of cell specialization is one of the least understood areas of biology.

In Chapter 6, we examined the fairly simple modes of reproduction in unicellular organisms. Here, we will concentrate on reproduction in multicellular organisms, which involves four main processes: *fertilization,* the union of an egg and a sperm to form the zygote, which occurs only in sexual reproduction; *morphogenesis*, the arrangement of cells into tissues and organs; *differentiation*, specialization in form and function of cells containing the same genetic information; and *growth*, the increase in the amount of living material. Except for fertilization, which must occur first, the other three processes are interrelated and may occur simultaneously. In this chapter, we will examine the general pattern of reproduction in two major groups of organisms—the flowering plants and the vertebrate animals. Chapter 23 will be devoted entirely to a detailed treatment of human reproduction and development.

Fertilization in Flowering Plants

In sexually reproducing organisms, the first step in the production of an individual occurs when two cells, each containing one-half of the number of chromosomes (the *haploid number*) characteristic of the species, unite to form a single cell, the zygote (or fertilized egg). The union of these two haploid cells provides the zygote with the full complement of chromosomes (the *diploid number*). Fertilization also initiates developmental changes in the egg and alterations of the egg surface that prevent other sperm from entering.

Multicellular plants form two types of haploid reproductive cells. In flowering plants, the *gametes* (eggs and sperm) are produced by meiosis in the male and female parts of the *flower*, as shown in Figure 22-1(a). In some plant forms, both male and female reproductive structures occur in each flower; in other forms, each flower contains either male or female parts. Moreover, in some plant species, each individual has only male flowers or only female flowers; in other plant species, each individual has both types of flowers.

The *pollen grains* produced by the male parts of a flower contain three haploid nuclei—two *sperm nuclei* and one *tube nucleus*. Pollen is transported by the wind or by an animal to the female part of a flower, where it becomes attached to the *stigma*. Part of a pollen grain, the *pollen tube*, then grows downward into the *ovary*, which lies deep inside the female part of a flower. The pollen tube penetrates one of the *ovules* inside the ovary and releases the two sperm nuclei as shown in Figure 22-1(b). (The tube nucleus, which controls the activities of the pollen grain, remains inside the pollen tube.) One sperm nucleus joins with the egg nucleus to form the *diploid zygote*, which becomes the *embryo*. The other sperm nucleus joins with two haploid nuclei in the ovule to form a single nucleus with three sets of chromosomes. This fertilized nucleus eventually develops into the *endosperm*—a special type of tissue that provides nutrients for the embryo. This form of union between male and female nuclei, called *double fertilization*, is unique to flowering plants. It ensures that nutrients do not develop until an egg is actually fertilized, so that structural materials are not wasted on eggs that never develop into embryos.

Plants that have both male and female flowers on the same individual or male and female parts in the same flower are usually capable of preventing *self-fertilization*—the fertilization of their own ovules by their own pollen. The pollen produced by some plants cannot penetrate the female parts of a flower on the same plant; in other plants, the pollen and the ovules of each flower reach maturity at different times during the reproductive season.

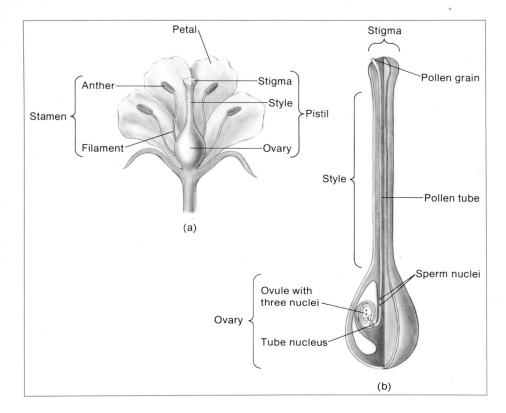

Figure 22-1
Fertilization in a Flowering
Plant

The production of gametes, fertilization, and early development all occur within the flower. (a) Reproductive cells formed by the male organs *(stamens)* are present in grains of pollen on the *anthers.* Reproductive cells formed by the female organs *(pistils)* are located within the *ovules* of the *ovary.* In the plant diagrammed here, which is similar to a lily or a tulip, each flower contains both male and female reproductive structures. In other plants, like the willow tree, all the flowers on each plant are only male or only female reproductive structures. (b) Enlarged diagram of the pistil, showing the growth of a *pollen tube* from the *stigma* to the ovule. One sperm nucleus joins with two nuclei in the ovule to form the *endosperm;* the other sperm nucleus joins with the egg nucleus in the ovule to form the *embryo.* The *tube nucleus* plays a role in development of the pollen tube but is not involved in the actual fertilization process.

Morphogenesis in Flowering Plants

Most flowering plants begin life as a zygote inside an ovule. *Morphogenesis* is the formation of the cells descended from this zygote into structural patterns of tissues and organs. Plants develop these patterns by (1) differential growth, during which certain groups of cells divide and enlarge rapidly while others do so slowly or not at all, and (2) modifying the shapes of cells.

Both fertilization and early development take place within the ovary of a flower (see Figure 22-1) while it is still attached to the parent plant. An ovary contains one or more ovules, and each ovule is capable of developing into an offspring.

Immediately after fertilization, the endosperm nucleus divides rapidly to form a large number of nuclei immersed in a fluid. Eventually, the growing endosperm completely surrounds the zygote, and a cell membrane and cell wall develop around each endosperm nucleus. These cells will provide nutrients for the growing plant embryo.

Once it is surrounded with endosperm, the zygote begins to undergo cell division. The first division produces two cells—a suspensor cell and an embryo cell. The *suspensor cell* divides repeatedly to form a stalk-like tissue, the *suspensor,*

Figure 22-2
Sequence of Changes in the Development of a Plant

(a) The first division of the zygote yields an *embryo cell* and a *suspensor cell*. (b) After several more cell divisions, the embryo is developing at one end, while an elongated tissue, the *suspensor*, is forming at the other end. (c) A later (less magnified) stage of development showing the developing embryo inside the ovule. The original suspensor cell has formed the suspensor, which attaches the embryo to the walls of the ovule. The embryo cell has formed the *shoot apical meristem* and *root apical meristem* tissues and *cotyledons* of the embryo.

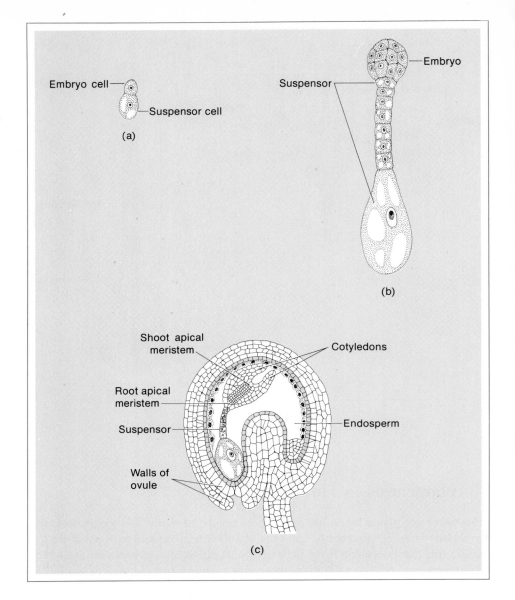

which eventually attaches the developing embryo to the inner wall of the ovule (Figure 22-2). At the same time the suspensor is developing, the embryonic cell divides repeatedly and becomes organized into three basic types of growing plant tissues—one or two *cotyledons* (embryonic leaves), the *root apical meristem,* and the *shoot apical meristem.* The cotyledons are strictly embryonic structures that absorb nutrients from the endosperm and make them available to the growing embryo. The meristem tissues remain present in certain parts of the plant throughout its life. These special tissues maintain the ability to undergo continued cell division, thereby permitting a plant to form new structures regularly long after embryonic development is complete. The root apical meristem cells divide rapidly to lengthen the embryonic root of the plant, and the shoot apical meristem cells divide to extend the embryonic stem of the plant in the opposite direction. Cells formed by the meristem tissues enlarge and become specialized in structure and function. Cell walls then develop around them and they become fixed in position, unable to move or to undergo further cell division.

A *seed coat* that completely encloses the embryo and the endosperm forms from the outer walls of the ovule in an early stage of embryonic development. The *seed* remains inside the ovary of the flower; each ovary may contain more than one seed.

While the embryo is developing, the ovary is undergoing changes that transform it into a *fruit*. The formation of a mature fruit is stimulated by growth hormones released from the pollen grain at the time of fertilization. The development of the ovary is similar to the development of other parts of the plant. A period of rapid cell division is followed by cell enlargement and then by cell differentiation. The coordination of these three processes eventually leads to the maturation and then separation, or *abscission*, of the fully ripe fruit from its parent plant. Abscission is controlled by *abscisic acid*—a hormone synthesized by the mature fruit itself.

The function of the fruit is to disperse the seeds, which contain the embryos, away from the parent plant to improve their chances of growing to maturity. The size and shape of fruits vary considerably, depending on the mode of seed dispersal. Some very small fruits contain only one seed and are shaped to facilitate movement through the air. The fruits of many coastal plants are large and float on water from the parent plant to their site of establishment. Other fruits are dispersed by animals. Some plants have sharp or sticky projections that permit their fruit to adhere to the outer surface of an animal. Many so-called *fleshy fruits*, including apples, tomatoes, and berries, are attractive and nourishing, which encourages their consumption by animals; when these fruits are eaten, their seeds usually pass through an animal's digestive tract and are deposited later in the feces some distance away from the parent plant. A variety of fruits are shown in Figure 22-3.

After the fruit and its seeds are released from the parent plant, the embryo remains dormant inside its seed coat until environmental conditions of temperature, light, and moisture become favorable for germination and subsequent development. *Germination*—the activation of an embryo that was previously dormant inside its seed coat—begins when water is absorbed into the seed, rupturing the seed coat and stimulating hormones and enzymes in the embryo and the endosperm. Cell division and enlargement of the cells within the meristem tissues of the embryo are then renewed. Finally, the *root* protrudes from the seed coat, exposing part of the embryo to the external environment for the first time.

As the root extends into the soil, the embryo becomes a *seedling* and grows rapidly. (Each day, the roots may lengthen as much as 7 centimeters and form as many as 14,000 new branches, or *feeder roots*.) After a root has become established, the *stem* and leaf tissues form (Figure 22-4). Strands of elongated cells develop within the stem and roots to form vascular structures (the *xylem* and the *phloem*), which transport fluids throughout the plant. All of the basic structures for survival are now fully developed and functioning, and the young plant enters into a period of growth until conditions are appropriate for it to reproduce.

Figure 22-3
The Fruit of a Flowering Plant Promotes Seed Dispersal

(a) The dandelion and maple produce fruits that are shaped to facilitate movement through the air. (b) The fruit of a coconut tree floats on the water, transporting its seeds from island to island. (c) Each blackberry fruit contains many seeds. These berries are usually eaten by animals, and their seeds are deposited later in the feces, often after the animals have traveled some distance from the parent plant. (d) The fruits of the cocklebur and sticktight have sharp projections that tend to adhere to the coats and feet of animals. These fruits are transported by the animal until they are rubbed off or removed in some other way.

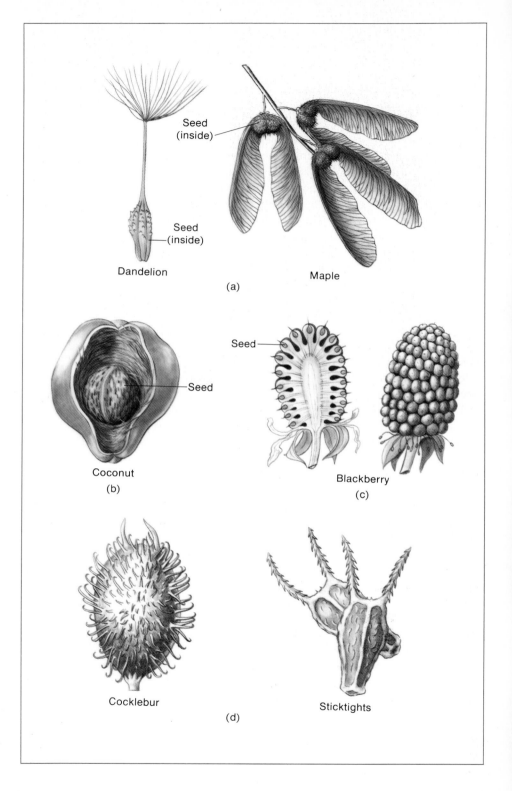

Seed (inside)

Seed (inside)

Dandelion

Maple

(a)

Seed

Coconut

(b)

Seed

Blackberry

(c)

Cocklebur

Sticktights

(d)

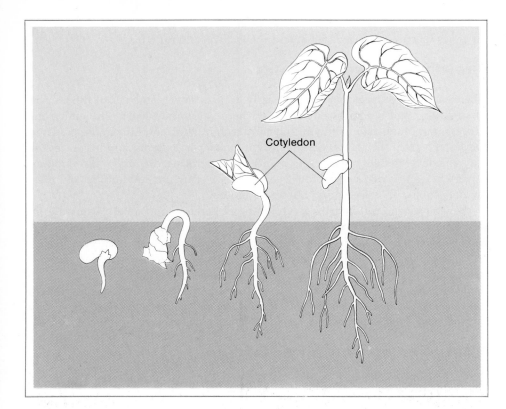

Cotyledon

Figure 22-4
Growth of a Bean Seedling
The first structure to emerge from the *seed coat* is the *root*. As the embryo unfolds and elongates, the *stem* and a pair of *leaves* appear above ground. Further development includes growth of the stem and of many *feeder roots* from the root. A bean plant is a *dicotyledon*, indicating that it has two cotyledons in the embryo. Other plants of this type include forest trees, fruit trees, potatoes, cabbages, roses, and snapdragons. Flowering plants with a single cotyledon are called *monocotyledons* and include grasses, cereals, palms, lilies, tulips, and orchids.

Cellular Differentiation in Flowering Plants

Cells produced by mitosis and descended from the original zygote eventually form all of the different structures of the adult plant. During *cellular differentiation,* these genetically identical cells become diversified and acquire the ability to perform specialized functions. All cells, whether they are differentiated or not, are capable of performing most of the basic functions of life, such as protein synthesis and cellular respiration. In addition to these basic processes, differentiated cells perform certain specialized functions that contribute to the survival of the multicellular individual. For example, the root cells of a plant contain structures that are specialized for the intake of water and minerals; the leaf cells of a plant are specialized to perform photosynthesis.

It is not known how plant cells become differentiated, but hormones certainly play an important role in coordinating developmental events. The construction of a new leaf, for example, requires many adjustments in the rest of the plant. The developing leaf produces a plant growth hormone, *auxin,* that stimulates the differentiation of adjacent shoot tissues into xylem and phloem tissues, so that the new leaf will have the necessary vascular connections when it begins to function.

Plant hormones are produced and released when certain conditions prevail in the external environment and within the developing plant. Their particular effects on cells depend on the condition of the cell at the time. A hormone probably influences cellular activity by activating or deactivating certain genes. The particular genes that are active in a cell determine its form and function.

Growth in Flowering Plants

The basic types of tissues are formed early in plant development, but the number of structures and the particular shape of the plant are not fixed. The embryronic processes of cell division, growth, and cell differentiation continue in the meristem tissues of a plant throughout its life, forming new roots, stems, and leaves. In this way, the development of flowering plants differs strikingly from the development of vertebrate animals; a vertebrate's general body structure and number of organs are determined early in life and cannot be altered.

Several hormones, including cytokinin, gibberellin, and auxin, stimulate the development of new plant structures from meristem tissues. Environmental conditions govern the amounts of these growth hormones that are produced as well as the time that hormonal production occurs. The amounts of nutrients and energy a plant acquires also greatly influence its growth rate (Figure 22-5).

Plants do not have the ability to regenerate lost parts. New branches, leaves, and roots are formed regularly, but do not appear in the same positions as the original, lost structures. Injury to one part of a plant stimulates growth elsewhere.

Maturation of Flowering Plants

An organism is considered an adult when it is capable of reproduction. Unlike animals, which carry their reproductive organs throughout life, plants produce their reproductive organs, flowers, only during one season of the year.

The shoot apical meristem of a plant is capable of several forms of new growth—stem elongation, branch and leaf production, or flower development. The time of the year and the size of the plant determine which developmental pathway the meristem tissues will take. When conditions are right, a hormone moves from the leaves, where it is synthesized, to the buds, where it initiates flower development. This hormone, as yet unidentified, is tentatively called *florigen.* A flower is composed of the *petals,* which attract animal pollinators, and the male and/or female organs. In each species of plant, the parts of the flower develop according to specific genetic instructions.

Annual plants reproduce during their first year of growth and then die. *Biennial* plants grow for two years, then reproduce and die. *Perennial* plants continue to grow and reproduce year after year. In some perennials, most of the adult structure endures throughout all seasons of the year, as is true of trees. In other perennials, the above-ground structures die after reproduction, but the roots and underground stems survive and provide nutrient stores for the plant's rapid growth prior to the next reproductive season.

Fertilization in Vertebrates

Two kinds of *haploid cells,* the gametes, are produced in the reproductive organs of adult multicellular animals by *meiosis*—the special form of cell division in which four haploid cells are formed from a *diploid cell. Haploid sperm* are produced from

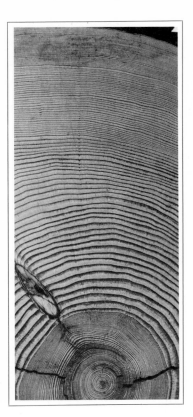

Figure 22-5
The Xylem of a Hemlock Tree
For the first 75 years of its life, this hemlock tree grew slowly (narrow inner rings) because it was shaded by other trees. Once the adjacent trees were cut down, the tree grew rapidly in subsequent years.

Figure 22-6
External Fertilization in
Amphibians
The female frog releases her eggs
when the male presses on her
abdomen. As the eggs are released, the
male sheds his sperm and fertilization
occurs in the surrounding water.

diploid cells in the *testes,* and *haploid eggs* are produced from diploid cells in the *ovaries.* In most animal groups, there are two types of reproducing individuals— males with testes and females with ovaries.

Sperm are always smaller and more mobile than eggs, which usually contain large quantities of nutrients for the developing embryo. For fertilization to occur, a sperm must find and penetrate an egg. In most animals, the probability that fertilization will occur is increased by courtship behavior, which guarantees that the two mating animals are of the same species and that the sperm and egg are released at about the same time and place. Courtship behavior is particularly important when fertilization occurs in water (Figure 22-6). Terrestrial animals release their gametes at a much more specific place; the sperm are usually deposited directly into the female's body, so that fertilization of the egg takes place within a small, enclosed area.

In a few animals, such as earthworms and snails, both testes and ovaries are found within each individual. Mating occurs when two of these *hermaphrodites* release their sperm into each other's bodies, simultaneously fertilizing each other's eggs. It is difficult for slow-moving organisms that exist in populations of low density to locate mates; if every individual in a species is both male and female, only two members of the population have to meet for reproduction to occur.

Morphogenesis in Vertebrates

Most vertebrate animals begin life as a zygote, or fertilized egg. The zygote cell is similar in appearance to other animal cells, but it is usually quite large to accommodate the nutrients required to support the animal's early development. Vertebrate *morphogenesis* involves migrations of cells from one part of the embryo to another. In this way, animal development differs markedly from plant development; plant cells become cemented together and are not free to move.

Early vertebrate development occurs in a variety of locations. In most fishes and amphibians, fertilization and all development take place in water. In reptiles and birds, fertilization occurs inside the female; a shell then forms around the zygote, producing an "egg" that is released into the external environment. Most of the developmental sequence takes place inside the shell. In mammals, both fertilization and a large portion of the developmental sequence occur inside the female's body.

After fertilization, cleavage begins to occur in the zygote. *Cleavage* is a series of mitotic cell divisions that increases the number of cells but does not change the size of the original mass. The large zygote is therefore subdivided into a mass of smaller cells. As the mass becomes organized into smaller and smaller units, the particular developmental pattern that emerges is influenced by the amount of *yolk,* or nutrients, stored in the original zygote. This amount is identical for all individuals of a species but varies greatly among different species. Developing birds and reptiles require large quantities of stored nutrients because they have no access to other resources while they are enclosed in their shells. In contrast, mammals require very little yolk because they establish a connection with their mother's circulatory system and begin to receive nutrients from her blood early in their development. The zygotes of fishes and amphibians contain just enough yolk to carry them through the early stages of their development until they become self-feeding larvae.

The patterns of cleavage in an amphibian (a frog) and in a bird (a chicken) are compared in Figure 22-7. The yolk is heavier than the rest of the material in the frog zygote and is concentrated in the lower half of the zygote. The rate at which cleavage occurs is always slower in areas where the concentration of yolk is high. The first two cleavages divide the frog egg vertically into four equal quarters. The third cleavage, which occurs along a horizontal plane, divides the zygote into eight cells—four smaller and more active cells on top, and four larger, yolk-filled cells on the bottom. Subsequent cleavages occur more rapidly in the top half, so that the developing embryo eventually contains tiny cells in the top half and larger cells in the bottom half.

A bird zygote provides nutrients to support the embryo during most of its developmental period. Cleavage occurs only in a tiny disk-shaped area that contains little yolk and never extends throughout the entire zygote. The first cleavage furrow starts in the middle of this disk of nonyolk material and extends almost to its edges. The second cleavage furrow is perpendicular to the first and divides the active material of the zygote into four equal parts. In subsequent cleavages, the flat disk is further subdivided into many small cells, which rest on a large volume of yolk below.

After a certain number of cells have been produced by cleavage, they become arranged in the form of a hollow structure, the *blastula,* which is usually no larger than the original zygote. A noncellular cavity, the *blastocoel,* inside the blastula is filled with fluid. The frog blastula, shown in Figure 22-7(a), is a hollow sphere of cells; the chicken blastula, shown in Figure 22-7(b), is a hollow disk with several layers of cells on top and a single layer of cells separating the embryo from the yolk. Although no organized structures are apparent in the vertebrate blastula, cleavage has provided the building blocks (cells) for the future organization of tissues and organs.

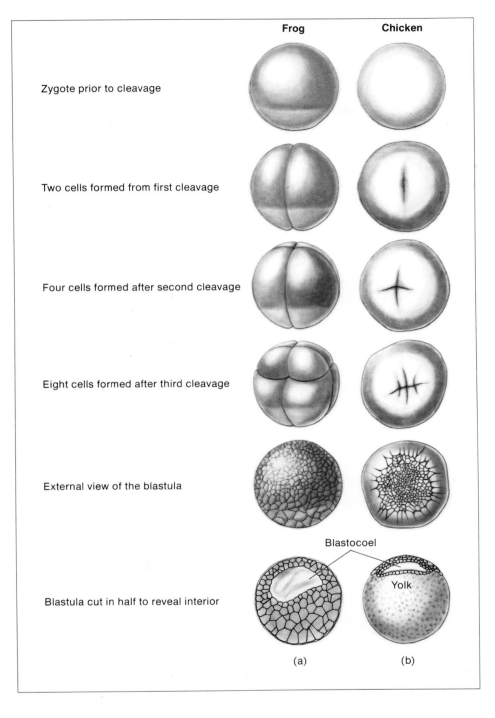

Frog **Chicken**

Zygote prior to cleavage

Two cells formed from first cleavage

Four cells formed after second cleavage

Eight cells formed after third cleavage

External view of the blastula

Blastula cut in half to reveal interior

Blastocoel

Yolk

(a) (b)

Figure 22-7
Patterns of Cleavage in Two Types of Vertebrates
Although the same types of developmental changes occur in these two forms of vertebrates, they appear to differ due to the varying amounts of yolk present in the zygote. The rate of cleavage is slower in areas where the concentration of yolk is higher.
(a) The amphibian (frog) egg contains relatively little yolk, which is concentrated in the lower half of the cleaving embryo. The top half, which contains almost no yolk, cleaves more rapidly than the bottom half. By the time the blastula is formed, the top half contains smaller cells than the bottom half. (b) The bird (chicken) egg contains a great quantity of yolk. Cleavage is restricted to a small, flat area containing little yolk where the cells of the embryo develop. The remainder of the egg contains yolk-filled cells and remains relatively inactive during cleavage.

Shortly after the blastula forms, a series of cellular divisions and major cellular rearrangements transforms it into a new structure, the *gastrula*. The onset of this change is indicated by the *involution* (turning inward) of the surface of the blastula,

Figure 22-8
The Direction of Cell Movement During Gastrulation in the Frog
Cells on the outside of the blastula move inward to form new tissue layers.

Figure 22-9
Changes Occurring Inside a Frog Embryo During Gastrulation
Only one-half of the frog embryo has been drawn here to reveal the interior. (a) The blastula, with its fluid-filled interior cavity, the *blastocoel.* (b) The indentation of one wall of the blastula to form a *blastopore.* (c) As cells move from the outside to the inside of the blastula through the blastopore, the blastocoel becomes smaller. (d) A new cavity, the *primitive gut,* begins to form. Two layers of cells are now evident—the ectoderm and the endoderm. (e) The gastrula with two layers of cells and a primitive gut. The blastopore has extended and is now blocked by yolk-filled cells. (f) Gastrulation is complete. Three layers of cells have formed—an outer layer of *ectoderm,* an intermediate layer of *mesoderm,* and an inner layer of *endoderm.*

during which about one-half of the cells move inside (Figure 22-8) to form a second layer of cells. This process, shown in Figure 22-9(a)–(d), is most clearly seen in embryos that contain little yolk. The dorsal (top) half of the organism now has two layers—an inner layer formed from the migrated cells and an outer layer formed from the cells that did not migrate inward. The opening through which the cells moved inward, the *blastopore,* eventually becomes the *anus* in vertebrate animals.

In the next stage of gastrula formation, more cells migrate to the area between the two layers of cells and establish themselves as a middle layer, as shown in Figure 22-9(e) and (f). The outer layer of cells, the *ectoderm,* will later form the nervous system, skin, and sense organs. The middle layer, the *mesoderm,* will develop into the notochord, muscles, bones, ovaries or testes, kidneys, and heart. The

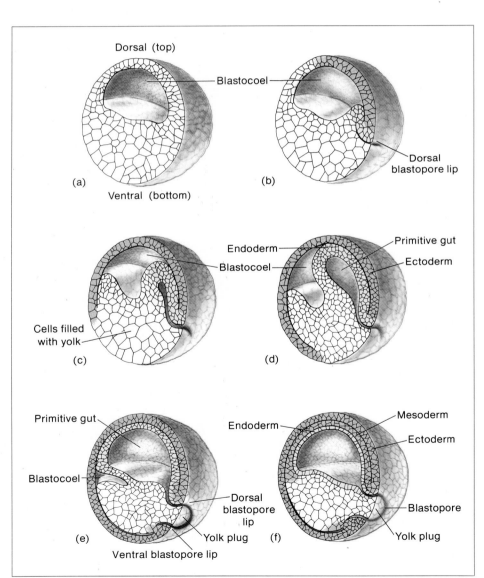

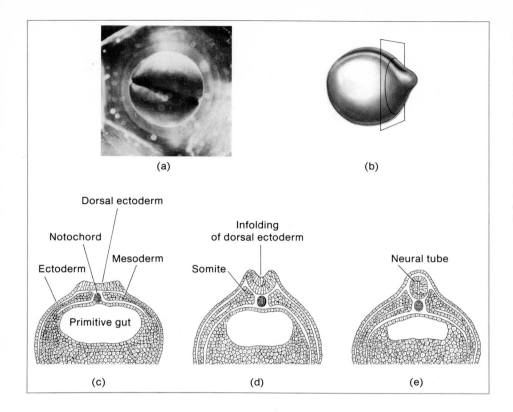

(a)

(b)

Dorsal ectoderm

Notochord

Infolding
of dorsal ectoderm

Mesoderm

Neural tube

Ectoderm

Somite

Primitive gut

(c)

(d)

(e)

Figure 22-10
Formation of a Frog Neurula
(a) Photograph of a neurula forming inside the embryonic membranes. (b) Diagram of a neurula, showing how the embryo was cut to reveal the structures shown in parts (c) through (e). (c) As the mesoderm cells form the notochord and somites, the ectoderm cells on the dorsal side of the embryo thicken. (d) The dorsal ectoderm cells fold inward. (e) The ectoderm cells form a hollow neural tube. More ectoderm cells then cover the top of the newly formed neural tube.

innermost layer, the *endoderm,* will form such internal structures as the lungs, liver, pancreas, bladder, thyroid gland, and digestive tract lining. The central cavity of the gastrula—the *primitive gut*—becomes the cavity of the digestive tract.

The formation of the gastrula changes the organism from a hollow sphere or disk of cells to an *embryo* with three separate layers of cells. The gastrula is then transformed into a *neurula*—an embryo possessing the rudiments of a central nervous system. The ectoderm cells on the dorsal side of the embryo sink inward, as shown in Figure 22-10(c) and (d), to form a hollow *neural tube* that extends from the front to the rear of the embryo and becomes covered with a layer of ectoderm cells, as shown in Figure 22-10(e). The front end of the neural tube (farthest from the blastopore) is broadest and develops into the brain; the remainder of the neural tube forms the spinal cord.

While the neural tube is forming, the mesoderm cells just beneath it become organized into a flexible rod—the *notochord,* as shown in Figure 22-10(c), (d), and (e). This structure gives support to the developing embryo and is eventually replaced by the bones of the vertebral column (the *backbone*). Most of the mesoderm cells that are not involved in the formation of the notochord become organized into blocks of tissue, the *somites,* which will form the muscles and bones of the individual.

Further development involves an elongation of the neurula and the appearance of a definite head, trunk, tail, and rudimentary internal organs. In fishes and

Figure 22-11
Three-Spined Stickleback Embryos Just Before and Just After Hatching
The embryo breaks out of its surrounding membranes and becomes a self-feeding larva.

amphibians, the embryo usually hatches from its surrounding membranes at this stage and becomes a self-feeding larva (Figure 22-11). In most amphibians, the larval stage persists until the individual reaches a certain size, at which point significant structural changes transform it into the adult form. In fishes, further development from the larval stage is a gradual transition to the mature form and involves no abrupt changes. Other vertebrate embryos continue to develop within the shell of an "egg" or the uterus of the mother until the adult pattern of tissues and organs is established and functionally integrated. When embryonic development is complete, young birds and reptiles hatch and young mammals are born.

Cellular Differentiation in Vertebrates

In vertebrate animals, genetically identical cells descended from the same zygote become differentiated in structure and function. *Cellular differentiation* is believed to be a process of selective gene expression during which specific genes in a cell are "turned on" or "off" at particular times during development. Once differentiation is accomplished, only certain genes within the cell continue to direct protein synthesis. For example, all of the cells in a vertebrate animal contain genes capable of directing the synthesis of actin and myosin and arranging these molecules into the structure of muscle fibrils; however, these genes are active only in the muscle cells.

It is not known how gene activity is regulated in animals, but the control systems are apparently passed on during cell division. When a differentiated cell divides, its descendants "inherit" the same capabilities for protein synthesis. Liver cells, for example, divide to form more liver cells; their structure and function are predetermined.

Cloning

Differentiated cells are not known to revert naturally to their embryonic, undifferentiated state, but they can be stimulated to do so under the appropriate conditions. A cell taken from the root of a mature carrot will undergo the complete developmental sequence to form another, fully formed mature carrot if environmental conditions are favorable. This process is called *cloning.*

There is also evidence that cloning can occur in animals. When the nucleus from a differentiated cell of an amphibian is placed inside a fertilized egg (of the same species) from which the nucleus has been removed, the cell can direct the development of a complete individual. This type of experiment is much more successful when nuclei are taken from younger animals than when nuclei are taken from adults. When the transplantation is effective, the nucleus of the differentiated cell behaves like the nucleus of a zygote; it undergoes cell division by mitosis, and its cells develop into all of the specialized cells found in a normal amphibian.

The psychological and moral implications of these experiments are enormous. The possibility that a large number of "offspring," genetically identical to one another and to the "parent," can be formed from the cells of an adult organism has led to increasing speculation about a future world in which genetically engineered organisms—even humans—could be mass-produced on demand.

Induction: Short-Range Interactions Among Cells

The particular specialization that each cell of a multicellular organism develops depends primarily on its location within the embryo. *Embryonic induction* is the process through which one group of cells influences the way in which another group of cells becomes differentiated.

The influence of developing cells on one another was first observed by a German experimental embryologist, Hans Spemann. Beginning in 1901, Spemann performed a variety of experiments on amphibian embryos in an attempt to determine which fragments of an embryo can produce all of the cell types necessary for the development of a complete individual. He found that if an embryo was separated into two parts prior to the gastrula stage, each part developed into a normal adult. If the embryo was divided after gastrulation, however, neither part developed normally. To determine at precisely what point during the formation of a gastrula the cells lost their ability to form individuals, Spemann divided embryos into fragments at various times during the process of gastrula formation and found that one part of a developing gastrula always produced a complete, normal individual and that the other parts yielded shapeless masses of cells. Further studies revealed that the part of the gastrula from which a normal individual developed always contained cells from the blastopore—the indented area where the cells migrate inward, shown in Figure 22-9(b).

As a demonstration of the special nature of the blastopore cells in directing differentiation, Spemann and his student, Hilde Mangold, removed the blastopore cells from one salamander blastula just prior to gastrulation and transplanted them onto another, normal salamander blastula. The host embryo, containing both its own blastopore and the foreign blastopore, developed into a sort of Siamese twin with two brains, spinal cords, and associated organs (Figure 22-12). Spemann and Mangold concluded that cells from the blastopore area cause other undifferentiated cells to begin the initial construction of an embryo. This phenomenon is now referred to as *induction.*

Another well-documented case of embryonic induction occurs during the development of a vertebrate eye. While the brain is developing as a swelling at the front end of the neural tube, rudiments of the eyes begin to form as bulges extending from the sides of the brain (Figure 22-13). The presence of a rudimentary eye (the *optic cup,* which later forms the retina) causes the surrounding ectoderm on the surface of the head to form the lens of the eye. The optic cup induces differentiation of the surface ectoderm cells into the specialized cells of a vertebrate lens. If the optic cup is cut out and placed in another position just beneath the ectoderm of the embryo, it causes a lens to develop in this new, inappropriate area.

Development is controlled not only by the particular genes the zygote receives but also by a series of inductive events. As certain cells differentiate, they influence the way in which adjacent cells differentiate, which explains how each type of tissue or organ develops within an organism in the appropriate place at a particular time. Clearly, some sort of communication among groups of cells controls which genes remain active and, therefore, the kinds of cells that develop. The exact form of communication that produces induction among groups of cells has yet to be identified.

Figure 22-12
Experimental Demonstration of Embryonic Induction in a Salamander Embryo

In this classic study by Spemann and Mangold, the dorsal lip of the blastopore of a salamander embryo in a stage of development just prior to gastrulation was transplanted onto a host embryo at the same stage of development. The presence of two blastopores on the host embryos initiated the development of two heads, spinal cords, and associated structures.

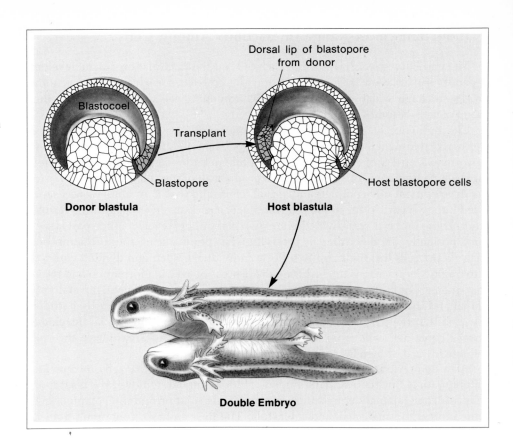

Figure 22-13
Development of a Vertebrate Eye

(a) The brain develops as a swelling at the anterior end of the neural tube. Eyes form as bulges extending from the developing brain. (b) As the rudimentary eye forms into an *optic cup,* it causes the surrounding ectoderm to fold inward and thicken. (c) A lens forms from the thickened ectoderm.

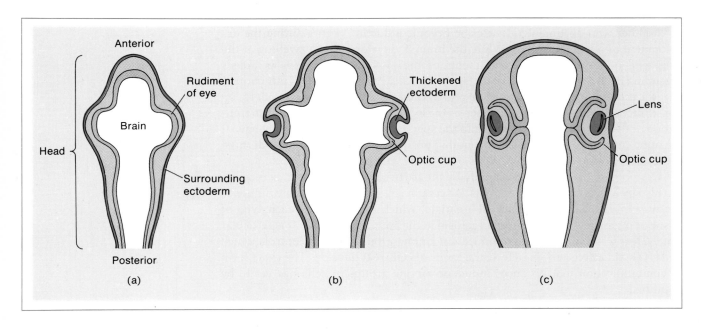

Growth in Vertebrates

The basic arrangement of cells into tissues and organs occurs at a very early stage of vertebrate development. The general structure and organization of a human is established within two months after fertilization, when the embryo is only a few centimeters in length. During the remaining months of development, these tissues and organs increase in size through cell growth and division. Vertebrate growth is regulated by the growth hormone released from the *anterior pituitary gland.*

Indeterminate Versus Determinate Growth

All swiftly moving terrestrial animals exhibit a *determinate growth* pattern, in which the individual increases in size until a characteristic age or size is reached. After this point, although some cells continue to divide, the rate at which new cells are added is approximately equal to the rate at which old cells die, and no further increase in body size occurs. These vertebrates inherit an adult body size that is critical to their way of life. Body size has a significant effect on speed of movement, the ability of an animal to capture prey and escape from predators, and the size range of prey organisms available to a predator.

In contrast, many aquatic animals exhibit an *indeterminate growth* pattern, in which the individual slowly increases in size throughout its lifetime. Large size is not a disadvantage for most aquatic animals; an individual moving through water is not as encumbered by its weight as an animal moving on land. If its size is not restricted, it is advantageous for an aquatic animal to continue to grow because larger individuals are less likely to be killed by predators and, in the case of females, are able to carry more eggs. Although unlimited size is theoretically possible and some very large aquatic animals have been found, accidents, starvation, and predation prevent most individuals from living long enough to reach an enormous size.

Rates of Growth

All parts of a developing organism do not grow at the same rate. Consequently, the percent of the body occupied by each part changes at each stage of development. The rate of growth of an organ is related to its degree of differentiation; cell division is slower in more differentiated organs. The changes in human head size with increasing age (Figure 22-14) provide an example. The brain and other features of the head grow and differentiate before other areas of the body, such as the legs, begin to develop. Once an infant is born, further growth of the head is slow, but the legs grow rapidly during the first 20 years of life.

A remarkable feature of animal development is that each organ only continues to grow until it occupies a certain percent of the adult body. The number of organs and the size of each organ are regulated in some way, so that little deviation occurs among the adult individuals of a species. For example, an adult fish may continue to increase to an enormous size as it grows older, but the number of its fins and their size relative to other body parts will remain the same.

Regeneration

Some animals have the ability to replace parts of their bodies that have been removed or destroyed. This phenomenon, called *regeneration,* is much more common among invertebrates than vertebrates. In general, less complex animals that

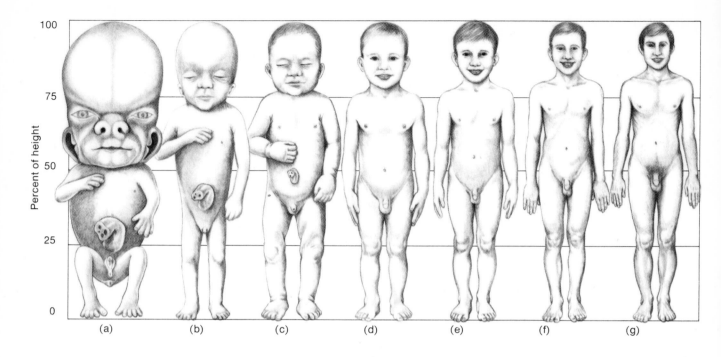

Figure 22-14
Body Proportions at Different Ages

All of these figures have been drawn as if they were of equal height, so that the percent of the total height contributed by each part of the body can be estimated. Age: (a) 8 weeks after conception; (b) 4 months after conception; (c) newborn; (d) 2 years old; (e) 6 years old; (f) 12 years old; (g) 25 years old. The percent of the body occupied by the head gradually declines with age, and the percent of the body occupied by the legs increases with age.

occupy lower positions on the evolutionary scale have a greater capacity for regeneration.

The processes involved in regeneration are similar to those involved in embryonic development. All embryonic processes, including regeneration, occur to a lesser degree in more specialized cells and tissues. Among the invertebrates, lobsters can regenerate new claws and spiders can regenerate new legs. Not only can a starfish regenerate a new ray, but if the entire body is fragmented, each part will develop into an entire starfish body (Figure 22-15). Among the lower vertebrates, salamanders are capable of regenerating an entirely new limb. Humans and other mammals have very little ability to regenerate lost or damaged parts. Any body damage larger than a small cut is often repaired imperfectly, leaving scar tissue that does not function as well as the original tissue.

Figure 22-15
Regeneration in a Starfish
Only three rays of this animal remain. It is in the process of regenerating two more rays to form a complete adult body with five rays.

Becoming an Adult Animal

Many animals change dramatically as they progress through the developmental stages to adulthood. In humans, anatomical, physiological, and behavioral alterations occur during puberty in preparation for reproduction, but these changes are relatively minor compared to the dramatic transformations that occur in other life forms. The bodies of most amphibians and invertebrates undergo major restructuring in preparation for the reproductive phase of their lives. Here, we will examine two important examples of how animals change during these final developmental stages—puberty in mammals and metamorphosis in amphibians.

Reproductive Maturity in Mammals

Newborn mammals are sexually and reproductively immature. Due to the limited space within the uterus and the small size of the birth canal, mammals are small at birth. The rate of growth in the young, which appears to be controlled by hormones in the mother's bloodstream, actually slows down just prior to birth, thereby preventing them from becoming too large to pass safely through the birth canal.

In mammals, the time from birth to reproductive maturity is devoted primarily to two developmental processes: (1) the coordinated growth of the body parts and (2) acquiring the ability to use these parts to move the body and to perform the complex behavior required for adult life.

Puberty—the time of life during which mammals attain sexual maturity—is initiated by the release of *gonadotrophic hormones* (follicle-stimulating hormone and luteinizing hormone) from the anterior pituitary gland. The age at which puberty occurs is genetically programmed and varies greatly among different types of mammals, although it may be modified to some extent by such environmental factors as stress and nutrition.

The gonads of mammals are formed very early in embryonic development but remain inactive until puberty. The release of gonadotrophic hormones stimulates further growth and maturation of the gonads at the time of puberty. The maturing *ovaries* (female reproductive organs) synthesize and release *estrogens* and *progesterone;* the maturing *testes* (male reproductive organs) synthesize and release *testosterone.* These sex hormones produce the gradual development of the *secondary sex characteristics,* or physical signs of femaleness or maleness. In humans, changes occur in the distribution of body hair, muscles, and fat deposits, as well as in the quality of the voice. The sex drive is also initiated, resulting in the emergence of sexual behavior.

Puberty and reproductive maturity do not necessarily occur at the same time. *Reproductive maturity*—the age at which a female is capable of carrying young to full term—usually occurs later than puberty. In human females, eggs may not be released from the ovaries until after the menstrual cycle is well established. Even then, the adolescent female may not be able to carry a child for the full nine months of development.

Metamorphosis in Amphibians

Many amphibians undergo *metamorphosis*—a period of major structural changes that abruptly transform an aquatic larva into a terrestrial adult. A frog embryo hatches from its embryonic membranes shortly after it reaches the neurula stage of development. A young frog larva, or *tadpole,* is a self-feeding animal, with a tail and gills, that is adapted to life in the water. Tadpoles feed on plant particles which they scrape off submerged objects with horny teeth. These frog larvae are entirely confined to water, and this developmental period is primarily one of growth.

At metamorphosis, the tadpole changes into an adult frog (Figure 22-16) with adaptations for a terrestrial life of walking on land, breathing air, and eating insects and worms. The transition involves changes in most parts of the body. The tail, fins, gills, and teeth disappear, while the legs, reproductive organs, and a tongue develop. Extensive alterations occur in the shape of the mouth, thickness and color of the skin, length of the digestive tract, and types of enzymes produced by the cells. Eyelids develop, and the eyes move from the sides of the head to the top of the head.

Metamorphosis in frogs is completed in only a few days. All of the physiological and anatomical changes involved are regulated by *thyroxine*—a hormone produced in the thyroid gland, which is, in turn, controlled by the anterior pituitary gland. Although metamorphosis is genetically programmed to occur during a particular stage of development, the timing of this period of dramatic structural change is also influenced by such environmental factors as temperature, food supply, and population density.

Figure 22-16
Developmental Changes in a
Frog After Hatching
(a) Newly hatched larva; (b) tadpole;
(c) growth of the hind legs; (d) adult
frog.

(a)

(b)

(c)

(d)

Summary

All organisms are capable of *reproduction* — the development of new individuals from existing individuals. Multicellular organisms that reproduce sexually begin life as a single cell — the *zygote* — which is formed from the union of an egg and a sperm. The development from zygote to adult involves cell division by mitosis and the specialization of the cells in the developing *embryo* in form and function according to the genetic instructions carried by the original egg and sperm. Reproduction in multicellular plants and animals can be divided into four processes: fertilization, morphogenesis, differentiation, and growth.

The *flower* — the reproductive organ in flowering plants — contains male and/or female parts that produce *haploid* nuclei. Two *sperm nuclei* are located in each *pollen grain* produced by the male parts of a flower. The *ovule*, or egg, is located in the *ovary*, which lies deep inside the female parts of a flower. The ovule contains three nuclei that are involved in *fertilization*; two nuclei become fused with one sperm nucleus to form the *endosperm*, and one nucleus becomes fused with the other sperm nucleus to form the *diploid zygote*. This form of union is called *double fertilization*.

Morphogenesis — the formation of tissues and organs from the cells descended from the zygote — begins shortly after fertilization within a flower. The endosperm, which will provide nutrients for the developing embryo, grows until it completely surrounds the zygote. The zygote then divides by mitosis to form the cells of the *suspensor* (a structure that holds the embryo in place) and the embryo itself. Three types of tissues develop in the plant embryo — the *root apical meristem*, the *shoot apical meristem*, and the *cotyledons*. Growth continues inside the ovule. At some point early in development, a *seed coat* forms from the walls of the ovule, the ovary develops into a *fruit*, and the fruit becomes separated from the parent plant. The function of the fruit is to disperse the *seeds*, which contain the embryos, away from the parent plant.

Further development of the plant embryo occurs only under certain environmental conditions, which initiate *germination*. The seed coat then breaks open, a *root* grows into the soil, and a *stem* grows upward. Leaves, additional roots, and vascular structures (the *xylem* and the *phloem*) appear. The meristem tissues produce new cells and the young plant continues to grow. A plant is capable of changing its shape throughout its life by the addition of new roots, stems, and leaves.

It is not known how *cellular differentiation*, or specialization, is controlled in plants. However, hormones appear to play a major role in coordinating developmental events in response to internal and external environmental conditions.

The appearance of flowers indicates that a plant has reached reproductive maturity. These structures develop from the shoot apical meristem in response to the presence of an unidentified hormone, tentatively called *florigen*. Flowers remain on the adult plant only until fertilization is complete and one or more embryos are developing. The ovary is then converted into a fruit, which leaves the parent plant, and the other parts of the flower disintegrate.

Plants are classified into three categories on the basis of their pattern of growth and reproduction. *Annual plants* grow, reproduce, and die within a single year; *biennial plants* grow for two years, reproduce and then die. *Perennial plants* have no defined life span and grow and reproduce year after year.

Multicellular animals that reproduce sexually develop from a single cell, the zygote, which is formed at fertilization from the union of an egg and a sperm (the *gametes*). In most animal species, there are two types of individuals—males, which have *testes* and produce *sperm*, and females, which have *ovaries* and produce *eggs*. In a few animal species, every individual has testes and ovaries and produces both sperm and eggs; these animals are called *hermaphrodites*.

The egg contains nutrients in the form of *yolk* for the developing embryo. Vertebrate development begins with *cleavage*, or cell division by mitosis, which increases the number of cells without increasing the size of the original mass. In time, a *blastula*—a hollow sphere or disk of cells about the size of the original zygote—forms. After *morphogenesis*, which involves a series of cell divisions and migrations, a new structure—the *gastrula*—emerges from the blastula. The gastrula contains the three basic types of embryonic tissues: *ectoderm, mesoderm,* and *endoderm.* The gastrula gradually develops into a *neurula*—an embryo possessing the rudiments of a central nervous system. Organs and specialized tissues then begin to form, and further development of the individual primarily involves growth.

Cellular differentiation in vertebrates apparently involves the activation or deactivation of certain genes and is governed largely by where the cell is located in the embryo. The influence of some cells on the differentiation of other cells is called *induction.* Under experimental conditions, a differentiated nucleus can be reversed to an undifferentiated nucleus by *cloning*—the process of placing the nucleus from a cell in an adult body inside an enucleated fertilized egg, where it directs the development of a complete individual.

Many aquatic animals exhibit an *indeterminate growth* pattern, in which the individual continues to grow throughout its life. Most terrestrial animals exhibit a *determinate growth* pattern, in which adult size is predetermined. Animal size is of great importance in terrestrial locomotion.

Growth takes place at different rates in different tissues and organs during development, so that the percent of the body occupied by a certain tissue or organ varies with time. The final proportions in the adult, however, are the same in all individuals of a species.

Some animals are capable of *regeneration*—the replacement of lost or injured parts. In general, this embryonic ability in adult animals is greater among invertebrates than vertebrates and greater among amphibians and reptiles than among birds and mammals.

Animals undergo changes during their transformation from fully grown immature individuals to reproductively mature adults. In mammals, this developmental stage is called *puberty* and involves anatomical, physiological, and behavioral alterations that are governed by sex hormones released from the mature *gonads*. Maturity of the gonads is controlled by *gonadotrophic hormones* released from the *anterior pituitary gland*. *Reproductive maturity*—the ability to give birth to live offspring—may occur at a later age than puberty.

Dramatic changes accompany the onset of reproductive maturity in other animals. Amphibians, for example, are transformed from aquatic to terrestrial animals. This change, or *metamorphosis*, involves a major restructuring of the body and is governed by the hormone *thyroxine*.

Human Reproduction, Development, and Maturation

23

D uring the approximately nine months from conception to birth, a child develops from a single cell into a wholly formed individual inside its mother's womb. Unfortunately, we cannot fully appreciate the precision of this orderly but remarkable transition because this initial stage of human development is an internal process.

Human reproduction begins with sexual intercourse to unite a sperm and an egg. Then, during the first nine months of human development, no special parental attention is required other than a few precautions regarding maternal diet and health. Parental care essentially begins with the birth of the infant. Human offspring are usually formed singly (a litter size of one), so that the parents only have to care for one infant at a time. As long as the mother breast-feeds her baby, her physiological condition often prevents another pregnancy.

This natural interval between successive pregnancies permits the parents to give each child the nourishment and attention it requires to survive to maturity. Unlike the young of most animals, a human infant is entirely helpless and can only cry to indicate its wants and needs.

During its long period of parental dependency, the human child develops in a protected environment, where learning begins. Learned behaviors have shaped the human life history pattern, naturally selecting for a litter size of one, a long interval between successive births, a reproductive period of approximately 30 years to permit the production of many offspring, a lasting pair-bond between mother and father, and extended lifetimes for the parents after the last child has been born.

In this chapter, we will consider the conception, development, and birth of a human child and trace the maturation of a new individual from birth to adulthood. We will then discuss various methods of birth control, genetic diseases, venereal diseases, and the aging process.

Formation of Eggs in the Human Female

The *female gametes,* or *eggs,* are formed in the *ovaries* of a woman prior to her own birth. Early in the development of the female embryo, the cells that will later

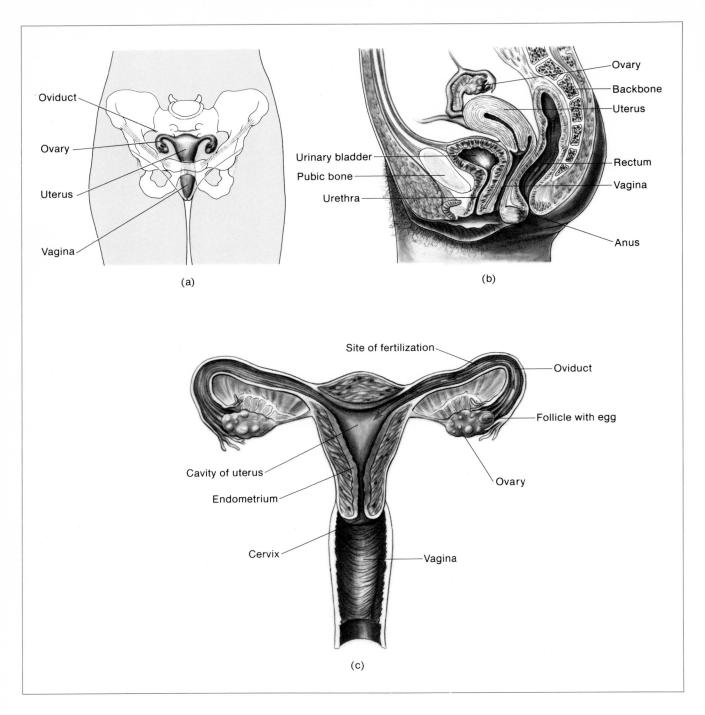

Figure 23-1
The Female Reproductive System

(a) Front view, showing the positions of the ovaries, oviducts, uterus, and vagina. (b) Side view, showing the position of the uterus between the urinary bladder and the rectum. (c) Enlarged diagram of the female reproductive structures viewed from the front.

become eggs are separated from the remaining embryonic cells and are stored where the ovaries will eventually form. At the time of birth, these cells have developed into approximately 400,000 immature eggs. No new egg cells are produced after the female is born; the number of eggs declines steadily with increasing age until the end of the reproductive period (*menopause*), when only a few eggs remain.

The female reproductive system is located in the lower abdominal cavity between the urinary bladder and the rectum (Figure 23-1). The female has two ovaries, which develop mature eggs in alternate months (only one egg normally matures each month). Inside an ovary, each egg resides within its own individual sac, or *follicle,* until it matures. Mature eggs are fairly large in size (about 1 millimeter in diameter) and contain sufficient materials to nourish the first cells of an embryo.

A mature egg is released each month from an ovary into one of the two *oviducts* leading to the uterus. These tubes are not directly connected to the ovaries, so that the egg must pass briefly through an open space between the organs of the abdominal cavity. The opening to each oviduct is expanded into a funnel that captures almost every egg released from an ovary. Fertilization of an egg by a sperm occurs in the segment of the oviduct closest to the funnel, as shown in Figure 23-1(c).

On rare occasions, more than one egg is released into one or both of the oviducts at the same time. If all of these eggs are fertilized, they may develop simultaneously and produce more than one infant. These *fraternal siblings* (twins, triplets, etc.) are as genetically similar as siblings from single births of the same parents; in either case, each child develops from a different fertilized egg. By contrast, *identical siblings* are formed from a *single* fertilized egg in which the developing embryo divides into two or more separate parts sometime during the first two weeks of development. To some extent, the tendency to produce fraternal siblings appears to be inherited; the occurrence of identical siblings is an accident of development.

The oviducts are directly connected with the *uterus*—the chamber in which the embryo develops. The uterus terminates in the *cervix.* As shown in Figure 23-1(c), the opening of the cervix projects a short distance into the *vagina*—a channel that extends to the exterior of the body where it exits between the *urethra* (a much narrower tube that carries urine) and the *anus.* The vagina receives sperm during intercourse and serves as a birth canal when the baby is delivered.

The Menstrual Cycle

From puberty until menopause, a woman's reproductive system undergoes a succession of cyclic changes. A period of *menstruation* marks the onset of each cycle, during which some cells and blood are loosened from the inner lining of the uterus and released through the cervix and vagina. Each *menstrual cycle* lasts an average of 28 days. Increased muscle contractions in the uterus and the large intestine during menstruation are caused by a group of hormone-like substances called the *prostaglandins.* If produced in excessive amounts, these hormones may be responsible for painful *cramps* during menstruation. The function of menstruation is not known.

The significant point in each reproductive cycle—the time that an egg is released (*ovulates*) from the ovary—usually occurs in the human female about two weeks prior to the onset of menstruation. All mammals except humans exhibit obvious physical and behavioral changes, known as *estrus*, during this period of *ovulation*. In other mammals, the female will accept the male for intercourse only during estrus and indicates her receptive condition to the male through various visual, auditory, and olfactory signs. The human female will, in principle, accept the male at any time and exhibits no obvious behavioral pattern when ovulation is occurring.

Regulation of the Menstrual Cycle

Four hormones regulate the menstrual cycle: follicle-stimulating hormone, luteinizing hormone, estrogens, and progesterone. The *anterior pituitary gland* produces two of these hormones. One—the *follicle-stimulating hormone* (FSH)—causes the ovarian follicle and its egg to mature. The maturing follicle produces *estrogens*—several steroid hormones of similar molecular structure that are constructed in part from cholesterol and require vitamin C to be synthesized. Estrogens stimulate cell division in the *endometrium* (the inner lining of the uterus) to repair the damage caused by menstruation and to increase the thickness of the tissue.

As the follicle grows and matures, it produces larger quantities of estrogens. High concentrations of estrogen molecules in the bloodstream stimulate the an-

Figure 23-2

Changes in the Amount of Luteinizing Hormone (LH) and Follicle-Stimulating Hormone (FSH) Produced by the Anterior Pituitary Gland During the Menstrual Cycle

Dramatic increases in the concentrations of both hormones in the bloodstream on about day 14 of the cycle causes ovulation to occur.

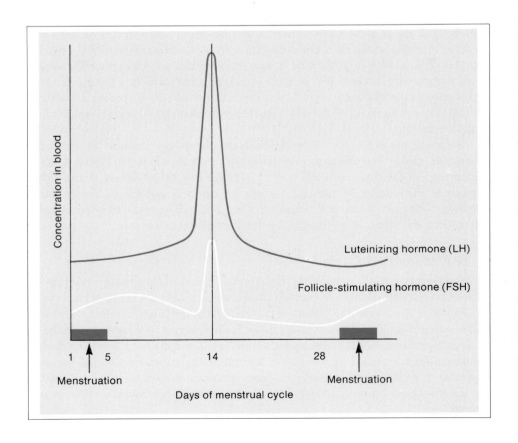

terior pituitary gland to produce a brief spurt of FSH and large amounts of *luteinizing hormone* (LH). This sudden increase in the production of FSH and LH causes ovulation to occur (Figure 23-2).

After ovulation, the anterior pituitary gland stops producing FSH but continues to release LH. The presence of LH stimulates the conversion of the ruptured follicle into another gland, the *corpus luteum*. Two types of hormones produced by the corpus luteum—estrogens and progesterone (Figure 23-3)—act together to prepare the reproductive organs for pregnancy and stimulate the endometrium to become soft, moist, and thick. *Progesterone* increases the volume of muscles and blood vessels in the walls of the uterus, stimulates the activity of mucous glands in the endometrium, and causes milk glands to develop in the breasts.

If the egg becomes fertilized and pregnancy occurs, the corpus luteum continues to produce estrogens and progesterone to maintain the uterus in a suitable condition to support the embryo. The presence of these two hormones also inhibits production of FSH by the anterior pituitary gland, so that no new follicles and eggs mature during pregnancy.

In the absence of pregnancy, the corpus luteum degenerates and the estrogen and progesterone levels decline. Without hormonal support to the endometrium, the arteries to this area constrict and the tissues are deprived of nutrients and sloughed off. Menstruation occurs, marking the end of one menstrual cycle and the beginning of another. The low levels of estrogen and progesterone no longer inhibit the production of FSH, and this hormone increases in concentration, causing

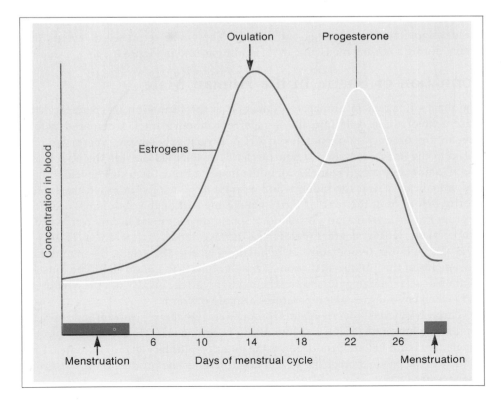

Figure 23-3
Concentrations of Estrogens and Progesterone in the Bloodstream on Different Days of the Menstrual Cycle

Hormones released from the anterior
pituitary gland regulate changes in
the ovary; hormones released from the
follicle and the corpus luteum in the
ovary regulate changes in the uterus.

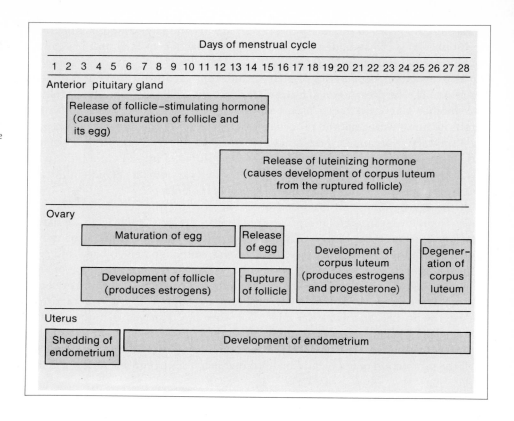

the development of a new egg and follicle. The entire sequence of events that
occurs during a normal menstrual cycle is summarized in Figure 23-4.

Formation of Sperm in the Human Male

The male's function in human reproduction is the formation of sperm and the
placement of these cells into the female reproductive tract. Compared with a
woman, the production of gametes in a man is relatively simple. Sperm are produced in the *testes,* or *testicles*—a pair of glands suspended outside the abdominal
cavity in human males (Figure 23-5). In the male embryo, the testes develop in the
abdominal cavity near the kidneys and migrate to a sac of skin called the *scrotum*
shortly before or at the time of birth. Unlike the rest of the abdominal area, the
scrotum has little or no fatty insulation. Sperm development occurs optimally at
temperatures a few degrees below the normal body temperature of 98.6°F
(37.3°C). A lower temperature is maintained inside the testes by involuntary
movements of the muscles that connect the scrotum to the body wall. The muscles
contract or relax, moving the scrotum nearer to or farther away from the heat of the
body when environmental temperatures are cold or warm.

Three hormones govern male reproductive functions: testosterone, follicle-
stimulating hormone (FSH), and luteinizing hormone (LH). In the male, as in the
female, FSH and LH are produced by the anterior pituitary gland. In males, FSH
stimulates the production of sperm and LH stimulates the production of *testosterone*
by the testes.

Although both FSH and LH are essential for reproduction, the structure and function of the entire male reproductive system depend on testosterone. Testosterone produces sperm (along with FSH) and is responsible for erection, ejaculation, and sex behaviors that result in the delivery of sperm into the female reproductive tract. A deficiency of this hormone leads to sterility.

Virtually all of the characteristics associated with being a human male are testosterone-dependent, including depth of voice and the amount and distribution of

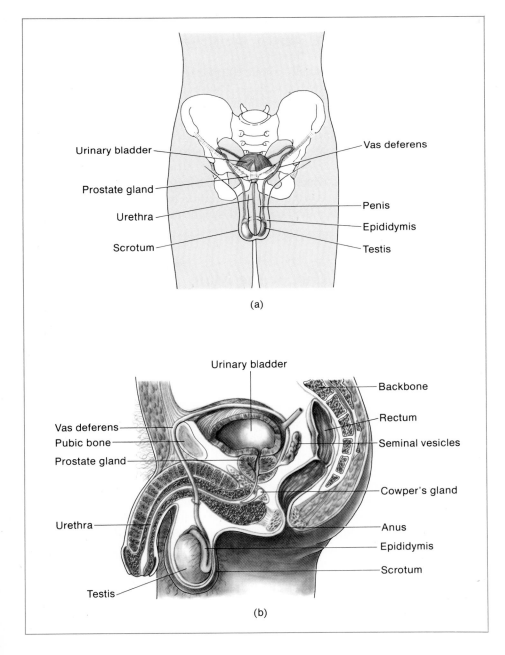

(a)

(b)

Figure 23-5
The Male Reproductive System

(a) Front view, showing the positions of the penis, testes, vas deferens, and prostate gland. (b) Side view, showing the pathway taken by the sperm after they leave the testes. Mature sperm move from the epididymis into the vas deferens, where they travel by a circuitous route around the urinary bladder to the urethra. Fluids are added to the semen from the seminal vesicles and the prostate and Cowper's glands.

Figure 23-6
The Structure of a Sperm

body hair, muscle, and fat. Although the concentration of testosterone in the blood of human males remains fairly constant, small increases may occur during sexual excitement, as evidenced by an increase in beard growth during such times. Testosterone controls male traits in other species as well, including the growth and shedding of antlers in male deer, the brilliant plumage of many male birds, the characteristic marking odor of tomcats, and even a dog's habit of lifting his leg to urinate.

There is no specific cycle of secretion of FSH, LH, and testosterone in human males; sperm essentially develop at an unchanging rate. Several hundred million sperm are produced each day by an adult male, and the average sperm develops to maturity in about 74 days. All stages of sperm formation and development are present in the testes at the same time.

A mature sperm is very small—about 100,000 times smaller than an egg, or about one-third the size of a red blood cell. The sperm cell has a head and a tail region (Figure 23-6). The *nucleus* (containing the *chromosomes*) is located in the head, which is partially covered by a *capsule*—a covering of materials that break down the cells surrounding the egg during fertilization. The region of the tail adjacent to the head contains the *mitochondria*—energy-generating structures that are important to the survival and movement of the sperm. The major structure for movement is the whiplike portion of the tail, which generates speeds that can move the sperm distances of 8–25 centimeters (3–10 inches) per minute in the female reproductive tract. Male fertility depends much more on the vigor than on the numbers of the sperm.

Delivery of Sperm Out of the Male

A copulatory organ, or *penis,* is found in all male mammals. Male bats, seals, dogs, and some other mammals have a bone in the penis to help make it rigid; in human males, rigidity is achieved solely by the circulatory system. The tissues of the penis contain a spongy mass of blood vessels. During an *erection,* arteries carrying blood to the penis enlarge and veins carrying blood away from the penis constrict, so that the penis is engorged with blood and becomes sufficiently rigid to be inserted into the female.

In human males, developing sperm move from each testis into a network of long, coiled tubes called the *epididymis,* which lies within the scrotum just outside each testis (see Figure 23-5). The sperm undergo the final stage of maturation in the epididymis and are stored there until ejaculation. Just prior to ejaculation, mature sperm move into the *vas deferens*—a long tube that extends from each epididymis up into the abdominal cavity. Inside the body cavity, each vas deferens circles around the bladder and connects with the single urethra, as shown in Figure 23-5(b). The urethra serves the dual function of transporting urine from the bladder and conducting sperm to the tip of the penis.

When stimulated by copulation or masturbation, or during spontaneous emission, the muscles in the walls of the vas deferens contract strongly, forcing the sperm to move rapidly through the vas deferens. Fluids from the seminal vesicles, the prostate gland, and Cowper's glands (see Figure 23-5) are added to the sperm to form the *semen.* About a teaspoonful of semen exits through the urethra during an *ejaculation.*

The *prostate gland* surrounds the urethra directly below the bladder and contributes fluid to the semen as well as substances that stimulate the mobility of the sperm. The *seminal vesicles,* two glands located near the bladder, secrete a fluid that contains sugar molecules to provide the sperm with an energy source. The seminal vesicles also produce and release prostaglandins (molecules similar to those that induce muscle contractions in the uterus during menstruation). The prostaglandins in the semen cause the muscles of the female reproductive tract to contract rhythmically, facilitating movement of the sperm through the uterus and toward the oviducts.

The two *Cowper's glands,* located near the urethra where it enters the penis, secrete an alkaline lubricant that is released onto the tip of the penis just prior to ejaculation. This clear, sticky material removes any urine that may be present in the urethra and also neutralizes the acidic environment of the woman's vagina, so that sperm activity will not be inhibited. Glands in the walls of the vagina also secrete an alkaline fluid that acts as a lubricant and neutralizes the vagina at this time.

Journey of the Sperm to the Egg

To reach the egg, sperm must move from the vagina through the uterus and enter the oviduct containing the egg—a journey that can take 30 minutes to 3 hours. Movement of the sperm into the uterus is chiefly dependent on the activity of the sperm themselves. Movement from the uterus to the oviducts is aided by the muscular contractions of the female reproductive tract caused by prostaglandins in the semen. Female orgasm—a poorly understood aspect of reproduction—is not essential to fertilization but contributes to sperm movement.

Millions of sperm are deposited in the vagina, but rarely more than a few hundred reach the site of fertilization. Once the sperm reach the outer third of the oviduct, where the egg is located, they must depend on their swimming movements to reach the egg.

Conception: Fertilization of an Egg by a Sperm

For fertilization to occur inside the oviduct, a sperm must pass through a layer of cells (originally part of the follicle) surrounding the egg (Figure 23-7). An enzyme contained in the capsule on the head of the sperm makes an opening in this layer, permitting the sperm to reach and attach itself to the outer membrane of the egg. The cell membranes of the egg and sperm soon fuse and become continuous, so that the sperm and the egg are enclosed within the same outer membrane. The sperm structure then degenerates, except for the nucleus which contains the chromosomes.

The egg reacts rapidly to the entry of a sperm. The outer membrane is altered immediately to make it impossible for other sperm to penetrate the egg. Protein and DNA synthesis are initiated, and the fertilized egg, or *zygote,* is chemically activated in preparation for its development into an embryo.

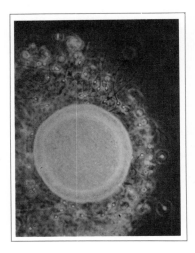

Figure 23-7
Sperm Approaching an Egg
Two of the sperm are beginning to penetrate the layer of cells surrounding the egg.

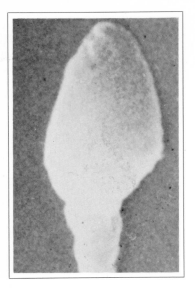

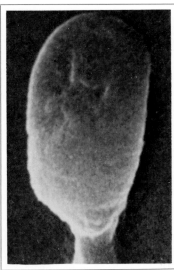

**Figure 23-8
Differences Between
X-Carrying and Y-Carrying
Sperm (× 5,600)**
The sperm on the top carries the Y
chromosome, which produces a male
offspring if it fertilizes an egg.
The sperm on the bottom carries an X
chromosome, which produces a
female offspring if it fertilizes an egg.
The X-carrying sperm is larger and is
shaped differently than the Y-carrying
sperm.

Sex Determination

Within 12 hours after sperm penetration, the nuclei of the sperm and the egg fuse together, restoring the appropriate number of chromosomes for body cells. All of the genes that will determine the characteristics of the offspring are now present in a single nucleus, and the sex of the offspring is established.

With the exception of the gametes (sperm and eggs), each human cell contains 46 chromosomes, two of which are involved in sex determination. A woman carries two X chromosomes, and a man carries an X and a Y chromosome. During gamete formation, the chromosomes of a reproductive cell separate and each gamete receives 23 chromosomes, one of which is a sex chromosome. A sperm may carry an X or a Y chromosome; an egg can only carry an X chromosome. The sex of the offspring depends on whether the sperm that fertilizes the egg is carrying an X or a Y chromosome. If the sperm carries an X chromosome, the fertilized egg will contain two X chromosomes and the offspring will be a girl; if the sperm carries a Y chromosome, the fertilized egg will contain an X and a Y chromosome and the offspring will be a boy.

Because the Y chromosome is smaller and lighter than the X chromosome, the Y-carrying sperm are also smaller and lighter (Figure 23-8) and are able to move slightly faster than the X-carrying sperm. The Y-carrying sperm also has a somewhat longer tail than the X-carrying sperm. Sperm carrying a Y chromosome therefore tend to reach the egg first, so that more males than females are conceived. At the time of *conception* (fertilization), the ratio of males to females in the human population is about 160 males to every 100 females. However, more males than females die during embryonic development, so that this ratio is reduced to about 106 males to every 100 females at birth. Throughout life, the ratio of males to females continues to decline. By the ages of 15–19, the sex ratio is equal; by the age of 85, females outnumber males by a ratio of about two to one. Genetic and environmental differences between men and women account for this discrepancy in the survival rates of the two sexes. However, evidence suggests that this ratio is changing as more women seek professional careers and encounter the high stress levels associated with many of these lifestyles.

Influencing the Sex of the Child

Although the X-carrying sperm move more slowly, they are more durable and survive longer in the female reproductive tract. These sperm also perform better in acidic environments, whereas the Y-carrying sperm function better in more alkaline environments.

Researchers have attempted to use their knowledge about the differences between X- and Y-carrying sperm to influence the type of sperm that will fertilize an egg, thereby controlling the sex of the offspring. If a daughter is desired, conditions favoring the slower, longer living sperm that carry the X chromosome should be generated by (1) intercourse no later than two days prior to ovulation, so that the shorter living sperm carrying the Y chromosome are no longer active when the egg is present in the oviduct; (2) no orgasm by the woman; (3) ejaculation when the penis is not very far into the vagina; and (4) an acid douche of vinegar and water by the woman prior to intercourse. To increase the chance of producing a son, conditions favoring the faster, shorter living sperm carrying the Y chromosome should be generated by (1) intercourse at the time of or the day following ovulation; (2) orgasm by the woman; (3) ejaculation when the penis is far into the vagina; and (4) an alkaline douche of water and baking soda by the woman prior to intercourse. These procedures are about 80% effective.

The Journey to the Womb

Once the egg is fertilized, it undergoes a series of mitotic cell divisions (*cleavage*) during the two to four days it takes to travel along the oviduct on its way to the uterus, or *womb.* The developing embryo is transported by the fluid currents produced by the wavelike actions of small, hairlike structures (*cilia*) lining the inner walls of the oviduct. Muscular contractions of the oviduct also help to move the embryo. By the time it reaches the uterus, the embryo is a cluster of approximately 30 cells.

The tiny ball of embryonic cells floats freely in the uterine fluid for several days while cell division continues. During this period of relative independence from the mother, the embryo is nourished by materials originally stored in the egg and by nutrients diffused into the fluids from the walls of the oviduct and the uterus.

In the meantime, hormones from the corpus luteum in the ovary stimulate the preparation of the endometrium for the attachment (*implantation*) of the embryo. Now a hollow sphere, or *blastula,* of about 100 cells, the embryo begins to attach itself to the inner wall of the uterus on approximately the sixth day after fertilization (about the twenty-first day since the onset of the mother's last menstrual period). Implantation fixes the embryo in place and establishes close physical contact between the tissues of the embryo and the mother.

The Placenta—Lifeline Between the Embryo and the Mother

After the embryo becomes implanted in the endometrium, the membranes surrounding the embryo secrete a hormone that suppresses menstruation. This hormone can be detected in the urine of pregnant women and can be an early indicator of pregnancy.

Five days after implantation, the embryo becomes deeply embedded in the inner lining of the uterus and begins to receive nutrients by diffusion from the endometrium. Membranes form around the embryo, and this sac fills with fluid. The embryo remains cushioned in a liquid environment in this membranous sac until birth.

By the age of two or three weeks, the embryo begins to receive maternal nutrients through the *placenta*—a system of blood vessels that develops between the embryo and the maternal circulatory system. This organ, which is found only in placental mammals, is composed of interlocking embryonic and maternal tissues that serve as a lung, digestive tract, and kidney for the developing embryo. Not all materials can move freely between the bloodstreams of the mother and the embryo, however, because the connecting vessels are separated by membranes. Food molecules, oxygen, carbon dioxide, and embryonic waste materials, as well as some drugs, anesthetics, alcohol, antibodies, and certain viruses (including the virus that causes German measles), pass between the mother and the developing child. However, larger materials, such as red blood cells and most bacteria, cannot pass across these membranes, which also serve as a barrier to most maternal hormones and thereby protect male embryos from the influence of female hormones.

In addition to providing a connection between the embryo and the mother's circulatory system, the placenta also serves as a gland and produces hormones that stimulate the corpus luteum to continue to produce estrogens and progesterone. These hormones help to maintain the uterus in a suitable condition for embryonic development. During the third month of pregnancy, the placenta itself begins to produce estrogens and progesterone and the corpus luteum gradually degenerates.

The Formation of the Fetus

The most dramatic changes in the human embryo occur during the first two months of development. As cell division continues, the cells differentiate and migrate to new positions and the specific tissues and organs that will form the basic pattern of the human individual gradually emerge. As early as the third week of development, a series of spectacular events occurs that leads to the formation of the major organ systems. The brain and spinal cord begin to develop first, followed by rudiments of the heart, blood vessels, digestive tract, and muscles.

By the fourth week, the heart begins to beat and outgrowths from the body begin to form the arms and legs (Figure 23-9). The cells that will eventually develop into sperm or eggs become separated. Branchial arches resembling a gill appear in the throat area and a definite tail develops at this time. Although the gill structures form parts of the jaw and ears and the tail eventually disappears, these embryonic formations are identical to those found in fish and other vertebrates. The fact that all vertebrates, including humans, share many features of early embryonic development is an indication of a common evolutionary origin.

Figure 23-9
A Human Embryo After Four Weeks of Development
The membranes around the embryo have been removed in this illustration to make the features more visible.

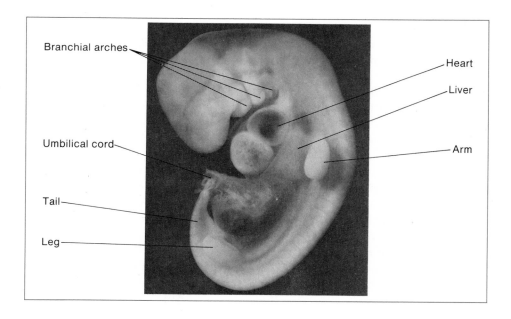

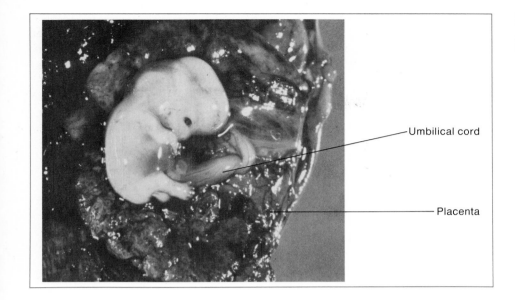

Figure 23-10
A Human Embryo After Eight
Weeks of Development

Umbilical cord

Placenta

After the fifth week of development, the major blood vessels are formed and an *umbilical cord* develops that connects the vessels of the embryo with the mother's circulatory system. During the next few weeks, the facial features emerge and the fingers and toes become evident.

By the end of the eighth week of pregnancy, the most significant aspects of development have taken place and the embryo has an unmistakable human appearance (Figure 23-10). The tail has disappeared, and the head is bulging and round and has a high forehead, a chin, and eyes. The embryo is now called a *fetus.*

The first two months of pregnancy—when the major organs form—is the most crucial period of development, but many women do not even know that they are pregnant at this time and do not take necessary precautions to ensure the normal formation of their child. If the embryo does not receive all the nutrients essential for development, certain organs may fail to develop and the fetus may be deformed. Noxious chemicals, drugs, alcohol, irradiation, major viral infections, and certain other materials that the mother may be exposed to during this critical period could also permanently damage the developing embryo. Many chemical substances can cause birth defects. Most of these chemicals, which are collectively referred to as *teratogens* ("monster producers"), are drugs or environmental contaminants such as insecticides and herbicides. Although few of these substances have been definitely linked to the incidence of human birth defects, the effects of many chemicals on the developing embryo are now being studied. Even aspirin, which is readily available and massively consumed, has been associated with some developmental malformations. When narcotics are taken by the mother, the fetus is in danger of becoming addicted as well as deformed, and an addicted newborn will suffer severe withdrawal symptoms.

At least 20% of all pregnancies end in *spontaneous abortions,* or *miscarriages,* during the first two months of development. Most naturally aborted embryos have gross abnormalities caused by genetic defects or the introduction of damaging materials into the fetus from the mother's bloodstream.

Figure 23-11
A Human Fetus After Four
Months of Development

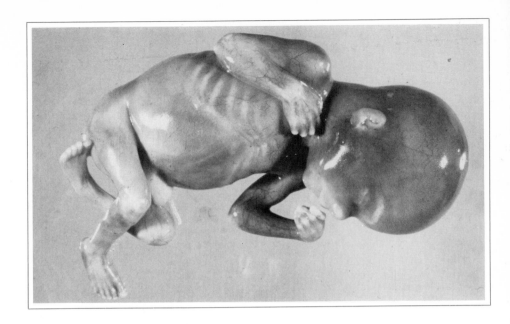

The Development of the Fetus

The remaining period of development within the mother is primarily one of growth and the perfection of the details of the body. During the third month, the ears form and paired eyelids (upper and lower) develop around each eye and fuse together. The fetus is now able to move its arms and kick its legs. By the end of the fourth month (Figure 23-11), most of the child's features are well developed. An oily coating (*vernix*) covers the skin at this time to protect the fetus from abrasions and to help maintain a constant body temperature. This coating remains on the skin until birth, when it is washed off.

By the time the fetus is seven months (28 weeks) old, it weighs about two pounds and has a fair chance of surviving outside the uterus. Primarily for this reason, a fetus is now legally considered to be an individual.

By the age of eight months, the eyelids have separated and, if premature birth occurs, the fetus is able to perceive light. At this stage of development, the fetus swallows about a pint of the fluid (giving it "hiccups," which the mother can feel) that surrounds it every day and may suck its thumb. These activities prepare the fetus for feeding after it is born. Most of the fluid that is swallowed is absorbed into the fetal bloodstream from the digestive tract. Some unusable materials in the fluid collect in the intestines and are defecated shortly after birth. Prior to birth, the fetus acquires antibodies from its mother's blood, so that the baby is born with some level of resistance to infections. Additional protective substances are later transferred through the mother's milk during breast-feeding.

Birth: The Appearance of the New Individual

About 266 days after conception (or 280 days from the onset of the mother's last menstrual period), the pituitary gland in the fetus produces hormones that induce the maternal tissues to release prostaglandins, and these molecules, in turn, stimulate the onset of *labor.* The later stages of labor, during which very strong contractions of the uterus occur, are controlled by hormones produced by the mother.

Just prior to labor, the entire contents of the uterus shift downward, bringing the fetus into contact with the cervix. The first rhythmic contractions of labor cause the cervix to dilate, and it eventually becomes large enough for the infant to pass through. Usually the baby's head pushes against the dilating cervix, acting as a wedge to increase the size of the opening to the vagina. During this stage, the membranes surrounding the fetus rupture and the fluids in the membranes are expelled through the vagina.

Uterine contractions gradually increase in strength and frequency from once every 20 minutes to once every minute near the end of labor. As the cervix continues to expand, the top of the baby's head becomes visible in the opening. Once the head emerges, the body, which is smaller in diameter, passes readily through the cervix and the vagina (Figure 23-12). The head takes much longer than the rest of the body to pass through the birth canal. If the birth is normal, the expulsion stage may last from a few minutes to an hour.

Delivery is easier when the baby's head is oriented downward and emerges first, which occurs in about 80% of all births. If the baby is upside down (*a breech delivery*), passage is much more difficult and the infant may be temporarily deprived of oxygen. During birth, the baby receives oxygen through the umbilical cord,

Figure 23-12
Childbirth

(a) and (b) The head turns sideways as it moves through the cervix.
(c) After one shoulder has emerged, the position of the body rotates.
(d) The rest of the body, which is smaller in diameter than the head, moves out of the cervix and the vagina easily.

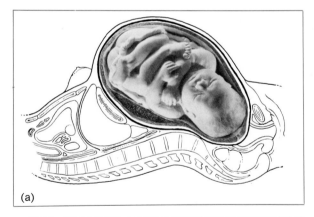

(a)

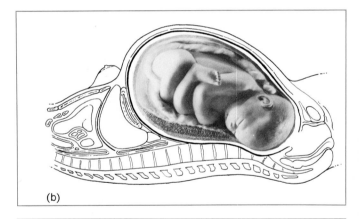

(b)

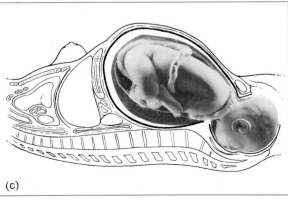

(c)

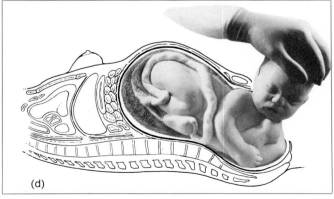

(d)

which is now about 50 centimeters (20 inches) long and is still attached to the placenta in the uterus. In a breech delivery, the umbilical cord is squeezed between the baby's head and the cervix and may be constricted long enough to interrupt the flow of oxygen to the baby. Oxygen deprivation can cause tissue damage, especially in the brain. A doctor or midwife can often reorient a baby in the early stages of labor so that the head emerges first.

After the infant is born and the umbilical cord is cut and tied (Figure 23-13), a series of adjustments within the newborn's circulatory system reorient the flow of blood to permit it to breathe through its lungs. The lungs begin to open and

Figure 23-13
A Child is Born
After birth, the umbilical cord is cut and tied.

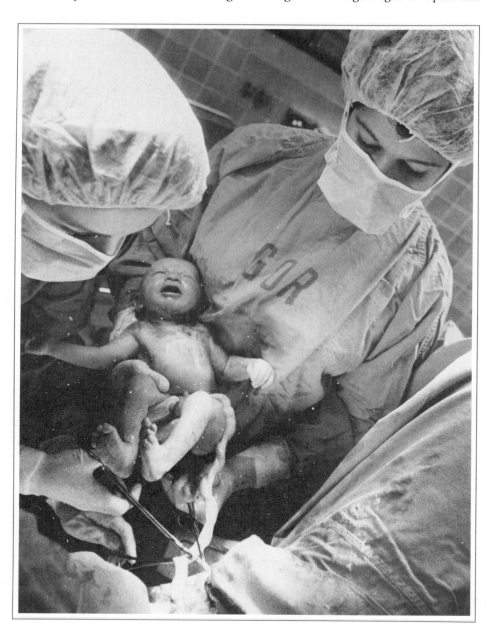

become functional when the baby takes its first breath, which happens once it begins to cry (spontaneously or when encouraged by a helpful slap).

After childbirth, the uterus continues to contract until blood, fluid, the placenta, and some maternal uterine tissues are expelled as the "afterbirth." The placenta is now a flat, circular tissue about 23 centimeters (9 inches) in diameter and almost 3 centimeters (1 inch) thick. After expulsion, the hormones produced by the placenta are no longer released into the mother's bloodstream. This sudden drop in progesterone and estrogen levels stimulates milk production in the breasts.

The removal of the placenta leaves a raw, bleeding wound on the inner wall of the uterus. The danger of the mother becoming infected just after childbirth is considerable, but sterile techniques and the administration of antibiotics have greatly reduced this danger. After several weeks, the wound heals.

Natural Childbirth

More and more women are choosing to have their babies by *natural childbirth* without relying on pain-killing drugs or anesthesia. In addition to protecting the infant from any detrimental side effects of drugs, these women feel that childbirth will be a richer and more satisfying experience if they remain alert. Painkillers and anesthetics administered to the mother during labor decrease the infant's heart rate and delay the onset of breathing. Both of these effects decrease the rate of oxygen flow to the tissues and may damage some of the infant's brain cells. The infant's sucking response and the mother's milk production are also depressed by drugs and anesthesia.

Of course, there is some pain associated with natural childbirth. However, psychological and muscular conditioning prior to giving birth teaches the mother to relax, thereby reducing her fear and permitting her to control the level of pain to some degree.

Nourishing the Infant

The natural way to nourish an infant is to feed it milk from its mother's breasts (Figure 23-14). Over the past several decades, however, doctors felt that there were no important differences between human milk and formula substitutes, and most mothers chose the more convenient method of bottle-feeding instead of breast-feeding their babies. We now know that no formula can provide a complete substitute for human milk, and breast-feeding in the United States is increasing as more and more young women return to the natural processes of reproduction. By contrast, women in other countries are changing from breast-feeding to bottle-feeding. In Chile, for example, 95% of all mothers breast-fed their babies 20 years ago; today 80% bottle-feed their babies.

Breast-feeding occurs in all mammals; the milk of each species is especially adapted to the nutritional needs of the young. The composition of human milk is uniquely suited to the needs of the growing human infant. It is easily digested and provides protection against food allergies and infections, particularly in the intestinal tract. Human milk is also inexpensive, hygienic, and requires no preparation.

The mother derives some physical benefits from nursing as well. Breast-feeding reduces the uterus to its pre-pregnancy size more quickly. The nursing mother

Figure 23-14
The Natural Way to Nourish
an Infant

also loses weight more rapidly after childbirth because she expends about 1,000 calories a day in milk production alone. Moreover, breast-feeding tends to inhibit ovulation and therefore acts as a natural—although not a highly reliable—contraceptive.

Growing Up

Biologically, an individual who is capable of producing offspring is an *adult*. The major physical changes that occur between birth and adulthood are the maturation of the nervous system, development of the reproductive organs, and growth of the entire body. The brain and spinal cord are not well developed at birth and it takes many years for these tissues to mature. A child gradually acquires the ability to walk, talk, and develop muscular control. Many aspects of the social system are also learned during childhood. Genetic and cultural forces continually interact to mold the life of the developing child.

We know that infants do not thrive if their mothers are hostile or merely indifferent to them. Emotional deprivation in infancy can stunt the physical as well as the psychological development of a child. Without adequate infant–parent interaction, a child exhibits abnormal sleep patterns that inhibit the secretion of the pituitary hormones, including the hormone that controls the individual's growth rate. Children in emotionally deprived environments can become physically stunted because their pituitary glands release insufficient amounts of this growth hormone.

Normal human growth is a regular process that continues until the individual's early twenties. However, a dramatic increase in the growth rate occurs in early adolescence. The reason why one child grows faster or to a greater adult height than another child is still unknown. We do know that both genetic and environmental factors (especially diet) are important in determining an individual's growth rate and that several hormones are essential for normal growth. On the average, the growth spurt occurs when a girl is about 11 and a boy is about 13 years of age.

At about the same time that the growth spurt is under way, males and females begin to undergo distinct physical changes. The development of the ovaries or testes greatly increases the production of sex hormones and alters the body profoundly. In boys, growth of the penis and testes accelerates; in girls, the breasts begin to develop. In both sexes, pubic hair appears for the first time and fat and muscle are redistributed throughout the body. This time of sexual development, called *puberty,* extends over a period of 2–4 years. At the end of puberty, *sexual maturity* is achieved and the individual can produce mature eggs or sperm.

The rates at which children mature to adulthood vary greatly. These differences are present at all ages, but their effects are most dramatically apparent during puberty. Both the ability to learn and the ability to perform athletic sports are much more dependent on the maturity level than on the age of an individual. Because children are educated according to their age, rather than their degree of maturity, being an early or a late maturer can have considerable and lasting effects on an individual's emotions and performance.

Methods of Birth Control

The explosive growth of human populations over the last few decades is primarily the result of our conquest of childhood diseases and improved standards of nutrition and sanitation. Most children in modern cultures survive to adulthood, whereas most children in previous cultures did not. If humans continue to reproduce at the present rate, our resources will not be sufficient to support our population densities. To stop the rapid increase in human populations, the death rate must increase or the birth rate must decrease.

Permitting the death rate to increase is an unthinkable solution to the problem of overpopulation, but most people view birth control as an acceptable way of limiting the human growth rate. Our knowledge of the human reproductive process allows us to prevent or arrest the formation or development of a child in several ways. These methods of birth control are listed in Table 23-1 and discussed here in the same sequence in which the reproductive process occurs—from gametes to fetus.

Table 23-1
Comparison of Some of the Most Commonly Used Methods of Birth Control

Method	Where obtained	Reliability	Advantages	Drawbacks
Rhythm	Available to anyone	Depends on regularity of menstrual cycle	Only method sanctioned by the Roman Catholic Church	Frequent failures; requires abstinence during the female's fertile periods
The pill	By prescription	99.5% effective	Affords sexual freedom; highly reliable	May cause cancer, blood clots, and liver and/or heart disease; must be taken regularly
Mini-pill	By prescription	97% effective	Does not contain estrogen, eliminating many side effects of the pill	Long-term effects unknown
Vasectomy	Doctor's office or hospital	Fail-proof once absence of sperm in the ejacula is established	Minor operation; sexual freedom; full reliability	Difficult to reverse; may contribute to arteriosclerosis
Condom	Drug counters	98% effective	No health hazards	Can slip or break; interrupts foreplay and may lessen male pleasure
Tubal ligation	Doctor's office or hospital	Almost 100% effective	Sexual freedom; full reliability	Operation usually requires hospitalization; difficult to reverse
Diaphragm	By prescription	98% effective when used consistently with a spermicide	No health hazards; does not diminish sexual pleasure	Must be inserted properly; may inhibit spontaneity
Aerosol foams	Drug counters	80% effective	No prescription required; no health hazards	Must be applied shortly before intercourse
Jellies, creams and suppositories	Drug counters	Varies, but less effective than foams	No health hazards	Not fully reliable
Intra-uterine device (IUD)	By prescription	98% effective	Sexual freedom; no side effects from drugs	May puncture uterus; may cause bleeding, cramps, and other discomforts; may be unknowingly expelled
Morning-after pill	By prescription	98% effective	Eliminates pregnancy after intercourse. *Prescribed only in cases of incest or rape.* Too dangerous for regular use; increases estrogen levels more than 50% above normal levels.	
Abortion by menstrual evacuation	Doctor's office or clinic	Not a preventative	Ends unwanted pregnancies; painless, relatively inexpensive procedure	Must be performed within three months (first trimester) of pregnancy

The Rhythm Method

The *rhythm method* of birth control was practiced for many centuries prior to any real understanding of the biological nature of the menstrual cycle. This method was based on the incorrect assumption that pregnancy occurred as a result of the union of semen from the man and menstrual blood in the woman. To avoid preg-

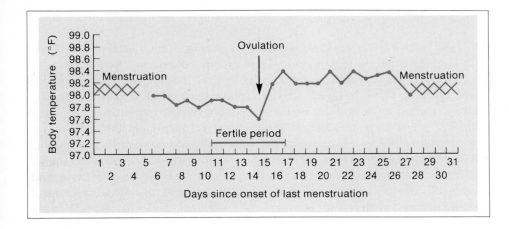

Figure 23-15
The Rhythm Method
of Contraception

In some women, body temperature
increases distinctly just after
ovulation. Intercourse immediately
prior to and during this temperature
change will probably result in
pregnancy. To be reasonably sure of
not becoming pregnant, a woman
should avoid intercourse from day 11
through day 16 of the menstrual cycle.

nancy, a woman was advised to abstain from intercourse during menstruation.

By the early 1930s, it was generally understood that ovulation does not occur during menstruation but midway between successive menstruations, and the modern rhythm method of contraception was developed. To use this method effectively, however, a woman must know precisely when ovulation is occurring. Although the time between the release of an egg and the onset of menstruation is quite exact (14 days), the length of time between menstruation and ovulation is variable and can be affected by many environmental factors. Severe stress, for example, can inhibit ovulation for weeks, months, or even years.

In some women, body temperature rises approximately 0.7°F just after ovulation (Figure 23-15). If this increase is charted for several months and occurs in a regular pattern, a woman can estimate when she normally ovulates to a fairly accurate degree. Intercourse can then be regulated so that sperm and egg are not present in an oviduct at the same time. An unfertilized egg survives about 24 hours after it is released from the ovary; a sperm can live as long as three days inside the woman, but can fertilize an egg only for the first day or two. If intercourse is avoided for several days prior to ovulation and for at least a day following ovulation, the egg will probably not be fertilized.

Unfortunately, only about 25% of all adult females exhibit a mid-cycle temperature rise, and even these women do not always ovulate on schedule. Emotions and changes in habit often delay ovulation, and some women may actually ovulate in response to intercourse. An average of 25 pregnancies per year can be expected to occur among every 100 women who use the rhythm method of contraception. Until a more precise indicator of the time of ovulation—or, better still, an indicator of the time just prior to ovulation—is found, the reliability of the rhythm method will continue to depend on the regularity of the individual woman's menstrual cycle.

Hormonal Regulation

Birth control pills were developed in the late 1950s and began to be used on a widespread basis in the 1960s. All of these pills contain synthetically produced estrogens and progesterone, but the relative amounts of these hormones vary in different brands of the pill. The dosages of estrogens and progesterone in the pill are equal to or less than the amounts that occur naturally, but the pill produces hormone levels that do not fluctuate the way naturally occurring hormone levels do.

These constant levels of estrogens and progesterone inhibit the production of FSH and LH, so that the eggs do not mature and are not released from the ovaries.

Progesterone is an effective contraceptive by itself, but it usually causes spontaneous bleeding from the uterus at times other than menstruation. Estrogen is added to the pill to prevent this excessive bleeding, but estrogen also causes most of the pill's undesirable side effects. Minor effects include weight gain, nausea, and skin blotches; more serious problems that may be associated with the birth control pill are the development of blood clots, cancer, and heart disease. Because the pill has been in use for such a short time, the long-term risks of developing these more serious disorders is not yet known. Evidence does indicate, however, that a woman should not take a birth control pill regularly for several successive years or smoke cigarettes when using the pill. The pill also suppresses milk production and cannot be used to prevent pregnancy in mothers who are breast-feeding.

A recently developed oral contraceptive—the *"mini" pill*—contains less than one-third the progesterone in other pills and no estrogen. Although the small amounts of progesterone in the mini-pills do not cause bleeding, they do not prevent conception as reliably as the progesterone–estrogen pills. The primary effect of the mini-pill is to increase the amount of mucus on the cervix, which hinders the migration of sperm into the uterus. However, any sperm that do enter the uterus may encounter a receptive egg.

Except for sterilization, the progesterone–estrogen pill is the most effective means of preventing conception. The failure rate of this birth control pill is one pregnancy per year per 500 woman.

Attempts are being made to develop a birth control pill for men that would prevent the production of mature sperm. However, any alteration of the male hormones would cause profound changes in male appearance and behavior—clearly undesirable side effects.

Preventing Sperm from Entering the Female

An almost totally effective method of birth control in the male is a *vasectomy*, in which the vas deferens is blocked on each side of the scrotum to prevent sperm from leaving the body at ejaculation. A vasectomy is a simple operation that can be completed in a few minutes. Many different methods of closing the vas deferens are used; they may be cut and tied or blocked with a metal clip or beads. A new technique—the insertion of a tiny plug fitted with a valve that can be turned on and off—makes the vasectomy reversible. The penis and the testes are not involved in this operation (only the tissues of the vas deferens and scrotum are cut), so that hormone production, erection, and ejaculation are unaffected. However, the ejaculated fluid does not contain sperm; apparently the sperm degenerate in the epididymis, and their components are absorbed into the bloodstream.

Males have practiced two other contraceptive methods for centuries—the withdrawal of the penis from the woman just prior to ejaculation and the use of condoms. The *withdrawal method* is unreliable because some sperm may enter the vagina before actual ejaculation. Moreover, the timing of withdrawal is critical, because the first drop of semen contains more sperm than later drops. A more reliable method is to place a *condom*—an "envelope" through which sperm cannot penetrate—around the penis. Because they do not permit direct contact between male and female tissues, condoms prevent the transmission of venereal diseases (to be discussed in a later section) as well as conception.

Preventing Sperm from Reaching the Egg

Mechanical, physical, or chemical barriers along the female reproductive tract can prevent sperm from reaching the egg. The most effective barrier is a *tubal ligation,* in which the oviducts are blocked, cut, or tied off. This operation is more serious than a vasectomy, because surgical incisions must be made in the abdominal wall or the vagina and a general anesthesia must be administered. A tubal ligation has no effect on hormone production, female behavior, or appearance.

A *diaphragm*—a flexible, dome-shaped, rubber hemisphere that covers the cervix—is a temporary barrier designed to prevent sperm from entering the uterus. The diaphragm is inserted through the vagina prior to intercourse and effectively stops sperm penetration when used in combination with a chemical that kills sperm (*spermicide*). Used alone, neither spermicides nor a diaphragm are particularly effective in preventing conception. A douche (liquid used to flush out the vagina) after intercourse is ineffective because the sperm reach the uterus within a few minutes after ejaculation.

Methods for Preventing Implantation

The implantation of the embryo (and, therefore, pregnancy) can be prevented by means of an *intra-uterine device* (IUD) or the *morning-after pill.* An IUD is a small plastic device that is inserted into the uterus by a physician to prevent embryonic implantation. Exactly how an IUD works is not known. More recent models (the "Copper T" and the "Copper-7") are designed to release small amounts of copper or progesterone continuously and are proving to be much more effective in preventing pregnancy than earlier models.

The morning-after pill contains large amounts of estrogens, which increase muscle contractions in the oviducts. Should fertilization occur, these contractions transport the developing embryo through the oviduct so rapidly that it reaches the uterus too soon. The fluids in the uterus at this stage of pregnancy are usually toxic to the embryo and the endometrium has not yet been prepared to accept implantation. When implantation fails to occur, the tiny embryo is discharged through the vagina.

Abortion

Induced abortions by *menstrual evacuation* during the first three months of pregnancy (the first *trimester*) are relatively simple, safe, and inexpensive. During this procedure, which can usually be conducted in a physician's office or a clinic, the inner lining of the uterus is sucked out by means of a vacuum device.

After the third month (the twelfth week) of pregnancy, abortions cannot be performed by menstrual evacuation and the fetus can only be aborted by inducing premature labor, which is accomplished by injecting a salt solution into the uterus or by administering prostaglandins to stimulate strong uterine contractions. These abortions are not as safe as menstrual evacuations and usually require a hospital stay of one or two days.

Abortions after the age of six months (24 weeks) are not legally permitted in the United States unless the mother's life or health is endangered or the baby is known to be grossly abnormal.

Genetic Diseases

Genetic defects can be detected as early as the middle of the third month (about the tenth week) of pregnancy by inserting a needle through the lower abdominal wall into the membranous sac that surrounds the fetus and withdrawing a small sample of the fluid surrounding the fetus (Figure 23-16). At this stage of development, an ample amount of fluid is present and the fetus is still quite small. The process of *amniocentesis* involves some risk and should not be done without a good reason, but there is less than a 1% chance that the procedure will cause a spontaneous abortion, injury to either mother or fetus, or an infection. Body cells that have been shed by the fetus are isolated from the collected fluid, grown in a laboratory culture, and analzyed for abnormal chromosomes and metabolic disorders. The sex of the fetus can also be determined by amniocentesis.

Figure 23-16
Amniocentesis—Examining the Genes of the Fetus
Genetic defects can be detected by employing this procedure to check the shape and the number of chromosomes and the products of cellular metabolism present in the cells within the fluid surrounding the fetus.

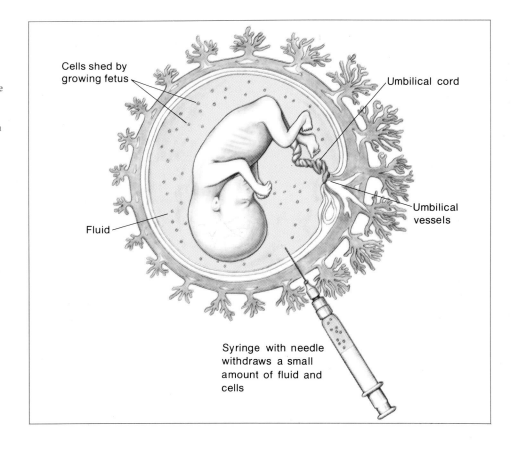

Cells shed by growing fetus

Umbilical cord

Umbilical vessels

Fluid

Syringe with needle withdraws a small amount of fluid and cells

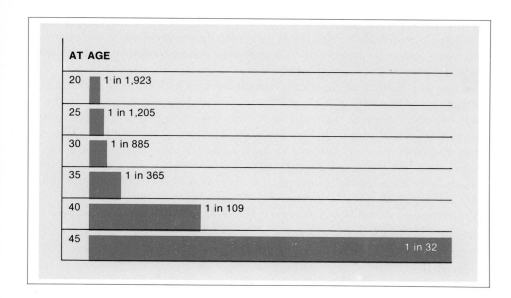

Figure 23-17
The Risk of Down's Syndrome, a Genetic Disease, in Relation to the Age of the Mother
Older women are more likely than younger women to give birth to a child with Down's syndrome—a serious genetic defect caused by an extra chromosome. A child with this disease suffers from severe mental retardation.

A genetic disease or abnormal embryonic development affects 5% of all newborns in the United States. *Genetic counseling* helps people to understand their genetic origins and what influence their genes may have on their children. This counseling procedure begins with a study of the health histories of both families. When a possible genetic disorder is revealed, cell samples from both potential parents are usually tested for the presence or absence of certain genes. If the woman is already pregnant and the initial analysis indicates that the child may be carrying defective genes, she may choose to undergo amniocentesis.

People who have a higher risk of producing children with genetic disorders and who should definitely seek genetic counseling include women over 35 (Figure 23-17), couples with a confirmed genetic disease in either family, couples who already have a child with a genetic disease, women with a history of two or more miscarriages, pregnant women who have had x-rays or a viral infection or who have taken harmful drugs early in their pregnancies, cousins or other blood relatives who plan to have children, and couples from ethnic groups known to have a high incidence of a particular disease.

Many babies who have genetic disorders appear to be normal at birth. In such cases, by the time the disease is recognized, it has usually caused irreversible damage. However, when a genetic disorder is detected by amniocentesis, treatments as simple as special diets, if started early enough, often ensure the normal development of the child.

Sexually Transmitted Diseases: Prevention and Treatment

The enormous increase in the occurrence of *venereal* or *sexually transmitted diseases* in the past two decades is largely the result of the use of birth control pills and intrauterine devices (IUDs), rather than condoms, as methods of contraception. Venereal diseases can be transmitted homosexually as well as heterosexually and can occur in the throat and the rectum as well as in the reproductive tissues. The transmission of all venereal diseases can be prevented by using a condom during sexual intercourse and, to a lesser extent, by washing the areas of contact thoroughly with medicated soap and water after intercourse.

Every year, millions of Americans become infected with syphilis, gonorrhea, or nongonococcal urethritis (NGU). All three of these venereal diseases are contagious bacterial infections that are passed from person to person through contact between the sexual organs. In rare cases, syphilis can be transmitted by kissing when one of the partners happens to have a syphilis sore on the mouth. In the United States, NGU is now the most common infectious disease, gonorrhea is second, and syphilis is fourth.* However, it is likely that all sexually transmitted diseases are more widespread than statistics indicate, because many cases are probably not reported. About 50% of all reported cases of venereal disease occur between the ages of 15 and 24.

Syphilis

The first symptoms of *syphilis* appear 10–90 days after exposure to an infected person. Painless sores develop on the skin at the point of physical contact between the sexual organs. They do not itch and may be so small that they are unnoticed, ignored, or mistaken for measles, chicken pox, or poison ivy. These sores usually disappear in about two weeks, even without medical treatment, but the syphilitic bacteria move from the skin into the bloodstream.

The second stage of infection occurs 3–6 weeks after the sores disappear. By that time, a large population of the bacteria has formed in the bloodstream, causing a fever, the loss of small patches of hair, a sore throat, a skin rash (usually on the soles of the feet and the palms of the hands), and fatigue. During the third and final stage of a syphilitic infection, which may occur years later, the bacteria invade the internal organs and blindness, heart attack, bone and liver ailments, brain damage, and death may result.

Syphilis can be detected in its early stages by a blood test. Once diagnosed, the disease can be treated effectively with antibiotics; penicillin is usually prescribed. In the near future, researchers may develop a vaccine to combat this serious disease.

Gonorrhea

Gonorrhea is such a common venereal disease because about 15% of all infected men and 85% of all infected women exhibit no symptoms and unknowingly transmit the disease to others. The symptoms of gonorrhea in men are a yellowish discharge from the penis and a painful, burning sensation during urination. Infected women may observe a thick, white vaginal discharge, but this symptom is common of many other, less serious, gynecological conditions. The scar tissue that remains after a gonorrheal infection can result in sterility, difficulties in urinating, or a painful form of arthritis.

Two methods of detecting gonorrhea are currently available—a blood test or an analysis of fluid from the vagina or discharge from the penis for the presence of

*The third most common infectious disease is the common cold.

gonorrhea bacteria. Like syphilis, gonorrhea is treated with antibiotics. Recent strains of this bacteria have evolved a resistance to penicillin but can still be effectively treated with other antibiotics.

Nongonococcal Urethritis

Both the symptoms of *nongonococcal urethritis* (NGU) and its effects on the body are similar to those of gonorrhea. *Urethritis* is an infection of the urethra that usually occurs in males who have NGU. This venereal disease acquired the label "nongonococcal" because no gonococcus bacteria (those causing gonorrhea) were found in the blood or discharge of many persons believed to have gonorrhea. The only symptom known to distinguish NGU from gonorrhea is that the discharge from the penis is clear, rather than yellow in color.

Although the specific type of bacteria causing NGU is not always identified, in about 50% of all cases it has been found to be *Chlamydia trachomatis*—the same bacteria that causes a serious eye disease. An infection of these bacteria in the reproductive tract of a pregnant woman may cause eye and lung infections in her newborn.

The incidence of NGU in the United States is increasing rapidly. Almost all sexually transmitted diseases treated by college health services in recent years have been NGU infections. Diagnosis of this common disease remains difficult. At present, when the patient exhibits the symptoms of gonorrhea but no gonococcal bacteria can be found, the disease is diagnosed as NGU. The bacteria causing NGU are not affected by penicillin. The disease is treated with other kinds of antibiotics, particularly tetracycline.

Herpes

This sexually transmitted disease is caused by a virus, rather than a bacteria. *Herpes sores* (or *fever blisters*) may develop on the face, especially around the mouth and nose, or in the genital area. Herpes sores on the face are usually (but not always) caused by a different strain of virus than herpes sores in the genital area.

In men, venereal herpes sores appear on the penis and in the urethra; in women, the sores can be external or may appear on the walls of the vagina and cervix. Herpes sores begin as tiny, fluid-filled blisters that eventually erupt and spread the infection throughout the body. The more widespread infection is painful; symptoms include fever, chills, and a swelling of the lymph nodes. Once an individual has been infected with herpes, the virus never leaves the body but remains dormant in the nerve endings between eruptions.

Pregnant women who have active venereal herpes usually miscarry. If the sores remain dormant throughout pregnancy but become active at the time of childbirth, the infant usually dies or suffers from brain damage unless delivery is by *Cesarean section* (surgical removal through an opening in the abdomen), so that the baby does not pass through the infected cervix and vagina.

No current medication is effective against a herpes infection. Antibiotics cannot combat a viral infection, and patients can only be medicated to reduce the pain. Women who have genital herpes are advised not to become pregnant for at least a year after the sores disappear.

Aging—Deterioration of the Individual

All things deteriorate with time, but the same mechanisms do not cause deterioration in living and nonliving materials. Biological material has a unique capacity for *self-renewal* and *self-maintenance*—for preserving its structural organization by utilizing a continual input of energy and nutrients to reform and replace its parts according to specific genetic instructions. Biological organization is also maintained by *homeostasis*—the promotion of a constant internal environment.

Natural Selection and Mortality

Senility, or the progressive decline of biological functions with age, may not be the universal fate of all living matter. Many life forms do not physically deteriorate as they grow older. These apparently immortal organisms are not confined to a particular taxonomic group but are scattered throughout the world of life. (Examples include the bristlecone pine tree, the flatworm, and the sea anemone.) In all of these organisms, reproduction continues throughout life.

The processes of aging and death are not well understood. We still do not know why biological functions decline with increasing age or whether or not death is an inevitable consequence of life. Most organisms are *mortal;* they live for a limited period of time and then die. Both genetic and environmental factors contribute to determining the length of life. When environmental conditions are optimal for survival, the *maximum life span* of an individual appears to be approximately the same in all populations of a given species, suggesting that length of life, like puberty and menopause, is genetically programmed.

The onset of the aging process is closely associated with the cessation of the reproductive function (including parental care) in mortal organisms. Once reproduction ceases, the forces of natural selection can no longer operate to remove detrimental genes from a population. Traits that appear late in life do not affect reproductive performance and cannot be selected for or against. Tissues may be genetically programmed to function and renew themselves only for a set period of time—from birth through the reproductive ages. When individuals of a mortal species are no longer reproductively functional, the adaptations that promote their survival gradually begin to deteriorate and they become increasingly vulnerable to death from starvation, predation, accidents, diseases, and adverse weather conditions.

The association between the termination of reproduction and the onset of senility is most evident in organisms that reproduce only once. Annual plants deteriorate and die soon after releasing their seeds. The adult forms of many insects do not have feeding structures; they are destined to die of starvation after reproduction. Among the vertebrate species, a dramatic example is provided by the Pacific salmon, which spends several years in the ocean growing into a large and powerful fish. When it is four or five years old, the salmon swims up freshwater streams and, after reproducing, becomes senile virtually overnight (Figure 23-18) and dies within two weeks.

The relationship between reproduction and aging is less pronounced in species that reproduce more than once in a lifetime. However, the reproductive outputs of

Figure 23-18
Senility in a Pacific Salmon
The condition of this fish changes from healthy and vital (top) to the total degeneration of all tissues (bottom) within two weeks after it reproduces.

species that do become senile generally decrease with increasing age, whereas the reproductive outputs of immortal organisms do not. Selection for traits that could maintain life becomes less strong after an individual has produced most of its offspring. Consequently, adaptations are less effective in promoting an individual's survival after peak reproductive age has been reached.

Aging and the Functional Decline of Tissues

Many tissues gradually decline in functional ability during the aging process. This decline in tissue function appears to originate within the cells. As aging progresses, more and more of the cells within a tissue stop functioning normally. Exactly what happens to an aging cell to interfere with its normal activities has not yet been identified.

In humans, the rate of decline of tissue function with age is about 1% of original physiological capacity per year after the age of 30 (Figure 23-19). At this rate, the

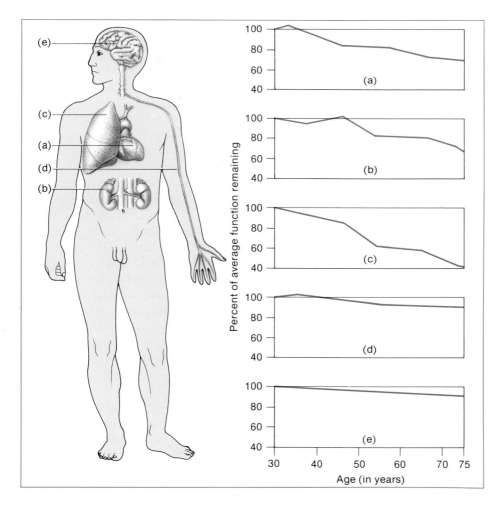

Figure 23-19
Aging and the Functional Decline of Human Tissues

These graphs indicate the loss of function of various tissues in percentage form, based on a 100% level of function at age 30. Decline in function is shown for: (a) the amount of blood pumped from the heart during each contraction while the body is at rest, (b) the rate at which the kidneys process fluids, (c) maximum breathing capacity, (d) velocity of nerve conduction, and (e) the weight of the brain.

ability of the human body to maintain physiological homeostasis would be exhausted by about the age of 120. (The greatest age attained by any human for whom accurate birth records are available is 118.) The deterioration of tissues in the circulatory, nervous, and immune systems produce the most frequent and profound symptoms of aging.

Changes in the heart muscles and blood vessels affect many of the other tissues in the body. The muscles of the heart and of the large arteries gradually lose their elasticity as the body becomes older, so that the heart pumps less blood during each contraction. The blood moves through the tissues at an increasingly slower rate, and the body becomes less and less able to respond to conditions that require greater blood flow. Due to the lower rate of blood flow, materials are exchanged more slowly between the bloodstream and cells, preventing the cells from carrying out their chemical activities at normal rates and resulting in the functional decline of the brain, heart, kidney, and other organs.

A prominent symptom of senility is *hypertension,* or high blood pressure, which results from a decrease in the diameter and flexibility of the arteries due to *arteriosclerosis.* In all individuals, blood pressure tends to increase with age; in general, a higher blood pressure level indicates a greater probability of death. Arteriosclerosis is a progressive disease that develops slowly over years or decades of an individual's life. All humans eventually suffer from this disease, and it occurs in most other vertebrate species. Arteriosclerosis is generally considered to be an inevitable manifestation of aging, and the majority of human deaths result from a breakdown of the circulatory system due to this disease. The two most common causes of human death are: (1) *heart attacks,* which occur when the heart muscles cease to function because passage of blood through the coronary arteries is obstructed, and (2) *strokes,* which occur when an artery in the brain is ruptured or blocked. Death from heart attack or stroke is so common that we have come to regard these circulatory failures as the natural end of the human life span.

The deterioration of the nervous system also contributes to many conspicuous symptoms of aging. The capacities for precise coordination, initiation of movement, and response to environmental stimuli gradually decline as an individual grows older. Some of the decline in these abilities results from a decrease in the threshold at which sensory organs such as the eyes and ears respond to environmental changes. Another contributing factor is that electrochemical impulses are transmitted through the nervous system 5–10% more slowly in older individuals (over 60) than in younger individuals (under 30). Therefore, an event takes longer for the nervous system to process as the individual grows older, which increases reaction and decision times. A high incidence of accidents among the elderly, especially in urban environments, may be due to the fact that life is too rapidly paced to give their declining neuromuscular systems time enough to adjust appropriately to environmental events.

After the age of 30, the number of brain cells that are active gradually begins to decline (see Figure 23-19). Once brain cells lose their ability to function, they are broken down and their components are absorbed into the blood. Brain cells are never replaced because nerve cells do not divide in an adult organism. However, the functions of many areas of the brain appear to overlap, and many brain cells can be lost without impairing normal mental functions. Scores on intelligence tests indicate that older people readily learn new tasks, especially when they are not hurried. In some cases, the apparent loss of mental capacity reflected by such test results is actually due to the fact that older people are less accustomed than younger people to the pressure and competition of examinations.

A decline in intellectual abilities due to blood-vessel damage in the brain is more likely to occur in the elderly. A blocked or ruptured artery may prevent a large population of cells from receiving nutrients and oxygen, and these cells will eventually die. An inadequate blood supply to the brain is the primary cause of *senile dementia*—the progressive deterioration of mental faculties with increasing age that occurs in some individuals.

The immune system also exhibits a functional decline with age. The elderly are more susceptible to infections and require longer periods of time to recover from an infection once it becomes established. The deterioration of the immune system also makes the elderly more vulnerable to *autoimmune diseases,* which result from the immune system's attack on the body's own molecules. Many of the long-term degenerative diseases associated with aging may be due to autoimmunity.

Aging and the Decline of Homeostasis

Homeostasis requires the coordination of many physiological adjustments of the body. If some systems are not functioning optimally, this entire mechanism may fail. When homeostasis fails, the internal environment can deviate greatly from its optimal state, resulting in serious damage or death. Older people are often unable to cope with the stress of environmental change as readily as they could when they were younger. All environmental changes are threatening to an individual's well-being when homeostatic mechanisms are not functioning optimally. Death frequently occurs simply because the body is unable to adjust to such a seemingly trivial environmental change as a heavy meal, a mild case of pneumonia, a fall, or a walk in the sun on a hot day.

Human Longevity and Lifestyle

A marked decline in physical and psychological capabilities is not necessarily a characteristic of advanced age. People age at different rates, and the aging rate is apparently determined by genetic and environmental factors. Perhaps the best assurance of a vigorous, long life is the fact that your parents and grandparents lived long, healthy lives. Most of the people who live beyond the age of 70 have parents who also lived at least that long. However, it is difficult to determine how much of the correlation between *longevity* (a long life span) in children and their parents is due to genes and how much is the result of similar lifestyles within family lines.

We are not yet able to alter our genes at will, but we can alter many of the environmental factors that influence health and longevity. Studies of the lifestyles of the elderly indicate that certain habits—particularly the kinds of diets, exercise, and social life—contribute to a long and healthy life.

Excessive food intake shortens an individual's life span. Extra fat burdens the heart, requiring it to pump additional amounts of blood over longer distances. Ideally, body weight should decline gradually as age increases. Healthy older people typically consume fewer than 2,000 calories per day, only about 25% of which is in the form of fats. By contrast, the average adult in the United States consumes 3,300 calories per day, almost 50% of which is in the form of fats.

Figure 23-20
A Healthy, Vigorous Life at Age 117

This remarkable gentleman lives in the Soviet Republic of Georgia. A farmer all his life, he continues to work in the fields every day. For a long life, he advocates "active physical work, and a moderate interest in alcohol and the ladies."

Many of the symptoms of aging, including stiffness of the joints and a tottering gait, result from a lack of exercise. Muscle tissue can be easily rejuvenated, and proper exercise can produce strong, flexible muscles regardless of chronological age. Exercise also increases the number of small arteries and capillaries that branch from the coronary arteries, providing more alternative pathways through which blood can flow to the heart. A blockage of any one of these vessels is then less serious, and the individual is less likely to suffer from a heart attack. Physical fitness is the inevitable reward of an active life. Most of the very old people in the world (Figure 23-20) live in mountainous areas, where the primary means of transportation is walking and even a short walk requires movement over steep terrain. Healthy old people also tend to live in agricultural societies, where physical labor usually continues throughout life.

Many degenerative changes formerly attributed to aging are now known to be the result of *social withdrawal*. In our modern society, it is often assumed that old people have no further contributions to make, and the elderly are excluded from many aspects of life. Many older people share this view and withdraw completely from an active life. Often people who are judged to be mentally and legally incompetent due to advanced age are actually suffering from malnutrition or depression. Some older people suffer from psychological problems that are treated and cured in younger people.

Another factor contributing to the symptoms of aging—the loss of *social status* which accompanies retirement at a fixed age—often leads to depression, irritability, loss of interest in life, lack of energy, increased consumption of alcohol, and psychosomatic illnesses. All of these behaviors tend to accelerate the aging process. Retirement may also lead to poverty. Individuals who do not have sufficient funds for a good diet or proper housing tend to become ill, depressed, and withdrawn.

Most old people who continue to lead rewarding and interesting lives are actively engaged in living—in pursuing business or personal goals that give their lives a purpose that, in turn, promotes health, longevity, and a feeling of well-being.

Summary

Each human begins life as a single cell, the *zygote*, which is formed from the union of an *egg* (the female gamete) and a *sperm* (the male gamete) inside the mother's body. Eggs mature within *follicles* inside the two ovaries of a woman. In an adult woman, one mature egg is released *(ovulates)* from an *ovary* each month and moves into an *oviduct*, which is connected directly to the *uterus*. After intercourse has occurred and sperm are present inside the female reproductive tract, the egg is fertilized by a sperm inside the oviduct.

Menstruation begins in a woman during *puberty* and occurs repeatedly throughout her reproductive years until *menopause*. Each *menstrual cycle* lasts about 28 days. A woman can become pregnant two weeks prior to the onset of each menstrual period, when an egg is released from one of her two ovaries. The sequence of changes that take place in the ovaries and the uterus during the menstrual cycle are controlled by four hormones. Two of these hormones—the *follicle-stimulating hormone* (FSH) and the *luteinizing hormone* (LH)—are produced by the anterior pituitary gland. The other two hormones—*estrogens* and *progesterone*—are produced by the maturing follicle and the *corpus luteum*.

Sperm are produced continuously in the *testes* of a man. Three hormones govern sperm production: *testosterone* (which greatly influences many male characteristics), FSH, and LH. At a certain stage of development, the sperm are moved into the *epididymis*—a collection of coiled tubes that lies outside each testis. Just prior to *ejaculation*, mature sperm are moved into a pair of long tubes, called the *vas deferens*, inside the abdomen. The vas deferens eventually join the single *urethra*, which carries the sperm through the *penis* to the outside of the body. As they are transported through the vas deferens and the urethra, fluids and materials from the *seminal vesicles*, the *prostate gland*, and *Cowper's glands* are added to the sperm to form the *semen*.

Sperm released into a woman travel from the *vagina* through the *cervix* and the uterus to the oviducts, where fertilization takes place. Once fertilization of an egg by a sperm *(conception)* has occurred, the zygote undergoes a series of mitotic cell divisions *(cleavage)* as it travels down the oviduct toward the uterus. The sex of the developing child is already determined at this point. If the egg has fused with a sperm carrying a Y chromosome, a boy has been conceived; if the egg has united with an X-carrying sperm, a girl has been conceived.

The tiny embryo reaches the uterus within four days after fertilization and begins to attach itself to the inner wall of the uterus on about the sixth day after fertilization. It is now a hollow sphere, or *blastula*, of approximately 100 cells. If *embryonic implantation* occurs, menstruation is suppressed—usually the first indication that a woman is pregnant.

Membranes form a sac around the developing embryo, and it is immersed in fluids until birth. By the age of two or three weeks, the embryo begins to receive oxygen and nutrients from its mother's circulatory system through a special organ, the *placenta.*

Most stages of human development are completed within the first two months after conception. By this time, the basic structures of all major tissues and organs have been formed and the embryo's circulatory system is operative. The blood vessels of the embryo are connected to the mother's circulatory system by an *umbilical cord;* at this point, the embryo has become a *fetus.* The remaining period of development within the mother is primarily one of growth and the completion of the details of the body.

The birth of the infant occurs about 266 days after conception. Hormones produced in the infant's pituitary gland stimulate the onset of strong contractions of the uterus, or *labor*—the main force that causes the expulsion of the baby from the mother's womb. After the infant is born, it begins to breathe through its lungs for the first time. The placenta is expelled from the uterus as the "afterbirth."

The most nutritious food for a newborn baby is the milk from its mother's breasts, because it contains the correct balance of nutrients as well as antibodies to protect the infant from infectious diseases. The mother also derives benefits from breast-feeding, including a more rapid reduction of the uterus to pre-pregnancy size and a more rapid weight loss.

After birth, humans undergo a prolonged period of development that involves both major physical changes and an immense amount of learning. Adolescents reach physical or *sexual maturity* at varying ages, and the rate at which a child matures to adulthood strongly influences his or her behavior and ability to learn.

Various methods of birth control can be employed to prevent or arrest the formation or development of the embryo or fetus. The *rhythm method* of contraception is practiced by abstaining from sexual intercourse for several days prior to and following ovulation. The *birth control pill* contains estrogens and progesterone and prevents the release of eggs from the ovaries by maintaining a constant level of progesterone in the body at all times. Methods of contraception that prevent sperm from reaching the egg during intercourse include a *vasectomy, withdrawal* of the penis at ejaculation, and the use of a *condom* on the part of the man and a *tubal ligation* or the use of a *diaphragm* in conjunction with a *spermicide* on the part of the woman. Embryonic implantation can be prevented by the use of an *intra-uterine device* (IUD) or the *morning-after pill.*

The *induced abortion* of the embryo or fetus by *menstrual evacuation* during the first three months of pregnancy (the first *trimester*) is a relatively simple procedure. Abortion later in pregnancy is more serious because premature labor and birth must be induced. Abortions after the fetus is six months old cannot be legally performed in the United States unless the mother's life or health is endangered or the fetus is known to be grossly abnormal. A genetic defect in the fetus can be detected as early as the middle of the third month by a process called *amniocentesis.*

Several serious *venereal diseases* can be passed from one partner to the other during sexual intercourse. The three major bacterial infections—*syphilis, gonorrhea,* and *nongono-coccal urethritis* (NGU)—can all be treated with antibiotics. Another sexually transmitted disease, *herpes,* is caused by a virus and has no known cure. Proper use of a condom can prevent the transmission of venereal diseases.

What happens to the human body during aging is not well understood. However, the same mechanisms do not cause deterioration in living and nonliving things. Unlike nonliving material, biological material has the capacity for *self-renewal* and *self-maintenance*.

Many organisms appear to be immortal. They continue to reproduce throughout their lives and may have the potential to live forever. However, most organisms are *mortal*; they have a defined period of reproduction and a *maximum life span*. The symptoms of aging appear shortly after reproductive ability begins to decline. Natural selection cannot produce adaptations to conditions that occur after an organism has ceased to reproduce.

The maximum human life span appears to be about 120 years. The symptoms of aging or *senility*, begin at about 30, when the body cells no longer function optimally. The deterioration of the tissues of the circulatory, nervous, and immune systems produce the major symptoms of aging. A decline in the function of these and other tissues causes a decline in *homeostasis*, so that changes in the external environment can seriously alter the individual's internal environment and lead to death. Two common causes of death, *heart attack* and *stroke*, are due to a deterioration of the arteries *(arteriosclerosis)*.

A correlation appears to exist between the severity of the aging process and an individual's lifestyle and genes. People in certain family lines tend to live longer than people in other family lines. Aspects of lifestyle that promote *longevity* (a long life span) and a vigorous old age include a low-calorie, low-fat diet, regular physical exercise, and remaining socially active.

Biology and the Challenges of Modern Life

5

Modern Life: Challenges and Opportunities

24

Through natural selection, organisms acquire characteristics that enable them to utilize their environment to benefit themselves and to form progeny. Like other species, the human species has inherited a variety of traits that contribute to maximizing reproductive output; unlike other species, humans have also developed *culture*, through which knowledge about life is accumulated. This combination of biological and cultural inheritance provides the human race with an ever-increasing power to exploit nature—to utilize greater and greater amounts of environmental materials to sustain human life.

According to one criterion of evolution—the abundance of a species—we are a great biological success. If we use a different test of evolutionary success—the length of time that a species exists prior to extinction—we may turn out to be a relatively unsuccessful species. We depend on the environment to nourish us and to absorb our wastes. The production of increasing amounts of human biomass at the expense of environmental degradation may eventually destroy us. Unlike other organisms, however, humans are not compelled to respond blindly to the dictates of their genes and continue to exploit the environment until it can no longer support life. To some degree, we can predict the long-term consequences of our activities and make changes that will influence our future. In this final chapter, we will examine some of the challenges we must face if the human species is to survive and some of the opportunities available to us to create lifestyles in which human biology can be adjusted to improve the quality and ensure the continuance of life on earth.

The Human Predicament: Limits to Adaptability

The history of human populations differs strikingly from the histories of other life forms due to the enormous impact of cultural changes on our lifestyle and environment. The different rates of change in the two aspects of human history—incredibly fast-paced cultural changes compared with the slower changes brought about by biological evolution—have led to our current predicament: we may not be biologically adapted to our present, culturally modified environment.

We inherited our genes from stone-age hunters and gatherers and probably possess no greater innate intelligence, artistic abilities, physical capacities, or emotional feelings than our prehistoric ancestors. Cro-Magnon humans, who existed about 35,000 years ago, were physically very similar to twentieth-century men and women but were more closely adapted to their environment due to two important factors. First, Cro-Magnon culture was less developed and communication of new ideas among different populations was minimal. Changes in the Cro-Magnon environment caused by cultural modifications occurred more slowly, permitting biological evolution to keep pace with cultural change. Second, natural selection acted continuously to modify the biological characteristics of Cro-Magnon individuals, removing genes that made them less able to cope with direct environmental challenges to their survival and reproduction.

By contrast, technological and cultural innovations cause rapid changes in the environments and lifestyles of modern human populations and shelter individuals from natural selection. Housing, agriculture, education, and medicine promote the survival and successful reproduction of all human beings today, regardless of the genes they carry. Because natural selection is much less effective due to the many changes we have imposed on our environment, our biological attributes are less closely matched to the environment than those of our prehistoric ancestors. From a purely biological viewpoint, the most optimal relationship between the human species and the environment may have been achieved by the Cro-Magnon peoples. Although this is not an indication that we should return to the hunter–gatherer lifestyle, we should recognize our possible genetic limitations when planning new ways of life in the future.

The Pace of Life

One aspect of modern life that has changed dramatically in the past 35,000 years is its pace. Human beings may have reached the limits of their capacity to adjust to the pace of life. For example, consider the speed of movement from place to place (Figure 24-1). The speed of Cro-Magnon travel was restricted to the rate at which a human could run. Camels were not domesticated for transportation until 6,000 B.C.; the top speed of long-distance travel by camel was then only 8 miles per hour. The horse-drawn chariot, which appeared 4,000 years later, increased human speed to a maximum of 20 miles per hour. The invention of the steam locomotive 3,500 years later in the 1880s increased the top speed that a human could travel to 100 miles per hour. In today's world of the automobile and the airplane, even the old and the sick can travel at very fast speeds simply by pressing down on a gas pedal or buying a ticket, and astronauts hurtle through space in their capsules at thousands of miles per hour.

These new speeds make unusual demands on the human nervous system—particularly on its capacity for decision making. When the human brain evolved more than 100,000 years ago, the tempo of life was much slower. When we walk or run (for which we are biologically adapted), the information from the passing scene moves slowly through the mind, providing ample time for us to integrate it, make decisions, and handle emergencies. But when we guide a car through fast traffic

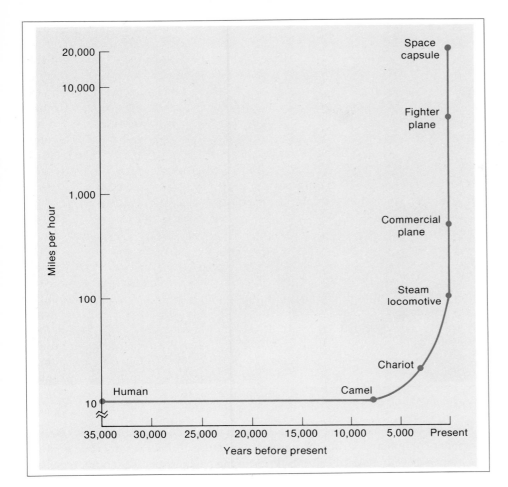

Figure 24-1
The Speed of Human Travel Over the Past 35,000 Years
Rapid travel is a recent innovation to which humans may not be biologically adapted. The vertical axis of this graph gives speed in miles per hour on a logarithmic scale.

(for which we are not biologically adapted), information from the environment flashes through the mind and decisions that could prove fatal must be made instantaneously.

Speed of movement is only one example of how our capacities for processing information and making decisions are being stretched to their limits. The sheer amount of information that an individual must possess to participate meaningfully in society has become formidable. As our knowledge increases and society becomes more complex, the consequences of a particular action are exceedingly difficult to predict. The environmental effects of new industries or the international consequences of specific political and military acts cannot be known with certainty. The dimensions of human relationships are also being taxed; we interact with greater numbers of people, largely due to improved long-distance travel and communication and the intensity of urban life. To keep ourselves from being emotionally and socially overloaded, many of these relationships must be fleeting or superficial. This type of social activity is new to the human organism and was not part of the biology of our hunter–gatherer ancestors.

Figure 24-2
Escaping the Stresses
of Modern Life

Periods of high anxiety or intense concentration have probably always been a component of human life, but the unrelenting nature of this stressful state of mind and body is new to modern cultures. Prior to the past few decades, human culture had developed snug harbors in which the nervous system rested between emergencies, including the comforts of tradition, close family relationships, long-lasting marriages and friendships, and ties to the land and its seasonal changes (Figure 24-2). The number of these restful harbors is rapidly decreasing in our modern culture of continuous and rapid change, transcience, and "progress."

Without protection from this flood of information and change, the human mind and body are vulnerable to incapacitating stresses. Both the autonomic and the endocrine control systems react to environmental danger signals, preparing the body for "fight or flight." Unless these stress reactions are carried through and released in strenuous activity, a variety of mental and physical disabilities can develop. Unfortunately, although the stresses of modern life require us to be active to maintain physical and mental health, humans are becoming increasingly sedentary. Both unrelenting stress and a decrease in physical activity are biologically unnatural.

Dealing with Biologically Foreign Materials

The limits of human adaptability are most evident in terms of the central nervous system and behavior, but other environmental changes present difficulties for our old-fashioned bodies. We are not able to tolerate the increasing levels of unusual materials in our food, water, and air. If the unnatural molecules in pesticides and food additives had been present in our food throughout our evolutionary past, we would probably now possess digestive enzymes that would break them apart and even make them useful sources of energy and nutrients. Instead, these toxic foreign molecules accumulate in our cells, where they interfere with many critical processes, including normal cell division (that is, they cause cancer) and the immune system's natural mechanisms for fighting diseases, as well as neuromuscular function and behavior. Similarly, the delicate tissues of human lungs are not adapted to deal effectively with air pollutants and gradually decline in function after long-term exposure to smog, tobacco smoke, and other sources of unnatural air-borne materials.

The evolution of mechanisms that prevent tissue damage from toxic materials has been demonstrated in other organisms. Many insect populations are no longer detrimentally affected by insecticides, and many bacterial populations are no longer destroyed by antibiotics. However, the evolution of such resistant populations involves strong natural selection in which the death rates among nonresistant types are exceedingly high. Only given such high mortality rates, could humans expect to become biologically adjusted to the high levels of damaging materials in our environment.

Primitive Rates of Reproduction

The single most pervasive cause of environmental deterioration in the world today is *overpopulation*. The number of humans is rapidly and continuously increasing; each year the number of babies who are born exceeds the number of people who die (Figure 24-3). Like our physical adaptations, our rate of reproduction was established thousands of years ago when humans were hunters and

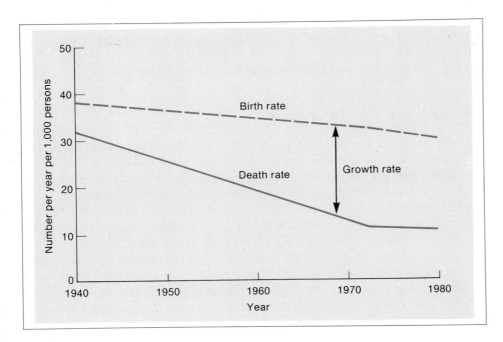

Figure 24-3
World Birth and Death Rates
Both birth and death rates have declined in the past 40 years, but the drop in the death rate has exceeded the drop in the birth rate. In 1980, the birth rate was 30 per year per 1,000 people and the death rate was 11 per year per 1,000 people. The population growth rate—the difference between these two rates—is 19 per year per 1,000 people, or 1.9% per year. As this graph indicates, the growth rate has been positive over the past 40 years and is now greater than it was in 1940.

gatherers. Just as her ancestors before her, a modern woman has the capacity to give birth to a child every year or two during her reproductive span of approximately 30 years. Despite this high rate of reproduction in Cro-Magnon women, the populations did not grow rapidly because many offspring died from disease before their second or third year of life. In modern populations, most infants survive to adulthood, and natural reproductive rates can produce much larger families than occurred in ancestral populations. The tremendous change in the survival rate of offspring has made our reproductive ability a biological relict that is no longer adaptive.

The Human Predicament:
Environmental Deterioration

It is becoming increasingly clear that we must modify our activities to achieve harmony with the natural systems of the earth. The organisms in each ecosystem are highly interdependent, and a small change in one species can magnify and eventually disrupt the existing biological equilibrium. Species become extinct when changes take place too rapidly to permit evolutionary adjustments, and the disappearance of each species jeopardizes the stability of other populations.

Most of the organisms with which we share the earth are endangered because humans are destroying their habitats so that farms, mines, and cities can be developed. It is difficult to protect species and natural ecosystems that have no economic value or demonstrate no potential value (as food or in medical research for example) in human societies that have such strong exploitative relationships with nature. Natural prairies—and with them, all of the plants and animals that evolved in a prairie habitat—have essentially disappeared from the United States. Lands that were formerly prairies now produce a significant portion of the world's grains. Another habitat that will soon disappear from the world is wet, tropical forest land. Most of the trees on this land will be cut by the year 2,000 so that agricultural lands can expand into these areas. About 25% of all living species are dependent on tropical forests; virtually all of these organisms will become extinct.

Possibilities for the Future

The ancient Greeks consulted oracles when they wanted to know about the future. At such shrines, like the famous Oracle of Apollo at Delphi shown in Figure 24-4(a), the gods would answer the pilgrims' questions through the medium of resident priests and priestesses, who would fall into trances and spew out unintelligible words that interpreters would then translate.

Today, we use a similar procedure to look into the future, only now our oracles are *computers.* Masses of information from the past are fed into a computer like the one shown in Figure 24-4(b). It falls into a deep electronic trance, and then spews out an almost endless stream of numbers, graphs, and statistics. The resident priests, called *programmers,* periodically gather up a pile of computer output and ship it off to experts, called *futurologists,* who translate the computer data into predictions about the future (Figure 24-5).

(a)

Figure 24-4
Modern and Ancient Methods of Making Predictions
Humans have the ability to make predictions about the future. (a) The ancient Greeks consulted oracles to learn about the future. This is a photograph of the sacred precincts in the temple of Apollo at Delphi, where the gods occasionally revealed information to pilgrims about what was to happen. (b) In the 1980s, future trends are obtained from computer assimilations of past trends.

(b)

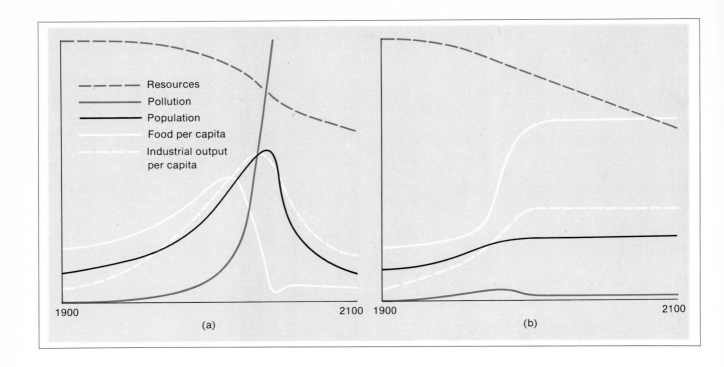

Figure 24-5
Predictions from the Computer

The computer uses information from the past and certain assumptions about the future to make very precise predictions. (a) Predictions about food, industry, resources, population, and pollution, based on assumptions of unlimited nuclear energy, an increased availability of resources, and the recycling of materials. In this model, population growth is stopped by rising pollution. (b) Predictions about the same human activities, based on assumptions of birth control, a reduced utilization of resources, recycling of materials, pollution control, and an emphasis on the provision of food and services rather than industrial output.

The ancient oracles were often wrong, and the computer's version of the future may be just as inaccurate. We do not know why the oracles failed, but we do know that computers can only be completely accurate if human behavior in the future is identical to human behavior in the past, which is highly improbable. In some of the key problem areas of modern life—population growth, use of resources, continuous and rapid technological change, and violence—predictions of future disasters will come true only if we persist in the perverse, self-destructive behavior patterns of the past. As you are reading this sentence, hundreds of thousands of people are starving to death or trying to kill one another. But human behavior can change. In the past, we have always been able to master the challenges of our environment. Humans can make choices that transcend genetic and environmental determinism. If we reorder our priorities, focus creative energies on these problems, and make necessary changes, future disasters and eventual human extinction will not be inevitable.

We will now briefly examine some of the possible directions in which the human spirit could carry us in the near future. If we maintain our essential humanity—a mutual respect for all human beings—these technological and biological avenues of change may ultimately help us to solve our existing problems.

Basic Changes in Human Biology

In the past, we have used tools to enhance our biological adaptations. Now, such cultural innovations as computers, machinery, and various entertainment mediums have extended our biological capacities to increase our calculating capabilities, achieve greater physical power, and greatly expand the limits of our experience through books, television, and movies. Moreover, we have acquired entirely new abilities, such as flight (airplanes) and underwater dives of long duration (scuba diving and submarines). Technological advancements continue to extend human capabilities in new and exciting ways.

We can now enhance our biological adaptations by altering our basic structure, chemistry, and genetic repertoire, which provides us with the potential to re-design our own framework—to initiate and control a wholly new path of evolution. Given our newly found capabilities for altering human biology, it is important and timely for us to develop an appropriate image of the desirable human being. It is even more important to determine whether we want to influence the development of the human species at all. We must seriously consider how humans differ from other species, how we should interrelate, and other basic moral and ethical issues. For the first time in biological history, changes occurring within a species can be goal-oriented. We must approach this opportunity with as much knowledge, wisdom, and care as possible.

Perhaps the most dramatic change in human biology is our ability to manipulate our genetic inheritance. The odds of producing a child with specific traits have been increased by the establishment of *sperm banks,* where men with certain recorded abilities and physical characteristics leave their semen to be frozen and, perhaps, later requested by a woman to father her offspring. *Egg banks* will probably develop in the future, thereby enabling a prospective "parent" to choose both gametes; after being united within a test tube, the fertilized egg can then be placed in a woman's uterus for development.

Before the year 2,000 it will be possible to modify the physical traits and bodily processes of our children by making changes in their genes. Segments of DNA could be inserted into a fertilized egg to replace defective genes as well as to direct the development of traits that are considered desirable. Because the genetic code is universal, the inserted DNA segments would not have to be taken from humans, and they could contain genetic instructions for traits never before present in the human species, or in an animal of any kind. For example, genes coding for the synthesis of vitamin C and the essential amino acids might be desirable additions to the human genetic repertoire. These and similar additions could eliminate the need for a diverse diet and reduce our dietary requirements to calories and certain inorganic materials, which would greatly simplify our present problems of food production.

We now have the biological expertise to reproduce asexually—to develop offspring that are exact genetic copies of a single adult. Obtaining desirable traits through normal sexual reproduction is difficult due to the random nature of chromosome assortment and crossovers during meiosis. Reproduction without this process of genetic recombination would permit the offspring to inherit the genotype intact. Once an individual had acquired a desirable set of genes (through sexual reproduction or genetic engineering), the genetic program could be preserved and perpetuated through asexual reproduction, which occurs in a large variety of organisms. Although it is rare in vertebrates and is not known to occur naturally in any mammal, the process of asexual reproduction is understood and could be generated in humans.

Humans are already being improved physiologically and anatomically by medical breakthroughs and biological alterations. A recent technological invention, *computerized axial tomography* (CAT scanning), combines the techniques of X-rays, computers, and television to provide such a clear view of the internal anatomy of a patient that physicians can detect tumors and determine whether they are *benign* (localized and not apt to recur after removal) or *malignant* (liable to recur and spread throughout the body). CAT scanning also reveals blood-vessel blockage, bile-duct obstructions, broken bones, foreign objects, and other anatomical aspects that may create health problems. Other medical wonders include artificial parts, such as metal bones, Dacron tendons and arteries, plastic and stainless-steel joints, and kidney dialysis machines, as well as such surgical feats as the organ and blood-vessel transplantations from one individual to another. An artificial pancreas developed recently to cure people who suffer from diabetes continuously monitors levels of blood-sugar concentration and releases appropriate amounts of insulin into the bloodstream. Future medical breakthroughs will probably occur in the area of organ and limb regeneration, in which the missing part is reformed through growth and development. Regeneration occurs naturally in many animals, although it is limited in humans to the reformation of a complete liver from a fragment and the repair of minor wounds and bone fractures. Once regeneration is understood, the renewal of other human organs and the growth of new limbs from severed stumps can probably be initiated. Even now, researchers are meeting with limited success in stimulating this basically embryonic process in the amputated limbs, damaged heart muscles, and brain tissues of lizards, rats, and dogs. The medical benefits that can be derived from this knowledge will be profound.

Each year, our understanding of the human brain becomes more complete. The possibilities for *behavioral engineering* are expanding rapidly, and it may soon be possible to make precise adjustments in human behavior through such methods as drug therapy, neurosurgery, conditioning, and sleep teaching. Although these techniques present potential dangers in that they could be used to manipulate people in ways that are not in their best interests, behavioral engineering could also eliminate a patient's pain or self-destructive tendencies. An exciting development has been the recent discovery that the human brain and pituitary gland contain substances—the *endorphins*—that act very much like morphine and heroin. The genetic code for the synthesis of these natural drugs can be inserted into bacterial cells, where the DNA segment directs the synthesis of these molecules, which can then be harvested for use as natural pain killers. Endorphins may also prove valuable in the treatment of certain mental illnesses and in understanding drug addiction.

Before we can decide whether or not to tamper with our genetic programming, we must determine how the techniques of genetic engineering would be used. These techniques could be of enormous benefit to most individuals with obvious biological handicaps, but they could also be used to deprive most humans of their basic freedoms.

The Genetic and Ecological Manipulation of Food Organisms

A new area of research is the production of strains of crop plants that do not require nitrogen fertilizer. A significant agricultural cost is the commercial synthesis of nitrogen fertilizers. About 3% of the annual consumption of natural gas in the United States is used for this purpose.

Green plants can be divided into two categories—plants that require the addition of nitrogen fertilizers (such as corn, wheat, and rice) and plants that manufacture their own nitrogen fertilizer (the *legumes,* such as peas, soybeans, alfalfa, clover, and peanuts). Actually, no plant is capable of using atmospheric nitrogen (N_2) to construct nitrogen-containing molecules from glucose. However, special tissues in the roots of legumes provide an environment suitable for certain soil-dwelling bacteria, which convert atmospheric nitrogen into a form that can be used by the plant. This bacterial process is referred to as *nitrogen fixation.* In return, the plant produces energy-rich molecules that the bacteria use for their own maintenance and reproduction. Natural selection has favored this symbiotic relationship because it benefits both species.

A set of about 15 genes that govern nitrogen fixation has been identified and isolated. A current line of biological research is the transfer of these bacterial genes into the cells of multicellular crop plants, such as corn and wheat. If this could be accomplished, the crops could produce their own nitrogen fertilizer, using energy from the sun rather than fossil fuels to drive the synthesis.

New kinds of plants may also be developed by uniting material from two or more different species within a single cell. Such forced hybrids could contain chromosomes from several species or the chromosomes from one species and the organelles from another. It may also be possible to create new symbiotic associations—for example, by incorporating nitrogen-fixing bacteria directly into plant cells. This process may have occurred naturally to produce such organelles as the mitochondria and chloroplasts within eukaryotic cells. The ability to unite various organelles or even whole prokaryotic cells within a single eukaryotic cell would make it possible to create new strains of food plants with desirable characteristics.

Today, many people throughout the world are experimenting with new kinds of agricultural ecosystems. They feel that commercial foods are nutritionally deficient, filled with toxic materials, and expensive due to unnecessary packaging, handling, and transport. In an effort to develop inexpensive, ecologically sound farming techniques that would permit each family or community to produce their own fresh food, people are experimenting with renewable forms of energy, such as the sun and wind, establishing small farms that produce a variety of crops, and taking advantage of the natural cycling of nutrients as well as natural means of controlling pests.

The *solar tube*—a cylindrical aquarium (Figure 24-6) that utilizes solar energy—is one of the simpler new ways to produce food. Algae placed inside the tube transform solar light, carbon dioxide, and water into biological molecules, oxygen, and heat. The water is continually circulated by the movements of swimming fish, usually a species of *Tilapia* from Africa, which feeds on algae and prefers warm water. Conditions within the solar tube can be adjusted so that these fish survive and reproduce well despite a high population density. Under natural conditions, the fish release a chemical in their urine that becomes so concentrated when the fish are crowded together that it inhibits their reproduction. To prevent reproductive inhibition in the solar tube, oysters are added. Bacteria growing on the oyster shells convert the fish urine into nitrites and nitrates, which the algae use in the synthesis of nitrogen-containing molecules such as amino acids and nucleic acids.

Figure 24-6
A Solar Tube

This new invention, which is basically a cylindrical aquarium, can be used to produce large quantities of fish protein in a small space. An almost self-contained ecosystem is maintained within the solar tube.

Reproduction in the fish is enhanced by both the breakdown of their urine and the growth of their food plant, algae. The fish, in turn, are a source of high-quality protein for humans.

Solar tubes and other kinds of farm aquaria and greenhouses are particularly useful in crowded urban areas, because they can be used to produce food in alleys or vacant lots, on rooftops, or anywhere else that there is sunlight. They are inexpensive to build and maintain and will probably play an important role in future lifestyles.

Ecolibrium—Balance in Our Earthly Home

Human beings have and will continue to alter the surface of the earth profoundly. Past alterations have often been haphazard, exploitative, and destructive. Future alterations, however, can be carefully planned, so that consequences of change are fully understood and the continuance of a healthy, life-promoting environment is ensured.

A major threat to the earthly balance is *urban sprawl*—the continual replacement of farmlands and wildlife habitats with residences and shopping centers. Our expanding centers of human activity should be redesigned in accordance with clearly defined priorities for global food production and wildlife preservation.

The Design of Human Settlements

The human population must attempt to coexist harmoniously with the re-mainder of the earth's organisms. One solution might be the *decentralization* of human activities—the development of small, self-sufficient communities in rural settings like the one shown in Figure 24-7(a), where each village would rely on

Figure 24-7
Communities of the Future
Two very different kinds of human settlements, both of which would help to solve our problems of farmland shortage and wilderness loss. (a) Life in a self-sufficient, small community appeals to many of us. This lifestyle could meet our basic biological needs. (b) It may become necessary to construct urban centers to support large future populations. Centers like this one would free vast areas of land for farming and wilderness preservation.

(a)

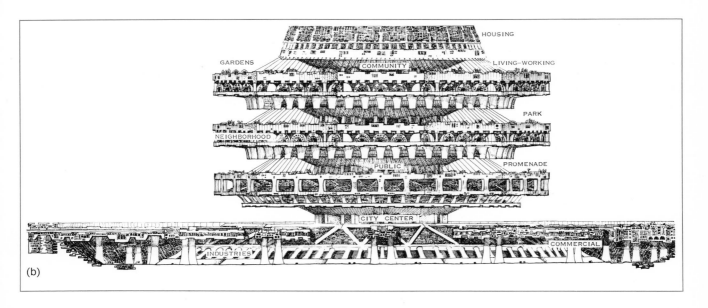

(b)

locally available sources of food, materials, and fuel. The success of such settlements would depend on the treatment of the local environment. An advantage of this approach to living is that all residents would participate fully in local politics, economy, and culture; the daily activities of each individual would include close personal relationships, physical labor, and contributions to a dynamic culture. In short, the lifestyle would meet our basic biological needs. These communities would not necessarily be independent of modern technology. Instead, new technologies designed specifically for small community industries, food production, and recreation could be employed.

Another approach to achieving harmonious coexistence with nature would be the *centralization* of human activities—the establishment of multi-level cities like the model shown in Figure 24-7(b) that would house hundreds of thousands of people but occupy only a small portion of the earth's surface. Such structures would free most of the land for food production and wilderness habitats and would bring large numbers of people together for the exchange of ideas and services as well as cultural enrichment. These new cities would provide a new form of human environment designed and built to be as close to perfection as modern technology could achieve. All transportation systems, for example, could be confined to underground tubes, which would free even more land for food production and recreation.

Of these two future lifestyles, centralization is more realistic in view of the present population growth rates and the pressing need for increasing quantities of food. In recognition that different lifestyles appeal to different people, however, an optimum plan for human settlement might include a variety of communities ranging from pastoral villages to well-designed, highly technological urban centers.

Conservation: Will We Leave Room on the Ark?

Billions of years of natural selection have produced an enormous variety of organisms, each exquisitely adapted to its own unique way of life. Many of these evolutionary achievements are familiar aspects of your own life. However, the accumulation of small but rapid changes in their environments, caused by human activities, is gradually eliminating many of these life forms from the face of the earth. We all feel that these creatures should be saved from extinction in some way. No one wants to deprive them willfully of their right to life. Yet their existence is now at our mercy.

Agruments supporting the preservation of wildlife can be developed on the basis of their future use to humans. But the fundamental argument—if one is needed—is that we have an obligation to pass on all of the components of our living world to future generations. This sentiment is eloquently expressed in the words of the American ecologist Archie Carr, who wrote:

> A reverence for original landscape is one of the humanities. It was the first humanity. Reckoned in terms of human nerves and juices, there is no difference in the value of a work of art and a work of nature. There is this difference, though. . . . Any art might somehow, some day be replaced—the full symphony of the savanna landscape, never.

Will Humans Learn to Share the Earth's Bounty with These Evolutionary Achievements?

Summary

The human species is a biological success in terms of its ability to utilize environmental resources to produce offspring, but our continued existence is threatened by our exploitation of the environment. Unlike other organisms, however, humans possess the unusual capacity to predict the future consequences of their activities and to prevent their own extinction.

Modern humans are biologically adapted to the hunter–gatherer lifestyle of our Cro-Magnon ancestors, who existed about 35,000 years ago, but our environment has changed profoundly as a consequence of recent rapid cultural modifications. The pace of life — including the speed of travel, the quantity of information to be processed, and the intensity and level of human interaction — has increased to such an extent that we may be at the limits of our biological capacities to adjust to stress and change. We are exposed to chemicals that were not present during our evolution and that present a threat to our health. Our reproductive rate is a biological trait that is no longer adapted to our modern environment. The human population is now so large and environmental changes are occurring so rapidly that many of the earth's organisms are threatened with extinction.

The key to the survival of life on this planet is our ability to make accurate predictions about the future by analyzing computer assimilations of the effects of past trends in human behavior. These assimilations indicate that future disasters will occur unless we reorder our present priorities and modify our behavior.

A knowledge of biology can be used to make future modifications in the human body. We have already the capacity to alter our genetic framework, to reproduce asexually, and to substitute artificial parts for human organs. In the near future, we may even be able to stimulate the regeneration of organs and limbs. Our increasing understanding of the brain provides us with the enormous and dangerous potential to alter human behavior. A knowledge of biology can also be used to modify the genes of our food organisms through techniques of genetic recombination and forced symbiosis. We are just beginning to explore ecosystems that will provide new methods of food production.

Ecolibrium, or environmental stability, should be our primary objective in the near future. An obvious move toward achieving this goal is the reduction of urban sprawl. Two possible designs for future human settlements are small, self-sufficient rural communities that would make use of appropriate new technologies and large, multi-level cities that would provide an entirely new form of human environment and that would occupy a relatively small portion of land, thereby freeing more of the earth's surface for food production, wilderness habitats, and recreation.

The far-reaching challenge in the future appears to be the perpetuation of the human species and the maintenance of satisfying lifestyles for all individuals. But in our attempts to achieve this goal, we must not relinquish our ultimate responsibility to preserve all existing life on earth — to preserve our past and our present heritage for future generations.

Illustration and Photo Credits (*continued*)

McGraw-Hill, Figure 17–14, p. 521.   **13–19(a)** Adapted from C. A. Villee and V. G. Dethier, *Biological Principles and Processes*, 1971, W. B. Saunders, Figure 22–18, p. 719. Copyright to this text has been transferred to Holt, Rinehart and Winston. **(b)** Adapted from N. R. Carlson, *Physiology of Behavior*, 1977, Allyn and Bacon, Figure 8–27, p. 194.   **13–20(a)** Modified from *Biology*, by G. C. Stephens and B. B. North, John Wiley & Sons, New York, 1974, p. 315, Figure 11–5.   **(b)** Modified from *Structure and Function in the Nervous System of Invertebrates*, Vol. I, by Theodore Holmes Bullock and G. Adrian Horridge, W. H. Freeman and Company. Copyright © 1965.   **13–21** Adapted from Weinberg, S. L., *Biology: An Inquiry into the Nature of Life*, Allyn and Bacon, Inc., 1974, p. 295, Figure 15–8.   **13–22(b)** Adapted from W. Keeton, 1973, *Elements of Biological Science*, Second Edition, W. W. Norton & Company, Inc., p. 203, Figure 10–3.   **13–23** Adapted from R. E. Snodgrass, *Anatomy of the Honey Bee*, copyright © 1965 by Cornell University. Used by permission of the publisher, Cornell University Press.   **13–24** From *Human Physiology: The Mechanisms of Body Function* by A. J. Vander, et al. Copyright © 1970, McGraw-Hill. Used with permission of the McGraw-Hill Book Company.   **13–25(a)** Adapted and reproduced by permission of Harcourt Brace Jovanovich, Inc., from *Life: An Introduction to Biology*, by G. G. Simpson, C. S. Pittendrigh and L. H. Tiffany, © 1957.   **(b)** Adapted and reproduced by permission of Harcourt Brace Jovanovich, Inc., from *Psychology: An Introduction*, Second Edition, by J. Kagan and E. Havemann, © 1972.   **13–26** Adapted and reproduced by permission of Harcourt Brace Jovanovich, Inc., from *Life: An Introduction to Biology*, Second Edition, by G. G. Simpson and W. S. Beck, © 1957, 1965.   **13–27** Adapted from G. C. Stephens and B. B. North, *Biology*, John Wiley & Sons, Inc., New York, 1974, p. 335, Figure 11–24.   **13–28** Kimball, *Biology*, © 1974, Addison–Wesley, Reading, MA, Figure 27.17. Reprinted with permission.   **13–29** Adapted from G. C. Stephens and B. B. North, *Biology*, John Wiley & Sons, New York, 1974, p. 327, Figure 11–19b.   **13–30** From "Brain Function and Blood Flow" by Niels A. Lassen, David H. Ingvar and Erik Shinhøj. Copyright © October, 1978 by *Scientific American*. All rights reserved.   **13–31** Reprinted with permission of Macmillan Publishing Co., Inc., from *The Cerebral Cortex of Man* by Wilder Penfield and Theodore Rasmussen. Copyright 1950 by Macmillan Publishing Co., Inc. Renewed 1978 by Theodore Rasmussen.   **13–32** Adapted from Vander, A. J., et al., *Human Physiology: The Mechanisms of Body Function*, McGraw-Hill, Inc., New York, 1970, Figure 19–7, p. 576.   **13–33(a)** From C. Sagan, *The Dragons of Eden*, Ballantine Books, Inc., 1977, a division of Random House, Inc.   **(b)** Illustration from *The Brain: The Last Frontier* by Richard M. Restak, M.D. Copyright © 1979 by Richard M. Restak. Reprinted by permission of Doubleday and Company, Inc.   **13–34** Adapted from *Psychology: An Introduction*, Third Edition, by P. Mussen and M. Rosenzweig, p. 67, Figure 3–2, D. C. Heath and Co.

Chapter 14
14–1(a) Scanning electron micrograph of stomata on the leaf surface of *Dianthus* sp. (Caryophyllaceae), J. Heslop–Harrison, University College of Wales.   **14–1(b)** John H. Troughton.   **14–4** Thomas Eisner, Cornell University.   **14–5** William H. Amos.   **14–11** Photo by E. Bernstein and E. Kairinen, *Science*, Vol. 173, cover, © August 27, 1971. American Association for the Advancement of Science.   **14–17** Adapted and reproduced by permission of Harcourt Brace Jovanovich, Inc., from *Health*, Second Edition, by B. A. Kogan, © 1970.   **14–18** AP Laserphoto.   **14–23** © Reprinted with permission American Heart Association.

Chapter 15
15–1 Russ Kinne/Photo Researchers, Inc.   **15–5** Adapted from John Henry Comstock: *The Spider Book*. Revised and edited by W. J. Gertsch. Copyright 1912, 1940 by Doubleday, Doran and Company, Inc. Copyright assigned 1948 to Comstock Publishing Company, Inc. Used by permission of the publisher, Cornell University Press.   **15–9** Dr. Robert W. Guimond, Boston State College, Department of Biology; Attorney at Law.

Chapter 16
16–2 Wellcome Museum of Medical Science.   **16–3** Wellcome Museum of Medical Science.   **16–7(a)** and **(b)** Photograph by Thomas Eisner, Cornell University.   **(c)** Photograph by Penelope Jenkins, Birmingham, England.   **16–7(a)**, **(b)**, and **(c)** Line drawing adapted from *Life on Earth* by E. O. Wilson, T. Eisner, et al. Sinauer Associates, Stamford, CT, p. 407, Figure 23, 1973.   **16–11(a)** and **(b)** Jeanne M. Riddle, Ph.D., Henry Ford Hospital.

Chapter 17
17–14(a) © Animals Animals/Irene Vandermolen.   **(b)** © Animals Animals/ M. Ansterman.

Chapter 18
18–1(a) Centers for Disease Control, Atlanta, GA.   **18–1(b)** Carolina Biological Supply Company.   **18–2** Centers for Disease Control, Atlanta, GA/Biological Photo Service.   **18–3(a)**, **(b)**, and **(c)** Centers for Disease Control, Atlanta, GA.

18–3(d) Runk Shoenberger/Grant Heilman.   **18–4** Centers for Disease Control, Atlanta, GA.   **18–6** From R. W. Horne and J. Nagington, *Journal of Molecular Biology*, 1959, Academic Press.   **18–7** From R. J. Dubos and J. G. Hirsch, *Bacterial and Mycotic Infections of Man*, Fourth Edition, J. B. Lippincott Company, 1965.

Chapter 19
19–3 Illustration by Birgit H. Satir.

Chapter 20
20–1 Adapted from M. H. A. Keenleyside, "Some Aspects of Schooling in Fish," *Behavior*, Vol. 8, 183–248, 1955.   **20–2** Photo from Kenneth D. Roeder, *Nerve Cells and Insect Behavior*, Harvard University Press, 1967.   **20–3** From E. Peter Volpe, *Man, Nature and Society: An Introduction to Biology*, Second Edition, © 1975, 1979, William C. Brown Company, Publishers, Dubuque, Iowa. Reprinted by permission.   **20–4** Art adapted from J. Alcock, *Animal Behavior*, Second Edition, 1979, Figure 7A, p. 63.   **20–5** V. G. Dethier, Eliot Stellar, *Animal Behavior*, Third Edition, © 1970, p. 91. Adapted by permission of Prentice-Hall, Inc., Englewood Cliffs, New Jersey.   **20–6** Adapted from D. P. Barash, *Sociobiology and Behavior*, 1977, Figure 1–1, p. 4.   **20–7** From *Biology: An Inquiry Into the Nature of Life* by Stanley L. Weinberg. Copyright © 1971 by Allyn and Bacon, Inc. Used by permission.   **20–8** Adapted from A. Silverstein, 1974, *The Biological Sciences*, Holt, Rinehart and Winston. Figure 13–19, p. 365.   **20–11** Graph, figure 7, page 35, from Chapter 4, "Evolutionary Interpretation of Neural and Behavioral Studies of Living Vertebrates" by William Hodos from *The Neurosciences, Second Study Program* (Francis O. Schmitt, Editor-in-Chief), published by the Rockefeller University Press, New York, 1970.   **20–12** Adapted from S. E. Glickman, R. W. Sroges, "Curiosity in Zoo Animals," *Behavior*, Vol. 26, 151–188, 1966.   **12–13** Permission granted by Harlow & Mears.   **20–14** Thomas McAvon, *Life Magazine*, © 1955, Time, Inc.   **20–15** Permission granted by Harlow & Mears.   **20–16(b)** Adapted from S. W. Maters, "The Changing Earth," *National Geographic*, 1973, Vol. 143, p. 1–37.   **20–17(a)** © Bob Clay/Jeroboam, Inc.   **20–18** © Animals Animals/M. A. Chappell.   **20–20** © Animals Animals/John Stern.   **20–21** Patricia Caulfield.   **20–22** Adapted from *Animal Behavior*, Dr. Paul A. Johnsgard, 1967, Figure 3, p. 65, William C. Brown Company, Publishers.

Chapter 21
21–2(a) FPG/Leonard Lee Rue III.   **21–2(b)** Hal Harrison/Grant Heilman.   **21–3** Karl H. Maslowski/Photo Researchers, Inc.   **21–4** Paul R. Johnson/Photophile.   **21–6** Reprinted from *Population Growth and Land Use* by Colin Clark. © 1967, 1977 by Colin Clark. Reprinted by permission of St. Martin's Press.

Chapter 22
22–5 Photo by William M. Harlow.   **22–6** © Lynwood M. Chace/Photo Researchers, Inc.   **22–10(a)** © Photo Researchers, Inc.   **22–12** Adapted from *The Biological Sciences* by A Silverstein, Rinehart Press/Holt, Rinehart and Winston, p. 458, Figure 16–19.   **22–15** Runk/ Schoenberger from Grant Heilman Photography.   **22–16** Carolina Biological Supply Company.

Chapter 23
23–7 Excerpted from the book *A Child is Born* by Lennart Nilsson. English translation copyright © 1966, 1977 by Dell Publishing Co., Inc. Originally published in Swedish under the title *Ett Barn Blir Till* by Albert Bonniers Forlag. Copyright © 1965 by Albert Bonniers Forlag, Stockholm. Revised edition copyright © 1976 by Lennart Nilsson, Mirjam Furuhjelm, Axel Ingerlman-Sundberg, Cales Wirson. Used by permission of Delacorte Press/Seymour Lawrence.   **23–8** Landrum B. Shettles, M.D.   **23–9** From Omikron/Photo Researchers, Inc.   **23–10** Martin M. Rotker/Taurus Photos.   **23–11** Martin M. Rotker/Taurus Photos.   **23–13** © Ken Love/Black Star.   **23–14** Shirley Zeiberg/Taurus Photos.   **23–16** Adapted and reproduced by permission of Harcourt Brace Jovanovich, Inc. from *Health*, Second Edition, by B. A. Kogan © 1970.   **23–17** Copyright 1979 by Consumers Union of United States, Inc., Mount Vernon, NY 10550. Reprinted by permission from *Consumer Reports*, May, 1979.   **23–18** Andrew A. Benson.   **23–19** From "Getting Old" by Alexander Leaf. Copyright © September, 1973 by Scientific American, Inc. All rights reserved.   **23–20** John Launois/Black Star.

Chapter 24
24–2 Stephen Bernstein.   **24–4(a)** Marion H. Levy/Photo Researchers, Inc.   **(b)** Bill Pierce/*Time Magazine*.   **24–5** *The Limits To Growth*: a report for The Club of Rome's Project on the Predicament of Mankind, by Donella H. Meadows, Dennis L. Meadows, Jørgen Randers, William W. Behrens III. A Potomac Associates book published by Universe Books, NY, 1972. Graphics by Potomac Associates.   **24–6** Sandor Acs.   **24–7(a)** Stephen Bernstein.   **(b)** From Paolo Soleri, *Arcology: The City in the Image of Man;* © 1970, MIT Press, p. 57.   **24–8** Top left: Miriam Weinstein. Top right: © Animals Animals/Harry Engles.   Middle left: Charlie Ott from National Audubon Society/Photo Researchers, Inc.   Bottom left: © Animals Animals/Leonard Lee Rue III.   Bottom right: © Animals Animals/ Leonard Lee Rue III.

Glossary/Index

This index has been designed so that it can also be used as a *glossary*. Whenever definitions of terms and concepts appear in the text, the relevant page numbers are printed in **boldface** in the index. An italicized letter in parentheses after a page number indicates that an illustration (*i*) or table (*t*) also appears on the page.

A 2
B 3
C 4
D 5
E 6
F 7
G 8
H 9
I 0
J 1